LONDON MATHEMATICAL SOCIETY LECTURE NOTE SERIES

Managing Editor: Professor M. Reid, Mathematics Institute,
University of Warwick, Coventry CV4 7AL, United Kingdom

The titles below are available from booksellers, or from Cambridge University Press at www.cambridge.org/mathematics

London Mathematical Society Lecture Note Series: 380

Real and Complex Singularities

Edited by

M. MANOEL
Universidade de São Paulo, Brazil

M. C. ROMERO FUSTER
Universitat de València, Spain

C. T. C. WALL
University of Liverpool, UK

CAMBRIDGE
UNIVERSITY PRESS

CAMBRIDGE UNIVERSITY PRESS
Cambridge, New York, Melbourne, Madrid, Cape Town, Singapore,
São Paulo, Delhi, Dubai, Tokyo, Mexico City

Cambridge University Press
The Edinburgh Building, Cambridge CB2 8RU, UK

Published in the United States of America by Cambridge University Press, New York

www.cambridge.org
Information on this title: www.cambridge.org/9780521169691

© Cambridge University Press 2010

First published 2010

Printed in the United Kingdom at the University Press, Cambridge

A catalogue record for this publication is available from the British Library

Library of Congress Cataloging in Publication Data
International Workshop on Real and Complex Singularities (10th : 2008 : São Carlos,
São Paulo, Brazil)
Real and complex singularities / edited by M. Manoel, M. C. Romero Fuster, C. T. C Wall.
p.   cm. – (London Mathematical Society lecture note series ; 380)
Includes bibliographical references.
ISBN 978-0-521-16969-1 (pbk.)
1. Singularities (Mathematics) – Congresses.   I. Manoel, M., 1966–   II. Romero Fuster,
M. C.   III. Wall, C. T. C. (Charles Terence Clegg)   IV. Title.   V. Series.
QA614.58.I527   2008
516.3′5 – dc22   2010030522

ISBN 978-0-521-16969-1 Paperback

This volume is dedicated to
Maria Aparecida Ruas
and
Terence Gaffney
on their 60th birthdays

Maria Aparecida Ruas

Terence Gaffney

Contents

Preface

The *Workshops on Real and Complex Singularities* form a series of biennial meetings organized by the Singularities group at Instituto de Ciências Matemáticas e de Computação of São Paulo University (ICMC-USP), Brazil. Their main purpose is to bring together world experts and young researchers in singularity theory, applications and related fields to report recent achievements and exchange ideas, addressing trends of research in a stimulating environment.

These meetings started in 1990 following two pioneer symposia on Singularity Theory (in fact the very first talks on Singularities held in Brazil) organized respectively by G. Loibel and L. Favaro in 1982 and 1988 at ICMC-USP. Since then, with Maria Aparecida Ruas as a driving force, meetings have taken place every two years between singularists from around the world who find in São Carlos a centre to interact and develop new ideas.

The meeting held from the 27th of July to the 2nd of August 2008 was the tenth of these workshops. This was a special occasion, for it was also dedicated to Maria Aparecida Ruas (Cidinha) and Terence Gaffney on their 60th birthdays.

Cidinha and Terry started their scientific connection in 1976 when she was a Ph.D. student at Brown University in the U.S.A. At that time Terry held a position as instructor at that university. Their common interest in singularity theory brought them together and he became her (very young) thesis supervisor. Cidinha returned to Brazil in 1980 and joined the Singularities group created by G. Loibel at the ICMC-USP. Her great capacity and notable enthusiasm has brought the group to a leading position in the Brazilian mathematical community.

The mathematical interaction between Cidinha and Terry has had an important influence on the development of Singularity Theory at the ICMC-USP. From the beginning Terry has attended almost all workshops organized by the members of Cidinha's group, contributing to their research with stimulating discussions and seminars.

This workshop had a total of about 170 participants from about 15 different countries. The formal proceedings consisted of 27 plenary talks, 27 ordinary sessions and 3 poster sessions, with a total of 19 posters. The topics were divided into six categories: real singularities, classification of singularities, topology of singularities, global theory of singularities, singularities in geometry, and dynamical systems.

The Scientific Committee was composed of Lev Birbrair (Universidade Federal do Ceará, Brazil), Jean-Paul Brasselet (Institut de Mathématiques de Luminy, France), Goo Ishikawa (Hokkaido University, Japan), Shyuichi Izumiya (Hokkaido University, Japan), Steven Kleiman (Massachusetts Institute of Technology, USA), David Massey (Northeastern University, USA), David Mond (University of Warwick, UK), Maria del Carmen Romero Fuster (Universitat de València, Spain), Marcio Gomes Soares (Universidade Federal de Minas Gerais, Brazil), Marco Antonio Teixeira (Universidade Estadual de Campinas, Brazil), David Trotman (Université de Provence, France) and Terry Wall (University of Liverpool, UK).

Thanks are due to many people and institutions crucial in the realization of the workshop. We start by thanking the Organizing Committee: Roberta Wik Atique, Abramo Hefez, Isabel Labouriau, Miriam Manoel, Ana Claudia Nabarro, Regilene Oliveira and Marcelo José Saia. We also thank the members of the Scientific Committee for their support. The workshop was funded by FAPESP, CNPq, CAPES, USP and SBM, whose support we gratefully acknowledge. Finally, it is a pleasure to thank the speakers and the other participants whose presence was the real success of the tenth Workshop.

The editors

Introduction

This book is a selection of papers submitted for the proceedings of the *10th Workshop on Real and Complex Singularities*. They are grouped into three categories: singularity theory (7 papers), singular varieties (8 papers) and applications to dynamical systems, generic geometry, singular foliations, etc. (10 papers). Among them, four are survey papers: *Local Euler obstruction, old and new, II*, by N. G. Grulha Jr. and J.-P. Brasselet, *Global classifications and graphs*, by J. Martínez-Alfaro, C. Mendes de Jesus and M. C. Romero-Fuster, *Pairs of foliations on surfaces*, by F. Tari, and *Gaffney's work on equisingularity*, by C. T. C. Wall.

We thank the staff members of the London Mathematical Society involved with the preparation of this book. All papers presented here have been refereed.

1

On a conjecture by A. Durfee

E. ARTAL BARTOLO, J. CARMONA RUBER

AND A. MELLE-HERNÁNDEZ

Abstract

We show how *superisolated surface singularities* can be used to find a counterexample to the following conjecture by A. Durfee [8]: for a complex polynomial $f(x, y, z)$ in three variables vanishing at 0 with an isolated singularity there, "the local complex algebraic monodromy is of finite order if and only if a resolution of the germ $(\{f = 0\}, 0)$ has no cycles". A Zariski pair is given whose corresponding superisolated surface singularities, one has complex algebraic monodromy of finite order and the other not (answering a question by J. Stevens).

1. Introduction

In this paper we give an example of a *superisolated surface singularity* $(V, 0) \subset (\mathbb{C}^3, 0)$ such that a resolution of the germ $(V, 0)$ has no cycles and the local complex algebraic monodromy of the germ $(V, 0)$ is not of finite order, contradicting a conjecture proposed by Durfee [8].

For completeness in the second section we recall well known results about monodromy of the Milnor fibration, about normal surface singularities and state the question by Durfee.

In the third section we recall results on superisolated surface singularities and with them we study in detail the counterexample.

In the last section we show a Zariski pair (C_1, C_2) of curves of degree d given by homogeneous polynomials $f_1(x, y, z)$ and $f_2(x, y, z)$ whose

2000 *Mathematics Subject Classification* 14B05 (primary), 32S05, 32S10 (secondary).
The first author is partially supported through grant MEC (Spain) MTM2007-67908-C02-01. The last two authors are partially supported through grant MEC (Spain) MTM2007-67908-C02-02.

corresponding superisolated surface singularities $(V_1, 0) = (\{f_1(x, y, z) + l^{d+1} = 0\}, 0) \subset (\mathbb{C}^3, 0)$ and $(V_2, 0) = (\{f_2(x, y, z) + l^{d+1} = 0\}, 0) \subset (\mathbb{C}^3, 0)$ (l is a generic hyperplane) satisfy: 1) $(V_1, 0)$ has complex algebraic monodromy of finite order and 2) $(V_2, 0)$ has complex algebraic monodromy of infinite order (answering a question proposed to us by J. Stevens).

2. Invariants of singularities

2.1. Monodromy of the Milnor fibration

Let $f : (\mathbb{C}^{n+1}, 0) \to (\mathbb{C}, 0)$ be an analytic function defining a germ $(V, 0) := (f^{-1}\{0\}, 0) \subset (\mathbb{C}^{n+1}, 0)$ of a hypersurface singularity. The *Milnor fibration* of the holomorphic function f at 0 is the C^∞ locally trivial fibration $f| : B_\varepsilon(0) \cap f^{-1}(\mathbb{D}_\eta^*) \to \mathbb{D}_\eta^*$, where $B_\varepsilon(0)$ is the open ball of radius ε centered at 0, $\mathbb{D}_\eta = \{z \in \mathbb{C} : |z| < \eta\}$ and $\mathbb{D}_\eta^*$ is the open punctured disk ($0 < \eta \ll \varepsilon$ and ε small enough). Milnor's classical result also shows that the topology of the germ $(V, 0)$ in $(\mathbb{C}^{n+1}, 0)$ is determined by the pair $(S_\varepsilon^{2n+1}, L_V^{2n-1})$, where $S^{2n+1} = \partial B_\varepsilon(0)$ and $L_V^{2n-1} := S_\varepsilon^{2n+1} \cap V$ is the *link* of the singularity.

Any fiber $F_{f,0}$ of the Milnor fibration is called the *Milnor fiber* of f at 0. The *monodromy transformation* $h : F_{f,0} \to F_{f,0}$ is the well-defined (up to isotopy) diffeomorphism of $F_{f,0}$ induced by a small loop around $0 \in \mathbb{D}_\eta$. The *complex algebraic monodromy of f at 0* is the corresponding linear transformation $h_* : H_*(F_{f,0}, \mathbb{C}) \to H_*(F_{f,0}, \mathbb{C})$ on the homology groups.

If $(V, 0)$ defines a germ of isolated hypersurface singularity then $\tilde{H}_j(F_{f,0}, \mathbb{C}) = 0$ but for $j = 0, n$. In particular the non-trivial complex algebraic monodromy will be denoted by $h : H_n(F_{f,0}, \mathbb{C}) \to H_n(F_{f,0}, \mathbb{C})$ and $\Delta_V(t)$ will denote its characteristic polynomial.

2.2. Monodromy Theorem and its supplements

The following are the **main properties of the monodromy operator**, see e.g. [11]:

(a) $\Delta_V(t)$ is a product of cyclotomic polynomials.
(b) Let N be the maximal size of the Jordan blocks of h, then $N \leq n + 1$.
(c) Let N_1 be the maximal size of the Jordan blocks of h for the eigenvalue 1, then $N_1 \leq n$.
(d) The monodromy h is called of *finite order* if there exists $N > 0$ such that $h^N = Id$. Lê D.T. [12] proved that the monodromy of an irreducible plane curve singularity is of finite order.
(e) This result was extended by van Doorn and Steenbrink [7] who showed that if h has a Jordan block of maximal size $n + 1$ then

$N_1 = n$, i.e. there exists a Jordan block of h of maximal size n for the eigenvalue 1.

Milnor proved that the link L_V^{2n-1} is $(n-2)$-connected. Thus the link is an *integer (resp. rational) homology $(2n-1)$-sphere* if $H_{n-1}(L_V^{2n-1}, \mathbb{Z}) = 0$ (resp. $H_{n-1}(L_V^{2n-1}, \mathbb{Q}) = 0$). These can be characterized considering the natural map $h - id : H_n(F_{f,0}, \mathbb{Z}) \to H_n(F_{f,0}, \mathbb{Z})$ and using Wang's exact sequence which reads as (see e.g. [19, 21]):

$$0 \to H_n\big(L_V^{2n-1}, \mathbb{Z}\big) \to H_n(F_{f,0}, \mathbb{Z}) \to H_n(F_{f,0}, \mathbb{Z}) \to H_{n-1}\big(L_V^{2n-1}, \mathbb{Z}\big) \to 0.$$

Thus rank $H_n(L_V^{2n-1}) = $ rank $H_{n-1}(L_V^{2n-1}) = \dim \ker(h - id)$ and:

- L_V^{2n-1} is a rational homology $(2n-1)$-sphere $\Longleftrightarrow \Delta_V(1) \neq 0$,
- L_V^{2n-1} is an integer homology $(2n-1)$-sphere $\Longleftrightarrow \Delta_V(1) = \pm 1$.

2.3. Normal surface singularities

Let $(V, 0) = (\{f_1 = \ldots = f_m = 0\}, 0) \subset (\mathbb{C}^N, 0)$ be a normal surface singularity with link $L_V := V \cap S_\varepsilon^{2N-1}$, L_V is a a connected compact oriented 3-manifold. Since $V \cap B_\varepsilon$ is a cone over the link L_V then L_V characterizes the topological type of $(V, 0)$. The link L_V is called a *rational homology sphere* ($\mathbb{Q}$HS) if $H_1(L_V, \mathbb{Q}) = 0$, and L_V is called an *integer homology sphere* ($\mathbb{Z}$HS) if $H_1(L_V, \mathbb{Z}) = 0$. One of the main problems in the study of normal surfaces is to determine which analytical properties of $(V, 0)$ can be read from the topology of the singularity, see the very nice survey paper by Nemethi [20].

The resolution graph $\Gamma(\pi)$ of a resolution $\pi : \tilde{V} \to V$ allows to relate analytical and topological properties of V. W. Neumann [22] proved that the information carried in any resolution graph is the same as the information carried by the link L_V. Let $\pi : \tilde{V} \to V$ be a *good* resolution of the singular point $0 \in V$. Good means that $E = \pi^{-1}\{0\}$ is a normal crossing divisor. Let $\Gamma(\pi)$ be the dual graph of the resolution (each vertex decorated with the genus $g(E_i)$ and the self-intersection E_i^2 of E_i in $\tilde{V}$). Mumford proved that the intersection matrix $I = (E_i \cdot E_j)$ is negative definite and Grauert proved the converse, i.e., any such graph comes from the resolution of a normal surface singularity.

Considering the exact sequence of the pair $(\tilde{V}, E)$ and using I is non-degenerated then

$$0 \longrightarrow \operatorname{coker} I \longrightarrow H_1(L_V, \mathbb{Z}) \longrightarrow H_1(E, \mathbb{Z}) \longrightarrow 0$$

and rank $H_1(E) = $ rank $H_1(L_V)$. In fact L_V is a $\mathbb{Q}$HS if and only if $\Gamma(\pi)$ is a tree and every E_i is a rational curve. If additionally I has determinant ± 1 then L_V is an $\mathbb{Z}$HS.

2.4. Number of cycles in the exceptional set E and Durfee's conjecture

In general one gets

$$\text{rank } H_1(L_V) = \text{rank } H_1(\Gamma(\pi)) + 2\sum_i g(E_i),$$

where rank $H_1(\Gamma(\pi))$ is the number of independent cycles of the graph $\Gamma(\pi)$. Let $n : \tilde{E} \to E$ be the normalization of E. Durfee showed in [8] that the *number of cycles $c(E)$ in E*, i.e. $c(E) = \text{rank } H_1(E) - \text{rank } H_1(\tilde{E})$, does not depend on the resolution and in fact it is equal to $c(E) = \text{rank } H_1(\Gamma(\pi))$. Therefore, E contains cycles only when the dual graph of the intersections of the components contains a cycle. Durfee in [8] proposed the following

Conjecture. *For a complex polynomial $f(x, y, z)$ in three variables vanishing at 0 with an isolated singularity there, "the local complex algebraic monodromy h is of finite order if and only if a resolution of the germ $(\{f = 0\}, 0)$ has no cycles".*

He showed that the conjecture is true in the following two cases:

(1) if f is weighted homogeneuos (the resolution graph is star-shaped and therefore its monodromy is finite)
(2) if $f = g(x, y) + z^n$. Using Thom-Sebastiani [27], the monodromy of f is finite if and only if the monodromy of g is finite. Theorem 3 in [8] proves that the monodromy of f is of finite order if and only a resolution of f has no cycles.

2.5. Example (main result)

In this paper we show that the conjecture is not true in general, and for that we use superisolated surface singularities. Let $(V, 0) \subset (\mathbb{C}^3, 0)$ be the germ of normal surface singularity defined by $f := (xz - y^2)^3 - ((y - x)x^2)^2 + z^7 = 0$. Then the minimal good resolution graph Γ_V of (the superisolated singularity) $(V, 0)$ is

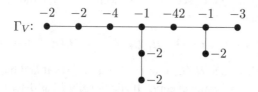

where every dot denotes a rational non-singular curve with the corresponding self-intersection. Thus the link L_V is a rational homology sphere and in particular this graph is a tree, i.e. it has no cycles. But the complex algebraic monodromy of f at 0 does not have finite order because there exists a Jordan block of size 2×2 for an eigenvalue $\neq 1$.

3. Superisolated surface singularities

Definition 3.1. A hypersurface surface singularity $(V, 0) \subset (\mathbb{C}^3, 0)$ defined as the zero locus of $f = f_d + f_{d+1} + \cdots \in \mathbb{C}\{x, y, z\}$ (where f_j is homogeneous of degree j) is *superisolated*, SIS for short, if the singular points of the complex projective plane curve $C := \{f_d = 0\} \subset \mathbb{P}^2$ are not situated on the projective curve $\{f_{d+1} = 0\}$, that is $\operatorname{Sing}(C) \cap \{f_{d+1} = 0\} = \emptyset$. Note that C must be reduced.

The SIS were introduced by I. Luengo in [17] to study the μ-constant stratum. The main idea is that for a SIS the embedded topological type (and the equisingular type) of $(V, 0)$ does not depend on the choice of f_j's (for $j > d$, as long as f_{d+1} satisfies the above requirement), e.g. one can take $f_j = 0$ for any $j > d + 1$ and $f_{d+1} = l^{d+1}$ where l is a linear form not vanishing at the singular points [18].

3.1. The minimal resolution of a SIS

Let $\pi : \tilde{V} \to V$ be the monoidal transformation centered at the maximal ideal $\mathfrak{m} \subset \mathcal{O}_V$ of the local ring of V at 0. Then $(V, 0)$ is a SIS if and only if $\tilde{V}$ is a non-singular surface. Thus π is the *minimal resolution* of $(V, 0)$. To construct the resolution graph $\Gamma(\pi)$ consider $C = C_1 + \cdots + C_r$ the decomposition in irreducible components of the reduced curve C in $\mathbb{P}^2$. Let d_i (resp. g_i) be the degree (resp. genus) of the curve C_i in $\mathbb{P}^2$. Then $\pi^{-1}\{0\} \cong C = C_1 + \cdots + C_r$ and the self-intersection of C_i in $\tilde{V}$ is $C_i \cdot C_i = -d_i(d - d_i + 1)$, [17, Lemma 3]. Since the link L_V can be identified with the boundary of a regular neighbourhood of $\pi^{-1}\{0\}$ in $\tilde{V}$ then the topology of the tangent cone determines the topology of the abstract link L_V [17].

3.2. The minimal good resolution of a SIS

The minimal good resolution of a SIS $(V, 0)$ is obtained after π by doing the minimal embedded resolution of each plane curve singularity $(C, P) \subset (\mathbb{P}^2, P)$, $P \in \operatorname{Sing}(C)$. This means that the support of the minimal good resolution graph

Γ_V is the same as the minimal embedded resolution graph Γ_C of the projective plane curve C in $\mathbb{P}^2$. The decorations of the minimal good resolution graph Γ_V are as follows:

1) the genus of (the strict transform of) each irreducible component C_i of C is a birational invariant and then one can compute it as an embedded curve in $\mathbb{P}^2$. All the other curves are non-singular rational curves.

2) Let C_j be an irreducible component of C such that $P \in C_j$ and with multiplicity $n \geq 1$ at P. After blowing-up at P, the new self-intersection of the (strict transform of the) curve C_j in the (strict transform of the) surface $\tilde{V}$ is $C_j^2 - n^2$. In this way one constructs the minimal good resolution graph Γ of $(V, 0)$.

In particular the theory of hypersurface superisolated surface singularities "contains" in a canonical way the theory of complex projective plane curves.

Example 3.2. If $(V, 0) \subset (\mathbb{C}^3, 0)$ is a SIS with an irreducible tangent cone $C \subset \mathbb{P}^2$ then L_V is a rational homology sphere if and only if C is a rational curve and each of its singularities (C, p) is locally irreducible, i.e a cusp.

Example 3.3. For instance, if $f = f_6 + z^7$ is given by the equation $f_6 = (xz - y^2)^3 - ((y - x)x^2)^2$. The plane projective curve C defined by $f_6 = 0$ is irreducible with two singular points: $P_1 = [0 : 0 : 1]$ (with a singularity of local singularity type $u^3 - v^{10}$) and $P_2 = [1 : 1 : 1]$ (with a singularity of local singularity type $\mathbb{A}_2$) which are locally irreducible. Let $\pi : X \to \mathbb{P}^2$ be the minimal embedded resolution of C at its singular points P_1, P_2. Let $E_i, i \in I$, be the irreducible components of the divisor $\pi^{-1}(f^{-1}\{0\})$.

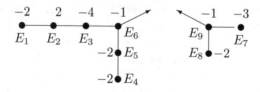

The minimal good resolution graph Γ_V of the superisolated singularity $(V, 0)$ is given by

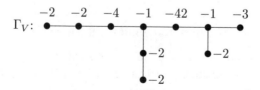

3.3. The embedded resolution of a SIS

In [2], the first author has studied, for SIS, the Mixed Hodge Structure of the cohomology of the Milnor fibre introduced by Steenbrink and Varchenko, [28], [29]. For that he constructed in an effective way an embedded resolution of a SIS and described the MHS in geometric terms depending on invariants of the pair $(\mathbb{P}^2, C)$.

The first author determined the Jordan form of the complex monodromy on $H_2(F_{f,0}, \mathbb{C})$ of a SIS. Let $\Delta_V(t)$ be the corresponding characteristic polynomial of the complex monodromy on $H_2(F_{f,0}, \mathbb{C})$. Denote by $\mu(V, 0) = \deg(\Delta_V(t))$ the Milnor number of $(V, 0) \subset (\mathbb{C}^3, 0)$.

Let $\Delta^P(t)$ be the characteristic polynomial (or Alexander polynomial) of the action of the complex monodromy of the germ (C, P) on $H_1(F_{g^P}, \mathbb{C})$, (where g^P is a local equation of C at P and F_{g^P} denotes the corresponding Milnor fiber). Let μ^P be the Milnor number of C at P. Recall that if $n^P : \tilde{C}^P \to (C, P)$ is the normalization map then $\mu^P = 2\delta^P - (r^P - 1)$, where $\delta^P := \dim_{\mathbb{C}} n_*^P(\mathcal{O}_{\tilde{C}^P})/\mathcal{O}_{C,P}$ and r^P is the number of local irreducible components of C at P.

Let H be a $\mathbb{C}$-vector space and $\varphi : H \to H$ an endomorphism of H. The i-th Jordan polynomial of φ, denoted by $\Delta_i(t)$, is the monic polynomial such that for each $\zeta \in \mathbb{C}$, the multiplicity of ζ as a root of $\Delta_i(t)$ is equal to the number of Jordan blocks of size $i + 1$ with eigenvalue equal to ζ.

Let Δ_1 and Δ_2 be the first and the second Jordan polynomials of the complex monodromy on $H_2(F_{f,0}, \mathbb{C})$ of V and let Δ_1^P be the first Jordan polynomial of the complex monodromy of the local plane singularity (C, P). After the Monodromy Theorem these polynomials joint with $\Delta_V(t)$ and Δ^P, $P \in \text{Sing}(C)$, determine the corresponding Jordan form of the complex monodromy. Let us denote the Alexander polynomial of the plane curve C in $\mathbb{P}^2$ by $\Delta_C(t)$, it was introduced by A. Libgober [13, 14] and F. Loeser and Vaquié [16].

Theorem 3.4 [2]. *Let $(V, 0)$ be a SIS whose tangent cone $C = C_1 \cup \ldots \cup C_r$ has r irreducible components and degree d. Then the Jordan form of the complex monodromy on $H_2(F_{f,0}, \mathbb{C})$ is determined by the following polynomials*

(i) *The characteristic polynomial $\Delta_V(t)$ is equal to*

$$\Delta_V(t) = \frac{(t^d - 1)^{\chi(\mathbb{P}^2 \setminus C)}}{(t - 1)} \prod_{P \in \text{Sing}(C)} \Delta^P(t^{d+1}).$$

(ii) *The first Jordan polynomial is equal to*

$$\Delta_1(t) = \frac{1}{\Delta_C(t)(t-1)^{r-1}} \prod_{P \in \mathrm{Sing}(C)} \frac{\Delta_1^P(t^{d+1})\Delta_{(d)}^P(t)}{\Delta_{1,(d)}^P(t)^3},$$

where $\Delta_{(d)}^P(t) := \gcd(\Delta^P(t), (t^d-1)^{\mu^P})$ *and* $\Delta_{1,(d)}^P(t) := \gcd(\Delta_1^P(t), (t^d-1)^{\mu^P})$.

(iii) *The second Jordan polynomial is equal to*

$$\Delta_2(t) = \prod_{P \in \mathrm{Sing}(C)} \Delta_{1,(d)}^P(t).$$

Corollary 3.5 [2, Corollaire 5.5.4]. *The number of Jordan blocks of size 2 for the eigenvalue 1 of the complex monodromy h is equal to*

$$\sum_{P \in \mathrm{Sing}(C)} (r^P - 1) - (r - 1). \tag{3.1}$$

Let $\tilde{D}_i$ be the normalization of D_i and $\tilde{C}$ the disjoint union of the $\tilde{D}_i$ and $n : \tilde{C} \to C$ be the projection map. Thus the first Betti number of $\tilde{C}$ is $2g :=$ $2\sum_i g(D_i)$ and the first Betti number of C is $2g + \sum_{P \in \mathrm{Sing}(C)}(r^P - 1) - r + 1$. Then $\sum_{P \in \mathrm{Sing}(C)}(r^P - 1) - (r - 1)$ is exactly the difference between the first Betti numbers of C and $\tilde{C}$. In fact this non-negative integer is equal to the first Betti number of the minimal embedded resolution graph Γ_C of the projective plane curve C in $\mathbb{P}^2$, which is nothing but rank $H_1(\Gamma_V)$.

Corollary 3.6. *Let $(V, 0)$ be a SIS whose tangent cone $C = C_1 \cup \ldots \cup C_r$ has r irreducible components. Then the number of independent cycles $c(E) =$ rank $H_1(\Gamma_V) = \sum_{P \in \mathrm{Sing}(C)}(r^P - 1) - (r - 1)$.*

In particular E has no cycles if and only if $\sum_{P \in \mathrm{Sing}(C)}(r^P - 1) = (r - 1)$ if and only if the complex monodromy h has no Jordan blocks of size 2 for the eigenvalue 1.

Corollary 3.7 [2, Corollaire 4.3.2]. *If for every $P \in \mathrm{Sing}(C)$, the local monodromy of the local plane curve equation g^P at P acting on the homology $H_1(F_{g^P}, \mathbb{C})$ of the Milnor fibre F_{g^P} has no Jordan blocks of maximal size 2 then the corresponding SIS has no Jordan blocks of size 3.*

Corollary 3.8. *Let $(V, 0) \subset (\mathbb{C}^3, 0)$ be a SIS with a rational irreducible tangent cone $C \subset \mathbb{P}^2$ of degree d whose singularities are locally irreducible. Then:*

(i) *the link L_V is a $\mathbb{Q}HS$ link and E has no cycles,*
(ii) *the complex monodromy on $H_2(F_{f,0}, \mathbb{C})$ has no Jordan blocks of size 2 for the eigenvalue 1,*

(iii) *the complex monodromy on $H_2(F_{f,0}, \mathbb{C})$ has no Jordan blocks of size 3.*

(iv) *The first Jordan polynomial is equal to*

$$\Delta_1(t) = \frac{1}{\Delta_C(t)} \prod_{P \in \text{Sing}(C)} \gcd(\Delta^P(t), (t^d - 1)^{\mu^P}).$$

The proof follows from the previous description and the fact that if every $P \in \text{Sing}(C)$ is locally irreducible then by Lê D.T. result (see 2.2) the plane curve singularity has finite order and $\Delta_1^P(t) = 1$.

Corollary 3.9. *Let $(V, 0) \subset (\mathbb{C}^3, 0)$ be a SIS whose tangent cone $C = C_1 \cup \ldots \cup C_r$ has r irreducible components. Assume that $\sum_{P \in \text{Sing}(C)} (r^P - 1) = (r - 1)$, then:*

(i) *E has no cycles,*

(ii) *the complex monodromy on $H_2(F_{f,0}, \mathbb{C})$ has no Jordan blocks of size 2 for the eigenvalue 1,*

(iii) *the complex monodromy on $H_2(F_{f,0}, \mathbb{C})$ has no Jordan blocks of size 3.*

(iv) *The first Jordan polynomial is equal to*

$$\Delta_1(t) = \frac{1}{\Delta_C(t)(t - 1)^{r-1}} \prod_{P \in \text{Sing}(C)} \gcd(\Delta^P(t), (t^d - 1)^{\mu^P}).$$

The proof follows from Corollary 3.6 and the part (*e*) Monodromy Theorem 2.2.

3.4. The first Jordan polynomial in Example 3.3

As we described above, the plane projective curve C defined by $f_6 = (xz - y^2)^3 - ((y - x)x^2)^3 = 0$ is irreducible, rational and with two singular points: $P_1 = [0 : 0 : 1]$ (with a singularity of local singularity type $u^3 - v^{10}$) and $P_2 = [1 : 1 : 1]$ (with a singularity of local singularity type $\mathbb{A}_2$) which are unibranched. Let $\pi : X \to \mathbb{P}^2$ be the minimal embedded resolution of C at its singular points P_1, P_2. Let $E_i, i \in I$, be the irreducible components of the divisor $\pi^{-1}(f^{-1}\{0\})$. For each $j \in I$, we denote by N_j the multiplicity of E_j in the divisor of the function $f \circ \pi$ and we denote by $v_j - 1$ the multiplicity of E_j in the divisor of $\pi^*(\omega)$ where ω is the non-vanishing holomorpic 2-form $dx \wedge dy$ in $\mathbb{C}^2 = \mathbb{P}^2 \setminus L_\infty$. Then the divisor $\pi^*(C)$ is a normal crossing divisor. We attach to each exceptional divisor E_i its numerical data (N_i, v_i).

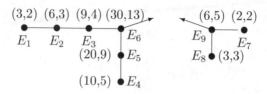

Thus $\Delta^{P_1}(t) = \frac{(t-1)(t^{30}-1)}{(t^3-1)(t^{10}-1)} = \phi_{30}\phi_{15}\phi_6$ and $\Delta^{P_2}(t) = \frac{(t-1)(t^6-1)}{(t^3-1)(t^2-1)} = \phi_6$, where ϕ_k is the k-th cyclotomic polynomial. Thus, by Corollary 3.8, the only possible eigenvalues of with Jordan blocks of size 2 are the roots of the polynomial $\Delta_1(t) = \frac{\phi_6^2}{\Delta_C(t)}$.

The proof of our main result will be finished if we show that the Alexander polynomial $\Delta_C(t) = \phi_6$. The Alexander polynomial, in particular of sextics, has been investigated in detail by Artal [1], Artal and Carmona [3], Degtyarev [6], Oka [24], Pho [25], Zariski [30] among others. In [23] Corollary 18, I.2, it is proved that $\Delta_C(t) = \phi_6$.

Consider a generic line L_∞ in $\mathbb{P}^2$, in our example the line $z = 0$ is generic, and define $f(x, y) = f_6(x, y, 1)$. Consider the (global) Milnor fibration given by the homogeneous polynomial $f_6 : \mathbb{C}^3 \to \mathbb{C}$ with Milnor fibre F. Randell [26] proved that $\Delta_C(t)(t - 1)^{r-1}$ is the characteristic polynomial of the algebraic monodromy acting on $F : T_1 : H_1(F, \mathbb{C}) \to H_1(F, \mathbb{C})$.

Lemma 3.10 (Divisibility properties) [13]. *The Alexander polinomial $\Delta_C(t)(t-1)^{r-1}$ divides $\prod_{P \in \text{Sing}(C)} \Delta^P(t)$ and the Alexander polynomial at infinity $(t^d - 1)^{d-2}(t - 1)$. In particular the roots of the Alexander polynomial are d-roots of unity.*

To compute the Alexander polynomial $\Delta_C(t)$ we combined the method described in [1] with the methods given in [13], [16] and [9].

Consider for $k = 1, \ldots, d - 1$ the ideal sheaf $\mathcal{I}^k$ on $\mathbb{P}^2$ defined as follows:

- If $Q \in \mathbb{P}^2 \setminus \text{Sing}(C)$ then $\mathcal{I}_Q^k = \mathcal{O}_{\mathbb{P}^2, Q}$.
- If $P \in \text{Sing}(C)$ then $\mathcal{I}_P^k$ is the following ideal of $\mathcal{O}_{\mathbb{P}^2, P}$: if $h \in \mathcal{O}_{\mathbb{P}^2, P}$ then $h \in \mathcal{I}_P^k$ if and only if the vanishing order of $\pi^*(h)$ along each E_i is, at least, $-(\nu_i - 1) + [\frac{kN_i}{d}]$ (where [.] stands for the integer part of a real number).

For $k \geq 0$ the following map

$$\sigma_k : H^0(\mathbb{P}^2, \mathcal{O}_{\mathbb{P}^2}(k - 3)) \to \bigoplus_{P \in \text{Sing}(C)} \mathcal{O}_{\mathbb{P}^2, P}/\mathcal{I}_P^k : h \mapsto (h_P + \mathcal{I}_P^k)_{P \in \text{Sing}(C)}$$

is well defined (up to scalars) and the result of [13] and [16] reinterpreted in this language as [1] and [9] reads as follows:

Theorem 3.11 (Libgober, Loeser-Vaquié).

$$\Delta_C(t) = \prod_{k=1}^{d-1}(\Delta^k(t))^{l_k}, \tag{3.2}$$

where $\Delta^k(t) := (t - \exp(\frac{2k\pi i}{d}))(t - \exp(\frac{-2k\pi i}{d}))$ *and* $l_k = \dim \operatorname{coker} \sigma_k$

In our case, by the Divisibility properties (Lemma 3.10), $\Delta_C(t)$ divides $\Delta^{P_1}(t)\Delta^{P_2}(t) = \phi_{30}\phi_{15}\phi_6^2$. Thus, by Theorem 3.11, we are only interested in the case $k = 1$ and 5, $\Delta^1(t) = \Delta^5(t) = \phi_6 = (t^2 - t - 1)$. In case $k = 1$, we have $l_1 = 0$.

In case $k = 5$, the ideal $\mathcal{I}_{P_1}^5$ is the following ideal of $\mathcal{O}_{\mathbb{P}^2, P_1}$:

$$\mathcal{I}_{P_1}^5 = \{h \in \mathcal{O}_{\mathbb{P}^2, P_1} : (\pi^*h) \geq E_1 + 3E_2 + 4E_3 + 4E_4 + 8E_5 + 13E_6\}$$

and with the local change of coordinates $u = x - y^2$, $w = y$, the generators of the ideal are $\mathcal{I}_{P_1}^5 = <uw, u^2, w^5>$ and the dimension of the quotient vector space $\mathcal{O}_{\mathbb{P}^2, P_1}/\mathcal{I}_{P_1}^5$ is 6. A basis is given by $1, u, w, w^2, w^3, w^4$. The ideal

$$\mathcal{I}_{P_2}^5 = \{h \in \mathcal{O}_{\mathbb{P}^2, P_2} : (\pi^*h) \geq 0E_7 + 0E_8 + E_9\} = \mathfrak{m}_{\mathbb{P}^2, P_2}$$

and the dimension of the quotient vector space $\mathcal{O}_{\mathbb{P}^2, P_2}/\mathcal{I}_{P_2}^5$ is 1. A basis is given by 1.

If we take as a basis for the space of conics $1, x, y, x^2, y^2, xy$, the map σ_5

$$\sigma_5 : H^0(\mathbb{P}^2, \mathcal{O}_{\mathbb{P}^2}(2)) \rightarrow \mathcal{O}_{\mathbb{P}^2, P_1}/\mathcal{I}_{P_1}^5 \times \mathcal{O}_{\mathbb{P}^2, P_2}/\mathcal{I}_{P_2}^5$$

$$= \mathbb{C}^6 \times \mathbb{C} : h \mapsto \left(h + \mathcal{I}_{P_1}^5, h + \mathcal{I}_{P_2}^5\right)$$

is given in such coordinates by (using $u = x - y^2$): $\sigma_5(1) = (1, 0, 0, 0, 0, 0, 1)$, $\sigma_5(x) = (0, 1, 0, 1, 0, 0, 1)$, $\sigma_5(y) = (0, 0, 1, 0, 0, 0, 1)$, $\sigma_5(x^2) = (0, 0, 0, 0, 0, 1, 1)$, $\sigma_5(y^2) = (0, 0, 0, 1, 0, 0, 1)$ and $\sigma_5(xy) = (0, 0, 0, 0, 1, 0, 1)$.

Therefore σ_5 is injective and $\dim \operatorname{coker} \sigma_5 = 7 - 6 + 0 = 1$. The key point is that $u \notin \mathcal{I}_{P_1}^5$.

4. Zariski pairs

Let us consider $C \subset \mathbb{P}^2$ a reduced projective curve of degree d defined by an equation $f_d(x, y, z) = 0$. If $(V, 0) \subset (\mathbb{C}^3, 0)$ is a SIS with tangent cone C, then the link L_V of the singularity is completely determined by C. Let us recall, that L_V is a Waldhausen manifold and its plumbing graph is the dual graph of the good minimal resolution. In order to determine L_V we do not need the embedding of C in $\mathbb{P}^2$, but only its embedding in a regular neighborhood. The needed data can be encoded in a combinatorial way.

Definition 4.1. Let $\mathrm{Irr}(C)$ be the set of irreducible components of C. For $P \in \mathrm{Sing}(C)$, let $B(P)$ be the set of local irreducible components of C. The *combinatorial type* of C is given by:

- A mapping deg : $\mathrm{Irr}(C) \to \mathbb{Z}$, given by the degrees of the irreducible components of C.
- A mapping top : $\mathrm{Sing}(C) \to \mathrm{Top}$, where Top is the set of topological types of singular points. The image of a singular point is its topological type.
- For each $P \in \mathrm{Sing}(C)$, a mapping $\beta_P : T(P) \to \mathrm{Irr}(C)$ such that if γ is a branch of C at P, then $\beta_\gamma(\gamma)$ is the global irreducible component containing γ.

Remark 1. There is a natural notion of isomorphism of combinatorial types. It is easily seen that combinatorial type determines and is determined by any of the following graphs (with vertices decorated with self-intersections):

- The dual graph of the preimage of C by the minimal resolution of $\mathrm{Sing}'(C)$. The set $\mathrm{Sing}'(C)$ is obtained from $\mathrm{Sing}(C)$ by forgetting ordinary double points whose branches belong to distinct global irreducible components. We need to mark in the graph the r vertices corresponding to $\mathrm{Irr}(C)$.
- The dual graph of the minimal good resolution of V. Since the minimal resolution is unique, it is not necessary to mark vertices.

Note also that the combinatorial type determine the characteristic polynomial $\Delta_V(t)$ of V (see Theorem 3.4).

Definition 4.2. A *Zariski pair* is a set of two curves $C_1, C_2 \subset \mathbb{P}^2$ with the same combinatorial type but such that $(\mathbb{P}^2, C_1)$ is not homeomorphic to $(\mathbb{P}^2, C_2)$. An *Alexander-Zariski pair* $\{C_1, C_2\}$ is a Zariski pair such that the Alexander polynomials of C_1 and C_2 do not coincide.

In [2], (see here Theorem 3.4) it is shown that the Jordan form of complex monodromy of a SIS is determined by the combinatorial type and the Alexander polynomial of its tangent cone. The first example of Zariski pair was given by Zariski, [30, 31]; there exist sextic curves with six ordinary cusps. If these cusps are (resp. not) in a conic then the Alexander polynomial equals $t^2 - t + 1$ (resp. 1). Then, it gives an Alexander-Zariski pair. Many other examples of Alexander-Zariski pairs have been constructed (Artal [1], Degtyarev [6]).

We state the main result in [2].

Theorem 4.3. *Let V_1, V_2 be two SIS such that their tangent cones form an Alexander-Zariski pair. Then V_1 and V_2 have the same abstract topology and characteristic polynomial of the monodromy but not the same embedded topology.*

Recall that the Jordan form of the monodromy is an invariant of the embedded topology of a SIS (see Theorem 3.4); since it depends on the Alexander polynomial $\Delta_C(t)$ of the tangent cone.

4.1. Zariski pair of reduced sextics with only one singular point of type $\mathbb{A}_{17}$

Our next Zariski-pair example (C_1, C_2) can be found in [1, Théorème 4.4]. The curves $C_i, i = 1, 2$ are reduced sextics with only one singular point P of type $\mathbb{A}_{17}$, locally given by $u^2 - v^{18}$.

(I) the irreducible componentes of C_1 are two non-singular cubics. These cubics meet at only one point P which moreover is an inflection point of each of the cubics, i.e. the tangent line to the singular point P goes through the infinitely near points P, P_1 and P_2 of C_1. The equations of C_1 are given for instance by $\{f_1(x, y, z) := (zx^2 - y^3 - ayz^2 - bz^3)(zx^2 - y^3 - ayz^2 - cz^3) = 0\}$, with $a, b, c \in \mathbb{C}$ generic.

(II) the irreducible componentes of C_2 are two non-singular cubics. These cubics meet at only one point P which is not an inflection point of any of the cubics, i.e. the tangent line to the singular point P goes through the infinitely near points P, P_1 of C_1 but it is not going through P_2. The equations of C_2 are given for instance by $\{f_2(x, y, z) := (zx^2 - y^2x - yz^2 - a_1(z^3 - y(xz - y^2)))(zx^2 - y^2x - yz^2 - a_2(z^3 - y(xz - y^2))) = 0\}$ with $a_1, a_2 \in \mathbb{C}$ generic.

Consider the superisolated surface singularities $(V_1, 0) = (\{f_1(x, y, z) + l^7 = 0\}, 0) \subset (\mathbb{C}^3, 0)$ and $(V_2, 0) = (\{f_2(x, y, z) + l^7 = 0\}, 0) \subset (\mathbb{C}^3, 0)$ (l is a generic hyperplane). In both cases the tangent cone has two irreducible components and it has only one singular point P of local type $u^2 - v^{18}$ and therefore $\Delta^P(t) = (t^{18} - 1)(t - 1)/(t^2 - 1) = \phi_{18}\phi_9\phi_6\phi_3\phi_1$, where ϕ_k is the k-th cyclotomic polynomial. Thus the number of local branches is 2 and $\sum_{P \in \text{Sing}(C)}(r^P - 1) = (r - 1)$. By Corollary 3.9, for $(V_i, 0), i = 1, 2$, the complex monodromy has no Jordan blocks of size 2 for the eigenvalue 1, and it has no Jordan blocks of size 3. Moreover the first Jordan polynomial is equal to

$$\Delta_1(t) = \frac{\gcd(\Delta^P(t), (t^6 - 1)^{\mu^P})}{\Delta_{C_i}(t)(t - 1)} = \frac{\phi_6\phi_3}{\Delta_{C_i}(t)}. \tag{4.1}$$

To compute $\Delta_{C_i}(t)$ we use the same ideas as in Theorem 3.11.

Lemma 4.4. *For the point P at the curve C_1 the ideals $\mathcal{I}_P^k = \mathcal{O}_{\mathbb{P}^2,P}$ if $k \leq 3$, $\mathcal{I}_P^4 = <y^3, z> \mathcal{O}_{\mathbb{P}^2,P}$ and $\mathcal{I}_P^5 = <y^6, z - y^3 - ay^4 - by^5> \mathcal{O}_{\mathbb{P}^2,P}$.*

Lemma 4.5. *For the point P at the curve C_2 the ideals $\mathcal{I}_P^k = \mathcal{O}_{\mathbb{P}^2,P}$ if $k \leq 3$, $\mathcal{I}_P^4 = <y^3, z - y^2> \mathcal{O}_{\mathbb{P}^2,P}$ and $\mathcal{I}_P^5 = <y^6, z - y^2 - y^5> \mathcal{O}_{\mathbb{P}^2,P}$.*

Thus $\Delta_{C_i}(t) = \phi_6^{\dim \operatorname{coker} \sigma_5} \phi_3^{\dim \operatorname{coker} \sigma_4}$.

Therefore the map σ_4 is

$$\sigma_4 : H^0(\mathbb{P}^2, \mathcal{O}_{\mathbb{P}^2}(1)) \simeq \mathbb{C}^3 \to \mathcal{O}_{\mathbb{P}^2,P}/\mathcal{I}_P^4 \simeq \mathbb{C}^3,$$

and if we choose as basis of the first space $\{1, y, z\}$ and of the second $\{1, y, y^2\}$ then

(1) by Lemma 4.4, for C_1 the dimension $\dim \operatorname{coker} \sigma_4 = \dim \ker \sigma_4 = 1$.
(2) by Lemma 4.5, for C_2 the dimension $\dim \operatorname{coker} \sigma_4 = \dim \ker \sigma_4 = 0$.

On the other hand for the map σ_5

$$\sigma_5 : H^0(\mathbb{P}^2, \mathcal{O}_{\mathbb{P}^2}(2)) \simeq \mathbb{C}^6 \to \mathcal{O}_{\mathbb{P}^2,P}/\mathcal{I}_P^5 \simeq \mathbb{C}^6,$$

if we choose as basis of the first space $\{1, y, z, y^2, yz, z^2\}$ and of the second $\{1, y, y^2, y^3, y^4, y^5\}$ then we can compute

(3) by Lemma 4.4, for C_1 the dimension $\dim \operatorname{coker} \sigma_5 = \dim \ker \sigma_5 = 1$.
(4) by Lemma 4.5, for C_2 the dimension $\dim \operatorname{coker} \sigma_5 = \dim \ker \sigma_5 = 0$.

Therefore, $\Delta_{C_1}(t) = \phi_6 \phi_3$ and $\Delta_{C_2}(t) = 1$ and by (4.1) we have proved that the pair (C_1, C_2) is a Alexander-Zariski pair.

Example 4.6. Consider the superisolated surface singularities $(V_1, 0) = (\{f_1(x, y, z) + l^7 = 0\}, 0) \subset (\mathbb{C}^3, 0)$ and $(V_2, 0) = (\{f_2(x, y, z) + l^7 = 0\}, 0) \subset (\mathbb{C}^3, 0)$ (l is a generic hyperplane). Then the complex algebraic monodromy of $(V_1, 0) \subset (\mathbb{C}^3, 0)$ has finite order and the complex algebraic monodromy of $(V_2, 0) \subset (\mathbb{C}^3, 0)$ has not finite order

This answers a question proposed to us by J. Stevens: find a Zariski pair C_1, C_2 such that for the corresponding SIS surface singularities $(V_1, 0) \subset (\mathbb{C}^3, 0)$ and $(V_2, 0) \subset (\mathbb{C}^3, 0)$ one has a finite order monodromy and the other it does not.

There are also examples of Zariski pairs which are not Alexander-Zariski pairs (see [23], [3], [4]). Some of them are distinguished by the so-called characteristic varieties introduced by Libgober [15]. These are subtori of $(\mathbb{C}^*)^r$, $r := \# \operatorname{Irr}(C)$, which measure the excess of Betti numbers of finite Abelian coverings of the plane ramified on the curve (as Alexander polynomial does it for cyclic coverings).

Problem 1. How can one translate characteristic varieties of a projective curve in terms of invariants of the SIS associated to it?

References

1. E. Artal, Sur les couples des Zariski, *J. Algebraic Geom.*, **3** (1994) 223-247.
2. E. Artal, Forme de Jordan de la monodromie des singularités superisolées de surfaces, *Mem. Amer. Math. Soc.* **525** (1994).
3. E. Artal and J. Carmona, Zariski pairs, fundamental groups and Alexander polynomials, *J. Math. Soc. Japan* **50(3)** (1998) 521–543.
4. E. Artal, J. Carmona, J. I. Cogolludo, and H.O. Tokunaga, Sextics with singular points in special position, *J. Knot Theory Ramifications* **10** (2001) 547–578.
5. E. Artal, J. I. Cogolludo, and H. O. Tokunaga, A survey on Zariski pairs, *Algebraic geometry in East Asia—Hanoi 2005*, 1–100, Adv. Stud. Pure Math. **50** Math. Soc. Japan, Tokyo, 2008.
6. A. Degtyarev, Alexander polynomial of a curve of degree six, *Knot theory and Ramif.* **3(4)** (1994) 439–454.
7. M. G M. van Doorn and J. H. M. Steenbrink, A supplement to the monodromy theorem, *Abh. Math. Sem. Univ. Hamburg* **59** (1989) 225–233.
8. A. Durfee, The monodromy of a degenerating family of curves, *Inv. Math.* **28** (1975) 231–241.
9. H. Esnault, Fibre de Milnor d'un cône sur une courbe plane singulière, *Inv. Math.* **68** (1982) 477–496.
10. G.-M. Greuel, G. Pfister, and H. Schönemann. Singular 3.0 — A computer algebra system for polynomial computations. In M. Kerber and M. Kohlhase: *Symbolic computation and automated reasoning, The Calculemus-2000 Symposium* (2001), 227–233.
11. A. Landman, On the Picard-Lefschetz transformation for algebraic manifolds acquiring general singularities, *Trans. Amer. Math. Soc.* **181** (1973) 89–126.
12. D. T. Lê, Sur les noeuds algébriques, *Compositio Math.* **25** (1972) 281–321.
13. A. Libgober, Alexander polynomial of plane algebraic curves and cyclic multiple planes, *Duke Math. J.* **49(4)** (1982) 833–851.
14. A. Libgober, Alexander invariants of plane algebraic curves, *Singularities, Part 2 (Arcata, Calif., 1981)*, 135–143, Proc. Sympos. Pure Math. **40** Amer. Math. Soc., Providence, RI, 1983.
15. A. Libgober, Characteristic varieties of algebraic curves, *Applications of algebraic geometry to coding theory, physics and computation (Eilat, 2001)*, Kluwer Acad. Publ., Dordrecht (2001) 215–254.
16. F. Loeser and M. Vaquié, Le polynôme d'Alexander d'une courbe plane projective, *Topology* **29(2)** (1990) 163–173.
17. I. Luengo, The μ-constant stratum is not smooth, *Invent. Math.*, **90(1)** (1987) 139–152.
18. I. Luengo and A. Melle Hernández, A formula for the Milnor number *C.R. Acad. Sc. Paris*, **321**, Série I. (1995) 1473–1478.
19. A. Némethi, Some topological invariants of isolated hypersurface singularities, *Low dimensional topology (Eger, 1996/Budapest, 1998)*, 353–413, Bolyai Soc. Math. Stud., **8**, JÛŠnos Bolyai Math. Soc., Budapest, 1999.

20. A. Némethi, Invariants of normal surface singularities, *Real and complex singularities*, 161–208, Contemp. Math. **354** AMS, Providence, RI, 2004.

21. A. Némethi and J. H. M. Steenbrink, On the monodromy of curve singularities, *Math. Z.* **223(4)** (1996) 587–593.

22. W.D. Neumann, A calculus for plumbing applied to the topology of complex surface singularities and degenerating complex curves, *Trans. Amer. Math. Soc.* **268(2)** (1981) 299–344.

23. M. Oka, A new Alexander-equivalent Zariski pair, Dedicated to the memory of Le Van Thiem (Hanoi, 1998). *Acta Math. Vietnam.* **27** (2002) 349–357.

24. M. Oka, Alexander polynomials of sextics, *J. Knot Theory Ramifications* **12(5)** (2003) 619–636.

25. D.T. Pho, Classification of singularities on torus curves of type (2, 3), *Kodai Math. J.*, **24** (2001) 259–284.

26. R. Randell, Milnor fibers and Alexander polynomials of plane curves, *Singularities, Part 2 (Arcata, Calif., 1981)*, 415–419, Proc. Sympos. Pure Math. **40** AMS, Providence, RI, 1983.

27. M. Sebastiani and R. Thom, Un résultat sur la monodromie, *Invent. Math.* **13** (1971) 90–96.

28. J. H. M. Steenbrink, Mixed Hodge structure on the vanishing cohomology, *Real and Complex Singularities (Proc. Nordic Summer School, Oslo, 1976)* Alphen a/d Rijn: Sijthoff & Noordhoff (1977) 525–563.

29. A. Varchenko, Asymptotic Hodge structure on vanishing cohomology, *Math. USSR Izvestija* **18** (1982) 469–512.

30. O. Zariski, On the problem of existence of algebraic functions of two variables possessing a given branch curve, *Amer. J. Math.* **51** (1929) 305–328.

31. O. Zariski, The topological discriminant group of a Riemann surface of genus *p*, *Amer. J. Math.* **59** (1937) 335–358.

E. Artal Bartolo
Departamento de Matemáticas, IUMA
Universidad de Zaragoza
Campus Pza. San Francisco s/n
E-50009 Zaragoza SPAIN
artal@unizar.es

A. Melle-Hernández
Departamento de Álgebra,
Universidad Complutense,
Plaza de las Ciencias s/n,
Ciudad Universitaria,
28040 Madrid, SPAIN
amelle@mat.ucm.es

J. Carmona Ruber
Departamento de Sistemas
Informáticos y Computación,
Universidad Complutense,
Plaza de las Ciencias s/n,
Ciudad Universitaria,
28040 Madrid, SPAIN
jcarmona@sip.ucm.es

2

On normal embedding of complex algebraic surfaces

L. BIRBRAIR, A. FERNANDES AND W. D. NEUMANN

Abstract

We construct examples of complex algebraic surfaces not admitting normal embeddings (in the sense of semialgebraic or subanalytic sets) whose image is a complex algebraic surface.

1. Introduction

Given a closed and connected subanalytic subset $X \subset \mathbb{R}^m$ the *inner metric* $d_X(x_1, x_2)$ on X is defined as the infimum of the lengths of rectifiable paths on X connecting x_1 to x_2. This metric defines the same topology on X as the Euclidean metric on $\mathbb{R}^m$ restricted to X (also called *"outer metric"*). This follows from the famous Lojasiewicz inequality and the subanalytic approximation of the inner metric [6]. But the inner metric is not necessarily bi-Lipschitz equivalent to the Euclidean metric on X. To see this it is enough to consider a simple real cusp $x^2 = y^3$. A subanalytic set is called *normally embedded* if these two metrics (inner and Euclidean) are bi-Lipschitz equivalent.

Theorem 1.1 [4]. *Let $X \subset \mathbb{R}^m$ be a connected and globally subanalytic set. Then there exist a normally embedded globally subanalytic set $\tilde{X} \subset \mathbb{R}^q$, for some q, and a global subanalytic homeomorphism $p\colon \tilde{X} \to X$ bi-Lipschitz with respect to the inner metric. The pair $(\tilde{X}, p)$ is called a normal embedding of X.*

2000 *Mathematics Subject Classification* 14P10 (primary), 32S99 (secondary).
The authors acknowledge research support under the grants: CNPq grant no 301025/2007-0 (Lev Birbriar), FUNCAP/PPP 9492/06, CNPq grant no 30685/2008-4 (Alexandre Fernandes) and NSF grant no. DMS-0456227 (Walter Neumann)

The original version of this theorem (see [4]) was formulated in a semialgebraic language, but it is easy to see that this result remains true for a global subanalytic structure. The proof remains the same as in [4].

Complex algebraic sets and real algebraic sets are globally subanalytic sets. By the above theorem these sets admit globally subanalytic normal embeddings. T. Mostowski asked if there exists a complex algebraic normal embedding when X is a complex algebraic set, i.e., a normal embedding for which the image set $\tilde{X} \subset \mathbb{C}^n$ is a complex algebraic set. In this note we give a negative answer for the question of Mostowski. Namely, we prove that a Brieskorn surface $x^b + y^b + z^a = 0$ does not admit a complex algebraic normal embedding if $b > a$ and a is not a divisor of b. For the proof of this theorem we use the ideas of the remarkable paper of A. Bernig and A. Lytchak [3] on metric tangent cones and the paper of the authors on the (b, b, a) Brieskorn surfaces [2]. We also briefly describe other examples based on taut singularities.

2. Proof

Recall that a subanalytic set $X \subset \mathbb{R}^n$ is called *metrically conical* at a point x_0 if there exists an Euclidean ball $B \subset \mathbb{R}^n$ centered at x_0 such that $X \cap B$ is bi-Lipschitz homeomorphic, with respect to the inner metric, to the straight cone over its link at x_0. When such a bi-Lipschitz homeomorphism is subanalytic we say that X is *subanalytically metrically conical* at x_0.

Example 2.1. The Brieskorn surfaces in $\mathbb{C}^3$

$$\{(x, y, z) \mid x^b + y^b + z^a = 0\}$$

$(b > a)$ are subanalytically metrically conical at $0 \in \mathbb{C}^3$ (see [2]).

We say that a complex algebraic set admits a *complex algebraic normal embedding* if the image of a subanalytic normal embedding of this set can be chosen complex algebraic.

Example 2.2. Any complex algebraic curve admits a complex algebraic normal embedding. This follows from the fact that the germ of an irreducible complex algebraic curve is bi-Lipschitz homeomorphic with respect to the inner metric to the germ of $\mathbb{C}$ at the origin (e.g., [8], [5]).

Theorem 2.3. *If* $1 < a < b$ *and* a *is not a divisor of* b, *then no neighborhood of* 0 *in the Brieskorn surface in* $\mathbb{C}^3$

$$\{(x, y, z) \in \mathbb{C}^3 \mid x^b + y^b + z^a = 0\}$$

admits a complex algebraic normal embedding.

We need the following result on tangent cones:

Theorem 2.4. *If* (X_1, x_1) *and* (X_2, x_2) *are germs of subanalytic sets which are subanalytically bi-Lipschitz homeomorphic with respect to the induced Euclidean metric, then their tangent cones* $T_{x_1} X_1$ *and* $T_{x_2} X_2$ *are subanalytically bi-Lipschitz homeomorphic.*

This result is a weaker version of the results of Bernig-Lytchak ([3], Remark 2.2 and Theorem 1.2). We present here an independent proof.

Proof of Theorem 2.4. Let us denote

$$S_x X = \{v \in T_x X : |v| = 1\}.$$

Since $T_x X$ is a cone over $S_x X$, in order to prove that $T_{x_1} X_1$ and $T_{x_2} X_2$ are subanalytically bi-Lipschitz homeomorphic, it is enough to prove that $S_{x_1} X_1$ and $S_{x_2} X_2$ are subanalytically bi-Lipschitz homeomorphic.

By Corollary 0.2 in [9], there exists a subanalytic bi-Lipschitz homeomorphism with respect to the induced Euclidean metric

$$h \colon (X_1, x_1) \to (X_2, x_2),$$

such that $|h(x) - x_2| = |x - x_1|$ for all x. Let us define

$$dh \colon S_{x_1} X_1 \to S_{x_2} X_2$$

as follows: given $v \in S_{x_1} X_1$, let $\gamma \colon [0, \epsilon) \to X_1$ be a subanalytic arc such that

$$|\gamma(t) - x_1| = t \ \forall \, t \in [0, \epsilon) \quad \text{and} \quad \lim_{t \to 0^+} \frac{\gamma(t) - x_1}{t} = v \, ;$$

we define

$$dh(v) = \lim_{t \to 0^+} \frac{h \circ \gamma(t) - x_2}{t}.$$

Clearly, dh is a subanalytic map. Define $d(h^{-1}) \colon S_{x_2} X_2 \to S_{x_1} X_1$ in the same way. Let $k > 0$ be a Lipschitz constant of h. Let us prove that k is a Lipschitz constant of dh. In fact, given $v_1, v_2 \in S_{x_1} X_1$, let $\gamma_1, \gamma_2 \colon [0, \epsilon) \to X_1$ be subanalytic arcs such that

$$|\gamma_i(t) - x_1| = t \ \forall \, t \in [0, \epsilon) \quad \text{and} \quad \lim_{t \to 0^+} \frac{\gamma_i(t) - x_1}{t} = v_i \quad \text{for } i = 1, 2.$$

Then

$$|dh(v_1) - dh(v_2)| = \left| \lim_{t \to 0^+} \frac{h \circ \gamma_1(t) - x_2}{t} - \lim_{t \to 0^+} \frac{h \circ \gamma_2(t) - x_2}{t} \right|$$

$$= \lim_{t \to 0^+} \frac{1}{t} |h \circ \gamma_1(t) - h \circ \gamma_2(t)|$$

$$\leq k \lim_{t \to 0^+} \frac{1}{t} |\gamma_1(t) - \gamma_2(t)|$$

$$= k |v_1 - v_2|.$$

Since $d(h^{-1})$ is Lipschitz (by the same argument) and dh and $d(h^{-1})$ are mutual inverses, we have proved the theorem.   □

Corollary 2.5. *Let $X \subset \mathbb{R}^n$ be a normally embedded subanalytic set. If X is subanalytically metrically conical at a point $x \in X$, then the germ (X, x) is subanalytically bi-Lipschitz homeomorphic to the germ $(T_x X, 0)$.*

Proof. The tangent cone of the straight cone at the vertex is the cone itself. So the result is a direct application of Theorem 2.4.   □

Proof of Theorem 2.3. Let $X \subset \mathbb{C}^3$ be the complex algebraic surface defined by

$$X = \{(x, y, z) \mid x^b + y^b + z^a = 0\}.$$

We are going to prove that the germ $(X, 0)$ does not have a normal embedding in $\mathbb{C}^N$ which is a complex algebraic surface. In fact, if $(\tilde{X}, 0) \subset (\mathbb{C}^N, 0)$ is a complex algebraic normal embedding of $(X, 0)$ and $p \colon (\tilde{X}, 0) \to (X, 0)$ is a subanalytic bi-Lipschitz homeomorphism, since $(X, 0)$ is subanalytically metrically conical [2], then $(\tilde{X}, 0)$ is subanalytically metrically conical and, by Corollary 2.5, $(\tilde{X}, 0)$ is subanalytically bi-Lipschitz homeomorphic to $(T_0 \tilde{X}, 0)$. Now, the tangent cone $T_0 \tilde{X}$ is a complex algebraic cone, thus its link is an S^1-bundle. On the other hand, the link of X at 0 is a Seifert fibered manifold with b singular fibers of degree $\frac{a}{\gcd(a,b)}$. This is a contradiction because the Seifert fibration of a Seifert fibered manifold (other than a lens space) is unique up to diffeomorphism.   □

The following result relates the metric tangent cone of X at x and the usual tangent cone of the normally embedded sets. See [3] for a definition of a metric tangent cone.

Theorem 2.6 [3], Section 5. *Let $X \subset \mathbb{R}^m$ be a closed and connected subanalytic set and $x \in X$. If $(\tilde{X}, p)$ is a normal embedding of X, then $T_{p^{-1}(x)} \tilde{X}$ is bi-Lipschitz homeomorphic to the metric tangent cone.*

Remark 1. We showed that the metric tangent cones of the above Brieskorn surface singularities are not homeomorphic to any complex cone.

2.1. Other examples

We sketch how taut surface singularities give other examples of complex surface germs without any complex analytic normal embeddings.

Both the inner metric and the outer (Euclidean) metric on a complex analytic germ (V, p) are determined up to bi-Lipschitz equivalence by the complex analytic structure (independent of a complex embedding). This is because $(f_1, \ldots, f_N) \colon (V, p) \hookrightarrow (\mathbb{C}^N, 0)$ is a complex analytic embedding if and only if the f_i's generate the maximal ideal of $\mathcal{O}_{(V,p)}$, and adding to the set of generators gives an embedding which induces the same metrics up to bi-Lipschitz equivalence.

A *taut* complex surface germ is an algebraically normal germ (V, p) (to avoid confusion we say "algebraically normal" for the algebro-geometric concept of normality) whose complex analytic structure is determined up to isomorphism by its topology. So if its inner and outer metrics are not bi-Lipschitz equivalent then it has no complex analytic normal embedding with algebraically normal image. Taut complex surface singularities were classified by Laufer [7] and include, for example, the simple singularities. A simple singularity (V, p) of type B_n, D_n, or E_n has non-reduced tangent cone, from which follows easily that it has non-equivalent inner and outer metrics. Thus (V, p) admits no complex algebraic normal embedding as an algebraically normal germ.

If we drop the requirement that the image be algebraically normal, (V, p) still has no complex analytic normal embedding. Indeed, suppose we have a subanalytic embedding $(V, p) \to (Y, 0) \subset (\mathbb{C}^n, 0)$ whose image Y is complex analytic but not necessarily algebraically normal (see also [1]). By tautness, the normalization of Y is isomorphic to V, which has non-reduced tangent cone. Hence Y also has non-reduced tangent cone, so it is not normally embedded.

References

1. L. Birbrair, A. Fernandes and W. D. Neumann, Bi-Lipschitz geometry of weighted homogeneous surface singularities, *Math. Ann.* **342** (2008) 139–144.
2. L. Birbrair, A. Fernandes and W. D. Neumann, Bi-Lipschitz geometry of complex surface singularities, *Geom. Dedicata.* **139** (2009) 259–267.
3. A. Bernig and A. Lytchak, Tangent spaces and Gromov-Hausdorff limits of subanalytic spaces, *J. Reine Angew. Math.* **608** (2007), 1–15.

4. L. Birbrair and Tadeusz Mostowski, Normal embeddings of semialgebraic sets. *Michigan Math. J.* **47** (2000), 125–132.

5. A. Fernandes, Topological equivalence of complex curves and bi-Lipschitz maps. *Michigan Math. J.* **51** (2003) 593–606.

6. K. Kurdyka and P. Orro, Distance géodésique sur un sous-analytique, *Rev. Mat. Univ. Complut. Madrid.* **10** (1997) supplementary, 174–182.

7. H. B. Laufer, Taut two-dimensional singularities. *Math. Ann.* **205** (1973) 131–164.

8. F. Pham and B. Teissier, Fractions Lipschitziennes d'une algebre analitique complexe et saturation de Zariski, *Prépublications Ecole Polytechnique* No. M17.0669 (1969).

9. G. Valette, The link of the germ of semi-algebraic metric space. *Proc. Amer. Math. Soc.* **135** (2007) 3083–3090.

L. Birbrair and A. Fernandes
Dep. de Matemática
Universidade Federal do Ceará (UFC)
Campos do Pici, Bloco 914
60455-760 Fortaleza-CE
Brazil
birb@ufc.br
alexandre.fernandes@ufc.br

W. D. Neumann
Dep. of Mathematics,
Barnard College
Columbia University
New York, NY 10027
USA
neumann@math.columbia.edu

3

Local Euler obstruction, old and new, II

JEAN-PAUL BRASSELET AND NIVALDO G. GRULHA JR.

Abstract

This paper is a continuation of the first author's survey *Local Euler Obstruction, Old and New* (1998). It takes into account recent results obtained by various authors, in particular concerning extensions of the local Euler obstruction for frames, functions and maps and for differential forms and collections of them.

1. Introduction

The local Euler obstruction was first introduced by R. MacPherson in [34] as a key ingredient for his construction of characteristic classes of singular complex algebraic varieties. Then, an equivalent definition was given by J.-P. Brasselet and M.-H. Schwartz in [7] using vector fields. This new viewpoint brought the local Euler obstruction into the framework of "indices of vector fields on singular varieties", though the definition only considers radial vector fields. There are various other definitions and interpretations in particular due to Gonzalez-Sprinberg, Verdier, Lê-Teissier and others, and there is a very ample literature on this topic, see for instance [3] and also [1, 7, 9, 12, 13, 17, 23, 33, 34].

A survey was written by the first author [2]. Then, the notion of local Euler obstruction developed mainly in two directions: the first one comes back to MacPherson's definition and concerns differential forms. That is

2000 *Mathematics Subject Classification* 14F, 32S, 57R, 55S.
The authors acknowledge the financial support given by the program USP-COFECUB, grant 07.1.12081.1.7. The second author also acknowledges the financial support from FAPESP, grant 2009/08774-0.

developed by W. Ebeling and S. Gusein-Zade in a series of papers. The second one relates local Euler obstruction with functions defined on the variety [5, 6] and with maps [23]. That approach is useful to relate local Euler obstruction with other indices. The aim of this paper is to update the previous survey in order to present together the new features on the subject. We use (and abuse) the monographies [3, 9] for the most classical features. Let us provide a brief "history" of the subject.

The local Euler obstruction at a point p of an algebraic variety V, denoted by $Eu_V(p)$, was defined by MacPherson. It is one of the main ingredients in his proof of Deligne-Grothendieck conjecture concerning existence of characteristic classes for complex algebraic varieties [34]. An equivalent definition was given in [7] by J.-P. Brasselet and M.-H. Schwartz, using stratified vector fields.

Independently of MacPherson, M. Kashiwara [29] introduced a local invariant of singular complex spaces in relation to his famous local index theorem for holonomic $\mathcal{D}$-modules. It was later observed by Dubson to be the same as MacPherson's local Euler obstruction [11, 12, 13].

The computation of local Euler obstruction is not so easy by using the definition. Various authors propose formulae which make the computation easier. G. Gonzalez-Sprinberg and J.L. Verdier give a formula in terms of Chern classes of the Nash bundle [17]. Lê D.T. and B. Teissier provide a formula in terms of polar multiplicities [33]. V. H. Jorge-Perez, D. Levcovitz and M. J. Saia [27], use the Lê-Teissier result to emphasize interest of local Euler obstruction in the context of maps.

In the paper [5], J.-P. Brasselet, D. T. Lê and J. Seade give a Lefschetz type formula for the local Euler obstruction. The formula shows that the local Euler obstruction, as a constructible function, satisfies the Euler condition relatively to generic linear forms. A natural continuation of the result is the paper by J.-P. Brasselet, D. Massey, A. J. Parameswaran and J. Seade [6], whose aim is to understand what is the obstacle for the local Euler obstruction to satisfy the Euler condition relatively to analytic functions with isolated singularity at the considered point. That is the role of the so-called local Euler obstruction of f, denoted by $Eu_{f,V}(0)$.

The relation between local Euler obstruction of f and the number of Morse points of a Morsification of f is described, for particular germs of singular varieties, in [40] by J. Seade, M. Tibar and A. Verjovsky. They compare $Eu_{f,V}(0)$ with two different generalizations of the Milnor number for functions with isolated singularities on singular varieties: one is the notion of the Milnor number given by Lê D. T. [30], the other is the one given by D. Mond, and D. van Straten [37] and by V. Goryunov [22] for curves, and considered by

T. Izawa and T. Suwa [26] for functions defined on complete intersections in general.

Another generalization of the Milnor number, denoted by $\mu_{BR}(f)$, is given by J. W. Bruce and R. M. Roberts in [10] for functions defined on a singular algebraic variety. In [24, 25], N. Grulha establishes a relation between $Eu_{V,f}(0)$ and $\mu_{BR}(f)$, for a function f with isolated singularity and V a hypersurface with isolated singularity such that the associated logarithmic characteristic variety, denoted by $LC(V)$, is Cohen-Macaulay.

Bruce and Roberts' Milnor number can be computed in an easier way [37]; however, its behavior relatively to deformations of the variety V can be complicated, in particular because tangent vector fields do not necessarily lift on deformations of the variety. An important result is that Bruce and Roberts' Milnor number is a topological invariant for families of functions with isolated singularities defined on hypersurfaces with isolated singularities [24, 25].

The natural generalization of the notion of local Euler obstruction of a function is the one of local Euler obstruction of a map $f : (V, 0) \rightarrow (\mathbb{C}^k, 0)$, where $(V, 0)$ is a germ of an equidimensional complex analytic variety with dimension $n \geq k$. Such a notion can be defined using the local Euler obstruction associated to a k-frame on an analytic variety, as defined and studied by J.-P. Brasselet, J. Seade and T. Suwa in [9]. That is performed by N. Grulha in [23, 24], where is introduced the notion of local Euler obstruction for maps defined on singular varieties, a priori depending on a particular choice of a cell.

Another way to extend the notion of local Euler obstruction is to define it by using differential forms instead of vector fields. That is in fact the original way that R. MacPherson introduced the local Euler obstruction. That is developed by W. Ebeling and S.M. Gusein Zade in various papers, in particular [14, 15], see also [9]. The link between the definitions by vector fields and differential forms is provided. That notion is also closely related to the one introduced by C. Sabbah in [38].

In the case of collections of differential forms, W. Ebeling and S.M. Gusein Zade introduce a notion of Chern obstruction that generalizes the local Euler obstruction [14]. The Chern obstruction can be characterized as an intersection number. In a paper in preparation [21] T. Gaffney and N. Grulha compute the Chern obstruction of a collection of 1-forms on a variety with isolated singularity not necessarily ICIS using the Multiplicity Polar Theorem [20]. The results in [21] generalize some results of [20].

In a recent paper, [4], J.-P. Brasselet, N. Grulha and M. Ruas studied the link between the Chern obstruction and the notion of local Euler obstruction for maps. In particular, using invariance results obtained for the Chern obstruction [14] one shows that local Euler obstruction for maps is independent of the (generic) choice of the cell (see above).

Another new development about the Euler obstruction is the extension of this local invariant to a corresponding global Euler obstruction for affine algebraic varieties. This starts with the work [42] in the language of stratified vector fields, and was further extended in the language of 1-forms in [44, 46].

2. Definition of the Euler obstruction

2.1. The Nash transformation

In the following, we will denote by$(V, 0)$ an equidimensional complex analytic singularity germ of (complex) dimension d. We will denote also by V a representative $V \subset U$ where U is an open subset in $\mathbb{C}^m$.

Let $G(d, m)$ denote the Grassmanian of complex d-planes in $\mathbb{C}^m$. On the regular part $V_{reg} = V \setminus \mathrm{Sing}(V)$ of V the Gauss map $\phi : V_{reg} \to U \times G(d, m)$ is well defined by $\phi(x) = (x, T_x(V_{reg}))$.

Definition 2.1. The *Nash transformation* (or *Nash blow up*) $\widetilde{V}$ of V is the closure of the image $\mathrm{Im}(\phi)$ in $U \times G(d, m)$. It is a (usually singular) complex analytic space endowed with an analytic projection map $v : \widetilde{V} \to V$ which is a biholomorphism away from $v^{-1}(\mathrm{Sing}(V))$.

That means that each point $y \in \mathrm{Sing}(V)$ is being replaced by all limits of planes $T_{x_i}(V_{reg})$ for sequences $\{x_i\}$ in V_{reg} converging to y.

The fiber of the tautological bundle $\mathcal{T}$ over $G(d, m)$, at the point $P \in G(d, m)$, is the set of the vectors v in the d-plane P. We still denote by $\mathcal{T}$ the corresponding trivial extension bundle over $U \times G(d, m)$. Let $\widetilde{\mathcal{T}}$ be the restriction of $\mathcal{T}$ to $\widetilde{V}$, with projection map π. The bundle $\widetilde{\mathcal{T}}$ on $\widetilde{V}$ is called *the Nash bundle* of V.

An element of $\widetilde{\mathcal{T}}$ is written (x, P, v) where $x \in U$, P is a d-plane in $\mathbb{C}^m$ based at x and v is a vector in P. We have a diagram:

$$
\begin{array}{ccc}
\widetilde{\mathcal{T}} & \hookrightarrow & \mathcal{T} \\
\pi \downarrow & & \downarrow \\
\widetilde{V} & \hookrightarrow & U \times G(d, m) \\
v \downarrow & & \downarrow \\
V & \hookrightarrow & U.
\end{array}
$$

Let us consider a complex analytic stratification $(V_\alpha)_{\alpha \in A}$ of V satisfying the Whitney conditions. Adding the stratum $U \setminus V$ we obtain a Whitney stratification of U. Let us denote by $TU|_V$ the restriction to V of the tangent bundle of U. We know that a stratified vector field v on V means a continuous section of $TU|_V$ such that if $x \in V_\alpha \cap V$ then $v(x) \in T_x(V_\alpha)$. The following proposition is a direct application of the Whitney condition (a) [7]:

Proposition 2.2. *Every stratified vector field v on a subset $A \subset V$ has a canonical lifting to a section $\tilde{v}$ of the Nash bundle $\tilde{T}$ over $v^{-1}(A) \subset \tilde{V}$.*

2.2. The local Euler obstruction

Let us firstly recall the original definition, due to R. MacPherson [34]:

Let $z = (z_1, \ldots, z_m)$ be local coordinates in $\mathbb{C}^m$ around $\{0\}$, such that $z_i(0) = 0$, we denote by $\mathbb{B}_\varepsilon$ and $\mathbb{S}_\varepsilon$ the ball and the sphere centered at $\{0\}$ and of radius ε in $\mathbb{C}^m$. Let us consider the norm $\|z\| = \sqrt{z_1 \bar{z}_1 + \cdots + z_m \bar{z}_m}$. Then the differential form $\omega = d\|z\|^2$ defines a section of the real vector bundle $T(\mathbb{C}^m)^*$, cotangent bundle on $\mathbb{C}^m$. Its pull back and, restricted to $\tilde{V}$, becomes a section denoted by $\tilde{\omega}$ of the dual bundle $\tilde{T}^*$. For small enough ε, the section $\tilde{\omega}$ is nonzero over $v^{-1}(z)$ for $0 < \|z\| \le \varepsilon$. The obstruction to extend $\tilde{\omega}$ as a nonzero section of $\tilde{T}^*$ from $v^{-1}(\mathbb{S}_\varepsilon)$ to $v^{-1}(\mathbb{B}_\varepsilon)$, denoted by $Obs(\tilde{T}^*, \tilde{\omega})$ lies in $H^{2d}(v^{-1}(\mathbb{B}_\varepsilon), v^{-1}(\mathbb{S}_\varepsilon); \mathbb{Z})$. Let us denote by $\mathcal{O}_{v^{-1}(\mathbb{B}_\varepsilon), v^{-1}(\mathbb{S}_\varepsilon)}$ the orientation class in $H_{2d}(v^{-1}(\mathbb{B}_\varepsilon), v^{-1}(\mathbb{S}_\varepsilon); \mathbb{Z})$.

Definition 2.3. the local Euler obstruction of V at 0 is the evaluation of $Obs(\tilde{T}^*, \tilde{\omega})$ on $\mathcal{O}_{v^{-1}(\mathbb{B}_\varepsilon), v^{-1}(\mathbb{S}_\varepsilon)}$, i.e.

$$Eu_V(0) = \langle Obs(\tilde{T}^*, \tilde{\omega}), \mathcal{O}_{v^{-1}(\mathbb{B}_\varepsilon), v^{-1}(\mathbb{S}_\varepsilon)} \rangle.$$

The local Euler obstruction is independent of all choices involved.

The following interpretation of the local Euler obstruction has been given by Brasselet-Schwartz [7].

Let us consider a stratified radial vector field $v(x)$ in a neighborhood of $\{0\}$ in V, i.e., there is ε_0 such that for every $0 < \varepsilon \le \varepsilon_0$, $v(x)$ is pointing outwards the ball $\mathbb{B}_\varepsilon$ over the boundary $\mathbb{S}_\varepsilon = \partial \mathbb{B}_\varepsilon$.

Definition 2.4. Let v be a radial vector field on $V \cap \mathbb{S}_\varepsilon$ and $\tilde{v}$ the lifting of v on $v^{-1}(V \cap \mathbb{S}_\varepsilon)$ to a section of the Nash bundle. The *local Euler obstruction* (or simply the Euler obstruction) $Eu_V(0)$ is defined to be the obstruction to extending $\tilde{v}$ as a nowhere zero section of $\tilde{T}$ over $v^{-1}(V \cap \mathbb{B}_\varepsilon)$.

More precisely, let $\mathcal{O}(\tilde{v}) \in H^{2d}(v^{-1}(V \cap \mathbb{B}_\varepsilon), v^{-1}(V \cap \mathbb{S}_\varepsilon))$ be the obstruction cocycle to extending $\tilde{v}$ as a nowhere zero section of $\tilde{T}$ inside $v^{-1}(V \cap \mathbb{B}_\varepsilon)$. The local Euler obstruction $Eu_V(0)$ is defined as the evaluation of the cocycle $\mathcal{O}(\tilde{v})$ on the fundamental class of the pair $(v^{-1}(V \cap \mathbb{B}_\varepsilon), v^{-1}(V \cap \mathbb{S}_\varepsilon))$. The Euler obstruction is an integer.

$Eu_V(x)$ is a constructible function on V, in fact it is constant along the strata of a Whitney stratification.

Remark 1. The Euler obstruction is not a topological invariant [24]. Let us provide the example of Briançon-Speder, which gives a counter example of topological invariance. Let us consider the family

$$\phi_t = x^5 + y^7z + z^{15} + txy^6,$$

that is a quasi-homogeneous family of hypersurfaces in $\mathbb{C}^3_{(x,y,z)}$ of the type $(3, 2, 1; 15)$, with isolated singularity, therefore topologically trivial. If we consider a generic section $z = ax + by$, we find a family $x^5 + y^7(ax + by) + (ax + by)^{15} + txy^6$ of planar curves. To compute the Milnor number of the elements of this family, it is enough to compute the Milnor number of the terms of the type $x^5 + by^8 + txy^6$.

From Lê and Teissier's results [33] we know that

$$Eu_{V_t}(0) = m_0(V_t) - m_1(V_t),$$

where $m_0(V_t)$ and $m_1(V_t)$ are polar multiplicities of the hypersurface V_t at the origin. The polar multiplicities are related with the Milnor number of the hypersurfaces [45] by the equation $m_i(V_t) = \mu_i(\phi_t) + \mu_{i+1}(\phi_t)$ where μ_i is the Milnor number of the intersection of the hypersurface with a hyperplane of dimension i. Therefore we have $Eu_{V_t}(0) = m_0(V_t, 0) - \mu_1(\phi_t) - \mu_2(\phi_t)$.

But we know that $m_0(V_t, 0) - 1 = \mu_1(\phi_t)$, therefore $Eu_{V_t}(0) = 1 - \mu_2(\phi_t)$. Looking the Newtons's polygon below we can see that $\mu_2(\phi_0) < \mu_2(\phi_t)$, so, the Euler obstruction is not constant for the family.

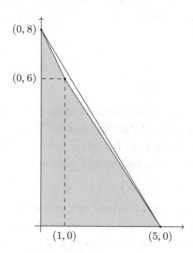

2.3. Proportionality Theorem for vector fields

The Proportionality Theorem for vector fields proved in [7] is an important property of local Euler obstruction.

Let us consider a vector field v_α defined on the stratum V_α with an isolated singularity at x and Poincaré-Hopf index on V_α denoted by $\mathrm{Ind}_{PH}(v_\alpha, x; V_\alpha)$. The vector field v_α can be lifted in a stratified vector field v defined in a neighborhood of x in $\mathbb{C}^m$ by the radial extension process described by M.-H. Schwartz (see [39, 7, 3]) in such a way that the extension v has an isolated singularity at x and satisfies

$$\mathrm{Ind}_{PH}(v, x; \mathbb{C}^m) = \mathrm{Ind}_{PH}(v_\alpha, x; V_\alpha).$$

One says that v is obtained by radial extension from v_α.

Definition 2.5. Let v be a stratified vector field obtained by radial extension of v_α with an isolated singularity at $x \in V_\alpha$. Let $\tilde{v}$ be the canonical lifting of v to a section of the Nash bundle $\tilde{T}$ over the boundary of $v^{-1}(V \cap \mathbb{B}_\varepsilon(x))$, where $\mathbb{B}_\varepsilon(x)$ is a small ball around x in $\mathbb{C}^m$. Let $\mathcal{O}(\tilde{v}) \in H^{2n}\big(v^{-1}(V \cap \mathbb{B}_\varepsilon(x)), v^{-1}(V \cap \mathbb{S}_\varepsilon(x))\big)$ be the obstruction cocycle to extending $\tilde{v}$ as a nowhere zero section of $\tilde{T}$ inside $v^{-1}(V \cap \mathbb{B}_\varepsilon(x))$. One defines the *local Euler obstruction of V relatively to the vector field v, at x*, and one denotes by $\mathrm{Eu}_V(v, x)$ the evaluation of $\mathcal{O}(\tilde{v})$ on the fundamental class of the pair $\big(v^{-1}(V \cap \mathbb{B}_\varepsilon(x)), v^{-1}(V \cap \mathbb{S}_\varepsilon(x))\big)$. That is:

$$\mathrm{Eu}_V(v, x) = \big\langle \mathcal{O}(\tilde{v}), \big[v^{-1}(V \cap \mathbb{B}_\varepsilon(x)), v^{-1}(V \cap \mathbb{S}_\varepsilon(x))\big]\big\rangle.$$

One has the following theorem:

Theorem 2.6. *Let v be a stratified vector field obtained by radial extension of v_α with an isolated singularity at $x \in V_\alpha$ with index $\mathrm{Ind}_{PH}(v_\alpha, x; V_\alpha)$. One has:*

$$\mathrm{Eu}_V(v, x) = \mathrm{Ind}_{PH}(v, x; V_\alpha) \cdot \mathrm{Eu}_V(x).$$

3. Local Euler obstruction for a frame

A k-field is a collection $v^{(k)} = (v_1, \ldots, v_k)$ of k vector fields. A singular point of the k-field $v^{(k)}$ is a point in which the vectors v_i fail to be linearly independent. A k-frame is a k-field without singularity. One says that the k-field $v^{(k)}$ is stratified if each vector v_i is a stratified vector field in the previous sense.

Let $(V, 0) \subset (\mathbb{C}^m, 0)$ be an equidimensional germ of complex analytic variety, of (complex) dimension d. Let us consider a Whitney stratification $\{V_\alpha\}$ of $\mathbb{C}^m$ compatible with V. Let us consider also a triangulation (K) of $\mathbb{C}^m$

compatible with the stratification $\{V_\alpha\}$. One denotes by σ a cell with (real) dimension $2(m - k + 1)$ in a dual cell decomposition (D) of M defined by duality from the triangulation (K). Such a cell σ is transverse to all strata of the given stratification.

Let $v^{(k)}$ be a stratified k-field on $\sigma \cap V$ with an isolated singularity at the barycenter a of σ. The k-field $v^{(k)}$ does not have singularity on $\partial\sigma \cap V$. Each vector v_i in $v^{(k)}$ admits a lifting $\tilde{v}_i$ a a section of $\widetilde{T}$ on $v^{-1}(\partial\sigma \cap V)$. The k-frame $v^{(k)}$ can be lifted as a set $\tilde{v}^{(k)}$ of k linearly independent sections of $\widetilde{T}$ on $v^{-1}(\partial\sigma \cap V)$ [9].

The cohomology class, in $H^{2(d-k+1)}(v^{-1}(\sigma \cap V), (v^{-1}(\partial\sigma \cap V)))$ of the obstruction cocycle to extending $\tilde{v}^{(k)}$ as a set of k linearly independent sections of $\widetilde{T}$ on $v^{-1}(\sigma \cap V)$ is denoted by $Obs(\tilde{v}^{(k)}, \sigma \cap V)$. The following definition is adapted from [9]:

Definition 3.1. The local Euler obstruction $Eu(v^{(k)}, V, \sigma)$ of the stratified k-field $v^{(k)}$ defined on $\sigma \cap V$ with an isolated singularity at the barycenter of σ is defined as the evaluation of the obstruction cocycle $Obs(\tilde{v}^{(k)}, \sigma \cap V)$ on the fundamental class of the pair $[v^{-1}(\sigma \cap V), v^{-1}(\partial\sigma \cap V)]$. In other words, that is:

$$Eu(v^{(k)}, V, \sigma) = \langle Obs(\tilde{v}^{(k)}, \sigma \cap V), [v^{-1}(\sigma \cap V), v^{-1}(\partial\sigma \cap V)]\rangle.$$

4. Euler obstruction and hyperplane sections

The idea of studying the Euler obstruction "à la" Lefschetz, using hyperplane sections, appears in the works of Dubson and Kato. The approach we follow here is that of [5, 6], which is topological.

We start with the following lemma, which is a special case of well known results about Lefschetz pencils. Let us denote by $\mathcal{L}$ the space of complex linear forms on $\mathbb{C}^m$. As before let $(V, 0) \subset (\mathbb{C}^m, 0)$ an equidimensional germ of complex analytic variety, of (complex) dimension d. Let us fix a Whitney stratification of V. There are a finite number of strata of this Whitney stratification which contain 0 in their closure, and we assume that the representative of $(V, 0)$ is chosen small enough so that these are the only strata of V.

Lemma 4.1 [5]. *There exists a non-empty Zariski open set Ω in $\mathcal{L}$ such that for every $l \in \Omega$, there exists a representative V of $(V, 0)$ so that:*

(i) *for each $x \in V$, the hyperplane $l^{-1}(0)$ is transverse in $\mathbb{C}^m$ to every limit of tangent spaces in TV_{reg} of points in V_{reg} converging to x;*

(ii) *for each y in the closure $\overline{V}_\alpha$ in V of each strata V_α, $\alpha = 1, \ldots, \ell$, the hyperplane $l^{-1}(0)$ is transverse in $\mathbb{C}^m$ to every limit of tangent spaces in TV_α of points converging to y.*

In particular, for each $l \in \Omega$ the Nash transformation $\widetilde{V}$ satisfies

$$\widetilde{V} \subset \mathbb{C}^m \times (G(d, m) \setminus H^*),$$

where $H^* := \{T \in G(d, m) \text{ such that } l(T) = 0\}$.

Then we can state the Theorem:

Theorem 4.2 [5]. *Let $(V, 0)$ be a germ of an equidimensional complex analytic space in $\mathbb{C}^m$. Let V_α, $\alpha = 1, \ldots, \ell$, be the (connected) strata of a Whitney stratification of a small representative V of $(V, 0)$ such that 0 is in the closure of every stratum. Then for each $l \in \Omega$ as in Lemma 4.1 there is ε_0 such that for any ε, $\varepsilon_0 > \varepsilon > 0$ and $t_0 \neq 0$ sufficiently small, the Euler obstruction of $(V, 0)$ is equal to:*

$$\mathrm{Eu}_V(0) = \sum_{\alpha=1}^{\ell} \chi(V_\alpha \cap \mathbb{B}_\varepsilon \cap l^{-1}(t_0)) \cdot \mathrm{Eu}_V(V_\alpha),$$

where χ denotes the Euler-Poincaré characteristic and $\mathrm{Eu}_V(V_\alpha)$ is the value of the Euler obstruction of V at any point of V_α, $\alpha = 1, \ldots, \ell$.

Theorem 4.2 has been proved in [5], an alternative proof is given by Schürmann in [42]. We notice that the formula above is somehow in the spirit of the formula by Lê-Teissier in [33].

Let us give some consequences of the theorem. We notice that the generic slice $V \cap \mathbb{B}_\varepsilon \cap l^{-1}(t_0)$ in 4.2 is by definition (see [18]) the *complex link* of 0 in V. In the case of an isolated singularity the complex link is smooth and there is only one stratum appearing in the sum in Theorem 4.2. In this case the theorem gives:

Corollary 4.3 [5]. *Let V be an equidimensional complex analytic subspace of $\mathbb{C}^m$ with an isolated singularity at 0. The Euler obstruction of V at 0 equals the GSV index (see [16]) of the radial vector field on a general hyperplane section $V \cap H$.*

Corollary 4.4 [12, 33]. *Let V be an equidimensional complex analytic space of dimension d in $\mathbb{C}^m$ whose singular set $\mathrm{Sing}(V)$ is 1-dimensional at 0. Let l be a general linear form defined on $\mathbb{C}^m$ and denote by $\mathbf{F}_l$ the local Milnor fiber at 0 of the restriction of l to V. The singularities of $\mathbf{F}_l$ are the points*

$\mathbf{F}_t \cap \mathrm{Sing}(V) =: \{x_1, \ldots, x_k\}$. *Then,*

$$\mathrm{Eu}_V(0) = \chi(\mathbf{F}_t) - k + \sum_1^k \mathrm{Eu}_V(x_i).$$

5. The local Euler obstruction of a function

In this section we define an invariant introduced by J.P. Brasselet, D. Massey, A. J. Parameswaran and J. Seade in [6], which measures in a way how far the equality given in Theorem 4.2 is from being true if we replace the generic linear form l by some other function on V with at most an isolated stratified critical point at 0. For this it is convenient to think of the local Euler obstruction as defining an index for stratified vector fields. To be precise, let $(V, 0)$ be again a complex analytic germ contained in an open subset U of $\mathbb{C}^m$ and endowed with a complex analytic Whitney stratification $\{V_\alpha\}$. We assume further that every stratum contains 0 in its closure. For every point $x \in V$, we will denote by $V_\alpha(x)$ the stratum containing x. We recall first some well known concepts about singularity theory which originate in the work of R. Thom.

Let $f : V \to \mathbb{C}$ be a holomorphic function which is the restriction of a holomorphic function $F : U \to \mathbb{C}$. We recall [18] that a *critical point* of f is a point $x \in V$ such that $dF(x)(T_x(V_\alpha(x))) = 0$. We say, following [31], [18], that f has an isolated singularity at $0 \in V$ relative to the given Whitney stratification if f does not have critical points in a punctured neighborhood of 0 in V.

Let us denote by $\overline{\nabla}F(x)$ the gradient vector field of F at a point $x \in U$, defined by $\overline{\nabla}F(x) := (\frac{\overline{\partial F}}{\partial x_1}, \ldots, \frac{\overline{\partial F}}{\partial x_m})$, where the bar denotes complex conjugation. From now on we assume that f has an isolated singularity at $0 \in V$. This implies that the kernel $\mathrm{Ker}(dF)$ is transverse to $T_x(V_\alpha(x))$ at any point $x \in V \setminus \{0\}$. Therefore at such a point, we have:

$$\mathrm{Angle}\langle \overline{\nabla}F(x), T_x(V_\alpha(x))\rangle < \pi/2,$$

so the projection of $\overline{\nabla}F(x)$ on $T_x(V_\alpha(x))$, denoted by $\zeta_\alpha(x)$, does not vanish.

Let V_β be a stratum such that $V_\alpha \subset \overline{V}_\beta$, and let $\pi : \mathcal{U}_\alpha \to V_\alpha$ be a tubular neighborhood of V_α in U. Following the construction of M.-H. Schwartz in [39, §2] we see that the Whitney condition (a) implies that at each point $y \in V_\beta \cap \mathcal{U}_\alpha$, the angle of $\zeta_\beta(y)$ and of the parallel extension of $\zeta_\alpha(\pi(y))$ is small. This property implies that these two vector fields are homotopic on the boundary of $\mathcal{U}_\alpha$. Therefore, we can glue together the vector fields ζ_α to obtain a stratified vector field on V, denoted by $\overline{\nabla}_V f$. This vector field is homotopic to $\overline{\nabla}F|_V$ and one has $\overline{\nabla}_V f \neq 0$ unless $x = 0$.

Intuitively, what we are doing in the construction of $\overline{\nabla}_V f$ is to take, for each stratum V_α of V, the gradient vector field of the restriction of f to V_α, and then gluing all these vector fields together.

Definition 5.1. Let $\nu : \widetilde{V} \to V$ be the Nash transform of V. We define *the local Euler obstruction of f on V* at 0, denoted $\mathrm{Eu}_{f,V}(0)$, to be the Euler obstruction $\mathrm{Eu}(\overline{\nabla}_V f, V, 0)$ (see Definition 2.5) of the stratified vector field $\overline{\nabla}_V f$ at $0 \in V$.

In other words, let $\tilde{\zeta}$ be the lifting of $\overline{\nabla}_V f$ as a section of the Nash bundle $\widetilde{T}$ over $\widetilde{V}$ without singularity over $\nu^{-1}(V \cap \mathbb{S}_\epsilon)$, where $\mathbb{S}_\epsilon = \partial \mathbb{B}_\epsilon$ is the boundary of a small sphere around 0. Let $\mathcal{O}(\tilde{\zeta}) \in H^{2n}\left(\nu^{-1}(V \cap \mathbb{B}_\varepsilon), \nu^{-1}(V \cap \mathbb{S}_\varepsilon)\right)$ be the obstruction cocycle to the extension of $\tilde{\zeta}$ as a nowhere zero section of $\widetilde{T}$ inside $\nu^{-1}(V \cap \mathbb{B}_\varepsilon)$. Then the local Euler obstruction $\mathrm{Eu}_{f,V}(0)$ is the evaluation of $\mathcal{O}(\tilde{\zeta})$ on the fundamental class of the pair $(\nu^{-1}(V \cap \mathbb{B}_\varepsilon), \nu^{-1}(V \cap \mathbb{S}_\varepsilon))$.

We notice that all these definitions and constructions also work when f is the restriction to V of a real analytic function on the ambient space. For instance, we can take f to be the function distance to 0 on V, then $\overline{\nabla}_V f$ is a radial vector field and the invariant $\mathrm{Eu}_{f,V}(0)$ is the usual local Euler obstruction of V at 0.

6. The Euler obstruction and the Euler defect

The following result [6] compares the Euler obstruction of the space V with that of a function on V.

Theorem 6.1. *Let $f : (V, 0) \to (\mathbb{C}, 0)$ have an isolated singularity at $0 \in V$. One has:*

$$\mathrm{Eu}_{f,V}(0) = \mathrm{Eu}_V(0) - \left(\sum_\alpha \chi(V_\alpha \cap \mathbb{B}_\varepsilon \cap f^{-1}(t_0)) \cdot \mathrm{Eu}_V(V_\alpha) \right).$$

In other words, the invariant $\mathrm{Eu}_{f,V}(0)$ can be regarded as the "defect" for the local Euler obstruction of V to satisfy the Euler condition with respect to the function f. In this way one can generalize the definition of the Euler obstruction to functions with non-isolated singularities and one gets the *Euler defect* introduced in [6]. This arises as a natural application of Massey's work [35, 36] on intersections of characteristic cycles and derived categories.

7. The Euler defect at general points

By definition, if 0 is a smooth point of V and a regular point of f then $\mathrm{Eu}_{f,V}(0) = 0$ since in this case $\mathrm{Eu}_{f,V}(0)$ is the Poincaré-Hopf index of a vector field at a non-singular point. Proposition 7.2 below shows that this is the case in a more general situation.

Definition 7.1. Let $(V, 0) \subset (U, 0)$ be a germ of analytic set in $\mathbb{C}^m$ equipped with a Whitney stratification and let $f : (V, 0) \to (\mathbb{C}, 0)$ be a holomorphic function, restriction of a regular holomorphic function $F : (U, 0) \to (\mathbb{C}, 0)$. We say that 0 *is a general point of* f if the hyperplane $\mathrm{Ker}\, dF(0)$ is transverse in $\mathbb{C}^m$ to every generalized tangent space at 0, *i.e.* to every limit of tangent spaces $T_{x_i}(V_\alpha)$, for every V_α and every sequence $x_i \in V_\alpha$ converging to 0.

We notice that for every f as above the general points of f form a nonempty open set on each (open) stratum of V, essentially by Sard's theorem. We also remark that this definition provides a coordinate free way of looking at the general linear forms considered in Theorem 4.2. In fact the previous definition is equivalent to saying that with an appropriate local change of coordinates F is a linear form in U, and it is general with respect to V.

Proposition 7.2. *Let* 0 *be a general point of* $f : (V, 0) \to (\mathbb{C}, 0)$. *Then*

$$\mathrm{Eu}_{f,V}(0) = 0\,.$$

8. The Euler obstruction via Morse theory

This section is taken from [41], by J. Seade, M. Tibăr and A. Verjovsky. Here we show how stratified Morse theory yields to a clear understanding of what the invariant $\mathrm{Eu}_{f,V}(x)$ is for arbitrary functions with an isolated singularity. These results can also be deduced from Schürmann's book [43], and also from the work of D. Massey, as for instance [35, 36]. For this we recall the definition of complex stratified Morse singularities (see Goresky-MacPherson [18], p. 52).

Definition 8.1. Let V_α be a Whitney stratification of V and let $f : V \to \mathbb{C}$ be the restriction to V of a holomorphic function $F : \mathbb{C}^m \to \mathbb{C}$; assume for simplicity $0 = f(x)$. One says that $f : (V, x) \to (\mathbb{C}, 0)$ has a *stratified Morse critical point* at $x \in V$ if the dimension of the stratum V_α that contains x is ≥ 1, the restriction of f to V_α has a Morse singularity at x and f is general with respect to all other strata containing x in its closure, *i.e.*, Ker $dF(x)$ is transverse in $\mathbb{C}^m$ to every limit of tangent spaces $T_{x_i}(V_\beta)$, for every stratum V_β such that $V_\alpha \subset \overline{V}_\beta$ and every sequence $x_i \in V_\beta$ converging to x.

We recall that every map-germ f on $(V, 0)$ can be morsified, *i.e.*, approximated by Morse singularities. This is proved in [32] for f with an isolated singularity.

The theorem below is contained in [41].

Theorem 8.2. *Let f be a holomorphic function germ on $(V, 0)$ with an isolated singularity (stratified critical point) at 0, restriction of a function F on an open subset in $\mathbb{C}^m$. Let $V_\alpha \subset V$ be the stratum that contains 0. Then:*

(i) *If $\dim V_\alpha < \dim V$ and $\mathrm{Ker}\, dF$ does not vanish on any generalized tangent space of the regular stratum (in particular if f is Morse at 0), then* $\mathrm{Eu}_{f,V}(0) = 0$.

(ii) *If f has a stratified Morse singularity at $0 \in V_\alpha$ and $\dim V_\alpha = \dim V = n$, then* $\mathrm{Eu}_{f,V}(0) = (-1)^n$.

(iii) *In general, the number of critical points of a Morsification of f in the regular part of V is $(-1)^{n+1}\mathrm{Eu}_{f,V}(0)$.*

9. Generalizations of Milnor number

In D. T. Lê's work [30] a new notion of Milnor number arises, that is a generalization of the Milnor number for analytic functions defined on singular analytic spaces such that the so-called rectified homotopical depth of V at 0, denoted $\mathrm{rhd}(V, 0)$ satisfies $\mathrm{rhd}(V, 0) = \dim_{\mathbb{C}}(V, 0)$.

Let V be a sufficiently small representative of the germ $(V, 0)$. The Milnor fiber of the complex analytic function f, defined on V, with an isolated singularity at 0 (in the stratified way), has the homotopy type of a bouquet of spheres. The Lê's Milnor number, denoted by $\mu_L(f)$, is defined as the number of spheres in the bouquet.

The relations between this invariant and the local Euler obstruction of f were obtained in [40]. In particular one has:

Theorem 9.1. *Let V be a sufficiently small representative of the germ $(V, 0)$ of a complex analytic space. Let us consider a complex analytic function defined on V with a stratified isolated singularity at 0. Then, if $\mathrm{rhd}(V, 0) = \dim_{\mathbb{C}}(V, 0)$ we have that*

$$\mu_L(f) \geq (-1)^{\dim_{\mathbb{C}}(V,0)} Eu_{f,V}(0).$$

The condition rhd$(V, 0) = \dim_{\mathbb{C}}(V, 0)$ is satisfied for a complete intersection with isolated singularity (ICIS). In this case the following holds (see [40]):

Theorem 9.2. *Let V be a sufficiently small representative of an ICIS germ, $(V, 0)$, f an analytic function on V with stratified isolated singularity at 0, and l a generic linear form. Then, we have*

$$Eu_{f,V}(0) = (-1)^{\dim_{\mathbb{C}}(V,0)}[\mu_L(f) - \mu_L(l)].$$

Proposition 9.3. *Let $(V, 0) \subset (\mathbb{C}^m, 0)$ be an ICIS, and $F : (\mathbb{C}^m \times \mathbb{C}^r, 0) \to \mathbb{C}$ a family of functions with isolated singularity, we denote $F(x, u) = f_u(x)$. Then, the following are equivalent,*

(i) $Eu_{f_u,V}(0)$ *is constant for the family.*
(ii) $\mu_L(f_u)$ *is constant for the family.*

10. Bruce and Roberts' Milnor Number

Bruce and Roberts gave in [10] an alternative generalization for the notion of Milnor number for a function on a singular variety. One of the main goals in [10] is to characterize germs of diffeomorphisms preserving V. The usual technique is the integration of germs of vector fields tangent to V.

Let us denote by $\mathcal{O}_m$ the ring of germs of holomorphic germs of functions $f : (\mathbb{C}^m, 0) \to \mathbb{C}$ at 0, V a sufficiently small representative of the germ $(V, 0)$ and $\mathcal{I}(V)$ denote the ideal in $\mathcal{O}_m$ consisting of the germs of functions vanishing on V.

Definition 10.1. For $x \in \mathbb{C}^m$, let $Der_x \mathbb{C}^m$ denote the $\mathcal{O}_m$-module of germs of analytic vector fields on $\mathbb{C}^m$ at x. A vector field δ in $Der_x \mathbb{C}^m$ is said to be logarithmic for (V, x) if, when considered as a derivation $\delta : \mathcal{O}_m \to \mathcal{O}_m$, we have $\delta(h) \in \mathcal{I}(V)$ for all $h \in \mathcal{I}(V)$. The $\mathcal{O}_m$-module of such vector fields is denoted by $\Theta_{(V,x)}$. When $x = 0$, we denote it by Θ_V.

Definition 10.2. Let $f : (\mathbb{C}^m, 0) \to (\mathbb{C}, 0)$ be a function and $J_V(f)$ the ideal $\{\delta f : \delta \in \Theta_{V,0}\}$ of $\mathcal{O}_m$. The Bruce-Roberts' Milnor number of f on V at 0, denoted by $\mu_{BR}(f)$ is defined by $\dim_{\mathbb{C}} \mathcal{O}_m/J_V(f)$.

Definition 10.3 [10]. Let us suppose that the vector fields $\delta_1, \ldots, \delta_m$ generate Θ_V for some neighborhood U at $0 \in \mathbb{C}^m$. Then if $T_U^* \mathbb{C}^m$ is the restriction of the cotangent bundle of $\mathbb{C}^m$ in U, we define

$$LC_U(V) = \{(x, \xi) \in T_U^* \mathbb{C}^m : \xi(\delta_i(x)) = 0, i = 1 \ldots, m\}.$$

$LC(V)$ is defined as the germ of $L\mathcal{C}_U(V)$ in $T_U^*\mathbb{C}^n$, and it can be shown that it is independent of the choice of generators of Θ_V.

Let V_α be a stratum of the logarithmic stratification (see [10]) of V. The conormal space of V_α is the subspace of $T_0^*\mathbb{C}^n$ given by all forms vanishing on the tangent bundle TV_α. We denote it by $C(V_\alpha)$.

Then we have,

$$LC(V) = \bigcup_\alpha \overline{C(V_\alpha)},$$

with multiplicities [10].

In [24, 25] the author prove the following result that relates the Euler obstruction and the Bruce-Roberts' Milnor number.

Theorem 10.4. *Let $(V, 0)$ be the germ of a reduced equidimensinal analytic variety and $f : (\mathbb{C}^n, 0) \to (\mathbb{C}, 0)$ a function with isolated singularity at the origin in such that f has also a isolated singularity in the stratified way. If $LC(V)$ is Cohen-Macaulay we have,*

$$\mu_{BR}(f) = \sum_{\alpha=0}^{d} m_\alpha(-1)^{\dim_\mathbb{C} V_\alpha} Eu_{f,\overline{V}_\alpha}(0),$$

where m_α denotes the multiplicity of T^*V_α in $LC(V)$.

An important class of examples is the case of V being the discriminant of an analytic stable map-germ $F : (\mathbb{C}^n, 0) \to (\mathbb{C}^p, 0)$, $n \geq p$, with (n, p) in nice dimensions of Mather [19].

The next result is one of the main results of [25].

Theorem 10.5. *Let $V \subset \mathbb{C}^n$ be a hypersurface with isolated singularity with $LC(V)$ Cohen-Macaulay, and $F : (\mathbb{C}^n \times \mathbb{C}^r, 0) \to \mathbb{C}$ a family of functions with isolated singularity, then:*

(a) *$\mu_{BR}(f_u)$ constant for the family implies $\mu(f_u)$, $\mu_L(f_u)$ and $Eu_{f_u,V}(0)$ constant for the family.*
(b) *$\mu(f_u)$ is constant for the family, and either $Eu_{f_u,V}(0)$ or $\mu_L(f_u)$ constant for the family implies $\mu_{BR}(f_u)$ is constant for the family.*

Corollary 10.6. *Let $V \subset \mathbb{C}^n$ a hypersurface with isolated singularity at the origin with $LC(V)$ Cohen-Macaulay, and $F : (\mathbb{C}^n \times \mathbb{C}^r, 0) \to \mathbb{C}$, a family of functions with isolated singularity, such that is a topological trivial family in V, then $\mu_{BR}(f_u)$ is constant for the family.*

The converse of Corollary 10.6, *i.e.* the topological invariance of μ_{BR}, is an open problem in this theory.

11. Local Euler obstruction of an analytic map

Let us fix an integer p, $1 \le p \le d$. Let us consider a germ of analytic map $f : (V, 0) \to (\mathbb{C}^p, 0)$, restriction of $F : (U, 0) \to (\mathbb{C}^p, 0)$, where U is a neighborhood of 0 in of $\mathbb{C}^m$ and

$$F(z) = (F_1(z), F_2(z), \ldots, F_p(z)).$$

One denotes by

$$f(z) = (f_1(z), f_2(z), \ldots, f_p(z))$$

the coordinate maps of f.

Following [6] let us denote by $\overline{\nabla} F_i(z) = (\overline{\frac{\partial F_i}{\partial z_1}}(z), \ldots, \overline{\frac{\partial F_i}{\partial z_n}}(z))$ the vector field associated to F_i on U, where the bar means that we take the complex conjugate vector field. Let us consider $z \in V \setminus \{0\}$ and denote by V_α the stratum containing z. The kernel $ker(d F_i(z))$ is transversal to $T_z(V_\alpha)$. So, we have, for $z \in V \setminus \{0\}$:

$$Angle\langle \overline{\nabla} F_i(z), T_z(V_\alpha) \rangle < \pi/2,$$

therefore the projection of $\overline{\nabla} F_i(z)$ on $T_z(V_\alpha)$, does not vanish. We denote it by $\overline{\nabla}_V f_i(z)$. The construction can be done in such a way that the vector fields $(\overline{\nabla}_V f_1(z), \overline{\nabla}_V f_2(z), \ldots, \overline{\nabla}_V f_p(z))$ are linearly independent (see [23]).

One obtains a stratified p-frame denoted by $\overline{\nabla}_V^{(p)} f$, without singularity on $(\partial \mathbb{B}_\varepsilon \cap V \setminus \{\Sigma f\})$, where $\mathbb{B}_\varepsilon$ is a ball in $\mathbb{C}^m$ centered at 0 and Σf the singular set of f.

The definition of the local Euler obstruction of a map at a singular point, given in [23], depends on the choice of a $2(m - p + 1)$-cell σ satisfying the following condition (δ). In the following, we will see (Corollary 13.5) that the local Euler obstruction of the map does not depend on the (generic) choice of the cell σ.

Definition 11.1. Let $(V, 0) \subset (\mathbb{C}^m, 0)$ a germ of equidimensional (complex) analytic variety with dimension d. Let us consider an analytic map $f : (V, 0) \to (\mathbb{C}^p, 0)$, defined by coordinate functions. One says that f satisfies the (δ) condition if there is a cell σ with (real) dimension $2(m - p + 1)$ and with barycenter 0, in a dual cell decomposition (D) of $\mathbb{C}^m$, such that:

$$\Sigma f \cap \partial \sigma = \emptyset. \tag{δ}$$

If f satisfies the (δ) condition for the cell σ, the p-frame $\overline{\nabla}_V^{(p)} f$ can be lifted up as a set of p linearly independent sections $\widetilde{\overline{\nabla}}_V^{(p)} f$ of $\widetilde{T}$ on $\nu^{-1}(V^\sigma)$ where $V^\sigma = V \cap \partial\sigma$. Let us denote by $\xi \in H^{2(d-p+1)}(\nu^{-1}(V^\sigma), \nu^{-1}(\partial V^\sigma))$ the obstruction cocycle to extend $\widetilde{\overline{\nabla}}_V^{(p)} f$ as a set of p linearly independent sections of $\widetilde{T}$ on $\nu^{-1}(V^\sigma)$.

Definition 11.2. With the previous hypothesis, one defines the local Euler obstruction of f, denoted by $Eu_{f,V}(\sigma)$, as the evaluation of the cocycle ξ on the fundamental class of the pair $[\nu^{-1}(V^\sigma), \nu^{-1}(\partial V^\sigma)]$. On other words, one defines

$$Eu_{f,V}(\sigma) = \langle \xi, [\nu^{-1}(V^\sigma), \nu^{-1}(\partial V^\sigma)] \rangle.$$

In the case $p = 1$, one has the local Euler obstruction of the function f defined in [6] (see Definition 5.1).

The following result generalizes Theorem 6.1 (see Corollary 13.5):

Theorem 11.3. *Let $(V, 0) \subset (\mathbb{C}^m, 0)$ a germ of equidimensional complex analytic variety of complex dimension d. Let $\{V_\alpha\}$ be a Whitney stratification of $\mathbb{C}^m$ compatible with V. For $0 < p \le d$, let $f : V \to \mathbb{C}^p$ be an analytic map satisfying the condition (δ) for a $2(m - p + 1)$-cell σ_f (that depends on f, [23]) in a cellular decomposition (D) and such that 0 is barycenter of σ. Let $v^{(p)} = (v^{(p-1)}, v_p)$ be a p-frame without singularity on ∂V^σ such that v_p is a radial vector field in the usual sense, then one has:*

$$Eu(v^{(p)}, V, \sigma) = \left(\sum_\alpha Eu_V(V_\alpha) \cdot \chi(V_\alpha \cap B_\epsilon \cap f^{-1}(z_0)) \right) + Eu_{f,V}(\sigma)$$

where $\{V_\alpha\}$ describes the set of strata in V such that $0 \in \overline{V}_\alpha$.

12. Local Euler obstruction of 1-forms

Let us consider the Nash bundle $\widetilde{T}$ on $\widetilde{V}$. The corresponding dual bundles of complex and real 1-forms are denoted by $\widetilde{T}^* \to \widetilde{V}$ and $\widetilde{T}_{\mathbb{R}}^* \to \widetilde{V}$, respectively. Observe that a point in $\widetilde{T}^*$ is a triple (x, P, ω) where x is in V, P is an d-plane in the tangent space $T_x\mathbb{C}^m$ which is limit of a sequence $\{T_{x_i}(V_{\text{reg}})\}$, where the x_i are points in the regular part of V converging to x, and ω is a $\mathbb{C}$-linear map $P \to \mathbb{C}$ (similarly for $\widetilde{T}_{\mathbb{R}}^*$).

Definition 12.1. Let $\{V_\alpha\}$ be a Whitney stratification of V. Let ω be a (real or complex) 1-form on V, *i.e.*, a continuous section of either $T_{\mathbb{R}}^* M|_V$ or $T^* M|_V$.

A singularity of ω in the stratified sense means a point x where the kernel of ω contains the tangent space of the corresponding stratum.

This means that the pull back of the form to V_α vanishes at x.

Given a section η of $T_\mathbb{R}^* \mathbb{C}^m|_A$, $A \subset V$, there is a canonical way of constructing a section $\tilde\eta$ of $\tilde{T}_\mathbb{R}^*|_{\tilde{A}}$, $\tilde{A} = \nu^{-1}A$, that we describe now. Notice that the same construction works for complex forms. First, taking the pull-back $\nu^*\eta$, we get a section of $\nu^* T_\mathbb{R}^* \mathbb{C}^m|_V$. Then $\tilde\eta$ is obtained by projecting $\nu^*\eta$ to a section of $\tilde{T}_\mathbb{R}^*$ by the canonical bundle homomorphism

$$\nu^* T_\mathbb{R}^* \mathbb{C}^m|_V \longrightarrow \tilde{T}_\mathbb{R}^*.$$

Thus the value of $\tilde\eta$ at a point (x, P) is simply the restriction of the linear map $\eta(x): (T_\mathbb{R}\mathbb{C}^m)_x \to \mathbb{R}$ to P. We call $\tilde\eta$ the *canonical lifting* of η.

By the Whitney condition (a), if $a \in V_\alpha$ is the limit point of the sequence $\{x_i\} \in V_{\text{reg}}$ such that $P = \lim(T_{x_i}(V_{\text{reg}}))$ exists and if the kernel of η is transverse to V_α, then the linear form $\tilde\eta$ will be non-vanishing on P. Thus, if η has an isolated singularity at the point $0 \in V$ (in the stratified sense), then we have a never-zero section $\tilde\eta$ of the dual Nash bundle $\tilde{T}_\mathbb{R}^*$ over $\nu^{-1}(\mathbb{S}_\varepsilon \cap V) \subset \tilde{V}$. Let $o(\eta) \in H^{2d}(\nu^{-1}(\mathbb{B}_\varepsilon \cap V), \nu^{-1}(\mathbb{S}_\varepsilon \cap V); \mathbb{Z})$ be the cohomology class of the obstruction cycle to extend this to a section of $\tilde{T}_\mathbb{R}^*$ over $\nu^{-1}(\mathbb{B}_\varepsilon \cap V)$. Then define (c.f. [6, 14]):

Definition 12.2. The *local Euler obstruction* of the real differential form η at an isolated singularity is the integer $\mathrm{Eu}_V(\eta, 0)$ obtained by evaluating the obstruction cohomology class $o(\eta)$ on the orientation fundamental cycle $[\nu^{-1}(\mathbb{B}_\varepsilon \cap V), \nu^{-1}\mathbb{S}_\varepsilon \cap V)]$.

MacPherson's local Euler obstruction $\mathrm{Eu}_V(0)$ corresponds to taking the differential $\omega = d\|z\|^2$ of the square of the function distance to 0.

In the complex case, one can perform the same construction, using the corresponding complex bundles. If ω is a complex differential form, section of $T^*\mathbb{C}^m|_A$ with an isolated singularity, one can define the local Euler obstruction $\mathrm{Eu}_V(\omega, 0)$. Notice that it is equal to the local Euler obstruction of its real part up to sign:

$$\mathrm{Eu}_V(\omega, 0) = (-1)^d \mathrm{Eu}_V(\mathrm{Re}\,\omega, 0).$$

This is an immediate consequence of the relation between the Chern classes of a complex vector bundle and those of its dual.

We note that the idea to consider the (complex) dual Nash bundle was already present in [38], where Sabbah introduces a local Euler obstruction $\mathrm{E\check{u}}_V(0)$ that satisfies $\mathrm{E\check{u}}_V(0) = (-1)^d \mathrm{Eu}_V(0)$. See also [42, sec. 5.2].

Just as for vector fields, (see [7]), one can define the Poincaré-Hopf index of an 1-form in a singular point and one has in this situation the following·

Theorem 12.3 [9]. *Let $V_\alpha \subset V$ be the stratum containing 0, $\mathrm{Eu}_V(0)$ the local Euler obstruction of V at 0 and ω a (real or complex) 1-form on V_α with an isolated singularity at 0. Then the local Euler obstruction of the radial extension ω' of ω and the Poincaré-Hopf index of ω at 0 are related by the following proportionality formula:*

$$\mathrm{Eu}_V(\omega', 0) = \mathrm{Eu}_V(0) \cdot \mathrm{Ind}_{\mathrm{PH}}(\omega, 0; V).$$

13. Local Chern obstruction of collections of 1-forms and special points

In this section we recall some ideas and notation from W. Ebeling and S. M. Gusein-Zade about indices of collections of 1-forms [14, 15].

The notion of local Chern obstruction extends the notion of local Euler obstruction in the case of collections of germs of 1-forms. More precisely, W. Ebeling and S. M. Gusein-Zade perfom the following construction.

Let $(V^d, 0) \subset (\mathbb{C}^m, 0)$ be the germ of a purely d-dimensional reduced complex analytic variety at the origin. Let $\{\omega_j^{(i)}\}$ be a collection of germs of 1-forms on $(\mathbb{C}^m, 0)$ with s fixed, $i = 1, \ldots, s$, k_i are integers such that $\sum k_i = d$, $j = 1, \ldots, d - k_i + 1$. Let $\varepsilon > 0$ be small enough so that there is a representative V of the germ $(V, 0)$ and representatives $\{\omega_j^{(i)}\}$ of the germs of 1-forms inside the ball $B_\varepsilon(0) \subset \mathbb{C}^m$.

Definition 13.1. A point $x \in V$ is called a *special* point of the collection $\{\omega_j^{(i)}\}$ of 1-forms on the variety V if there exists a sequence x_n of points on the non-singular part V_{reg} of the variety V such that the sequence $T_{x_n} V_{reg}$ of the tangent spaces at the points x_n has a limit L (in $G(d, m)$) and the restriction of the 1-forms $\omega_1^{(i)}, \ldots, \omega_{d-k_i+1}^{(i)}$ to the subspace $L \subset T_x \mathbb{C}^m$ are linearly dependent for each $i = 1, \ldots, s$. The collection $\{\omega_j^{(i)}\}$ of 1-forms has an *isolated special point* on $(V, 0)$ if it has no special point on V in a punctured neighborhood of the origin.

Notice that we require each set $\omega_j^{(i)}$ for i fixed in the collection to be linearly dependent when restricted to the same limit plane. Notice also, that if an element of the collection has less than maximal rank at a point, then it is linearly dependent on all planes passing through the point.

Let $\{\omega_j^{(i)}\}$ be a collection of germs of 1-forms on $(V, 0)$ with an isolated special point at the origin. The collection of 1-forms $\{\omega_j^{(i)}\}$ gives rise to a section $\Gamma(\omega)$ of the bundle

$$\widetilde{\mathbb{T}} = \bigoplus_{i=1}^{s} \bigoplus_{j=1}^{d-k_i+1} \widetilde{T}_{i,j}^*$$

where $\widetilde{T}_{i,j}^*$ are copies of the dual Nash bundle $\widetilde{T}^*$ over the Nash transform $\widetilde{V}$ numbered by indices i and j.

Let $\mathbb{D} \subset \widetilde{\mathbb{T}}$ be the set of pairs $(x, \{\alpha_j^{(i)}\})$ where $x \in \widetilde{V}$ and the collection $\{\alpha_j^{(i)}\}$ is such that $\alpha_1^{(i)}, \ldots, \alpha_{d-k_i+1}^{(i)}$ are linearly dependent for each $i = 1, \ldots, s$.

Definition 13.2. Let 0 be a special point of the collection $\{\omega_j^{(i)}\}$. *The local Chern obstruction $Ch_{V,0}\{\omega_j^{(i)}\}$ of the collection of germs of 1-forms $\{\omega_j^{(i)}\}$ on $(V, 0)$ at the origin is the obstruction to extend the section $\Gamma(\omega)$ of the fibre bundle $\widetilde{\mathbb{T}} \setminus \mathbb{D} \to \widetilde{X}$ from the preimage of a neighbourhood of the sphere $S_\varepsilon = \partial B_\varepsilon$ to $\widetilde{V}$. More precisely its value (as an element of the cohomology group $H^{2d}(\nu^{-1}(V \cap B_\varepsilon), \nu^{-1}(V \cap S_\varepsilon), \mathbb{Z}))$ on the fundamental class of the pair $(\nu^{-1}(V \cap B_\varepsilon), \nu^{-1}(V \cap S_\varepsilon))$.*

We can see that we have the correct obstruction dimension as follows. For each $\widetilde{\mathbb{T}}_i = \bigoplus_{j=1}^{d-k_i+1} \widetilde{T}_{i,j}^*$ let $\mathbb{D}_i \subset \widetilde{\mathbb{T}}_i$ be the set of pairs $(x, \{\alpha_j^{(i)}\})$ where $x \in \widetilde{V}$ and the collection $\{\alpha_j^{(i)}\}$ are such that $\alpha_1^{(i)}, \ldots, \alpha_{d-k_i+1}^{(i)}$ are linearly dependent. Then, the set $\widetilde{\mathbb{T}}_i \setminus \mathbb{D}_i$ is a Stiefel manifold, with associated obstruction dimension equal to k_i, therefore the obstruction dimension for $\widetilde{\mathbb{T}}$ is $\sum k_i = d$.

The following result is one of the main results in [14].

Proposition 13.3. *The local Chern obstruction $Ch_{V,0}\{\omega_j^{(i)}\}$ of a collection $\{\omega_j^{(i)}\}$ of germs of holomorphic 1-forms is equal to the number of special points on V of a generic deformation of the collection.*

In [4] J.-P. Brasselet, N. Grulha and M. Ruas proved that the Euler obstruction of a map, defined in [23, 24] (see Definition 11.2) is in fact independent of a generic choice of σ, as a consequence of the last proposition, more precisely:

Theorem 13.4 [4]. *Let $(V, 0)$ as above and $f : (V, 0) \to \mathbb{C}^p$ be a map-germ defined on V. Then there exists a collection $\{\omega_j^{(i)}\}$ as above such that*

$$Ch_{V,0}\{\omega_j^{(i)}\} = (-1)^{d-p+1} Eu_{f,V}(\sigma).$$

Corollary 13.5. *The local Euler obstruction of a map, defined in Definition 11.2 is independent of a generic choice of σ, we denote it by $Eu_{f,V}(0)$.*

References

1. J.-P. Brasselet, Existence des classes de Chern en théorie bivariante, *Astérisque* **101–102** (1983) 7–22.
2. J.-P. Brasselet, Local Euler obstruction, old and new. In: Proceedings of XI Brazilian Topology Meeting (Rio Claro, 1998), World Sci. Publishing, River Edge, NJ (2000) 140–147.
3. J.-P. Brasselet, *Characteristic Classes of Singular Varieties*. Book in preparation.
4. J.-P. Brasselet, N. G. Grulha Jr. and M. A. S. Ruas, The Euler obstruction and the Chern obstruction. Preprint 2009.
5. J.-P. Brasselet, D. T. Lê and J. Seade, Euler obstruction and indices of vector fields, *Topology*, **6** (2000) 1193–1208.
6. J.-P. Brasselet, D. Massey, A. J. Parameswaran and J. Seade, Euler obstruction and defects of functions on singular varieties, *Jornal London Math. Soc.* (2) **70(1)** (2004) 59–76.
7. J.-P. Brasselet et M.-H. Schwartz, Sur les classes de Chern d'un ensemble analytique complexe, *Astérisque* **82–83** (1981) 93–147.
8. J.-P. Brasselet, J. Seade and T. Suwa, A proof of the proportionality theorem. Preprint, 2005.
9. J.-P. Brasselet, J. Seade and T. Suwa, Vector fields on singular varieties. *Lecture Notes in Mathematics*, 1987, Springer-Verlag, Berlin 2009.
10. J. W. Bruce and R. M. Roberts Critical points of functions on analytic varieties, *Topology* **27(1)** (1988) 57–90.
11. J. L. Brylinski, A. Dubson and M. Kashiwara, Formule d'indice pour les modules holonomes et obstruction d'Euler locale *C.R. Acad. Sci. Paris* **293** (1981) 573–576.
12. A. Dubson, Classes caractéristiques des variétés singulières, *C.R. Acad. Sci. Paris* **287(4)** (1978) 237–240.
13. A. Dubson, Formule pour l'indice des complexes constructibles et des D-modules holonomes, *C. R. Acad. Sci. Paris*, Sér. I Math. **298(6)** (1984) 113–116.
14. W. Ebeling and S. M. Gusein-Zade, Chern obstruction for collections of 1-forms on singular varieties, *Singularity theory*, World Sci. Publ., Hackensack, NJ, (2007) 557–564.
15. W. Ebeling and S. M. Gusein-Zade, Indices of vector fields and 1-forms on singular varieties, *Global aspects of complex geometry*, Springer, Berlin (2006) 129–169.
16. X. Gómez-Mont, J. Seade and A. Verjovsky, The index of a holomorphic flow with an isolated singularity, *Math. Ann.* **291** (1991), 737–751.
17. G. Gonzalez-Sprinberg, L'obstruction Locale d'Euler et le Théorème de MacPherson, *Astérisque* **82–83** (1981) 7–32.
18. M. Goresky and R. MacPherson, *Stratified Morse theory*, Ergebnisse der Mathematik und ihrer Grenzgebiete. 3 Folge, Bd. 14. Berlin Springer-Verlag, 1988.
19. J. N. Mather, Stability of C^∞ mappings, VI: The nice dimensions, *Proc. of Liverpool Singularities-Symposium, I* (1969/70) LNM, Springer **192** (1971) 207–253.
20. T. Gaffney, The Multiplicity polar theorem and isolated singularities, *Journal of Algebraic Geometry* **18(3)** (2009) 547–574.
21. T. Gaffney and N. G. Grulha Jr, The Multiplicity Polar Theorem, Collections of 1-forms and Chern Numbers. In preparation.

22. V. Goryunov, Functions on space curves, *J. London Math. Soc.* **61(3)** (2000) 807–822.

23. N. G. Grulha Jr, L'Obstruction d'Euler Locale d'une Application, *Annales de la Faculté des Sciences de Toulouse* **17(1)** (2008) 53–71.

24. N. G. Grulha Jr, *Obstrução de Euler de Aplicações Analíticas*, PhD Thesis, ICMC - USP, 2007.

25. N. G. Grulha Jr, The Euler Obstruction and Bruce-Roberts' Milnor Number, *Quart. J. Math.* **60(3)** (2009) 291–302.

26. T. Izawa and T. Suwa, Multiplicity of functions on singular varieties, *Intern. J. Math.* **14(5)** (2003) 541–558.

27. V. H. Jorge Pérez, D. Levcovitz and M. J. Saia, Invariants, equisingularity and Euler obstruction of map germs from $\mathbb{C}^n$ to $\mathbb{C}^n$, *J. für die Reine und Angewandte Mathematik* **587** (2005) 145–167.

28. V. H. Jorge Pérez, Multiplicities and equisingularity of map germs from $\mathbb{C}^3$ to $\mathbb{C}^3$, *Houston J. Math.* **29(4)** (2003) 901–924.

29. M. Kashiwara, Index theorem for maximally overdetermined systems of linear differential equations *Proc. Japan Acad.* **49** (1973) 803–803.

30. D. T. Lê, Complex analytic functions with isolated singularities, *J. Algebraic Geometry* **1** (1992) 83–100.

31. D. T. Lê, Le concept de singularité isolée de fonction analytique, *Adv. Stud. Pure Math.* **8** (1986) 215–227.

32. D. T. Lê, *Singularités isolées des intersections complètes.* In Introduction à la théorie des singularités, Travaux en cours **36** Hermann, 1988.

33. D. T. Lê et B. Teissier, Variétés polaires locales et classes de Chern des variétés singulières, *Ann. of Math.* **114** (1981) 457–491.

34. R. D. MacPherson, Chern classes for singular algebraic varieties, Ann. of Math. **100** (1974) 423–432.

35. D. B. Massey, *Lê Cycles and Hypersurface Singularities*, Springer-Verlag, Lect. Notes in Math. **1615**, 1995.

36. D.B. Massey, Hypercohomology of Milnor fibers, *Topology* **35(4)** (1996) 969–1003.

37. D. Mond and D. van Straten, Milnor number equals Tjurina number for functions on space curves, *J. London Math. Soc.* **63** (2001) 177–187.

38. C. Sabbah, Quelques remarques sur la géométrie des espaces conormaux, *Astérisque* **130** (1985) 161–192.

39. M.-H. Schwartz, *Champs radiaux sur une stratification analytique*, Travaux en cours **39** Hermann, Paris, 1991.

40. J. Seade, M. Tibăr, and A. Verjovsky, Milnor Numbers and Euler obstruction, *Braz. Bull. Math. Soc.* (N.S.) **36(2)** (2005) 275–283.

41. J. Seade, M. Tibăr and A. Verjovsky, Global Euler obstruction and polar invariants, *Math. Ann.* **333(2)** (2005) 393–403.

42. J. Schürmann, *A short proof of a formula of Brasselet, Lê and Seade*, math. AG/0201316, 2002.

43. J. Schürmann, *Topology of singular spaces and constructible sheaves*, Mathematical Monographs-new series, Birkhäuser, 2003.

44. J. Schürmann and M. Tibar, *Index formula for MacPherson cycles of affine algebraic varieties* preprint: math.AG/0603338, 2006.

45. B. Teissier, *Variétés polaires. II. Multiplicités polaires, sections planes et conditions de Whitney*, Lect. Notes in Math. **961** Springer, Berlin, 1982.
46. M. Tibar, *Polynomials and vanishing cycles*, Cambridge Tracts in Mathematics **170**, 2007.

J.-P. Brasselet
CNRS-Institut de Mathématiques de Luminy
Université de la Méditerranée
Campus de Luminy, Case 907
13288 Marseille Cedex 9
France
jpb@iml.univ-mrs.fr

N. G. Grulha Jr.
Instituto de Ciências Matemáticas e de Computação
Av. Trabalhador São-Carlense, 400
13560-970 São Carlos
Brazil
njunior@icmc.usp.br

4

Branching of periodic orbits in reversible Hamiltonian systems

C. A. BUZZI, L. A. ROBERTO AND M. A. TEIXEIRA

Abstract

This paper deals with the dynamics of time-reversible Hamiltonian vector fields with 2 and 3 degrees of freedom around an elliptic equilibrium point in the presence of symplectic involutions. The main results discuss the existence of one-parameter families of reversible periodic solutions terminating at the equilibrium. The main techniques used are Birkhoff and Belitskii normal forms combined with the Liapunov-Schmidt reduction.

1. Introduction

The resemblance of dynamics between reversible and Hamiltonian contexts, probably first noticed by Poincaré and Birkhoff, has caught much attention since the 1960s. Since then many important results, e.g. KAM theory, Liapunov center theorems, etc, holding in the Hamiltonian context have been carried over to the reversible one (see [13, 20] and reference therein).

The concept of reversibility is linked with an involution R, i. e., a map $R : \mathbb{R}^N \to \mathbb{R}^N$ such that $R \circ R = Id$. Let X be a smooth vector field on R^N. The vector field is called R–reversible if the following relation is satisfied

$$X(R(x)) = -DR_x.X(x).$$

2000 *Mathematics Subject Classification* 37C27 (primary), 37C10 (secondary).
All authors are partially supported by a FAPESP–BRAZIL grant 07–06896–5 and by the joint project CAPES–MECD grants 071/04 and HBP2003–0017.

Reversibility means that $x(t)$ is a solution of X if and only if $Rx(-t)$ is also a solution. The set $Fix(R) = \{x \in \mathbb{R}^N : R(x) = x\}$ plays an important role in the reversible systems. We say that a singular point p is symmetric if $p \in Fix(R)$, and analogously we say that an orbit γ is symmetric if $R(\gamma) = \gamma$.

Many dynamical systems that arise in the context of applications possess robust structural properties, for instance, symmetries or Hamiltonian structure. In order to understand the typical dynamics of such systems, their structure needs to be taken into account, leading one to study phenomena that are generic among dynamical systems with the same structure. In the last decade there has been a surging interest in the study of systems with time-reversal symmetries (see [18] and [11]). Symmetry properties arise naturally and frequently in dynamical systems. In recent years, a lot of attention has been devoted to understanding and using the interplay between dynamics and symmetry properties. It is worthwhile to mention that one of the characteristic properties of Hamiltonian and reversible systems is that minimal sets appear in one-parameter families. So a number of natural questions can be formulated, such as: (i) how do branches of such minimal sets terminate or originate?; (ii) can one branch of minimal sets bifurcate from another such branch?; (iii) how persistent is such a branching process when the original system is slightly perturbed? Recently, there has been increased interest in the study of systems with time-reversal symmetries and we refer to [14] for a survey in reversible systems and related problems.

Our main concern, in this article, is to find conditions for the existence of one-parameter families of periodic orbits terminating at the equilibrium.

We present some relevant historical facts. In 1895 Liapunov published his celebrated center theorem, see Abraham and Marsden [1] p 498; This theorem, for analytic Hamiltonians with n degrees of freedom, states that if the eigenfrequencies of the linearized Hamiltonian are independent over $\mathbb{Z}$, near a stable equilibrium point, then there exist n families of periodic solutions filling up smooth 2-dimensional manifolds going through the equilibrium point. Devaney [6] proved a time-reversible version of the Liapunov center theorem. Recently this center theorem has been generalized to equivariant systems, by Golubitsky, Krupa and Lim [7] in the time-reversible case, and by Montaldi, Roberts and Stewart [16] in the Hamiltonian case. We recall that in [7] Devaney's theorem was extended and some extra symmetries were considered. Contrasting Devaney's geometrical approach, they used Liapunov-Schmidt reduction, adapting an alternative proof of the reversible Liapunov center theorem given by Vanderbauwhede [19]. In [16] the existence of families of periodic orbits around an elliptic semi-simple equilibrium is analyzed. Systems with symmetry, including time-reversal symmetry, which are anti-symplectic are studied.

Their approach is a continuation of the work of Vanderbauwhede, in [19], where the families of periodic solutions correspond bijectively to solutions of a variational problem.

Recently Buzzi and Teixeira in [3] have analyzed the dynamics of time-reversible Hamiltonian vector fields with 2 degrees of freedom around an elliptic equilibrium point in presence of $1 : -1$ resonance. Such systems appear generically inside a class of Hamiltonian vector fields in which the symplectic structure is assumed to have some symmetric properties. Roughly speaking, the main result says that under certain conditions the original Hamiltonian H is formally equivalent to another Hamiltonian $\widetilde{H}$ such that the corresponding Hamiltonian vector field $X_{\widetilde{H}}$ has two Liapunov families of symmetric periodic solutions terminating at the equilibrium. It is worthwhile to say that all the systems considered there have been derived from the expression of Birkhoff normal form.

In this paper we address the problem to systems with 2 and 3 degrees of freedom. Physical models of such systems were exhibited in [5, 12]. As usual the main proofs are based on a combined use of normal form theory and the Liapunov-Schmidt Reduction. It is important to mention that our results concerning the existence of Liapunov families generalize those in [3]. As a matter of fact we deal with C^∞ or C^ω.

We begin in Section 2 with an introduction of the terminology and basic concepts for the formulation of our results. In Section 3 the Belitskii normal form is discussed. In Section 4 the Liapunov-Schmidt reduction is presented. In Section 5 the usefulness of Birkhoff normal form in our approach is pointed out. In Section 6 we study the Hamiltonian with 2 degrees of freedom denoted by Ω^0, and we denote by Ω^0_B the set of vector fields in Ω^0 that satisfy the Birkhoff Condition and by Ω^0_ω the vector fields in Ω^0 that are analytic. We generalize some results presented in [3] by proving Theorem A. That result says that there exists an open set $\mathcal{U}^0 \subset \Omega^0_B$ (respec. Ω^0_ω), in the C^∞-topology, such that (a) $\mathcal{U}^0$ is determined by the 3–jet of the vector fields; and (b) each $X \in \mathcal{U}^0$ possesses two 1–parameter families of periodic solutions terminating at the equilibrium. In Section 7 we study the Hamiltonian with 3 degrees of freedom, and we prove Theorems B and C. In Theorem B we consider the involution associated to the system satisfying $dim(Fix(R)) = 2$, and in Theorem C satisfying $dim(Fix(R)) = 4$. We denote these spaces of reversible Hamiltonian vector fields by Ω^1 and Ω^2, respectively. Again Ω^2_B is the set of vector fields in Ω^2 that satisfy the Birkhoff Condition and Ω^2_ω is the set of vector fields in Ω^2 that are analytic. The conclusions are the following: In Theorem B there exists an open set $\mathcal{U}^1 \subset \Omega^1$, in the C^∞-topology, such that (a) $\mathcal{U}^1$ is determined by the 2–jet of the vector fields, and (b) for each $X \in \mathcal{U}^1$

there is no periodic orbit arbitrarily close to the equilibrium. In Theorem C there exists an open set $\mathcal{U}^2 \subset \Omega_B^2$ (respec. Ω_ω^2), in the C^∞–topology, such that (a) $\mathcal{U}^2$ is determined by the 3–jet of the vector fields, and (b) each $X \in \mathcal{U}^2$ has infinitely many one–parameter families of periodic solutions terminating at an equilibrium with the periods tending to $2\pi/\alpha$. In Section 8 we present an example that satisfies the hypotheses of Theorem A and comment that it is possible to accomplish the vector fields of Theorem C.

2. Preliminaries

Now we introduce some of the terminology and basic concepts for the formulation of our results.

We consider (germs of) smooth functions $H : \mathbb{R}^{2n}, 0 \to \mathbb{R}$ having the origin as an equilibrium point. The corresponding Hamiltonian vector field, to be denoted by X_H, has the origin as an equilibrium or singular point. We recall that $dH = \omega(X_H, \cdot)$, where $\omega = dx_1 \wedge dy_1 + dx_2 \wedge dy_2 + \cdots + dx_n \wedge dy_n$ denotes the standard 2-form on $\mathbb{R}^{2n}$. In coordinates X_H is expressed as:

$$\dot{x}_i = \frac{\partial H}{\partial y_i}, \quad \dot{y}_i = -\frac{\partial H}{\partial x_i}; \quad i = 1, \ldots, n.$$

In $\mathbb{R}^6$ we have

$$\begin{pmatrix} \dot{x}_1 \\ \dot{y}_1 \\ \vdots \\ \dot{x}_3 \\ \dot{y}_3 \end{pmatrix} = \begin{pmatrix} 0 & 1 & 0 & 0 & 0 & 0 \\ -1 & 0 & 0 & 0 & 0 & 0 \\ 0 & 0 & 0 & 1 & 0 & 0 \\ 0 & 0 & -1 & 0 & 0 & 0 \\ 0 & 0 & 0 & 0 & 0 & 1 \\ 0 & 0 & 0 & 0 & -1 & 0 \end{pmatrix} \begin{pmatrix} \frac{\partial H}{\partial x_1} \\ \frac{\partial H}{\partial y_1} \\ \vdots \\ \frac{\partial H}{\partial x_3} \\ \frac{\partial H}{\partial y_3} \end{pmatrix}.$$

Here,

$$J = \begin{pmatrix} 0 & 1 & 0 & 0 & 0 & 0 \\ -1 & 0 & 0 & 0 & 0 & 0 \\ 0 & 0 & 0 & 1 & 0 & 0 \\ 0 & 0 & -1 & 0 & 0 & 0 \\ 0 & 0 & 0 & 0 & 0 & 1 \\ 0 & 0 & 0 & 0 & -1 & 0 \end{pmatrix}$$

is the symplectic structure associated with the 2-form ω given above.

We say that an involution is *symplectic* when it satisfies the equation $\omega(DR_p(v_p), DR_p(w_p)) = \omega(v_p, w_p)$. If the involution R is linear then this definition is equivalent to $JR = R^T J$, where J is the symplectic structure and R^T is the transpose matrix of R.

The next proposition exhibits normal forms for linear symplectic involutions on $\mathbb{R}^6$.

Proposition 2.1. *Given the symplectic structure ω and an involution R there exists a symplectic change of coordinates that transforms R in one of the following normal forms*

(i) $R_0 = Id$,

(ii) $R_0(x_1, y_1, x_2, y_2, x_3, y_3) = (x_1, y_1, x_2, y_2, -x_3, -y_3)$,

(iii) $R_0(x_1, y_1, x_2, y_2, x_3, y_3) = (x_1, y_1, -x_2, -y_2, -x_3, -y_3)$,

(iv) $R_0 = -Id$.

Before giving the proof we observe that the mapping $\psi = (1/2)(R + L)$, where $L = DR(0)$, is a symplectic conjugacy between R and L, i. e., $R \circ \psi = \psi \circ L$. So we may and do assume, without loss of generality, that the involution R is linear.

Lemma 2.2. *If R is a linear symplectic involution, then we have that $\mathbb{R}^6 = \text{Fix}(R) \oplus \text{Fix}(-R)$ and $\omega(\text{Fix}(R), \text{Fix}(-R)) = 0$.*

Proof: For every $u \in \mathbb{R}^6$, we can write $u = ((u + R(u))/2) + ((u - R(u))/2)$. Notice that $(u + R(u))/2 \in \text{Fix}(R)$ and $(u - R(u))/2 \in \text{Fix}(-R)$. Now, let $u \in \text{Fix}(R)$ and $v \in \text{Fix}(-R)$, so we have that $\omega(u, v) = \omega(R(u), -R(v))$. By using that R is symplectic and R is linear, we have that $\omega(R(u), R(v)) = \omega(u, v)$. So $-\omega(u, v) = \omega(u, v)$, and we have proved that $\omega(\text{Fix}(R), \text{Fix}(-R)) = 0$. $\qquad \square$

A linear subspace $U \in \mathbb{R}^6$ is *symplectic* if ω is *non-degenerate* in U, i. e, if $\omega(u, v) = 0$ for all $u \in U$ then $v = 0$.

Lemma 2.3. *$\text{Fix}(R)$ and $\text{Fix}(-R)$ are symplectic subspaces.*

Proof: Suppose $u \in \text{Fix}(R)$ and $u \neq 0$ such that $\omega(u, \text{Fix}(R)) = 0$. By using Lemma 2.2, we have $\omega(\text{Fix}(R), \text{Fix}(-R)) = 0$, so $\omega(u, \text{Fix}(-R)) = 0$. Again by Lemma 2.2 ($\mathbb{R}^6 = \text{Fix}(R) \oplus \text{Fix}(-R)$) we have $\omega(u, \mathbb{R}^6) = 0$ and so ω is degenerate in $\mathbb{R}^6$ which is not true. Then $\text{Fix}(R)$ is a symplectic subspace. The proof for $\text{Fix}(-R)$ is analogous. $\qquad \square$

Proof of Proposition 2.1: Let $R : \mathbb{R}^6 \to \mathbb{R}^6$ be a linear involution and ω be a fixed symplectic structure. From Lemma 2.2, $\mathbb{R}^6 = \text{Fix}(R) \oplus \text{Fix}(-R)$ and as $\text{Fix}(R)$ is a symplectic subspace, then $\dim \text{Fix}(R) = 0, 2, 4,$ or 6.

– if $\dim \text{Fix}(R) = 0$, then we can find a coordinate system, using Darboux Theorem [10], such that $R_0 = -Id$;

- If $\dim \mathrm{Fix}(R) = 6$, then we can find a coordinate system, using Darboux Theorem [10], such that $R_0 = Id$;
- if $\dim \mathrm{Fix}(R) = 4$, we consider the bases $\beta_1 = \{e_1, e_2, e_3, e_4\}$ for $\mathrm{Fix}(R)$ and $\beta_2 = \{f_1, f_2\}$ for $\mathrm{Fix}(-R)$. So $\beta = \{e_1, e_2, e_3, e_4, f_1, f_2\}$ is a basis for $\mathbb{R}^6$. Let us show that β can be chosen such that $[\omega]_\beta = J$ and $[R]_\beta = R_0 =$

$$\begin{pmatrix} 1 & 0 & 0 & 0 & 0 & 0 \\ 0 & 1 & 0 & 0 & 0 & 0 \\ 0 & 0 & 1 & 0 & 0 & 0 \\ 0 & 0 & 0 & 1 & 0 & 0 \\ 0 & 0 & 0 & 0 & -1 & 0 \\ 0 & 0 & 0 & 0 & 0 & -1 \end{pmatrix}.$$ Here $[\omega]_\beta$ means the matrix of ω with respect

to the basis β.

Note that $\omega(e_i, e_i) = 0$ and $\omega(f_j, f_j) = 0$, $i = 1, 2, 3, 4$ and $j = 1, 2$. By Lemma 2.2 $\omega(e_i, f_j) = 0$, $i = 1, 2, 3, 4$ and $j = 1, 2$. And as ω is alternating, then $\omega(f_1, f_2) = 1$ and $\omega(f_2, f_1) = -1$.

Define $\omega(e_i, e_j)$ for $i \neq j$. From Darboux's Theorem there exists a coordinate system around 0 such that $\omega|_{\beta_1}$ in this coordinate system is the symplectic structure J.

- if $\dim \mathrm{Fix}(R) = 2$, in the same way as above, we get

$$R_0 = [R]_\beta = \begin{pmatrix} 1 & 0 & 0 & 0 & 0 & 0 \\ 0 & 1 & 0 & 0 & 0 & 0 \\ 0 & 0 & -1 & 0 & 0 & 0 \\ 0 & 0 & 0 & -1 & 0 & 0 \\ 0 & 0 & 0 & 0 & -1 & 0 \\ 0 & 0 & 0 & 0 & 0 & -1 \end{pmatrix}.$$

$\square$

Using the previous proposition we consider the following cases:

$6 : 2$–**Case:** $R_1(x_1, y_1, x_2, y_2, x_3, y_3) = (x_1, y_1, -x_2, -y_2, -x_3, -y_3)$,
$6 : 4$–**Case:** $R_2(x_1, y_1, x_2, y_2, x_3, y_3) = (x_1, y_1, x_2, y_2, -x_3, -y_3)$.

2.1. Linear part of a R_j–reversible Hamiltonian vector field in $\mathbb{R}^6$

Denote by Ω^j the space of all R_j–reversible Hamiltonian vector field, X_{H_j}, in $\mathbb{R}^6$ with 3 degrees of freedom where H_j is the associate Hamiltonian and $j = 1, 2$. Fix the coordinate system $(x_1, y_1, x_2, y_2, x_3, y_3) \in (\mathbb{R}^6, 0)$. We endow Ω^j with the C^∞–topology.

The symplectic structure given by J is:

$$J = \begin{pmatrix} 0 & 1 & 0 & 0 & 0 & 0 \\ -1 & 0 & 0 & 0 & 0 & 0 \\ 0 & 0 & 0 & 1 & 0 & 0 \\ 0 & 0 & -1 & 0 & 0 & 0 \\ 0 & 0 & 0 & 0 & 0 & 1 \\ 0 & 0 & 0 & 0 & -1 & 0 \end{pmatrix}.$$

Observe that the involution R_j is symplectic, i.e, $J.R_j - R_j^T.J = 0$, $j = 1, 2$.

As the involution is symplectic, then the vector field is R_j−reversible if and only if the Hamiltonian function H_j is R_j−anti-invariant, $j = 1, 2$. This is equivalent to say that $H_j \circ R_j = -H_j$. (See [3])

Define the polynomial function with constant coefficients $a_k \in \mathbb{R}$:

$$H_j(x_1, y_1, x_2, y_2, x_3, y_3) = a_{01}x_1^2 + a_{02}x_1y_1 + a_{03}x_1x_2 + a_{04}x_1y_2 + a_{05}x_1x_3$$
$$+ a_{06}x_1y_3 + a_{07}y_1^2 + a_{08}y_1x_2 + a_{09}y_1y_2 + a_{10}y_1x_3 + a_{11}y_1y_3 + a_{12}x_2^2$$
$$+ a_{13}x_2y_2 + a_{14}x_2x_3 + a_{15}x_2y_3 + a_{16}y_2^2 + a_{17}y_2x_3 + a_{18}y_2y_3$$
$$+ a_{19}x_3^2 + a_{20}x_3y_3 + a_{21}y_3^2 + h.o.t.$$

First of all we impose the R_j−reversibility on our Hamiltonian system, $j = 1, 2$. For each case we have:

a) **Case** $6 : 2$

From the reversibility condition, $H_1 \circ R_1 = -H_1$, and

$$R_1 = \begin{pmatrix} 1 & 0 & 0 & 0 & 0 & 0 \\ 0 & 1 & 0 & 0 & 0 & 0 \\ 0 & 0 & -1 & 0 & 0 & 0 \\ 0 & 0 & 0 & -1 & 0 & 0 \\ 0 & 0 & 0 & 0 & -1 & 0 \\ 0 & 0 & 0 & 0 & 0 & -1 \end{pmatrix},$$

we obtain

$$H_1 = a_{03}x_1x_2 + a_{04}x_1y_2 + a_{05}x_1x_3 + a_{06}x_1y_3$$
$$+ a_{08}x_2y_1 + a_{09}y_1y_2 + a_{10}x_3y_1 + a_{11}y_1y_3 + h.o.t.$$

Then, the linear part of Hamiltonian vector field X_{H_1} is

$$A_1 = \begin{pmatrix} 0 & 0 & a & b & c & d \\ 0 & 0 & e & f & g & h \\ -f & b & 0 & 0 & 0 & 0 \\ e & -a & 0 & 0 & 0 & 0 \\ -h & d & 0 & 0 & 0 & 0 \\ g & -c & 0 & 0 & 0 & 0 \end{pmatrix}.$$

Just to simplify the notation we replace $a_{03}, a_{04}, a_{05}, a_{06}, a_{08}, a_{09}, a_{10}, a_{11}$ by $a, b, c, d, -e, -f, -g, -h$, respectively. Note that A_1 is R_1−reversible (i. e, $R_1.A_1 + A_1.R_1 = 0$). The eigenvalues of A_1 are $\{0, 0, \pm\sqrt{be - af + dg - ch}, \pm\sqrt{be - af + dg - ch}\}$. We restrict our attention to those systems satisfying the inequality:

$$be - af + dg - ch < 0. \tag{2.1}$$

The case when $be - af + dg - ch > 0$ will not be considered because the center manifold of the equilibrium has dimension two with double zero eigenvalue. We shall use the Jordan canonical form from A_1. So we stay, for while, away from the original symplectic structure. We call $\alpha = \sqrt{-be + af - dg + ch}$, and so the transformation matrix is

$$P_1 = \begin{pmatrix} 0 & 0 & \frac{-d}{dg-ch}\alpha & 0 & \frac{-c}{dg-ch}\alpha & 0 \\ 0 & 0 & \frac{-h}{dg-ch}\alpha & 0 & \frac{-g}{dg-ch}\alpha & 0 \\ \frac{df-bh}{be-af} & \frac{cf-bg}{be-af} & 0 & \frac{-df+bh}{dg-ch} & 0 & \frac{-cf+bg}{dg-ch} \\ \frac{-de+ah}{be-af} & \frac{-ce+ag}{be-af} & 0 & \frac{de-ah}{dg-ch} & 0 & \frac{ce-ag}{dg-ch} \\ 0 & 1 & 0 & 0 & 0 & 1 \\ 1 & 0 & 0 & 1 & 0 & 0 \end{pmatrix}.$$

So

$$\widehat{A_1} = P_1^{-1}.A_1.P_1 = \begin{pmatrix} 0 & 0 & 0 & 0 & 0 & 0 \\ 0 & 0 & 0 & 0 & 0 & 0 \\ 0 & 0 & 0 & \alpha & 0 & 0 \\ 0 & 0 & -\alpha & 0 & 0 & 0 \\ 0 & 0 & 0 & 0 & 0 & \alpha \\ 0 & 0 & 0 & 0 & -\alpha & 0 \end{pmatrix},$$

where P_1^{-1} is the inverse matrix of the matrix P_1. Moreover, in this way, $\widehat{R_1} = P_1^{-1}.R_1.P$ takes the form

$$\widehat{R_1} = \begin{pmatrix} -1 & 0 & 0 & 0 & 0 & 0 \\ 0 & -1 & 0 & 0 & 0 & 0 \\ 0 & 0 & 1 & 0 & 0 & 0 \\ 0 & 0 & 0 & -1 & 0 & 0 \\ 0 & 0 & 0 & 0 & 1 & 0 \\ 0 & 0 & 0 & 0 & 0 & -1 \end{pmatrix}.$$

b) **Case** $6 : 4$

We proceed in the same way as in the previous case. The involution is

$$
R_2 = \begin{pmatrix}
1 & 0 & 0 & 0 & 0 & 0 \\
0 & 1 & 0 & 0 & 0 & 0 \\
0 & 0 & 1 & 0 & 0 & 0 \\
0 & 0 & 0 & 1 & 0 & 0 \\
0 & 0 & 0 & 0 & -1 & 0 \\
0 & 0 & 0 & 0 & 0 & -1
\end{pmatrix}
$$

and the Hamiltonian function in this case takes the form:

$$
\begin{aligned}
H_2 = {}& a_{05}x_1x_3 + a_{14}x_2x_3 + a_{10}x_3y_1 + a_{17}x_3y_2 \\
& + a_{06}x_1y_3 + a_{15}x_2y_3 + a_{11}y_1y_3 + a_{18}y_2y_3 + h.o.t..
\end{aligned}
$$

Then, the linear part of Hamiltonian vector field X_{H_2} is expressed by:

$$
A_2 = \begin{pmatrix}
0 & 0 & 0 & 0 & a & b \\
0 & 0 & 0 & 0 & c & d \\
0 & 0 & 0 & 0 & e & f \\
0 & 0 & 0 & 0 & g & h \\
-d & b & -h & f & 0 & 0 \\
c & -a & g & -e & 0 & 0
\end{pmatrix}.
$$

Again we change the notation. The eigenvalues of A_2 are given by $\{0, 0, \pm\sqrt{bc - ad + fg - eh}, \pm\sqrt{bc - ad + fg - eh}\}$. We consider the case

$$
bc - ad + fg - eh < 0. \tag{2.2}
$$

We call $\alpha = \sqrt{-bc + ad - fg + eh}$ and consider the transformation matrix

$$
P_2 = \begin{pmatrix}
\frac{be-af}{bc-ad} & \frac{-bg+ah}{bc-ad} & 0 & \frac{-b}{\alpha} & 0 & \frac{-a}{\alpha} \\
\frac{de-cf}{bc-ad} & \frac{-dg+ch}{bc-ad} & 0 & \frac{-d}{\alpha} & 0 & \frac{-c}{\alpha} \\
0 & 1 & 0 & \frac{-f}{\alpha} & 0 & \frac{-e}{\alpha} \\
1 & 0 & 0 & \frac{-h}{\alpha} & 0 & \frac{-g}{\alpha} \\
0 & 0 & 0 & 0 & 1 & 0 \\
0 & 0 & 1 & 0 & 0 & 0
\end{pmatrix},
$$

and the Jordan canonical form of A_2 is:

$$\widehat{A_2} = P_2^{-1}.A_2.P_2 = \begin{pmatrix} 0 & 0 & 0 & 0 & 0 & 0 \\ 0 & 0 & 0 & 0 & 0 & 0 \\ 0 & 0 & 0 & \alpha & 0 & 0 \\ 0 & 0 & -\alpha & 0 & 0 & 0 \\ 0 & 0 & 0 & 0 & 0 & \alpha \\ 0 & 0 & 0 & 0 & -\alpha & 0 \end{pmatrix}.$$

Moreover, in this way, $\widehat{R_2} = P_2^{-1} R_2 P_2$ takes the form

$$\widehat{R_2} = \begin{pmatrix} 1 & 0 & 0 & 0 & 0 & 0 \\ 0 & 1 & 0 & 0 & 0 & 0 \\ 0 & 0 & -1 & 0 & 0 & 0 \\ 0 & 0 & 0 & 1 & 0 & 0 \\ 0 & 0 & 0 & 0 & -1 & 0 \\ 0 & 0 & 0 & 0 & 0 & 1 \end{pmatrix}.$$

3. Belitskii normal form

In this section we present the Belitskii Normal Form. When a vector field is in this normal form we can write explicitly the resultant equation of Liapunov–Schmidt reduction.

Consider a formal vector field expressed by

$$\hat{X}(x) = Ax + \sum_{k \geq 2} X^{(k)}(x)$$

where $X^{(k)}$ is the homogeneous part of degree k. Let us look for a "simple" form of the formal vector field $\hat{Y} = \hat{\phi} * \hat{X}$ by means of formal transformation

$$\hat{\phi} = x + \sum_{k}^{\infty} \phi^{(k)}(x).$$

The proof of the next theorem is in [2].

Theorem 3.1. *Given a formal vector field*

$$\hat{X}(x) = Ax + \sum_{k \geq 2} X^{(k)}(x),$$

there is a formal transformation $\hat{\phi}(x) = x + \dots$ bringing $\hat{X}$ to the form $(\hat{\phi} * X)(x) = Ax + h(x)$ where h is a formal vector field with zero linear part *commuting with A^T, i.e*

$$A^T h(x) = h'(x) A^T x,$$

where A^T is the transposed matrix and h' is the derivative of h.

Here we call the normal form $(\hat{\phi}_* X)(x) = Ax + h(x)$ the Belitskii normal form. By abuse of the terminology, call $X_H = A + h$.

4. Liapunov–Schmidt reduction

In this section we recall the main features of the Liapunov–Schmidt reduction. As a matter of fact, we adapt the setting presented in [4, 21] to our approach. In this way consider the R-reversible system expressed by

$$\dot{x} = X_H(x); \; x \in \mathbb{R}^6 \tag{4.1}$$

satisfying $X_H(Rx) = -RX_H(x)$ with R a linear involution in $\mathbb{R}^6$. Assume that $X_H(0) = 0$ and consider

$$A = D_1 X_H(0), \tag{4.2}$$

the Jacobian matrix of X_H in the origin.

In our case the linear part of vector field has the following eigenvalues: 0 with the algebraic and geometric multiplicity 2, and $\pm \alpha i$, also with algebraic and geometric multiplicity 2, $\alpha \in \mathbb{R}$. Performing a time rescaling we may take $\alpha = 1$. We write the real form of the linear part of the vector field X_j:

$$A = \begin{pmatrix} 0 & 0 & 0 & 0 & 0 & 0 \\ 0 & 0 & 0 & 0 & 0 & 0 \\ 0 & 0 & 0 & 1 & 0 & 0 \\ 0 & 0 & -1 & 0 & 0 & 0 \\ 0 & 0 & 0 & 0 & 0 & 1 \\ 0 & 0 & 0 & 0 & -1 & 0 \end{pmatrix}.$$

Let $C_{2\pi}^0$ the Banach space of de 2π–periodic continuous mappings $x : \mathbb{R} \to \mathbb{R}^6$ and $C_{2\pi}^1$ the corresponding C^1–subspace. We define an inner product on $C_{2\pi}^0$ by

$$(x_1, x_2) = \frac{1}{2\pi} \int_0^{2\pi} < x_1(t), x_2(t) > dt$$

where $< \cdot, \cdot >$ denotes an inner product in $\mathbb{R}^6$.

The main aim is to find all small periodic solutions of (4.1) with period near 2π.

Define the map $F : C_{2\pi}^1 \times \mathbb{R} \to C_{2\pi}^0$ by

$$F(x, \sigma)(t) = (1 + \sigma)\dot{x}(t) - X_H(x(t)).$$

Note that if $(x_0, \sigma_0) \in C_{2\pi}^1 \times \mathbb{R}$ is such that

$$F(x_0, \sigma_0) = 0, \tag{4.3}$$

then $\tilde{x}(t) := x_0((1 + \sigma_0)t)$ is a $2\pi/(1 + \sigma_0)$−periodic solution of (4.1).

Our task now is to find the zeroes of F. Clearly, $(x_0, \sigma_0) = (0, 0)$ is one solution of $F(x_0, \sigma_0) = 0$. Let $L := D_x F(0, 0) : C_{2\pi}^1 \to C_{2\pi}^0$; explicitly L is given by

$$Lx(t) = \dot{x}(t) - Ax(t).$$

Consider the unique (S-N)-decomposition of A, $A = S + N$. Recall that in our case A is semi-simple, i. e, $A = S$. Define the subspace $\mathcal{N}$ of $C_{2\pi}^1$ as

$$\mathcal{N} = \{q; \dot{q}(t) = Sq(t)\}$$
$$= \{q; q(t) = exp(tS)x; \ x \in \mathbb{R}^6\}.$$

Observe that $\mathcal{N} \subset C_{2\pi}^1$ and a basis for the solutions of $\dot{q} = Sq$ is given by the set $\{(1, 0, 0, 0, 0, 0), (0, 1, 0, 0, 0, 0), (0, 0, \cos(t), \sin(t), 0, 0), (0, 0, -\sin(t), \cos(t), 0, 0), (0, 0, 0, 0, \cos(t), \sin(t)), (0, 0, 0, 0, -\sin(t), \cos(t))\}$.

In order to study certain properties of the operator L we introduce $\mathcal{N} \subset C_{2\pi}^1$ and the following definitions and notations.

We will put the solution of $F(x_0, \sigma_0) = 0$ in one-to-one correspondence with the solutions of an appropriate equation in $\mathcal{N}$. Define the subspaces

$$X_1 = \left\{x \in C_{2\pi}^1 : (x, \mathcal{N}) = 0\right\}$$

and

$$Y_1 = \left\{y \in C_{2\pi}^0 : (y, \mathcal{N}) = 0\right\}$$

as the orthogonal complements of $\mathcal{N}$ in $C_{2\pi}^1$ and $C_{2\pi}^0$, respectively.

Let $(q_1, q_2, q_3, q_4, q_5, q_6)$ with $q_i = exp(tS)u_i$ where u_i, $i = 1, \ldots, 6$, is a basis for $\mathbb{R}^6$. Then we define a projection

$$\mathcal{P} : C_{2\pi}^0 \to C_{2\pi}^0$$

by

$$P = \sum_{i=1}^{6} q_i^*(\cdot) q_i \ \in \ \mathcal{L}(C_{2\pi}^0)$$

with $q_i^*(x) = (q_i, x)$.

We have $\text{Im}(\mathcal{P}) = \mathcal{N}$ and $\text{Ker}(\mathcal{P}) = Y_1$. Hence,

$$C_{2\pi}^1 = X_1 \oplus \mathcal{N}, \quad C_{2\pi}^0 = Y_1 \oplus \mathcal{N}.$$

Now we consider

$$F(x, \sigma) = F(q + x_1, \sigma) =: \hat{F}(q, x_1, \sigma); \quad q \in \mathcal{N}, \ x_1 \in X_1.$$

The proof of next result can be found in [9].

Lemma 4.1 (Fredholm's Alternative). *Let $A(t)$ be a matrix in C_T^0 and let f be in C_T. Here C_T^0 is the space of the matrices with entries continuous and T–periodic, and C_T is the set of T-periodic maps from $\mathbb{R}$ to $\mathbb{R}^n$. Then the equation $\dot{x} = A(t)x + f(t)$ has a solution in C_T if, and only if,*

$$\int_0^T \ < y(t), g(t) > dt = 0$$

for all solution y of the adjoint equation

$$\dot{y} = -yA(t)$$

such that $y^t \in C_T$.

As $L(\mathcal{N}) \subset \mathcal{N}$ this lemma implies the following:

Lemma 4.2. *The mapping $\hat{L} := L|_{X_1} : X_1 \to Y_1$ is bijective.*

Let us study the solutions of $\hat{F}(q, x_1, \sigma) = 0$. These solutions are equivalent to the solutions of the system

$$(I - \mathcal{P}) \circ \hat{F}(q, x_1, \sigma) = 0,$$
$$\mathcal{P} \circ \hat{F}(q, x_1, \sigma) = 0.$$

With Lemma 4.2 and the Implicit Function Theorem we can solve the first equation as $x_1 = x_1^*(q, \sigma)$. Then, (4.3) is reduced to

$$\tilde{F}(q, \sigma) := \mathcal{P} \circ \hat{F}(q, x_1^*(q, \sigma), \sigma) = 0.$$

This equation is solved if, and only if,

$$q_i^*(\hat{F}(q, x_1^*(q, \sigma), \sigma) = 0, \quad i = 1, \dots, 6.$$

Notice that (u, σ) is a solution of (4.3) provided that

$$B(u, \sigma) = 0 \qquad (4.4)$$

with $B : \mathcal{N} \times \mathbb{R} \to \mathbb{R}^6$ defined by

$$B(u, \sigma) := \frac{1}{2\pi} \int_0^{2\pi} \exp(-tS) F(x^*(u, \sigma), \sigma) dt$$

and

$$x^*(u, \sigma) := \exp(tS)u + x_1^*(\exp(tS)u, \sigma).$$

Let us present some properties of the mapping B.

The proof of next lemma can be found in [13].

Lemma 4.3. *The following relations hold:*

i) $s_\phi B(u, \sigma) = B(s_\phi u, \sigma)$;

ii) $R B(u, \sigma) = -B(Ru, \sigma)$, *where s_ϕ is the S^1–action in $\mathbb{R}^6$ defined by $s_\phi u = exp(-\phi S_0)u$.*

Observe that under the condition i) the mapping B is S^1– equivariant whereas condition ii) states that the mapping B is R–anti-equivariant, i.e., B inherits the anti-symmetric properties of X_H.

Assume that (4.1) is in Belitskii normal form truncated at the order p. So $X_H(x) = Ax + h(x) + r(x)$ where $r(x) = \mathcal{O}(\| x \|^{p+1})$. The proof of next result is in [21].

Theorem 4.4. *The following relations hold:*

i) $x^*(u, \sigma) = \exp(tS)u + \mathcal{O}(\| x \|^{p+1})$,

ii) $B(u, \sigma) = (1 + \sigma)Su - Au - h(u) + \mathcal{O}(\| x \|^{p+1})$ *for σ near the origin.*

If (u, σ) is a solution of (4.4) then $x = x^*(u, \sigma)$ corresponds to a $2\pi/(1 + \sigma)$-periodic solution of (4.3).

Recall that the periodic solution of (4.4) is R-symmetric if and only if it intersects Fix(R) in exactly two points. In conclusion, we obtain all small symmetric periodic solutions of (4.4) by solving the equation

$$G(u, \sigma) = B(u, \sigma) \mid_{\text{Fix}(R)} = 0. \qquad (4.5)$$

5. Birkhoff normal form

In this section we briefly discuss some points concerning the Birkhoff normal form that will be useful in the sequel. The Birkhoff normal form is useful because it preserves the symplectic structure. In our cases if the vector field is in the Birkhoff normal form then it is in the Belitskii normal form, and so we can apply Theorem 4.4.

The function $\{f, g\} = \omega(X_f, X_g)$ is called the *Poisson bracket* of the smooth functions f and g. Let $\mathcal{H}_n$ be the set of all homogeneous polynomials of degree n. The adjoint map $Ad_{H_2} : \mathcal{H}_n \to \mathcal{H}_n$ is defined by

$$Ad_{H_2}(H) = \{H_2, H\} = \omega(X_{H_2}, X_H) = \langle -X_{H_2}, \nabla H \rangle. \qquad (5.1)$$

The *Birkhoff Normal Form Theorem* (cf. [17, 8, 22]) states that if we have a Hamiltonian $H = H_2 + H_3 + H_4 + \cdots$, where $H_i \in \mathcal{H}_i$ is the homogeneous part of degree i, and $\mathcal{G}_i \subset \mathcal{H}_i$ satisfies $\mathcal{G}_i \oplus Range(Ad_{H_2}) = \mathcal{H}_i$, then there exists a formal symplectic power series transformation Φ such that $H \circ \Phi = H_2 + \widetilde{H}_3 + \widetilde{H}_4 + \cdots$ where $\widetilde{H}_i \in \mathcal{G}_i$ $(i = 3, 4, \ldots)$. In particular, if Ad_{H_2} is semi-simple, as in our case, then $Ker(Ad_{H_2})$ is the complement of $Range(Ad_{H_2})$.

As R_j is symplectic, the change of coordinates Φ can be chosen in such a way that $H \circ \Phi$ satisfies $H \circ \Phi \circ R_j = -H \circ \Phi$. In order to see this, we can split $\mathcal{H}_i = \mathcal{H}_i^+ \oplus \mathcal{H}_i^-$, where $\mathcal{H}_i^+ = \{H \in \mathcal{H}_i : H \circ R_j = H\}$ and $\mathcal{H}_i^- = \{H \in \mathcal{H}_i : H \circ R_j = -H\}$. If R_j is symplectic, then $Ad_{H_2}(\mathcal{H}_i^{\pm}) = \mathcal{H}_i^{\mp}$. In this case, if $\mathcal{H}_i = \mathcal{G}_i \oplus Ad_{H_2}(\mathcal{H}_i)$, then $\mathcal{H}_i^- = (\mathcal{G}_i \cap \mathcal{H}_i^-) \oplus Ad_{H_2}(\mathcal{H}_i^+)$. Now we can perform the change of coordinates restricted to $\mathcal{H}_i^-$. It implies that all monomial terms in the image of the adjoint restricted to H_i^- can be removed and it will remain only monomials in the kernel of the adjoint restricted to H_i^-. And so, the normal form is also R_j–reversible.

Definition 5.1. We say that a Hamiltonian vector field X_H satisfies the Birkhoff Condition (BC) if $Ad_{H_2}(H) = 0$.

Observation 1. By the equalities (5.1), the condition of the Definition 5.1 is equivalent to $\omega(X_{H_2}, X_H) = 0$ or $\{H_2, H\} = 0$.

6. Two degrees of freedom

In [3] a Birkhoff normal form for each $X \in \Omega^0$ is derived and the following result is obtained:

Theorem 6.1. *Assume H is a Hamiltonian that is anti-invariant with respect to the involution and the associated vector field X_H has an elliptical equilibrium point. Then there exists another Hamiltonian $\widetilde{H}$, formally C^k–equivalent to*

H, such that the vector field $X_{\tilde{H}}$ has two one–parameter families of symmetric periodic solutions, with period near $2\pi/\sqrt{ad-bc}$, as in the Liapunov's Theorem, going through the equilibrium point.

Let Ω^0 be the space of the C^∞ R_0–reversible Hamiltonian vector fields with two degrees of freedom in $\mathbb{R}^4$ and fix the coordinate system $(x_1, y_1, x_2, y_2) \in \mathbb{R}^4$. We endow Ω^0 with the C^∞–topology. Let $\Omega^0_B \subset \Omega^0$ be the space of the vector fields that satisfy the Birkhoff condition and $\Omega^0_\omega \subset \Omega^0$ be the space of the analytic ones. We prove the following result, which generalizes the previous one.

Theorem A. *There exists an open set $\mathcal{U}^0 \subset \Omega^0_B$ (respec. Ω^0_ω) such that*

(a) *$\mathcal{U}^0$ is determined by the 3–jet of the vector fields.*
(b) *each $X \in \mathcal{U}^0$ possesses two 1–parameter families of symmetric periodic solutions terminating at the equilibrium point.*

Proof. Fix on $\mathbb{R}^4$ a symplectic structure as in the Proposition 2.1. So the normal form of an involution has one of the following form: $Id_{\mathbb{R}^4}$, or $-Id_{\mathbb{R}^4}$,

or $R_0 = \begin{pmatrix} 1 & 0 & 0 & 0 \\ 0 & 1 & 0 & 0 \\ 0 & 0 & -1 & 0 \\ 0 & 0 & 0 & -1 \end{pmatrix}$. We just work with R_0–reversible vector fields.

As in the cases in $\mathbb{R}^6$ we have that by the hypothesis the Hamiltonian H satisfies $H \circ R_0 = -H$, so the linear part of the vector field X_H is given by

$$A = \begin{pmatrix} 0 & 0 & a & b \\ 0 & 0 & c & d \\ -d & b & 0 & 0 \\ c & -a & 0 & 0 \end{pmatrix}, \tag{6.1}$$

and their eigenvalues are $\{\pm\sqrt{bc-ad}, \pm\sqrt{bc-ad}\}$. We are interested in the case with $bc - ad < 0$. We call $\alpha = \sqrt{ad-bc}$ and in order to obtain the Jordan canonical form of the matrix A we consider the transformation matrix

$$P = \begin{pmatrix} 0 & \frac{-b}{\alpha} & 0 & \frac{-a}{\alpha} \\ 0 & \frac{-d}{\alpha} & 0 & \frac{-c}{\alpha} \\ 0 & 0 & 1 & 0 \\ 1 & 0 & 0 & 0 \end{pmatrix}.$$

After this transformation we obtain

$$\widehat{A} = P^{-1}.A.P = \begin{pmatrix} 0 & \alpha & 0 & 0 \\ -\alpha & 0 & 0 & 0 \\ 0 & 0 & 0 & \alpha \\ 0 & 0 & -\alpha & 0 \end{pmatrix},$$

and

$$\widehat{R_0} = P^{-1}.R_0.P = \begin{pmatrix} -1 & 0 & 0 & 0 \\ 0 & 1 & 0 & 0 \\ 0 & 0 & -1 & 0 \\ 0 & 0 & 0 & 1 \end{pmatrix},$$

where P^{-1} is the inverse matrix of P.

Performing a time rescaling we can assume that $\alpha = 1$. We write the canonical real Jordan form of A as

$$\widehat{A} = \begin{pmatrix} 0 & 1 & 0 & 0 \\ -1 & 0 & 0 & 0 \\ 0 & 0 & 0 & 1 \\ 0 & 0 & -1 & 0 \end{pmatrix}. \qquad \square$$

First we obtain the Belitskii normal form of X_H, by considering $h : \mathbb{R}^4 \rightarrow \mathbb{R}^4$ up to 3rd order, which is given by $X_H(x_1, y_1, x_2, y_2) = A[x_1, y_1, x_2, y_2] + h(x_1, y_1, x_2, y_2)$; and after we require the condition that the Belitskii normal form is $\widehat{R_0}$−reversible, i. e, $X_H \widehat{R_0} = -\widehat{R_0} X_H$. Then the system obtained is given by

$$
\begin{aligned}
\dot{x}_1 &= y_1 + (e_{21}y_1 + e_{23}y_2)(x_1^2 + y_1^2) + e_{30}y_2(x_2^2 + y_2^2) \\
&\quad + (e_{16}x_1 + e_{24}x_2)(y_1x_2 - x_1y_2) + e_{26}y_2(x_1x_2 + y_1y_2), \\
\dot{y}_1 &= -x_1 + (-e_{21}x_1 - e_{23}x_2)(x_1^2 + y_1^2) - e_{30}x_2(x_2^2 + y_2^2) \\
&\quad + (e_{16}y_1 + e_{24}y_2)(y_1x_2 - x_1y_2) - e_{26}x_2(x_1x_2 + y_1y_2), \\
\dot{x}_2 &= y_2 + (-d_{15}y_1 - d_{22}y_2)(x_1^2 + y_1^2) - (d_{20}y_1 + d_{29}y_2)(x_2^2 + y_2^2) \\
&\quad - (d_{17}y_1 + d_{25}y_2)(x_1x_2 + y_1y_2), \\
\dot{y}_2 &= -x_2 + (d_{15}x_1 + d_{22}x_2)(x_1^2 + y_1^2) + (d_{20}x_1 + d_{29}x_2)(x_2^2 + y_2^2) \\
&\quad + (d_{17}x_1 + d_{25}x_2)(x_1x_2 + y_1y_2).
\end{aligned}
\tag{6.2}
$$

Now we use the fact that the vector field satisfies the Birkhoff Condition. First of all we observe that the canonical symplectic matrix

$$J = \begin{pmatrix} 0 & 1 & 0 & 0 \\ -1 & 0 & 0 & 0 \\ 0 & 0 & 0 & 1 \\ 0 & 0 & -1 & 0 \end{pmatrix},$$

after the linear change of coordinates P, is transformed into

$$\widehat{J} = P^{-1}JP = \begin{pmatrix} 0 & 0 & -1 & 0 \\ 0 & 0 & 0 & -1 \\ 1 & 0 & 0 & 0 \\ 0 & 1 & 0 & 0 \end{pmatrix}.$$

We take a general Hamiltonian function $H : \mathbb{R}^4 \to \mathbb{R}$ of 4th order, compute the kernel of Ad_{H_2} defined on (5.1), where H_2 is the homogeneous part of degree 2 of H, and require that H satisfies $H \circ \widehat{R_0} = -H$. The terms up to 3rd order is given by $h_b(x) = \widehat{J} \cdot \nabla H(x)$; its expression is

$$
\begin{aligned}
\dot{x}_1 &= y_1 + a_1 y_1 (x_1^2 + y_1^2) + a_2 (2x_1 x_2 y_1 - x_1^2 y_2 + y_1^2 y_2) \\
&\quad + a_3 (3x_2^2 y_1 - 2x_1 x_2 y_2 + y_1 y_2^2), \\
\dot{y}_1 &= -x_1 + (a_2 y_1 + 2a_3 y_2)(x_2 y_1 - x_1 y_2) - x_1 (a_1 (x_1^2 + y_1^2) \\
&\quad + a_2 (x_1 x_2 + y_1 y_2) + a_3 (x_2^2 + y_2^2)), \\
\dot{x}_2 &= y_2 + (2a_1 x_1 + a_2 x_2)(-x_2 y_1 + x_1 y_2) + y_2 (a_1 (x_1^2 + y_1^2) \\
&\quad + a_2 (x_1 x_2 + y_1 y_2) + a_3 (x_2^2 + y_2^2)), \\
\dot{y}_2 &= -x_2 + (2a_1 y_1 + a_2 y_2)(-x_2 y_1 + x_1 y_2) - x_2 (a_1 (x_1^2 + y_1^2) \\
&\quad + a_2 (x_1 x_2 + y_1 y_2) + a_3 (x_2^2 + y_2^2)).
\end{aligned}
\tag{6.3}
$$

Observation 2. We observe here that we can apply Theorem 4.4, when the vector field is in the Belitskii normal form. This is not a restriction because if the vector field satisfies the Birkhoff Condition then it is in the Belitskii Normal Form. It is easy to see that if $\{H_2, H\} = 0$ then $D(\{H_2, H\}) = 0$, and so $A_0^T X_H - DX_H A_0^T(x) = 0$. For example, in our case we have

$$
\widehat{J} = \begin{pmatrix} 0 & 0 & -1 & 0 \\ 0 & 0 & 0 & -1 \\ 1 & 0 & 0 & 0 \\ 0 & 1 & 0 & 0 \end{pmatrix}, \quad A_0 = \begin{pmatrix} 0 & 1 & 0 & 0 \\ -1 & 0 & 0 & 0 \\ 0 & 0 & 0 & 1 \\ 0 & 0 & -1 & 0 \end{pmatrix} \quad \text{and} \quad X_{H_2} = \begin{pmatrix} y_1 \\ -x_1 \\ y_2 \\ -x_2 \end{pmatrix}.
$$

The Birkhoff condition implies $-y_1 H_{x_1} + x_1 H_{y_1} - y_2 H_{x_2} + x_2 H_{y_2} = 0$. So

$$
\begin{aligned}
H_{y_1} - y_1 H_{x_1 x_1} + x_1 H_{y_1 x_1} - y_2 H_{x_2 x_1} + x_2 H_{y_2 x_1} &= 0, \\
-H_{x_1} - y_1 H_{x_1 y_1} + x_1 H_{y_1 y_1} - y_2 H_{x_2 y_1} + x_2 H_{y_2 y_1} &= 0, \\
H_{y_2} - y_1 H_{x_1 x_2} + x_1 H_{y_1 x_2} - y_2 H_{x_2 x_2} + x_2 H_{y_2 x_2} &= 0, \\
-H_{x_2} - y_1 H_{x_1 y_2} + x_1 H_{y_1 y_2} - y_2 H_{x_2 y_2} + x_2 H_{y_2 y_2} &= 0.
\end{aligned}
\tag{6.4}
$$

On the other hand if we compute $A_0^T X_H - DX_H A_0^T(x)$, we obtain

$$
- \begin{pmatrix} -H_{y_2} \\ H_{x_2} \\ H_{y_1} \\ -H_{x_1} \end{pmatrix} + \begin{pmatrix} -H_{x_2 x_1} & -H_{x_2 y_1} & -H_{x_2 x_2} & -H_{x_2 y_2} \\ -H_{y_2 x_1} & -H_{y_2 y_1} & -H_{y_2 x_2} & -H_{y_2 y_2} \\ H_{x_1 x_1} & H_{x_1 y_1} & H_{x_1 x_2} & H_{x_1 y_2} \\ H_{y_1 x_1} & H_{y_1 y_1} & H_{y_1 x_2} & H_{y_1 y_2} \end{pmatrix} \begin{pmatrix} y_1 \\ -x_1 \\ y_2 \\ -x_2 \end{pmatrix},
$$

and by (6.4) we have that $A_0^T X_H - D X_H A_0^T(x) = 0$, i.e. the system is in the Belitskii Normal Form.

The Liapunov-Schmidt reduction gives us all small $\widehat{R_0}$–symmetric periodic solutions by solving the equation

$$B(x, \sigma)|_{x \in \text{Fix}(\widehat{R_0})} = 0,$$

with

$$B(x, \sigma) = (1 + \sigma)Sx - \hat{A}x - h_b(x), \quad x \in \mathbb{R}^4, \tag{6.5}$$

where S is the semi-simple part of (unique) $S - N$–decomposition of $\hat{A}$. (See [15]).

In our case, $\hat{A}$ is semi-simple and $\text{Fix}(\widehat{R_0}) = \{(0, y_1, 0, y_2); \ y_1, y_2 \in \mathbb{R}\}$. Recall that the reduced equation, $B(x, \sigma) = 0$, is defined in $\mathcal{N} \times \mathbb{R}$, where $\mathcal{N} = \{\exp(\hat{A}t)x; x \in V\} \in C_{2\pi}^1$ and $V = \text{span}\{e_1, e_2, e_3, e_4\}$.

The symplectic structure J give us that X_H is written in the following form $h_b(x) = h_b(x_1, y_1, x_2, y_2) = (-H_{x_2}(x_1, y_1, x_2, y_2), -H_{y_2}(x_1, y_1, x_2, y_2), H_{x_1}(x_1, y_1, x_2, y_2), H_{y_1}(x_1, y_1, x_2, y_2))$. Using the fact that h_b satisfies the Birkhoff Condition we have that

$$y_1 H_{x_1}(x_1, y_1, x_2, y_2) - x_1 H_{y_1}(x_1, y_1, x_2, y_2) + y_2 H_{x_2}(x_1, y_1, x_2, y_2)$$

$$-x_2 H_{y_2}(x_1, y_1, x_2, y_2) = 0, \ \forall (x_1, y_1, x_2, y_2) \in \mathbb{R}^4.$$

Hence at the points $(0, 0, 0, y_2)$ we have $y_2 H_{x_2}(0, 0, 0, y_2) = 0$. It implies that $H_{x_2}(0, y_1, 0, y_2) = y_1 \bar{f}(y_1, y_2)$. Analogously we have that $H_{x_1}(0, y_1, 0, y_2) = y_2 \bar{g}(y_1, y_2)$. So

$$G(y_1, y_2, \sigma) = B(x, \sigma)|_{x \in \text{Fix}(\widehat{R_0})} = \begin{bmatrix} -y_1(a_1 y_1^2 + a_2 y_1 y_2 + a_3 y_2^2 - \sigma + \cdots) \\ -y_2(a_1 y_1^2 + a_2 y_1 y_2 + a_3 y_2^2 - \sigma + \cdots) \end{bmatrix}.$$

$$\tag{6.6}$$

For the analytic case we have that the equation

$$G(y_1, y_2, \sigma) = (0, 0)$$

is given by

$$-y_1(a_1 y_1^2 + a_2 y_1 y_2 + a_3 y_2^2 - \sigma) + H_1(y_1, y_2) = 0,$$

$$-y_2(a_1 y_1^2 + a_2 y_1 y_2 + a_3 y_2^2 - \sigma) + H_2(y_1, y_2) = 0,$$

and multiplying the first equation by $-y_2$ and the second by y_1 we get $y_2 H_1 = y_1 H_2$. Using the fact that H_1 and H_2 are analytic we have that there exists $\widetilde{H}$ such that $H_1 = y_1 \widetilde{H}$ and $H_2 = y_2 \widetilde{H}$ for all (y_1, y_2).

If $u_1 a_3 \neq 0$ in $(6,6)$, then we have two solutions for the equation $G(y_1, y_2, \sigma) = 0$. One solution is $y_1 = 0$ and $y_2(\sigma) = \pm\sqrt{\frac{\alpha}{a_3}} + \cdots$. And the second solution is $y_2 = 0$ and $y_1(\sigma) = \pm\sqrt{\frac{\sigma}{a_1}} + \cdots$.

We define $\mathcal{U}^0 = \mathcal{U}_1^0 \cap \mathcal{U}_2^0$ where

$$\mathcal{U}_1^0 = \left\{ X \in \Omega_B^0; \quad \text{the canonical form of } DX(0) \text{ satisfies } ad - bc > 0 \right\}$$

and

$$\mathcal{U}_2^0 = \left\{ X \in \Omega_B^0; \quad \text{the coefficients of (6.3) satisfies } a_1 a_3 \neq 0 \right\}.$$

In $\mathcal{U}^0 = \mathcal{U}_1^0 \cap \mathcal{U}_2^0 \subset \Omega_B^0$ for each σ the equation $G(y_1, y_2, \sigma) = 0$ has two nonzero solutions terminating at the origin when σ is tending to 0. So, in the original problem we have two one parameter families of periodic solutions terminating the origin (when $\sigma \to 0$). $\qquad\qquad\square$

7. Three degrees of freedom

As in the previous section, let Ω^1 (resp. Ω^2) be the space of the C^∞ R_1-reversible (resp. R_2-reversible) Hamiltonian vector fields with three degrees of freedom in $\mathbb{R}^6$ and fix a coordinate system $(x_1, y_1, x_2, y_2, x_3, y_3) \in \mathbb{R}^6$. We endow Ω^1 and Ω^2 with the C^∞-topology. Let Ω_B^2 (resp. Ω_ω^2) be the space of vector fields in Ω^2 that satisfy the Birkhoff Condition (resp. that are analytic).

7.1. Case 6:2

Theorem B. *There exists an open set* $\mathcal{U}^1 \subset \Omega^1$ *such that*

(a) $\mathcal{U}^1$ *is determined by the 2–jet of the vector fields.*
(b) *for each* $X \in \mathcal{U}^1$ *there is no symmetric periodic orbit arbitrarily close to the equilibrium point.*

Proof: First we obtain the Belitskii normal form of X_H, by considering $h : \mathbb{R}^6 \to \mathbb{R}^6$ up to 2nd order, and then we require that the Belitskii normal form is $\widehat{R_1}$-reversible, i. e, $X_H \widehat{R_1} = -\widehat{R_1} X_H$. After that we take the Birkhoff normal form. The new symplectic structure is $\widehat{J} = P_1^t J P_1$, where P_1 is the linear matrix that brings the linear part of the vector field to the Jordan canonical form. The Birkhoff normal form is obtained by taking a general Hamiltonian function $H : \mathbb{R}^6 \to \mathbb{R}$ of 3rd order, computing the kernel of Ad_{H_2} and requiring that H satisfies $H \circ \widehat{R_1} = -H$. The Birkhoff normal form up to 2nd order is given by $h_b(x) = \widehat{J} \cdot \nabla H(x)$. Finally, the Liapunov-Schmidt reduction gives us all small $\widehat{R}_1$–symmetric periodic solutions by solving the equation

$$B(x, \sigma)|_{x \in \text{Fix}(\widehat{R}_1)} = 0,$$

with

$$B(x, \sigma) = (1 + \sigma)Sx - \widehat{A}_1 x - h_b(x), \quad x \in \mathbb{R}^6,$$

and S is the semi-simple part of (unique) $S - N$–decomposition of $\widehat{A}_1$.
(See [15]). In our case, $\widehat{A}_1$ is semi-simple and $\text{Fix}(\widehat{R}_1) = \{(0, 0, x_2, 0, x_3, 0);$
$x_2, x_3 \in \mathbb{R}\}$. We recall that the reduced equation of the Liapunov-Schmidt,
$B(x, \sigma) = 0$, is defined in $\mathcal{N} \times \mathbb{R}$, where $\mathcal{N} = \{\exp(\widehat{A}_1 t)x; x \in V\} \in C^1_{2\pi}$ and
$V = \text{span}\{e_1, e_2, e_3, e_4, e_5, e_6\}$.

We derive the following expression

$$G(x_2, x_3, \sigma) = B(x, \sigma)|_{x \in \text{Fix}(\widehat{R}_1)} = \begin{bmatrix} b_1 x_2^2 + x_3(b_2 x_2 + b_3 x_3) + \cdots \\ b_4 x_2^2 + x_3(b_5 x_2 + b_6 x_3) + \cdots \\ x_2(-\sigma + \delta) + \cdots \\ x_3(-\sigma + \delta) + \cdots \end{bmatrix}.$$

$$(7.1)$$

Observe that the equation $b_1 x_2^2 + b_2 x_2 x_3 + b_3 x_3^2 = 0$, generically, either
has the solution $(x_2, x_3) = (0, 0)$, or has a pair of straight lines solutions given
by $(c_1 x_2 + d_1 x_3)(c_2 x_2 + d_2 x_3) = 0$. The equation $b_4 x_2^2 + b_5 x_2 x_3 + b_6 x_3^2 = 0$ is
analogous. We can conclude that if the two first components of (7.13) have no
common factor of the form $cx_2 + dx_3$ then we have just the solution $(x_2, x_3) = (0, 0)$ for the two previous equations.

We define the following open sets:

$$\mathcal{U}_1^1 = \left\{ X \in \Omega^1; \quad \text{the canonical form of } DX(0) \text{ satisfies (2.1)} \right\},$$
$$\mathcal{U}_2^1 = \left\{ \begin{array}{l} X \in \Omega^1; \quad \text{the 2–jet of the two first equations of (7.1)} \\ \text{have no common factor} \end{array} \right\}.$$

Then $\mathcal{U}^1 = \mathcal{U}_1^1 \cap \mathcal{U}_2^1$ is an open set in Ω^1.

The pair $(x_2, x_3) = (0, 0)$ is the unique solution of the equation $G = 0$. So,
near the origin there are no *symmetric* periodic orbits for this case.   □

7.2. Case 6:4

Theorem C. *There exists an open set $\mathcal{U}^2 \subset \Omega^2_B$ (respec. Ω^2_ω) such that*

(a) *$\mathcal{U}^2$ is determined by the 3–jet of the vector fields.*

(b) each $X \in \mathcal{U}^2$ has two 2–parameter families of periodic solutions $\gamma^1_{\sigma,\lambda}$ and $\gamma^2_{\sigma,\lambda}$ with $\sigma \in (-\epsilon, \epsilon)$ and $\lambda \in [0, 2\pi]$, such that, for each λ_0, $\lim_{\sigma \to 0} \gamma^j_{\sigma,\lambda_0} = 0$, for $j = 1, 2$, and the periods tend to $2\pi/\alpha$ when $\sigma \to 0$.

Proof: First of all we derive the reversible Belitskii normal form of X_H up to 2nd order. We observe that it coincides with the reversible Birkhoff normal form and is given by:

$$
X_{h_b} = \begin{bmatrix} -\dfrac{b(x_3 y_2 - x_2 y_3)\alpha^2}{\beta} \\[2mm] \dfrac{a(x_3 y_2 - x_2 y_3)\alpha^2}{\beta} \\[2mm] (-ax_1 - by_1)y_2 + \alpha y_2 \\[1mm] x_2(ax_1 + by_1) - \alpha x_2 \\[1mm] (-ax_1 - by_1)y_3 + \alpha y_3 \\[1mm] x_3(ax_1 + by_1) - \alpha x_3 \end{bmatrix}, \tag{7.2}
$$

where $a = b_{65}/\alpha$, $b = b_{71}/\alpha$ e $\alpha = \sqrt{-a_{06}a_{10} + a_{05}a_{11} - a_{15}a_{17} + a_{14}a_{18}}$.

As in the other cases, the Liapunov-Schmidt reduction gives us all small $\widehat{R_2}$–symmetric periodic solutions by solving the equation

$$
B(x, \sigma)|_{x \in \mathrm{Fix}(\widehat{R_2})} = 0,
$$

with

$$
B(x, \sigma) = (1 + \sigma)Sx - \widehat{A_2}x - h_b(x), \quad x \in \mathbb{R}^6.
$$

As before S is the semi-simple part of (the unique) $S - N$–decomposition of $\widehat{A_2}$. (See [15]). In our case, $\widehat{A_2}$ is semi-simple and $\mathrm{Fix}(\widehat{R_2}) = \{(x_1, y_1, 0, y_2, 0, y_3); x_1, y_1, y_2, y_3 \in \mathbb{R}\}$. We recall that the reduced equation of the Liapunov-Schmidt, $B(x, \sigma) = 0$, is defined on $\mathcal{N} \times \mathbb{R}$, where $\mathcal{N} = \{\exp(\widehat{A_2}t)x; x \in V\} \in C^1_{2\pi}$ and $V = \mathrm{span}\{e_1, e_2, e_3, e_4, e_5, e_6\}$.

Like in the proof of Theorem A, we derive the following expression

$$
\begin{aligned}
G(x_1, y_1, y_2, y_3, \sigma) &= B(x, \sigma)|_{x \in \mathrm{Fix}(\widehat{R_2})} \\
&= \begin{bmatrix} y_2(\sigma + a_1 x_1 + a_2 y_1 + a_3 x_1^2 + a_4 y_1^2 + a_5 y_2^2 + a_6 y_2 y_3 + a_7 y_3^2 + \cdots) \\ y_3(\sigma + a_1 x_1 + a_2 y_1 + a_3 x_1^2 + a_4 y_1^2 + a_5 y_2^2 + a_6 y_2 y_3 + a_7 y_3^2 + \cdots) \end{bmatrix}.
\end{aligned} \tag{7.3}
$$

If $a_5 a_7 \neq 0$ in (7.3), then for each (x_1, y_1) close to $(0, 0)$ we have two solutions for the equation $G(x_1, y_1, y_2, y_3, \sigma) = 0$. One solution is $y_2 = 0$ and

$y_3(x_1, y_1, \sigma) = \pm \sqrt{\frac{\sigma + a_1 x_1 + a_2 y_1}{a_7}} + \cdots$. And the second solution is $y_3 = 0$ and

$y_2(x_1, y_1, \sigma) = \pm \sqrt{\frac{\sigma + a_1 x_1 + a_2 y_1}{a_5}} + \cdots$.

We define the following open sets:

$$\mathcal{U}_1^2 = \left\{ X \in \Omega_B^2; \quad \text{the canonical form of } DX(0) \text{ satisfies (2.2)} \right\},$$
$$\mathcal{U}_2^2 = \left\{ X \in \Omega_B^2; \quad \text{the coefficients of (7.3) satisfies } a_5 a_7 \neq 0 \right\}.$$

Then $\mathcal{U}^2 = \mathcal{U}_1^2 \cap \mathcal{U}_2^2$ is an open set in Ω_B^2. For each $X \in \mathcal{U}^2$ and σ we consider $\gamma_\sigma^1 : (x_1, y_1) \mapsto (x_1, y_1, 0, y_3(x_1, y_1, \sigma))$ and $\gamma_\sigma^2 : (x_1, y_1) \mapsto (x_1, y_1, y_2(x_1, y_1, \sigma), 0)$. Now we take the parametrization $(x_1, y_1) \mapsto (a\sigma, b\sigma)$. We have $\gamma_{\sigma\lambda_0}^1 : (a\sigma, b\sigma) \mapsto (a\sigma, b\sigma, 0, y_3(a\sigma, b\sigma, \sigma))$ and $\gamma_{\sigma\lambda_0}^2 : (a\sigma, b\sigma) \mapsto (a\sigma, b\sigma, y_2(a\sigma, b\sigma, \sigma), 0)$ where $\lambda_0 = a/b$. Then, there exists two 2–parameter family of periodic orbits $\gamma_{\sigma\lambda}^1$ and $\gamma_{\sigma\lambda}^2$ such that for each $\lambda_0 \in \mathbb{R}$, the families of periodic orbits $\gamma_{\sigma\lambda_0}^j$, for $j = 1, 2$, are Liapunov families; i. e. $\lim_{\sigma \to 0} \gamma_{\sigma\lambda_0}^j = 0$ and the period tends to $2\pi/\alpha$. $\square$

8. Examples

This section is devoted to presenting a mechanical example for the Case $4 : 2$.

We consider two objects m_1 and m_2 with charge q and $-q$. They are at the position $(a, b) \in \mathbb{R}^2$ and $(-a, -b) \in \mathbb{R}^2$, respectively. We assume that the system does not have kinetic energy. So the total energy, i.e the Hamiltonian function is:

$$H(x, u, y, v) = \frac{-q}{\sqrt{(x - a)^2 + (y - b)^2}} + \frac{q}{\sqrt{(x + a)^2 + (y + b)^2}}.$$

Note that this Hamiltonian function satisfies the condition

$$H(\widehat{R}_0 \cdot (x, u, y, v)) = -H(x, u, y, v),$$

where $\widehat{R}_0 = \begin{pmatrix} -1 & 0 & 0 & 0 \\ 0 & 1 & 0 & 0 \\ 0 & 0 & -1 & 0 \\ 0 & 0 & 0 & 1 \end{pmatrix}$.

In another words, our system is a Hamiltonian $\widehat{R}_0$−reversible vector field.

Observation 3. It is worth to say that the system (7.2) (case $6 : 4$) can be considered, in a similar way as [23], a mathematical model of a theoretical electrical circuit diagram.

Acknowledgements. The authors thank the dynamical system research group of Universitat Autònoma de Barcelona for the hospitality offered to us during part of the preparation of this paper.

References

1. R. Abraham and J. E. Marsden, *Foundations of Mechanics*, Benjamin-Cummings, 2nd edn, 1978.
2. G. Belitskii, C^∞-normal forms of local vector fields, Symmetry and perturbation theory, *Acta Appl. Math.* **70** (2002) 23–41.
3. C. A. Buzzi and M. A. Teixeira, Time-reversible Hamiltonian vector fields with symplectic symmetries, *Journal of Dynamics and Differential Equations* **2** (2005) 559–574.
4. S. H. Chow and J. K. Hale, *Methods of bifurcation theory*. Grundlehren der Mathematischen Wissenschaften [Fundamental Principles of Mathematical Science] **251** Springer-Verlag, New York-Berlin, 1982.
5. S. N. Chow and Y. Kim, Bifurcation of periodic orbits for non-positive definite Hamiltonian systems', *Appl. Anal.* **31** (1998) 163–199.
6. R. Devaney, Reversible diffeomorphism and flows, *Transactions of the American Mathematical Society* **218** (1976) 89–113.
7. M. Golubitsky, M. Krupa and C. Lim, Time-reversibility and particle sedimentation, *SIAM J. Appl. Math.* **51** (1981) 49–72.
8. J. Guckenheimer and P. Holmes, *Nonlinear oscillations, dynamical systems, and bifurcations of vector fields*. Applied Mathematical Sciences, **42** Springer-Verlag, New York, 1983.
9. J. Hale, *Ordinary Differential Equations* Wilay-Interscience, New York, 1969.
10. G.R. Hall and K.R. Meyer, *Introduction to Hamiltonian dynamical system and the N-body problem* Springer–Verlag, 1992.
11. H. Hanβmann, The reversible umbilic bifurcation. In: Time-reversal symmetry in dynamical systems (Coventry, 1996). *Phys. D* **112** (1998) 81–94.
12. F. Heinz, Dynamik und Stabilität eines Hamiltonschen Systems, *Diplomarbeit Institut für Reine und Angewandte Mathematik*, RWTH Aachen, 1998.
13. J. Knobloch and A. Vanderbauwhede, A general reduction method for periodic solutions in conservative and reversible systems, *J. Dyn. Diff. Equations* **8** (1996) 71–102.
14. J. S. W. Lamb and J. A. G. Roberts, Time-reversal symmetry in dynamical systems: a survey, *Phys. D* **112** (1998) 1–39.
15. A. Jacquemard, M. Lima and M. A. Teixeira, Degenerate resonances and branching of periodic orbits, *Ann. Mat. Pura Appl.(4)* **187** (2008) 105–117.
16. J. Montaldi, M. Roberts and I. Stewart, Existence of nonlinear normal modes of symmetric Hamiltonian systems, *Nonlinearity* **3** (1990) 695–730.
17. F. Takens, Singularities of vector fields, *Publ. Math. IHES* **43** (1974) 47–100.
18. M. A. Teixeira, Singularities of reversible vector fields. *Phys. D* **100** (1997) 101–118.

19. A. Vanderbauwhede, *Local bifurcation and symmetry*, Res. Notes in Math. **75**, Pitman, Boston, 1982.
20. A. Vanderbauwhede and J.-C. van der Meer, A general reduction method for periodic solutions near equilibria in Hamiltonian systems. In: *Normal forms and homoclinic chaos* (Waterloo, ON, 1992), *Fields Inst. Commun.* **4** (1995) 273–294.
21. T. Wagenknecht, *An analytical study of a two degrees of freedom Hamiltonian System associated the Reversible Hyperbolic Umbilic,* Thesis, University Ilmenau, Germany, 1999.
22. S. Wiggins, *Introduction to applied nonlinear dynamical systems and chaos.* Texts in Applied Mathematics **2** Springer-Verlag, New York, 1990.
23. S. Yu and W. K. S. Tang, Tetrapterous butterfly attractors in modified Lorenz systems, *Chaos, Solitons and Fractals* (2008) doi:10.1016/j.chaos.2008.07.023.

C. A. Buzzi and L. A. Roberto
Dep. de Matemática
IBILCE-UNESP
R. C. Colombo, 2265
15054-000 S. J. Rio Preto
Brazil
buzzi@ibilce.unesp.br
lroberto@ibilce.unesp.br

M. A. Teixeira
Dep. de Matemática
IMECC-UNICAMP
13081-970 Campinas
Brazil
teixeira@ime.unicamp.br

5

Topological invariance of the index of a binary differential equation

L. S. CHALLAPA

Abstract

In this paper we prove the topological invariance of the index of a binary differential equation

$$a(x, y)dy^2 + 2b(x, y)dxdy + c(x, y)dx^2 = 0$$

at an isolated singular point at which the coefficients a, b and c do not all vanish. We also show the topological invariance when the coefficients vanish at the singular point, under the assumption that the zeros of the lifted vector field are isolated.

1. Introduction

Binary differential equations (BDEs) have been studied by several authors with applications to differential geometry of surfaces, partial differential equations and control theory. For example, lines of curvature, asymptotic and characteristic lines on a smooth surface in $\mathbb{R}^3$ are given by BDEs ([4]) and the characteristic lines of a general linear second-order differential equation are also given by BDEs ([14]).

A BDE is written locally, in a neighbourhood U of the origin in $\mathbb{R}^2$, in the form

$$a(x, y)dy^2 + 2b(x, y)dxdy + c(x, y)dx^2 = 0 \tag{1.1}$$

where the coefficients a, b, c are smooth functions. At a point (x, y) where $\delta = (b^2 - ac)(x, y) > 0$, equation (1.1) defines a pair of directions in the plane, no

2000 *Mathematics Subject Classification* 34A09(primary), 34A26, 58K05 (secondary).
The author was supported in part by FAPESP Grant 06/50075-3.

direction where $\delta < 0$ and a double direction on the set $\Delta = \{\delta = 0\}$ provided that the coefficients of the equation do not all vanish at the given point. At such points, every direction is a solution. A natural way to study these equations is to lift the bivalued direction fields to a single field on an associated double cover. Generically the singular points of the BDE are of type well folded saddle, node or focus [6].

J. W. Bruce and F. Tari introduced in [3] the multiplicity of a BDE when a, b and c are real analytic functions. It is defined as the maximum number of singularities of well folded saddle, node or focus type appearing in a generic perturbation of the BDE given by

$$a_t(x, y)dy^2 + 2b_t(x, y)dxdy + c_t(x, y)dx^2 = 0.$$

In [5] and [6] we defined the index of a binary differential equation in terms of generic perturbations of the BDE and showed that this index is independent of the choice of a generic perturbation. We also exhibit a formula that expresses the index in terms of the coefficients of the original equation. This definition extends Hopf definition of the index of a positive binary differential equation to a BDE not necessarily positive. One of the main results in [5] and [6] is the invariance of the index by smooth equivalences.

In this work we prove the topological invariance of the index of a binary differential equation at an isolated singular point at which the coefficients a, b and c do not all vanish. We also show the topological invariance when the coefficients vanish at the singular point, under the assumption that the zeros of the lifted vector field are isolated.

2. Index of a binary differential equation

A binary differential equation is of the form

$$E(x, y) = a(x, y)dy^2 + 2b(x, y)dxdy + c(x, y)dx^2 = 0, \qquad (2.1)$$

where a, b, c are smooth functions. The function $\delta : \mathbb{R}^2 \to \mathbb{R}$ defined by $\delta = b^2 - ac$ is called discriminant function and the zero set of this function is the *discriminant*. We say that the BDE (2.1) is positive if $\delta \geq 0$ and $\delta = 0$ if and only if $a = b = c = 0$. An integral curve of (2.1) is a smooth curve $\alpha : (-1, 1) \longrightarrow \mathbb{R}^2$ such that $E(\alpha(t))(\alpha'(t)) = 0$.

Definition 2.1. We say that $z_0 \in \mathbb{R}^2$ is a singular point of (2.1) if $\delta(z_0) = 0$ and $E(z_0)(\delta_y(z_0), -\delta_x(z_0)) = 0$.

A configuration of a BDE consists of its discriminant together with the pair of foliations it determines. We denote by (E, z_0) the germ of BDE (2.1) at a singular point z_0

Definition 2.2. Let (E_1, p_1), (E_2, p_2) be two germs of BDEs. We say that they are topologically equivalent (resp. equivalent) if there exists a germ of homeomorphism (resp. diffeomorphim) $h : (\mathbb{R}^2, p_1) \longrightarrow (\mathbb{R}^2, p_2)$ that sends the configuration of (E_1, p_1) to the configuration of (E_2, p_2).

The classification of BDEs with respect to the smooth equivalence presents modality ([9], [10]). The generic singularities are the *well folded singularities*, whose normal form is $(dy^2 + (-y + \lambda x^2)dx^2, 0)$, with $\lambda \neq 0, \frac{1}{16}$ (see [9]). There are three topological models, a well folded saddle if $\lambda < 0$, a well folded node if $0 < \lambda < \frac{1}{16}$ and a well folded focus if $\frac{1}{16} < \lambda$.

A 1-parameter perturbation E_t of BDE (2.1) is determined by the 1-parameter smooth perturbations $a_t(x, y) = \tilde{a}(x, y, t)$, $b_t(x, y) = \tilde{b}(x, y, t)$, $c_t(x, y) = \tilde{c}(x, y, t)$ of its coefficients,

$$E_t(x, y) = a_t(x, y)dy^2 + 2b_t(x, y)dxdy + c_t(x, y)dx^2 = 0. \qquad (2.2)$$

The discriminant function of E_t is given by $\delta_t = b_t^2 - a_t c_t$.

We say that E_t is a good perturbation of $(E, 0)$ if all the singular points of E_t are well folded singularities, for $t \neq 0$ sufficiently close to zero.

Theorem 2.3 ([6]). *If a, b, c are smooth functions then there exists a good perturbation E_t of $(E, 0)$.*

We say that a germ of smooth function $g : (\mathbb{R}^n, 0) \longrightarrow (\mathbb{R}, 0)$ is of Lojasiewicz type if there exists a germ of C^1-diffeomorphism $h : (\mathbb{R}^n, 0) \longrightarrow (\mathbb{R}^n, 0)$ such that $g \circ h$ is a germ of real analytic function.

We set $V = (\delta, a\delta_x^2 - b\delta_x\delta_y + c\delta_y^2)$. Then z_0 is a singular point of (2.1) if and only if z_0 is a zero of V.

Definition 2.4. We say that $(E, 0)$ is finite if δ is of Lojasiewicz type and 0 is an isolated zero of V.

If $dx = 0$ is not a solution of equation (2.1), we can set $p = \frac{dy}{dx}$ and reduce equation (2.1) to the IDE

$$F(x, y, p) = E(x, y)(1, p) = 0. \qquad (2.3)$$

Let $M = F^{-1}(0)$. Then, the vector field

$$\xi = F_p \frac{\partial}{\partial x} + pF_p \frac{\partial}{\partial y} - (F_x + pF_y)\frac{\partial}{\partial p}$$

induces a vector field $\xi_{|M}$ on M given by the restriction of ξ on M. One of the properties of this vector field is that the image by the natural projection π : $M \longrightarrow \mathbb{R}^2, (x, y, p) \longmapsto (x, y)$, of the integral curves of $\xi_{|M}$ on M, corresponds to integral curves of (2.1) (see [8]).

Definition 2.5. We say that z_0 is a non-degenerate singular point of (2.1) if z_0 is a singular point of (2.1) and a non-degenerate zero of the map $(\delta, a\delta_x - b\delta_y)$.

By Lemma 2.5 in [5] we obtain that if z_0 is a non-degenerate singular point of (2.1), then there exists $p_0 \in \mathbb{R}$ such that (z_0, p_0) is a hyperbolic singular point of the vector field $\xi_{|M}$, where $p_0 = -\frac{b(z_0)}{a(z_0)}$. Thus we can associate a number $K_\delta(z_0)$ to each non-degenerate singular point z_0 of (2.1), as follows:

(i) $K_\delta(z_0) = -\frac{1}{2}$ if (z_0, p_0) is a saddle point of the vector field $\xi_{|M}$.
(ii) $K_\delta(z_0) = \frac{1}{2}$ if (z_0, p_0) is a node or focus of the vector field $\xi_{|M}$.

It is not difficult to show that $K_\delta(z_0) = \frac{1}{2}\text{Ind}_{s_0}\xi_{|M}$, where $s_0 = (z_0, p_0)$.

Definition 2.6 ([5]). Let a, b, c be smooth functions and E_t a good perturbation of $(E, 0)$. Then the index of $(E, 0)$ at 0 is defined by

$$I(E, 0) = \sum_i K_{\delta_t}(z_i) + \sum_{\delta_t(u_i) < 0} \text{Ind}_{u_i} \nabla \delta_t,$$

where z_i are non-degenerate singular points of E_t and u_i are the critical points in the negative part of δ_t (i.e, $\nabla \delta_t(u_i) = 0$ and $\delta_t(u_i) < 0$).

The next theorem shows that $I(E, 0)$ can be calculated from data given by $(E, 0)$, therefore does not depend on the chosen perturbation.

Theorem 2.7 ([6]). *Let $(E, 0)$ be the germ of a finite binary differential equation. Then:*

(i) *If $(0, 0)$ is an isolated zero of (a, b) and (δ, δ_y) then*

$$I(E, 0) = \frac{1}{2}[Ind_0(\delta, (a\delta_x - b\delta_y)a\delta_y) - Ind_0(a, b) - Ind_0(\delta\delta_x, \delta_y)$$
$$+ Ind_0 \nabla \delta].$$

(ii) *If $(0, 0)$ is an isolated zero of (b, c) and (δ, δ_x) then*

$$I(E, 0) = \frac{1}{2}[Ind_0(\delta, (b\delta_x - c\delta_y)c\delta_x) + Ind_0(c, b) + Ind_0(\delta\delta_y, \delta_x)$$
$$+ Ind_0 \nabla \delta].$$

Theorem 2.8 ([6]). *Let $(E_1, 0)$ and $(E_2, 0)$ be finite and equivalent germs. Then $I(E_1, 0) = I(E_2, 0)$.*

Let $\omega = dy - pdx$ be a 1-form in $\mathbb{R}^3$ and let $\omega_{|M}$ be its restriction to M. We denote by $\mathrm{Ind}_{s_0}\omega_{|M}$ the index of the 1-form $\omega_{|M}$ at $s_0 \in M$, introduced by Ebeling and Gusein-Zade in [12].

Theorem 2.9 ([6]). *Let $(E, 0)$ be the germ of a finite binary differential equation and $a(0, 0) \neq 0$. Then,*

$$I(E, 0) = \mathrm{sign}\,[F_{pp}(s_0)] \cdot \frac{1}{2}\mathrm{Ind}_{s_0}\nabla F + \frac{1}{2}\mathrm{Ind}_{s_0}\omega_{|M},$$

where $s_0 = (0, 0, -\frac{b(0,0)}{a(0,0)})$.

3. Topological invariance of $I(E, 0)$

In this section, we prove the topological invariance of the index of a germ of binary differential equation in two cases. The first consists of cases where the coefficients do not all vanish at the origin, and the second where all the coefficients vanish at the origin but not all the 1-jets do.

Let $f, g : (\mathbb{R}^n, 0) \longrightarrow (\mathbb{R}, 0)$ be germs of smooth functions. We say that f and g are topologically V-equivalent if there is a germ of homeomorphism $\psi : (\mathbb{R}^n, 0) \longrightarrow (\mathbb{R}^n, 0)$ satisfying $\psi(f^{-1}(0)) = g^{-1}(0)$. We denote by B_r^n the closed ball of center 0 and radius r.

Lemma 3.1. *Let $f, g : (\mathbb{R}^2, 0) \longrightarrow (\mathbb{R}, 0)$ be germs of smooth functions and of Lojasiewicz type, with isolated critical points at the origin. If f and g are topologically V-equivalent then $\mathrm{Ind}_0\nabla f = \mathrm{Ind}_0\nabla g$.*

Proof. By hypothesis, the number of branches of $f^{-1}(0)$ and $g^{-1}(0)$ are equal. From [13] it follows that

$$\chi(f^{-1}(\varepsilon) \cap B_r^2) = 1 - \mathrm{Ind}_0\nabla f \quad \text{and} \quad \chi(g^{-1}(\varepsilon) \cap B_r^2) = 1 - \mathrm{Ind}_0\nabla g,$$

where $\varepsilon \neq 0$. Also from [1], we have that the number of branches of $f^{-1}(0)$ and $g^{-1}(0)$ coincides with $2\chi(f^{-1}(\varepsilon) \cap B_r^2)$ and $2\chi(g^{-1}(\varepsilon) \cap B_r^2)$, respectively. The result now follows. $\qquad\square$

Let $P, Q \subset \mathbb{R}^3$ be singular surfaces with an isolated singularity at 0. Let X and Y be vector fields on P and Q, respectively. We say that X and Y are topologically equivalent if there exists a germ of homeomorphism $H : (P, 0) \longrightarrow (Q, 0)$ that sends integral curves of X to integral curves of Y. We denote by $\mathrm{Ind}_0 X$ the index of the vector field X at 0 introduced by Ebeling and Gusein-Zade in [11]. This index is a generalization of the notion of the index of an isolated singular point of a vector field on a manifold with isolated singularities in such a way that the Poincar-Hopf theorem holds. If T is a subset of P we denote by $Int(T)$ the set of interior points of T in P.

Lemma 3.2. *If X and Y are topologically equivalent then $Ind_0 X = Ind_0 Y$.*

Proof. We set $U_r = B_r^3 \cap P$ and $U_s = B_s^3 \cap P$, where $s < r$. Let V be an open neighbourhood of ∂U_r in U_r such that $V \cap U_s = \emptyset$. By [11], there exist radial vector fields X_{rad}, Y_{rad} on U_s and $H(U_s)$, respectively. Furthermore,

$$Ind_0 X = 1 + \sum_{q \in U_r - \{0\}} Ind_q \tilde{X} \quad \text{and} \quad Ind_0 Y = 1 + \sum_{q \in H(U_r) - \{0\}} Ind_q \tilde{Y},$$

where $\tilde{X}, \tilde{Y}$ are, respectively, vector fields on U_r and $H(U_r)$, and $\tilde{X}_{|U_s} = X_{rad}$, $\tilde{X}_{|V} = X$, $\tilde{Y}_{|H(U_s)} = Y_{rad}$ and $\tilde{Y}_{|H(V)} = Y$. It is clear that $W = U_r - Int(U_s)$ is a compact smooth surface with boundary and the set of zeros of $\tilde{X}$ is contained in $Int(W)$. Note that the integral curves of $\tilde{X}$ tangent to ∂W correspond to integral curves of X. The result then follows by applying the generalized Poincar index formula given in [17]. $\square$

Lemma 3.3. *Let $(E, 0)$ be the germ of a finite binary differential equation and $a(0, 0) \neq 0$. Then, $Ind_0 \xi_{|M} = Ind_0 \omega_{|M}$.*

Proof. By Lemma 6.1 in [5], we have that $Ind_0 \omega_{|M} = Ind_0(FF_y, F_p, F_x + pF_y)$. Let $\gamma = F_p dx + pF_p dy - (F_x + pF_y)dp$ be a 1-form on $\mathbb{R}^3$. Then, by applying the same argument as in the proof of Lemma 6.1 in [5], we get that $Ind_0 \gamma_{|M} = Ind_0(FF_y, F_p, F_x + pF_y)$. It follows from [12] that $Ind_0 \xi_{|M} = Ind_0 \gamma_{|M}$, so the result is clear. $\square$

Let $(L, 0)$ be a germ of binary differential equation given by

$$L(x, y) = A(x, y)dy^2 + 2B(x, y)dxdy + C(x, y)dx^2 = 0. \qquad (3.1)$$

where A, B, C are smooth functions. We denote by $\hat{\delta}$ the discriminant function of (3.1).

Let $G(x, y, p) = L(x, y)(1, p) = 0$ and $\zeta = (G_p, pG_p, -G_x - pG_y)$. We set $N = G^{-1}(0)$.

Lemma 3.4. *Let $(E, 0)$ and $(L, 0)$ be finite and topologically equivalent germs with $a(0, 0) \neq 0$ and $A(0, 0) \neq 0$. Then there exists a germ of homeomorphism $H : (M, \pi_{|M}^{-1}(0)) \longrightarrow (N, \pi_{|N}^{-1}(0))$ which sends integral curves of $\xi_{|M}$ to integral curves of $\zeta_{|N}$.*

Proof. Let $h = (h_1, h_2) : (\mathbb{R}^2, 0) \longrightarrow (\mathbb{R}^2, 0)$ be the germ of homeomorphism that sends integral curves of $(E, 0)$ to integral curves of $(L, 0)$. Then, for each point, $(z_0, p_0) \in M$ and $\delta(z_0) > 0$, there exists an integral curve $\alpha = (\alpha_1, \alpha_2)$ of

$(E, 0)$ such that $\alpha(0) = z_0$, $\frac{\alpha_2'(0)}{\alpha_1'(0)} = p_0$ and $h \circ \alpha$ is an integral curve of $(L, 0)$. Thus we define $H(z_0, p_0) = (h(z_0), \frac{(h_2 \circ \alpha)'(0)}{(h_1 \circ \alpha)'(0)})$. Using the flows given by the BDE's, one can show that H is a homeomorphism that sends integral curves of $\xi_{|M}$ to integral curves of $\zeta_{|M}$. It clearly extends to a homeomorphism on $M - \pi_{|M}^{-1}(0)$. Since $\pi_{|M}^{-1}(0) = \{(0, 0, -\frac{a(0,0)}{b(0,0)})\}$ and $\pi_{|N}^{-1}(0) = \{(0, 0, -\frac{A(0,0)}{B(0,0)})\}$, the result follows. $\qquad\square$

The next theorem shows the topological invariance of the index of a germ of BDE when the coefficients do not all vanish at the origin.

Theorem 3.5. *If $(E, 0)$ and $(L, 0)$ are finite and topologically equivalent germs of BDE's with coefficients not all vanishing at 0 then $I(E, 0) = I(L, 0)$.*

Proof. Using Proposition 3.2 in [3] and Theorem 2.8, we can assume that $(E, 0)$ (resp. $(L, 0)$) is of the form

$$E(x, y) = dy^2 - f(x, y)dx^2 = 0 \quad \text{(resp. } L(x, y) = dy^2 - g(x, y)dx^2 = 0\text{).}$$

It is not difficult to show that

$$\mathrm{Ind}_0 \nabla F = \mathrm{Ind}_0 \nabla f \quad \text{and} \quad \mathrm{Ind}_0 \nabla G = \mathrm{Ind}_0 \nabla g.$$

By hypothesis, it is clear that f and g are topologically V-equivalent. Using Lemma 3.1, we obtain that $\mathrm{Ind}_0 \nabla F = \mathrm{Ind}_0 \nabla G$. From Lemmas 3.4 and 3.2, we get $\mathrm{Ind}_0 \xi_{|M} = \mathrm{Ind}_0 \zeta_{|N}$. Then the result follows from Lemma 3.3 and Theorem 2.9. $\qquad\square$

Let $P : \mathbb{R}^2 \times \mathbb{RP} \longrightarrow \mathbb{R}^2$ be the natural projection given by $P(x, y, [\alpha, \beta]) = (x, y)$, where $\mathbb{RP}$ is the real projective line.

Remark 1. A natural way to study BDEs is to consider in $\mathbb{R}^2 \times \mathbb{RP}$ the set $\tilde{M}$ of points $(x, y, [\alpha, \beta])$ where $\delta(x, y) \geq 0$ and the direction $[\alpha, \beta]$ is a solution of the BDE (2.1) at (x, y). When $\tilde{M}$ is a smooth surface and a, b, c all vanish at the origin, one can show that exists $B_r^2 \subset \mathbb{R}^2$ such that $\tilde{M} \cap P_{|\tilde{M}}^{-1}(B_r^2)$ is diffeomorphic to a cylinder (see [2]).

Lemma 3.6. *Let $(E, 0)$ and $(L, 0)$ be finite and topologically equivalent germs of BDEs with coefficients all vanishing at 0. Suppose that M is a smooth surface and the number of zeros of the vector field $\xi_{|M}$ is finite. Then*

$$\sum_{i=1}^{n_0} \mathrm{Ind}_{s_i} \xi_{|M} = \sum_{i=1}^{n_1} \mathrm{Ind}_{r_i} \zeta_{|N},$$

where $s_i = (0, 0, p_i)$ and $r_i = (0, 0, q_i)$.

Proof. By Remark 1, we may assume that $M = \tilde{M}$ and $N = \tilde{N}$. Then there exists $B_r^2 \subset \mathbb{R}^2$ such that $M \cap P_{|_M}^{-1}(B_r^2)$ and $N \cap P_{|_N}^{-1}(h(B_r^2))$ are diffeomorphic to the cylinder. From Lemma 3.4, we obtain that there exists a homeomorphism H from $M - P_{|_M}^{-1}(0)$ to $N - P_{|_N}^{-1}(0)$ which sends integral curves of $\xi_{|_M}$ to integral curves of $\zeta_{|_N}$. If we set $M_1 = M \cap P_{|_M}^{-1}(B_r^2)$ and $N_1 = N \cap P_{|_N}^{-1}(h(B_r^2))$, then H sends integral curves of $\xi_{|_{M_1}}$ tangent to $\partial(M_1)$ to integral curves of $\zeta_{|_{N_1}}$ tangent to $\partial(N_1)$. Since $\chi(M_1) = \chi(N_1)$, the result then follows by applying the generalized Poincar index formula given in [17]. $\qquad\square$

Note that a perturbation E_t of $(E, 0)$ induces a perturbation F^t of F given by $F^t(x, y, p) = E_t(x, y)(1, p)$. We denote by M^t the set of zeros de F^t and set

$$\xi^t = F_p^t \frac{\partial}{\partial x} + p F_p^t \frac{\partial}{\partial y} - (F_x^t + p F_y^t) \frac{\partial}{\partial p}.$$

Let C_n be the set of smooth functions $f : \mathbb{R}^n \longrightarrow \mathbb{R}$. We set $C_n^\varepsilon = \{f \in C_n \mid if \ \nabla f(z) = 0 \ then \ \varepsilon f(z) > 0\}$, where $\varepsilon = \pm 1$.

Definition 3.7. Let E_t be a good perturbation of $(E, 0)$. We say that E_t is an excellent perturbation if for all t sufficiently close to zero, $t \neq 0$, $\delta_t \in C_2^{+1}$ and $F^t \in C_3^{-1}$.

Note that if E_t is an excellent perturbation of $(E, 0)$, then

$$I(E, 0) = \sum_i K_{\delta_t}(z_i),$$

where z_i are non-degenerate singular points of E_t.

Remark 2. Let $(E, 0)$ be a finite germ with coefficients all vanishing at 0, and let 0 be a critical point of δ of Morse type. If 0 is a saddle point of δ, then there exists an excellent perturbation of $(E, 0)$ (see theorem 1.23 in [6]). Also, when the coefficients of the BDE do not all vanish at the origin, one can show that there exists an excellent perturbation of the BDE.

We denote by $\mathcal{F}^+$ the set of germs of BDE's which have an excellent perturbation.

Lemma 3.8. *Let $(E, 0) \in \mathcal{F}^+$ be a finite germ with coefficients all vanishing at 0. If the number of zeros of the vector field $\xi_{|_M}$ is finite, then*

$$I(E, 0) = \frac{1}{2} \sum_{i=1}^{n_0} Ind_{s_i} \xi_{|_M} + \frac{1}{2} \sum_{i=0}^{n_1} Ind_{r_i} \nabla F,$$

where $s_i = (0, 0, p_i)$ and $r_i = (0, 0, q_i)$.

Proof. Let E_t be an excellent perturbation of $(E, 0)$ and let $s_{ik} = (z_{ik}, p_{ik})$ be the zeros of $\xi_{t|M_t}$ in a neighbourhood of s_i, where $k \in \{1, \ldots, m_i\}$. Note that $\nabla F^t(s_{ik}) \neq 0$. Let r_{ik} be the zeros of ∇F^t in a neighbourhood of r_i, where $k \in \{1, \ldots, u_i\}$. It follows from Proposition 2.2 in [7], and Lemma 6.1 in [5] that

$$
\begin{aligned}
\mathrm{Ind}_{s_i}\xi_{|M} &= \sum_{k=1}^{m_i} \mathrm{Ind}_{s_{ik}}\xi_{t|M_t} + \sum_{k=1}^{u_i} \mathrm{sign}\,[F^t(r_{ik})] \cdot \mathrm{Ind}_{r_{ik}}\nabla F^t \\
&= \sum_{k=1}^{m_i} \mathrm{Ind}_{s_{ik}}\xi_{t|M_t} - \sum_{k=1}^{u_i} \mathrm{Ind}_{r_{ik}}\nabla F^t \\
&= \sum_{k=1}^{m_i} \mathrm{Ind}_{s_{ik}}\xi_{t|M_t} - \mathrm{Ind}_{r_j}\nabla F
\end{aligned}
$$

when $s_i = r_j$. When s_i is a regular point of M, we have $\mathrm{Ind}_{s_i}\xi_{|M} = \sum_{k=1}^{m_i} \mathrm{Ind}_{s_{ik}}\xi_{t|M_t}$. By hypothesis, $I(E, 0) = \frac{1}{2} \sum_{i=1}^{n_0} \sum_{k=1}^{m_i} \mathrm{Ind}_{s_{ik}}\xi_{t|M_t}$. The result now follows. $\qquad\square$

The following result shows the topological invariance of the index of a germ of BDE when M is a smooth surface and the coefficients all vanish at the origin.

Theorem 3.9. *Let $(E, 0), (L, 0) \in \mathcal{F}^+$ be finite and topologically equivalent germs with coefficients all vanishing at 0. If M is a smooth surface and the number of zeros of $\xi_{|M}$ is finite, then $I(E, 0) = I(L, 0)$.*

Proof. We observe that $\nabla F \neq 0$. The result then follows from Lemma 3.6 and 3.8. $\qquad\square$

4. Examples

Let $(E, 0)$ be a germ of BDE with coefficients not all vanishing at 0. Then using Proposition 3.2 in [3] we obtain that $(E, 0)$ is equivalent to $(\tilde{E}, 0)$, where

$$
\tilde{E}(x, y) = dy^2 - f(x, y)dx^2 = 0. \tag{4.1}
$$

We denote by $M(E, 0)$ the multiplicity of $(E, 0)$, and set $I(f, 0) = I(\tilde{E}, 0)$. In [18], F. Tari studied the singularities of codimension 2 of binary differential equations with coefficients not all vanishing at 0. He also obtained the normal forms of these singularities. We calculate in Table 5.1. the index of the singularities of codimension 2 in [18].

Table 5.1. *Normal forms for codimension 2 singularities.*

Normal forms of f	$I(f, 0)$	$M(\tilde{E}, 0)$	$\mu(f)$
$-y + x^4$	$-\frac{1}{2}$	3	0
$-y - x^4$	$\frac{1}{2}$	3	0
$xy + x^3$	$-\frac{1}{2}$	3	1
$x^2 + y^3$	$-\frac{1}{2}$	3	2
$-x^2 + y^3$	$\frac{1}{2}$	3	2

Table 5.2. *Normal forms of asymptotic curves at a cross-cap.*

Normal forms of $(E, 0)$	$I(E, 0)$	$M(E, 0)$
$ydy^2 + 2xdxdy - ydx^2$	$\frac{1}{2}$	∞
$ydy^2 + 2(-x + y^2)dxdy + ydx^2$	$-\frac{1}{2}$	3
$ydy^2 + 2(-x + xy)dxdy + ydx^2$	$\frac{1}{2}$	5

The local configurations of the lines of curvature and of the asymptotic curves around of a cross-cap which is the unique stable singularity for maps of surfaces into $\mathbb{R}^3$ are given by BDEs with coefficients vanishing at 0. The topological model of these configurations are given in [19]. We can apply Proposition 2.2 in [19] and Theorem 3.9 to show that the local topological configurations of the asymptotic curves around of a cross-cap are topologically distinct (Table 5.2).

Remark 3. Given the BDE $E(x, y) = ydy^2 + 4xdxdy - ydx^2 = 0$, we obtain that $M(E, 0) = 3$. From Theorem 4.1 in [2] we have that $(E, 0)$ is topologically equivalent to the normal form of multiplicity ∞ given in Table 5.2. This shows that the $M(E, 0)$ is not a topological invariant.

References

1. K. Aoki, T. Fukuda and T. Nishimura, An algebraic formula for the topological types of one parameter bifurcation diagrams, *Arch. Rational Mech. Anal.* **108** (1989) 247–265.

2. J. W. Bruce and F. Tari, On binary differential equations, *Nonlinearity.* **8** (1995) 255–271.

3. J. W. Bruce and F. Tari, On the multiplicity of implicit differential equations, *J. Differential Equations.* **148** (1998) 122–147.

4. J. W. Bruce and F. Tari, Dupin indicatrices and families of curve congruences, *Trans. Amer. Math. Soc.* **357(1)** (2005) 267–285.

5. L. S. Challapa, Invariants of Binary Differential Equations, *Journal of Dynamical and Control Systems.* **15** (2009) 157–176.

6. L. S. Challapa, Index of quadratic differential forms, *Contemporary Math. AMS* **459** (2008) 177–191.

7. A. Cima, A. Gasull and J. Torregrosa, On the relation between index and multiplicity, *J. Londo Math. Soc.* **57(2)** (1998) 757–768.

8. L. Dara, Singularités génériques des équations diffrentielles multiformes, *Bol. Soc. Brasil. Math.* **6** (1975) 95–128.

9. A. A. Davydov, Normal forms of differential equations unresolved with respect to derivatives in a neighbourhood of its singular point, *Funct. Anal. Appl.* **19** (1985) 1–10.

10. A. A. Davydov and L. Ortiz-Bobadilla, Smooth normal forms of folded elementary singular points, *J. Dynam. Control Systems.* **1(4)** (1995) 463–482.

11. W. Ebeling and S. M. Gusein-Zade, On the index of a vector field at an isolated singularity, *Arnoldfest, Fields Inst. Commun.* **24** (1999) 141–152.

12. W. Ebeling and S. M. Gusein-Zade, Indices of 1-forms on an isolated complete intersection singularity, *Mosc. Math. J.* **3** (2003) 439–455.

13. G. M. Khimshiashvili, On the local degree of a smoooth mapping, *Akad. Nauk Gruzin. SSR Trudy Tbiliss. Mat. Inst. Razmadze* **64** (1980) 105–124.

14. A. G. Kuz'min,Nonclassical equations of mixed type and their applications in gas dynamics, *International Series of Numerical Mathematics* **109**. Birkhauser Velag, Besel, 1992.

15. H. C. King, Topological type of isolated critical points, *Ann. Math.* **2(2)** (1978) 385–397.

16. H. C. King, Real analytic germs and their varieties at isolated singularities, *Invent. Math.* **37(3)** (1976) 193–199.

17. C. C. Pugh, A generalized Poincar index formula, *Topology.* **7** (1968) 217–226.

18. F. Tari, Two-parameter families of implicit differential equations, *Discrete Contin. Dyn. Syst.* **13** (2005) 139–162.

19. F. Tari, On pairs of geometric foliations on a cross-cap, *Tohoku Math. J. (2)* **59** (2007) 233–258.

L. S. Challapa
Dep. de Matemática
Universidade Federal da Paraíba
58051-900 João Pessoa
Brazil
challapa@mat.ufpb.br

6

About the existence of Milnor fibrations

J. L. CISNEROS-MOLINA AND R. N. ARAÚJO DOS SANTOS

Abstract

The aim of the present article is to review some results on the existence of Milnor fibrations for complex and real singularities and some generalizations in several settings that have been developed recently. After recalling the classical theorems by Milnor, we start with the complex case for germs of maps, then we continue with the real analytic cases and list some open questions.

1. Introduction

In his now classic book [21] Milnor proves a Fibration Theorem, which associates a locally trivial fibration to each singular point of a complex hypersurface. It is a very useful and fundamental tool to understand its local topological behavior.

Given a complex holomorphic map

$$f : (\mathbb{C}^n, 0) \to (\mathbb{C}, 0),$$

let $V(f) := f^{-1}(0)$, denote by $\sum(f)$ the **critical locus** of f and suppose $0 \in \sum(f)$. Let S_ϵ^{2n-1} be the sphere centered at $0 \in \mathbb{C}^n$ of radius $\epsilon > 0$. The set $K_\epsilon := V(f) \cap S_\epsilon^{2n-1}$ is called the **link** of the singularity at the origin.

2000 *Mathematics Subject Classification* 32S55 (primary), 32S60, 58K15 (secondary).

The first author is a Regular Associate of the International Centre for Theoretical Physics, Trieste, Italy.

The second author would like to thank the Brazilian agencies CNPq, grant PDE, number: 200643/2007-0 for supporting his stay in NEU/U.S. from 10/2007 to 08/2008 during the postdoc, and FAPESP grant num. 09/14383-3.

Theorem 1.1 (Milnor Fibration Theorem, [21]). *There exists a small $\epsilon_0 > 0$ such that for all $0 < \epsilon \leq \epsilon_0$, the map*

$$\frac{f}{|f|} : S_\epsilon^{2n-1} \setminus K_\epsilon \to S^1,$$

is the projection of a smooth locally trivial fibration. Furthermore, if $0 \in \mathbb{C}^n$ is an isolated critical point of f, then the fibers of this fibration have the homotopy type of a bouquet of spheres of dimension $n - 1$ and the topological closure of each fiber is the link K_ϵ.

Let us see one example in the case $n = 2$.

Example 1.2. Consider the holomorphic map $f : (\mathbb{C}^2, 0) \to (\mathbb{C}, 0)$, given by $f(z, w) = z^2 + w^3$. Since $\nabla f(z, w) = (0, 0)$ implies $(z, w) = (0, 0)$, we have that $\sum(f) = \{(0, 0)\}$, i. e., f has an isolated critical point at the origin. Since the map f is weighted-homogeneous of type $(3, 2; 6)$, we have that $V(f) \setminus \{(0, 0)\}$ is an analytic manifold that cuts transversely all spheres of any radius. Without loss of generality, we can choose the sphere $S_{\sqrt{2}}^3 = \{(z, w) \in \mathbb{C}^2 : \|z\|^2 + \|w\|^2 = 2\}$.

Using the transversality theorem, the link $K_{\sqrt{2}} = V(f) \cap S_{\sqrt{2}}^3$ is a real analytic submanifold in $S_{\sqrt{2}}^3$ of dimension one, i.e, a regular curve.

Let us describe this link.

If $(z, w) \in K_{\sqrt{2}}$ then we have the equations $|z|^2 = |w|^3$ and $|z|^2 + |w|^2 = 2$. This system provides the equation $X^3 + X^2 - 2 = 0$ on the variable $X = |w|$, which can be factorized as $X^3 + X^2 - 2 = (X - 1)(X^2 + 2X + 2)$. Hence, this equation has only one real solution $X = |w| = 1$ and two others which are complex conjugates.

Considering the real solution $|w| = 1$ and using the second equation $|z|^2 + |w|^2 = 2$, we get $|z| = 1$. This implies that the link $K_{\sqrt{2}}$ lies on the torus $S^1 \times S^1 = \{(e^{i\theta}, e^{i\phi}) \in S_{\sqrt{2}}^3 : (\theta, \phi) \in [0, 2\pi] \times [0, 2\pi]\}$.

Now consider the parametrization $\varphi : [0, 12\pi] \to V(f) \cap S_{\sqrt{2}}^3$ given by $\varphi(t) = (e^{\frac{it}{2}}, e^{\frac{it}{3}})$ (see the Figure 6.1 below). We can see clearly that the intersection is the well-known $(2, 3)$-torus knot also called the "trefoil knot".

According to Milnor's theorem this curve is the topological closure of the boundary of the fibers.

This example is a particular case of weighted-homogeneous polynomials in n complex variables called **Brieskorn-Pham polynomials**

$$z_1^{a_1} + \cdots + z_n^{a_n}, \quad n > 0, \ a_j \in \mathbb{N}, \ a_j \geq 2, \ j = 1, \ldots, n. \quad (1.1)$$

Figure 6.1. Fibre and link of $V(z^2 + w^3)$.

1.1. Milnor fibration for real singularities

In [20] Milnor proves that for a *real* polynomial map germ

$$f : (\mathbb{R}^n, 0) \to (\mathbb{R}^k, 0), \quad n \geq k \geq 2,$$

with $0 \in \mathbb{R}^n$ an *isolated critical point*, it is always possible to associate a locally trivial fibration. This article was never published but its results appear in §11 of [21].

In what follows we describe this fibration. Denote by B_ϵ the real closed ball of dimension n centered at $0 \in \mathbb{R}^n$ of radius ϵ and let S_η^{k-1} be the sphere centered at $0 \in \mathbb{R}^k$ of radius $\eta > 0$. Then Milnor's result is:

Theorem 1.3 ([20, 21]). *There exists a small $\epsilon_0 > 0$, such that for all $0 < \epsilon \leq \epsilon_0$ and all η, with $0 < \eta \ll \epsilon$, the map*

$$\frac{1}{\eta} f_| : B_\epsilon \cap f^{-1}\big(S_\eta^{k-1}\big) \to S_\eta^{k-1}$$

is the projection of a smooth locally trivial fibration.

Let $S_\epsilon^{n-1} := \partial B_\epsilon$ and, as before, let $V(f) := f^{-1}(0)$ and define the link by $K_\epsilon := V(f) \cap S_\epsilon^{n-1}$. Since $0 \in \mathbb{R}^n$ is an isolated singularity of f, then $0 \in \mathbb{R}^k$ is a regular value of $f_| : S_\epsilon^{n-1} \to \mathbb{R}^k$, if $\epsilon > 0$ is small enough. Let $N(K_\epsilon)$ stand for an open tubular neighborhood of K_ϵ in the sphere. Milnor used the flow of an appropriate vector field in $B_\epsilon \setminus V(f)$ to construct a diffeomorphism from $B_\epsilon \cap f^{-1}(S_\eta^{k-1})$ to $S_\epsilon^{n-1} \setminus N(K_\epsilon)$, getting in this way, up to diffeomorphism, a fibration $S_\epsilon^{n-1} \setminus N(K_\epsilon) \to S^{k-1}$. It is not difficult to see (for instance, [30, 29]) that one can always extend this fibration into the open tubular neighborhood $N(K_\epsilon)$ to get a fibration

$$S_\epsilon^{n-1} \setminus K_\epsilon \to S^{k-1}. \tag{1.2}$$

In his book [21, p. 100], Milnor comments that the major weakness of Theorem 1.3 is that the hypothesis is so strong that it is very difficult to find examples, except those that come from holomorphic maps. This raises the problem of finding dimensions $n \geq k \geq 2$ for which such examples exist (see [27]).

Returning to the fibration (1.2), we observe that the procedure explained in the paragraph following Theorem 1.3 only ensures the existence of a projection map giving a fibration, but unlike the complex case, it gives no explicit construction.

In other words, in the real setting, there is no *a priori* reason to expect this projection to be the canonical map $f/|f|$ as in the complex case. In fact, Milnor gave an example (see [21, p. 99]) in which the canonical map fails to be the projection map of a fibration (1.2). This motivates the following definition.

Definition 1.4 ([27], [32]). Let $n \geq 2$. Given a real analytic map germ $f: (\mathbb{R}^n, 0) \to (\mathbb{R}^2, 0)$, with isolated critical point at the origin, we say that f satisfies the **strong Milnor condition** at 0, if and only if, for all $\epsilon > 0$ sufficiently small the map

$$\frac{f}{|f|}: S_\epsilon^{n-1} \setminus K_\epsilon \to S^1$$

is the projection of a locally trivial fibration.

So, a natural question is: Are there real analytic maps which satisfy the **strong Milnor condition**?

By Theorem 1.1 we have a partial positive answer: consider any holomorphic map $f: (\mathbb{C}^n, 0) \to (\mathbb{C}, 0)$, with isolated singularity at the origin, as a pair of real analytic maps $f = (\Re(f), \Im(f))$, where $\Re(f)$ and $\Im(f)$ are, respectively, the real and imaginary part of f.

What about real analytic maps which do not come from holomorphic ones?

This introduces the problem of studying natural conditions under which a real analytic map germ $f = (f_1, f_2): (\mathbb{R}^n, 0) \to (\mathbb{R}^2, 0)$, $n \geq 2$, satisfies the strong Milnor condition, i.e,

How big is the class of real analytic maps which satisfy the strong Milnor condition?

As far as we know, this problem was first approached in [11, 12] by A. Jacquemard. There, the author gives two sufficient conditions, one geometric and one algebraic, to have the strong Milnor condition. Some years later, Seade in [31] and Seade, Ruas and Verjovsky in [27] and, Ruas and Araújo dos Santos in [26], used a pencil of real analytic hypersurfaces canonically associated to the corresponding map germs to study singularities that satisfies the strong Milnor condition (see §2.2). In [27], the authors provided a method to find an infinite family of singularities satisfying this condition, in particular, the

twisted Brieskorn-Pham polynomials, which are real analytic analogues to Brieskorn-Pham polynomials given in (1.1) (see Example 3.8).

In [26], using different tools of singularity theory and stratification theory, the authors refined the previous argument and proved that under a weaker condition (Bekka's (c)-regularity) on a stratification of an analytic set given by projections of the map f, it is also possible to guarantee the strong Milnor condition. See [26] for details. Afterwards, in [28] the author obtained a slight improvement of the previous result using only the so-called (m)-condition. Actually, this was inspired by the proof of K. Bekka and Koike that (c)-regularity implies the (m)-condition.

Recently, in [5, 30], for certain classes of real analytic maps $f : (\mathbb{R}^n, 0) \to (\mathbb{R}^k, 0)$, with $n \geq k \geq 2$, necessary and sufficient conditions were given for the map $\frac{f}{|f|} : S_\epsilon^{n-1} \setminus K_\epsilon \to S^{k-1}$ to be the projection of a locally trivial fibration. We will explain these results in more detail in §3.4 and §3.5.

In what follows, we present some interesting generalizations, from the authors' point of view, of Milnor fibrations in both complex and real cases, which have been developed in recent years. We will start with the complex case for germs of maps and after that we will move to the real analytic case and list some open questions. We also recommend the interesting survey article [33].

2. Milnor fibrations for complex maps

In this section we will be concerned with the complex case.

2.1. Milnor Fibration on complex analytic sets

We start with an improvement of Milnor Fibration Theorem [21] for complex maps, due to Lê Dũng Tráng in [14], which generalizes the existence of Milnor fibrations on complex analytic sets.

Let X be an analytic subset of an open neighborhood U of the origin 0 in $\mathbb{C}^n$. Let $f : (X, 0) \to (\mathbb{C}, 0)$ be holomorphic and set $V = f^{-1}(0)$. Let B_ϵ be a ball in U of sufficiently small radius $\epsilon > 0$, centred at $0 \in \mathbb{C}^n$ and $D_\eta - \{0\}$ be the punctured disc of radius η in $\mathbb{C}$.

Theorem 2.1 (Milnor-Lê Fibration [14]). *For all small enough $\epsilon > 0$, and all η with $0 < \eta \ll \epsilon$,*

$$f_| : B_\epsilon \cap X \cap f^{-1}(D_\eta - \{0\}) \to D_\eta - \{0\} \tag{2.1}$$

is a topological locally trivial fibration.

Proof (idea). Let $\mathcal{S}$ be a Whitney stratification of X, choose $\epsilon > 0$ small enough such that B_ϵ intersects only a finite number of strata of X and such that the sphere S_ϵ intersects such strata transversely. Moreover, according to [10], we can always choose this stratification in such a way that V is a union of strata and satisfies Thom's A_f-condition. This implies that for $0 < \eta \ll \epsilon$ the fibers of the map

$$f_| : B_\epsilon \cap X \cap f^{-1}(D_\eta - \{0\}) \to D_\eta - \{0\},$$

intersect transversely the strata of $X \cap S_\epsilon$ and that it is a stratified submersion. Now the result follows from the Thom-Mather First Isotopy Theorem [19]. $\square$

Lê also observed that H. Hamm in [8] proved that, if $X \setminus V$ is a non-singular analytic set in $\mathbb{C}^n$, this topological fibration is a smooth fibration.

As stated below, this result was generalized by E. Looijenga in [17] for complex analytic maps germs $f : (\mathbb{C}^n, 0) \to (\mathbb{C}^p, 0)$ that locally define an Isolated Complete Intersection Singularity (ICIS). This means that in a sufficient small ball $B_\epsilon \subset \mathbb{C}^n$ the intersection $(f^{-1}(0) - \{0\}) \cap B_\epsilon$ is an analytic manifold. Let $C(f) = f(\sum(f))$ be the discriminant set of f.

Theorem 2.2 ([17]). *There exists ϵ_0, such that for all $0 < \epsilon \leq \epsilon_0$, and $0 < \eta \ll \epsilon$, the projection map*

$$f_| : B_\epsilon \cap f^{-1}(D_\eta - C(f)) \to D_\eta - C(f)$$

is a smooth locally trivial fibration.

One has the following natural question:

Question 1. Find good conditions under which it is possible to guarantee the existence of a Milnor fibration for a complete intersection with non-isolated singularity.

More recently in [7], using the integral closure of modules, T. Gaffney presented a beautiful and interesting approach to this question.

There is also a generalization of Theorem 1.1 to complex analytic sets, implicit in the work of Lê Dũng Tráng [13]; a weaker form is given by Alan H. Durfee in [6, Thm. 3.9].

Let $L_{X_\epsilon} = X \cap S_\epsilon$ be the link of X and let $L_{f_\epsilon} = X \cap V \cap S_\epsilon$ be the link of f in X.

Theorem 2.3. *There exists $\epsilon_0 > 0$, such that, for all $0 < \epsilon \leq \epsilon_0$, the map*

$$\phi = \frac{f}{|f|} : L_{X_\epsilon} \setminus L_{f_\epsilon} \longrightarrow S^1. \tag{2.2}$$

is a locally trivial fibration.

A proof of this theorem and the fact that fibration (2.2) is equivalent to the restriction $f_|: B_\epsilon \cap X \cap f^{-1}(\partial D_\eta) \to \partial D_\eta$ of fibration (2.1) can be found in [5] (see Theorem 2.5 below).

2.2. Refinements of Milnor fibration theorems

In [5] the authors give some refinements of Theorems 2.1 and 2.3.

As in § 2.1, let X be an analytic subset of an open neighborhood U of the origin 0 in $\mathbb{C}^n$ and denote by $\text{Sing}(X)$ the set of singular points of X.

As we mentioned in the introduction, there is a canonical pencil of real analytic hypersurfaces associated to f (see also [31, 27]) defined as follows. For each $\theta \in [0, \pi)$, let $\mathcal{L}_\theta$ be the line through 0 in $\mathbb{R}^2$ with an angle θ (with respect to the x-axis) and set $X_\theta = f^{-1}(\mathcal{L}_\theta)$. Then each X_θ is a real analytic hypersurface and the family $\{X_\theta\}$ is called the **canonical pencil** of f. Its main properties are summarized in the following theorem.

Theorem 2.4 (Canonical Decomposition). *Let $\{X_\theta\}$ be the canonical pencil of f. Then,*

(i) *The X_θ are all homeomorphic real analytic hypersurfaces of X with singular set $\text{Sing}(V) \cup (X_\theta \cap \text{Sing}(X))$. Their union is the whole space X and they all meet at V, which divides each X_θ in two homeomorphic halves, i.e., $X_\theta = E_\theta^+ \cup V \cup E_\theta^-$ and $E_\theta^+ \cong E_\theta^-$.*

(ii) *If $\{S_\alpha\}$ is a Whitney stratification of X adapted to V (i.e., V is a union of strata), then the intersection of the strata with each X_θ determines a Whitney stratification of X_θ, and for each stratum S_α and each X_θ, the intersection $S_\alpha \cap X_\theta$ is transverse to every sphere with centre O and radius $\leq \epsilon_0$.*

(iii) *There exists a small $\epsilon > 0$ such that there is a uniform conical structure for all X_θ, i.e., there is a homeomorphism*

$$h: (X \cap B_{\epsilon_0}, V \cap B_{\epsilon_0}) \to (Cone(X \cap S_{\epsilon_0}), Cone(V \cap S_{\epsilon_0}))$$

whose restriction to each X_θ defines a homeomorphism

$$(X_\theta \cap B_{\epsilon_0}) \cong Cone(X_\theta \cap S_{\epsilon_0}).$$

Proof (idea). In (i) the fact that all the X_θ are homeomorphic follows from Theorem 2.8 below and the fact that all the $E_\theta^\pm$ are homeomorphic from Theorem 2.5. The rest of (i) is straightforward. To prove (iii) the authors construct a stratified vector field on $B_\epsilon \cap X$ tangent to each X_θ, and transverse to all spheres in B_ϵ centered at 0. We refer to [5] for further details of the proof. □

The following theorem combines Theorems 2.1 and 2.3.

Theorem 2.5 (Fibration Theorem). *There is a commutative diagram of fiber bundles*

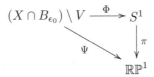

where $\Psi(x) = (Re(f(x)) : Im(f(x)))$, Ψ *has fiber* $(X_\theta \cap B_{\epsilon_0}) \setminus V$, $\Phi(x) = \frac{f(x)}{|f(x)|}$ *and* π *is the natural two-fold covering. The restriction of* Φ *to the link* $L_X \setminus L_f$ *is the Milnor fibration* ϕ *of Theorem 2.3, while the restriction of* Φ *to the* **Milnor tube** $f^{-1}(\partial D_\eta) \cap B_{\epsilon_0}$ *is the Milnor-Lê fibration of Theorem 2.1 (up to multiplication by a constant), and the two fibrations are equivalent.*

The following corollary gives the relation between the Milnor fibers of ϕ and the link of the X_θ.

Corollary 2.6. *Let the germ* $(X, 0)$ *be irreducible, and* $f : (X, 0) \to (\mathbb{C}, 0)$ *have Milnor fibration*

$$\phi = \frac{f}{|f|} : L_{X_\epsilon} \setminus L_{f_\epsilon} \longrightarrow S^1.$$

Then every pair of fibres of ϕ *over antipodal points of* S^1 *meet in the link* L_{f_ϵ}, *and their union is the link of a real analytic hypersurface* X_θ, *which is homeomorphic to the link of* $\{Re\ f = 0\}$. *Moreover, if both X and f have an isolated singularity at* 0, *then this homeomorphism is a diffeomorphism and the link of each* X_θ *is diffeomorphic to the double of the Milnor fiber of f regarded as a smooth manifold with boundary* L_{f_ϵ}.

Proof (idea). From Theorem 2.4(i) the link of X_θ is given by

$$X_\theta \cap S_\epsilon = (E_\theta^+ \cap S_\epsilon) \cup L_{f_\epsilon} \cup (E_\theta^- \cap S_\epsilon),$$

and $E_\theta^+ \cap US_\epsilon$ and $E_\theta^- \cap S_\epsilon$ are the Milnor fibers of ϕ over antipodal points of S^1. Also X_θ is homeomorphic to $X_\pi = \{Re\ f = 0\}$. $\qquad\square$

To prove Theorem 2.5 the authors introduce an auxiliary map called the **spherification** of f, defined by

$$\mathfrak{F}(x) = \|x\|\Phi(x) = \|x\| \frac{f(x)}{|f(x)|}.$$

Notice that given $z \in \mathbb{C} \setminus \{0\}$ with $\theta = \arg z$, the fiber $\mathfrak{F}^{-1}(z)$ is the intersection of $E_\theta^\pm$ with the sphere $S_{|z|}$ of radius $|z|$ centred at 0, and $\mathfrak{F}$ carries $S_{|z|} \setminus V$ into the circle around $0 \in \mathbb{R}^2$ of radius $|z|$.

Then they obtain the following fibration theorem.

Theorem 2.7. *For $\epsilon_0 > 0$ sufficiently small, one has a fiber bundle*

$$\mathfrak{F} \colon \big((X \cap B_{\epsilon_0}) \setminus V\big) \longrightarrow (D_{\epsilon_0} \setminus \{0\}),$$

where D_{ϵ_0} is the disc in $\mathbb{R}^2$ centred at 0 with radius ϵ_0. Furthermore, the restriction of $\mathfrak{F}$ to each circle around 0 of radius $\epsilon \leq \epsilon_0$ is a fiber bundle, and is the Milnor fibration ϕ in Theorem 2.3 up to multiplication by a constant.

Proof of Theorems 2.5 and 2.7 (idea). Using Theorem 2.4-(ii) one proves that $\mathfrak{F}$ is a submersion on each stratum in $X \setminus V$. This allows us to construct on $(X \cap B_{\epsilon_0}) \setminus V$ a complete, stratified, vector field $\hat{v}$ which is tangent to all the spheres in B_ϵ centred at 0, and whose orbits are transverse to the $X_\theta \setminus V$ and permute them. The vector field $\hat{v}$ gives the local triviality of the restriction of $\mathfrak{F}$ to any sphere S_ϵ, which is equivalent to the fibration ϕ since $\Phi = \frac{\mathfrak{F}(x)}{\|\mathfrak{F}(x)\|} = \frac{f(x)}{|f(x)|}$. Also $\hat{v}$ gives the local triviality of Φ and Ψ in Theorem 2.5. Using the uniform conical structure given in Theorem 2.4-(iii) one gets the local triviality of $\mathfrak{F}$ over $\mathbb{C} \setminus \{0\}$.

It is also possible to construct a vector field $\tilde{w}$ which is transverse to all spheres in B_{ϵ_0} centred at 0, transverse to all the tubes $f^{-1}(\partial D_\eta)$, and tangent to the strata of each $X_\theta \setminus V$. This vector field gives the equivalence of the fibration on the tube and that on the sphere. $\square$

Also a new Milnor-type fibration theorem is obtained in which it is not necessary to remove the zero locus of the map f.

Theorem 2.8. *Let $\tilde{X}$ be the space obtained by the real blow-up of V, i.e., the blow-up of $(Re(f), Im(f))$. The projection $\tilde{\Psi} \colon \tilde{X} \to \mathbb{RP}^1$ is a topological fiber bundle with fiber X_θ.*

This implies that all the hypersurfaces X_θ are homeomorphic and it can also be used to prove Theorem 2.5.

Proof (idea). The Whitney stratification of X induces a canonical Whitney stratification on $\tilde{X}$. The map $\tilde{\Psi}$ is a stratified submersion and by Theorem 2.4.(ii), the fibers of $\tilde{\Psi}$ are transverse to $\tilde{X} \cap (S_{\epsilon_0} \times \mathbb{RP}^1)$, the boundary of the compact set $\tilde{X} \cap (B_{\epsilon_0} \times \mathbb{RP}^1)$. Hence, one can apply the Thom-Mather First Isotopy Theorem to get the fibration. $\square$

3. Milnor fibrations for real maps

3.1. Fibrations for maps $f\bar{g}$ and meromorphic maps

It is natural to ask if, as in the holomorphic case, it is possible to have a fibration theorem for real analytic map germs $f: (\mathbb{R}^n, 0) \to (\mathbb{R}^k, 0)$ having an isolated critical value at $0 \in \mathbb{R}^k$, instead of having only $0 \in \mathbb{R}^n$ as an isolated critical point. This was first considered in [25] where the authors generalized Theorem 2.1 for the case of real analytic maps $f: X \to \mathbb{R}^k$ on a real analytic variety X of dimension $n > 0$, with $0 \in \mathbb{R}^k$ as an isolated critical value and satisfying Thom's A_f-condition. Using this generalization they prove the following fibration theorem.

Let X be an equidimensional complex analytic variety in $\mathbb{C}^N$ of dimension n with an isolated singularity at 0. Let $f, g: (X, 0) \to (\mathbb{C}, 0)$ be germs of holomorphic maps such that $f^{-1}(0)$ and $g^{-1}(0)$ have no common irreducible components. Denote by L_X the link of X and by $L_{f\bar{g}}$ the link of $f\bar{g}$ in X.

Theorem 3.1. *Suppose that $f\bar{g}: (X, 0) \to (\mathbb{C}, 0)$ has an isolated critical value at $0 \in \mathbb{R}^2 \cong \mathbb{C}$, and satisfies Thom's A_f-condition. Then one has a locally trivial fibration*

$$\Psi_{f\bar{g}}: L_X \setminus L_{f\bar{g}} \longrightarrow S_\eta^1.$$

Proof (idea). Using the fact that $f\bar{g}$ has an isolated critical value at 0 and that it satisfies Thom's A_f-condition, one can apply the Thom-Mather First Isotopy Theorem as in the proof of Theorem 2.1. $\square$

In the case that X has dimension 2, it is proved in [25] that for f and g with no common branch and $f\bar{g}$ with an isolated critical point at 0, $f\bar{g}$ satisfies Thom's A_f-condition. Also in [25] the authors generalize a fibration theorem in [2] for meromorphic maps f/g which are **semitame**. Notice that the meromorphic map f/g takes values in $\mathbb{P}^1$.

Definition 3.2. Set $V_{fg} = \{ fg = 0 \}$, and $h = f/g$. Define the set

$$M(f/g) = \left\{ x \in X \setminus V_{fg} \mid T_x\big((h^{-1}(h(x)))\big) \subset T_x S_{\|x\|}^{2N-1} \right\}.$$

The **bifurcation set** $B \subset \mathbb{P}^1$ of f/g is the union of $\{ 0, \infty \}$ and the set of $c \in \mathbb{P}^1$ such that there exists a sequence $(x_k)_{k \in \mathbb{N}}$ in $M(f/g)$ such that

$$\lim_{k \to \infty} x_k = 0 \qquad \text{and} \qquad \lim_{k \to \infty} (f/g)(x_k) = c.$$

The meromorphic map f/g is **semitame** at 0 if $B = \{ 0, \infty \}$.

Theorem 3.3 ([2]). *Let $f, g: (X, 0) \to (\mathbb{C}, 0)$ be holomorphic with no common branch. If f/g is semitame at 0, then the map*

$$\phi_{f/g} = \frac{f/g}{|f/g|}: \mathcal{L}_X \setminus L_{fg} \longrightarrow S^1$$

is a locally trivial C^∞ fiber bundle.

Proof (idea). The proof follows Milnor's proof [21, Chapter 4] with minor modifications in [21, Lemma 4.4] and in the Curve Selection Lemma to use a meromorphic map instead of a holomorphic one. See [2] for details. □

If $f, g: (X, 0) \to (\mathbb{C}, 0)$ are holomorphic with no common branch, such that $f\bar{g}$ has an isolated critical value at 0 and satisfies Thom's $A_{f\bar{g}}$-condition, and furthermore, f/g is semitame at 0, then the hypotheses of Theorems 3.1 and 3.3 are satisfied and we have two fibrations $\Psi_{f\bar{g}}$ and $\phi_{f/g}$. Since $\frac{f\bar{g}}{|f\bar{g}|} = \frac{f/g}{|f/g|}$ on $X \setminus V_{fg}$, the fibrations $\Psi_{f\bar{g}}$ and $\phi_{f/g}$ are topologically equivalent.

Recall that when X has dimension 2, the map $f\bar{g}$ automatically satisfies Thom's $A_{f\bar{g}}$-condition. In this case it is natural to compare the hypotheses: '$f\bar{g}$ has an isolated critical value at 0' and 'f/g is semitame at 0', which give the fibrations $\Psi_{f\bar{g}}$ and $\phi_{f/g}$ respectively. In the case $X = \mathbb{C}^2$ these two hypotheses are equivalent.

Theorem 3.4 ([25, Theorem 5.8]). *Let $f, g: (\mathbb{C}^2, 0) \to (\mathbb{C}, 0)$ be holomorphic germs such that $f^{-1}(0)$ and $g^{-1}(0)$ have no common component. The following are equivalent:*

  (i) *$f\bar{g}$ has an isolated critical value at 0,*
 (ii) *the map $\Psi_{f\bar{g}}: L_X \setminus L_{fg} \to S^1$ is a fibration,*
(iii) *the map $\phi_{f/g}: L_X \setminus L_{fg} \to S^1$ is a fibration,*
(iv) *f/g is semitame at 0.*

Proof (idea). The equivalence of (i) and (ii) is given by [25, Theorem 4.4] which uses a result of [15] which relates the determinantal ratios of the germ $(f, g): \mathbb{C}^2 \to \mathbb{C}^2$ to some topological invariants of the meromorphic germ f/g (see [25, Section 4] for details). The equivalence of (iii) and (iv) is proved in [2] and is given by Theorem 3.3 and its converse for the case $X = \mathbb{C}^2$. Both sets of equivalences are related by the fact that the fibrations $\Psi_{f\bar{g}}$ and $\phi_{f/g}$ are fibrations of the multilink $L_f \cup -L_g$. We refer to [25] for the definition of multilink as well as for details of the proof. □

3.2. Fibrations of polar weighted homogeneous polynomials

In [27] twisted Brieskorn-Pham polynomials were defined (see Example 3.8), which were the first examples of polar weighted homogeneous polynomials. Inspired by these examples, polar weighted homogeneous polynomials were defined in general in [4, 22]. These are **real** analytic maps which generalize complex weighted homogeneous polynomials. These polynomials have $0 \in \mathbb{C}$ as unique critical value, but do not necessarily have an isolated critical point. They have Milnor fibrations on the Milnor tube and on the sphere, the two fibrations being equivalent.

Consider $\mathbb{C}^n$ with coordinates $z_1, \ldots, z_n$; as usual, let $\bar{z}_j$ be the complex conjugate of z_j and write $z_j = x_j + i y_j$. Then considering $\mathbb{C}^n$ with coordinates $z_1, \ldots, z_n, \bar{z}_1, \ldots, \bar{z}_n$ is equivalent to considering it as a $2n$-dimensional *real* vector space with coordinates $x_1, y_1, \ldots, x_n, y_n$. To simplify notation write $\mathbf{z} = (z_1, \ldots, z_n), \bar{\mathbf{z}} = (\bar{z}_1, \ldots, \bar{z}_n)$; also set $\mathbb{C}^* = \mathbb{C} - \{0\}$.

Definition 3.5. Let p_j, u_j, with $j = 1, \ldots, n$, be positive integers such that

$$\gcd(p_1, \ldots, p_n) = 1, \qquad \gcd(u_1, \ldots, u_n) = 1.$$

Write $\tau \in \mathbb{C}^*$ in *polar form* $\tau = t\lambda$, with $t \in \mathbb{R}^+$ and $\lambda \in S^1$, that is, $t = |\tau|$ and $\lambda = \exp(i \arg \tau)$.

A **polar $\mathbb{C}^*$-action** on $\mathbb{C}^n$ with **radial weights** $(p_1, \ldots, p_n)$ and **polar weights** $(u_1, \ldots, u_n)$ is given by

$$t\lambda \cdot (\mathbf{z}) = (t^{p_1}\lambda^{u_1}z_1, \ldots, t^{p_n}\lambda^{u_n}z_n). \tag{3.1}$$

In fact, a polar $\mathbb{C}^*$-action is the combination of two actions: a $\mathbb{R}^+$-action given by the weights $(p_1, \ldots, p_n)$, and a S^1-action given by the weights $(u_1, \ldots, u_n)$.

Definition 3.6. Let $f: \mathbb{C}^n \to \mathbb{C}$ be a polynomial in the $2n$ variables $z_1, \ldots, z_n, \bar{z}_1, \ldots, \bar{z}_n$. Let a and c be positive integers. We say that f is **polar weighted homogeneous** with **radial weight type** $(p_1, \ldots, p_n; a)$ and **polar weight type** $(u_1, \ldots, u_n; c)$, if the following functional identity holds

$$f(t\lambda \cdot (\mathbf{z})) = t^a \lambda^c f(\mathbf{z}), \quad t \in \mathbb{R}^+, \ \lambda \in S^1, \tag{3.2}$$

where $t\lambda \cdot (\mathbf{z})$ is a polar $\mathbb{C}^*$-action.

In other words, it is weighted homogeneous of degree a with respect to the $\mathbb{R}^+$-action with weights $(p_1, \ldots, p_n)$ and it is weighted homogeneous of degree c with respect to the S^1-action with weights $(u_1, \ldots, u_n)$.

Example 3.7. Weighted homogeneous polynomials are a particular case of polar weighted homogeneous polynomials with no $\bar{z}_j$ for $j = 1, \ldots, n$ and with

$p_j = u_j$ and $a = c$. In particular, we have the Brieskorn-Pham polynomials given in (1.1).

Example 3.8. A polynomial in $\mathbb{C}^n$ of the form

$$v_1 z_1^{a_1} \bar{z}_{\sigma(1)} + \cdots + v_n z_n^{a_n} \bar{z}_{\sigma(n)},$$

was called in [27] a **twisted Brieskorn-Pham polynomial** of class $\{a_1, \ldots, a_n; \sigma\}$, where each $a_j \geq 2$, $j = 1, \ldots, n$, the v_j are non-zero complex numbers and σ is a permutation of the set $\{1, \ldots, n\}$ called the **twisting**.

In [27] it is proved that twisted Brieskorn-Pham polynomials are polar weighted homogeneous. It is also proved that they have isolated critical points and that they satisfy the strong Milnor condition.

Remark. The sum of two polar weighted homogeneous polynomials (with different variables) is again a polar weighted homogeneous polynomial.

Using a generalization of the Euler identities for weighted homogeneous polynomials, it is proved that $0 \in \mathbb{C}$ is the only critical value of polar weighted homogeneous polynomials. Then as for complex weighted homogeneous polynomials one has the following (global) fibration theorem (compare with [23]).

Proposition 3.9. *The restriction* $f \colon (\mathbb{C}^n - V) \to \mathbb{C}^*$ *is a locally trivial fibration. Its monodromy is given by the map*

$$h(\mathbf{z}) = e^{2\pi i/c} \cdot \mathbf{z}$$

Moreover, the map

$$\phi = \frac{f}{|f|} \colon \left(S_\epsilon^{2n-1} \setminus K_\epsilon \right) \to S^1,$$

is a locally trivial fibration for any $\epsilon > 0$.

Let $f_| \colon f^{-1}(S^1) \to S^1$ be the restriction of the fibration $f \colon (\mathbb{C}^n - V) \to \mathbb{C}^*$ of Proposition 3.9 to S^1.

Proposition 3.10. *The fibration $f_| \colon f^{-1}(S^1) \to S^1$ is equivalent to the Milnor fibration $\phi \colon S_\epsilon^{2n-1} \setminus K_\epsilon \to S^1$.*

Propositions 3.9 and 3.10 are proved using the $\mathbb{C}^*$-polar action in an analogous way as for complex weighted homogeneous polynomials.

In [4] is also proved a Join Theorem, which says that the Milnor fiber of the sum of two polar weighted homogeneous polynomials in disjoint sets of variables is homotopically equivalent to the join of the Milnor fibers of the summands.

Polar weighted homogeneous polynomials are **mixed analytic functions** which is a wider class of real analytic functions recently defined in [24]. There the author gives a condition to guarantee the existence of Milnor Fibrations for mixed analytic functions.

3.3. Milnor conditions (a) and (b) and Ł-analytic maps

An interesting improvement in the technique of proving the existence of Milnor fibrations for real analytic map germs $f : (\mathbb{R}^n, 0) \to (\mathbb{R}^k, 0)$, $m \geq k \geq 2$, is given in [18]. The author defines two conditions under which is possible to guarantee the existence of Milnor fibrations for non-isolated singularities, more precisely, for a map f with isolated critical value at $0 \in \mathbb{R}^k$.

To describe this result, following the author's notation in [18], let $f(x) = (f_1(x), \ldots, f_k(x))$ be a representative of the germ defined in a small neighborhood U of the origin, $\nabla f_i(x)$ the gradient map of the coordinate function f_i, for $i = 1, \ldots, k$. Denote by $\mathfrak{U} = \sum(f)$ the singular locus of f and by $\rho(x) = \|x\|^2$ the square of the distance function from the origin. Define

$$\mathfrak{B} := \{x \in U \mid \nabla f_i(x), \ldots, \nabla f_k(x), \nabla \rho(x) \text{ are linearly dependent}\}.$$

Definition 3.11. We say that a map germ f satisfies **Milnor condition (a) at the origin** 0, if $0 \notin \overline{\mathfrak{U} \setminus V(f)}$.

Definition 3.12. We say that a map germ f satisfies **Milnor condition (b) at the origin** 0, if 0 is an isolated point of (or, is not in) $V(f) \cap \overline{\mathfrak{B} \setminus V(f)}$.

It is easy to see that Milnor condition (a) means that in a small neighborhood of the origin $\sum(f) \subset V(f)$, i.e, $0 \in \mathbb{R}^k$ is an isolated critical value. Milnor condition (b) implies that there exists a small enough $\epsilon_0 > 0$, such that for each $0 < \epsilon \leq \epsilon_0$, $0 < \eta \ll \epsilon$, $f^{-1}(B_\eta^k - \{0\}) \cap S_\epsilon^{n-1} \cap \mathfrak{B} = \varnothing$, where $B_\eta^k - \{0\}$ stand for the punctured closed ball in $\mathbb{R}^k$ of radius η.

Definition 3.13. If f satisfies Milnor conditions (a) and (b) at the origin, we say that ϵ is a **Milnor radius for f at the origin**, if both conditions hold in $B_\epsilon \subset \mathbb{R}^n$.

Theorem 3.14 ([18]). *Suppose that the real analytic map f has Milnor radius $\epsilon_0 > 0$. Then, for all ϵ, $0 < \epsilon \leq \epsilon_0$, $0 < \eta \ll \epsilon_0$,*

$$f_| : B_\epsilon \cap f^{-1}(B_\eta^k - \{0\}) \to B_\eta^k - \{0\}$$

is the projection of a smooth locally trivial fibration.

Proof (idea). Since f has Milnor radius $\epsilon_0 > 0$, we have $\sum(f) \cap B_\epsilon \subset V(f) \cap B_\epsilon$ for all $0 < \epsilon \leq \epsilon_0$. Hence $0 \in \mathbb{R}^k$ is an isolated critical value of f, i.e, $f_| : B_\epsilon^\circ \setminus V(f) \to R^k$ is a smooth submersion, where B° stands for the open ball. It follows from Milnor condition (b) and the remark above that, for each ϵ and $0 < \eta \ll \epsilon_0$,

$$f_| : S_\epsilon^{n-1} \cap f^{-1}\left(B_\eta^\circ - \{0\}\right) \to B_\eta^\circ - \{0\}$$

is a submersion on the boundary of the closed ball B_ϵ. Now, combining these two conditions, it follows that for each ϵ and all small enough η,

$$f_| : B_\epsilon \cap f^{-1}\left(B_\eta^\circ - \{0\}\right) \to B_\eta^\circ - \{0\}$$

is a smooth, proper submersion, so by Ehresmann's Fibration Theorem is a locally trivial fibration.   □

In [18] the author gave the following analytic condition for existence of a Milnor radius of a map f.

Definition 3.15. We say that a real analytic map germ $f \colon (\mathbb{R}^n, 0) \to (\mathbb{R}^k, 0)$, $m \geq k \geq 2$, satisfies the **strong Łojasiewicz inequality at the origin**, or is **Ł-analytic** at the origin, if there exist an open neighborhood $U \ni 0$, and constants $c > 0$, $0 < \theta < 1$, such that, for all $x \in U$, the following holds:

$$|f(x)|^\theta \leq c. \min_{|(a_1,\ldots,a_k)|=1} \{|a_1.\nabla f_1(x) + \cdots + a_k.\nabla f_k(x)|\}.$$

It is easy to see that Ł-analytic maps always satisfy Milnor condition (a). In [18], using a technique going back to [9], the author proved the following result:

Theorem 3.16 ([18]). *If a real analytic map $f : (\mathbb{R}^n, 0) \to (\mathbb{R}^k, 0)$ is Ł-analytic at the origin, the Milnor condition (b) holds, so there exists a Milnor radius for the map f. Therefore by Theorem 3.14) there is a local Milnor fibration.*

In [18] the author posed the following open question:

Question 2. Let $f \colon (\mathbb{R}^n, 0) \to (\mathbb{R}^2, 0)$ be an Ł-analytic map that also satisfies the strong Milnor condition (see Definition 1.4). Are the two fibrations equivalent?

3.4. Real Milnor fibration and open book decompositions

As we have mentioned before, we cannot expect in general that given a real analytic map germ $f : (\mathbb{R}^n, 0) \to (\mathbb{R}^k, 0)$, with isolated critical point at the origin, the projection $\dfrac{f}{|f|} : S_\epsilon^{n-1} \setminus K_\epsilon \to S^{k-1}$ is the projection map of a smooth locally trivial fibration, i.e, for $k = 2$, that f satisfies the **strong Milnor condition** at the origin.

In [30] the second author and M. Tibar use the idea of open book decomposition in higher dimensions to get a characterization of these fibrations, for all $k \geq 2$, in the class of $\mathcal{K}-$finite map germs. Their main result is:

Theorem 3.17 ([30]). *Let $f : (\mathbb{R}^n, 0) \to (\mathbb{R}^k, 0)$ be a real analytic map germ and suppose that for all small enough radii $\epsilon > 0$, $\sum(f) \cap V(f) \cap B_\epsilon \subseteq \{0\}$. Then, the map $\dfrac{f}{|f|} : S_\epsilon^{n-1} \setminus K_\epsilon \to S^{k-1}$ is the map projection of a smooth locally trivial fibration if, and only if, it is a submersion for all ϵ. Moreover, if the map is weighted-homogeneous with isolated singularity at origin, this fibration is fiber-equivalent to the Milnor fibration given in Theorem 1.3.*

Proof (idea). Write $f(x) = (f_1(x), \ldots, f_k(x))$, $s = (s_1, \ldots, s_k) \in \mathbb{R}^k \setminus \{0\}$, and $[s] = (s_1 : \ldots : s_k) \in \mathbb{P}^{k-1}(\mathbb{R})$. Define the analytic set

$$X := \left\{ (x, [s]) \in B_{\epsilon_0} \times \mathbb{P}^{k-1}(\mathbb{R}) : \text{rank} \begin{bmatrix} f_1(x) & \cdots & f_k(x) \\ s_1 & \cdots & s_k \end{bmatrix} < 2 \right\}.$$

Write $\pi : X \to \mathbb{P}^{k-1}(\mathbb{R})$ and $p : X \to B_{\epsilon_0}$ for the projections, and denote by $X_{[s]}$ the fiber of π over $[s] \in \mathbb{P}^{k-1}(\mathbb{R})$. It is easy to see that $V \times \mathbb{P}^{k-1}(\mathbb{R}) \subset X$, so $V \times [s] \subset X_{[s]}$, for all $[s]$. Also, set

$$X_{[s]}^+ = \left(\frac{f}{|f|} \right)^{-1} \left(\frac{s}{|s|} \right) \text{ and } X_{[s]}^- = \left(\frac{f}{|f|} \right)^{-1} \left(-\frac{s}{|s|} \right).$$

The following three conditions hold:

1) $X_{[s]} \setminus V$ is a disjoint union of $X_{[s]}^+$ and $X_{[s]}^-$;
2) The projection $p : X \to B_{\epsilon_0}$ is a blow up along V, i.e,

$$p : X \setminus (V \times \mathbb{P}^{k-1}(\mathbb{R})) \to B_{\epsilon_0} \setminus V$$

is an analytic isomorphism.
3) For all $0 < \epsilon \leq \epsilon_0$, $X_{[s]} \setminus V$ is an analytic manifold for all $[s]$, transverse to all small spheres S_ϵ (this follows from the submersion condition in the hypothesis).

Now the equivalence follows by a construction similar to that given in [21], [26] (see also [29]). □

We observe that, if f has an isolated singular point at origin, i.e. if $\sum(f) = \{0\}$, our main hypothesis is satisfied, so the following result is an immediate consequence:

Corollary 3.18. *A necessary and sufficient condition for a pair of isolated singular map germs $f = (P, Q) : (\mathbb{R}^n, 0) \to (\mathbb{R}^2, 0)$ to satisfy the* **strong Milnor condition** *at the origin is that the projection map $\dfrac{f}{|f|} : S^{m-1}_\epsilon \setminus K_\epsilon \to S^1$ be a submersion for all small enough $\epsilon > 0$.*

3.5. Real Milnor fibrations and d-regularity

Using the ideas and constructions for holomorphic maps, the authors in [5] introduce a condition called d-regularity, for a class of real analytic maps $f : (\mathbb{R}^n, 0) \to (\mathbb{R}^2, 0)$ with an *isolated critical value* at $0 \in \mathbb{R}^2$, which is necessary and sufficient for the map $f/|f| : S^{n-1}_\epsilon \setminus K_\epsilon \to S^1$ to be a smooth locally trivial fibration.

As in §2.2 we define the **canonical pencil** of f, which is a family of real analytic spaces parametrized by $\mathbb{RP}^1$, as follows: for each $\ell \in \mathbb{RP}^1$, consider the line $\mathcal{L}_\ell \subset \mathbb{R}^2$ passing through the origin corresponding to ℓ and set

$$X_\ell = \{x \in U \mid f(x) \in \mathcal{L}_\ell\}.$$

Then each X_ℓ is a real analytic variety. It is easy to see that these varieties meet at V and away from it they are smooth submanifolds of U of dimension $n - 1$.

Definition 3.19. The map f is said to be *d-regular* at 0 if there exist a metric ρ induced by some positive definite quadratic form and an $\epsilon_0 > 0$ such that every sphere (for the metric ρ) of radius $\leq \epsilon_0$ centred at 0 meets every $X_\ell \setminus V$ transversely (whenever the intersection is non-empty). We shall also say that f is d-regular with respect to the metric ρ.

Example 3.20. The first four examples are d-regular at 0 for the usual metric:

(i) By [21, Lem. 5.9], every holomorphic germ $f : (\mathbb{C}^n, 0) \to (\mathbb{C}, 0)$ is d-regular (see [5]).

(ii) By [25], given $f, g : (\mathbb{C}^2, 0) \to (\mathbb{C}, 0)$ holomorphic germs (see §3.1):
 – If $f\bar{g}$ has an isolated critical value at $0 \in \mathbb{C}$, then it is d-regular.
 – If $f/g : (\mathbb{C}^n, 0) \to (\mathbb{C}, 0)$ is semitame at 0, then it is d-regular.

(iii) By [4] (see §3.2), polar weighted homogeneous polynomials are d-regular.

(iv) Assume $f: (\mathbb{R}^n, 0) \to (\mathbb{R}^2, 0)$ is real weighted homogeneous with isolated critical value at 0. Since the orbits of the action of $\mathbb{R}^*$ are tangent to each X_ℓ and transversal to all spheres centered at 0, it follows that f is d-regular.

(v) By [26], every map-germ $g: (\mathbb{R}^{n+2}, 0) \to (\mathbb{R}^2, 0)$ whose pencil is c-regular (in the sense of K. Bekka) with respect to the control function defined by a metric ρ in $\mathbb{R}^{n+2}$, is d-regular at 0 with respect to ρ.

The condition of d-regularity is equivalent to the condition given in §3.4 that the map

$$\phi = \frac{f}{|f|} : S_\epsilon^{n-1} \setminus K_\epsilon \to S^{k-1}$$

is a submersion for any sufficiently small ϵ. In fact, one has the following characterizations of d-regularity in terms of the spherification of f.

As in §2.2, define the map $\Phi: U \setminus V \to S^1$ by $\Phi(x) = \frac{f(x)}{|f(x)|}$ and the **spherification** $\mathfrak{F}: U \setminus V \to \mathbb{R}^2 \setminus \{0\}$ of f by $\mathfrak{F}(x) = \|x\| \Phi(x)$.

Proposition 3.21. *Let $f: (B_{\epsilon_0}, 0) \to (\mathbb{R}^2, 0)$. Then the following are equivalent:*

(i) *The map f is d-regular at 0.*

(ii) *For each sphere S_ϵ in $\mathbb{R}^n$ centred at 0 of radius $\epsilon \le \epsilon_0$, the restriction map $\mathfrak{F}_\epsilon: S_\epsilon \setminus V \to S_\epsilon^1$ of $\mathfrak{F}$ is a submersion.*

(iii) *The spherification map $\mathfrak{F}$ is a submersion at each $x \in B_{\epsilon_0} \setminus V$.*

(iv) *The map $\phi = \dfrac{f}{|f|} : S_\epsilon \setminus K_\epsilon \longrightarrow S^1$ is a submersion for every sphere S_ϵ with $\epsilon \le \epsilon_0$.*

Proof (idea). The equivalences are straightforward from the definitions of d-regularity and the spherification map. The important point is to notice that $\Phi(x) = \frac{f(x)}{|f(x)|} = \frac{\mathfrak{F}(x)}{|\mathfrak{F}(x)|}$. $\qquad\square$

If f is d-regular, the following proposition gives us a special vector field which will be used in the Fibration Theorem.

Proposition 3.22. *If f is d-regular for some metric ρ, then there exist $\varepsilon > 0$ and a C^∞ vector field ξ on $B_\varepsilon \setminus V$ such that :*

(i) *Each integral curve of ξ is contained in an element X_θ of the pencil;*

(ii) *The vector field ξ is transverse to all ρ-spheres around 0; and*

(i) *The vector field ξ is transverse to all Milnor tubes $f^{-1}(\partial D_\delta)$ for all sufficiently small discs D_δ centred at $0 \in \mathbb{R}^2$.*

Proof (idea).   Since f is d-regular, the spherification map $\mathfrak{F}$ is a submersion on $B_\varepsilon \setminus V$. We lift the radial vector field u in $\mathbb{R}^2$ given by $u(x, y) = (x, y)$, to vector fields $\omega_{\mathfrak{F}}$ and ω_f on $B_\varepsilon \setminus V$ using $\mathfrak{F}$ and f respectively. The desired vector field is given by adding $\omega_{\mathfrak{F}}$ and ω_f on $B_\varepsilon \setminus V$. $\square$

The main result is the following real analogue of Theorem 2.5.

Theorem 3.23 (Fibration Theorem).   *Let $f : (U, 0) \to (\mathbb{R}^2, 0)$ be a locally surjective real analytic map with an isolated critical value at $0 \in \mathbb{R}^2$ and U an open neighborhood of 0 in $\mathbb{R}^{n+2}$. Assume further that f has the Thom A_f property at 0, is d-regular, and that dim $V > 0$. Then:*

(i)   *One has a Milnor-Lê fibration (a fiber bundle)*

$$f : N(\epsilon, \eta) \longrightarrow \partial D_\eta \,,$$

   where $N(\epsilon, \eta) = B_\epsilon \cap f^{-1}(\partial D_\eta)$ is a Milnor tube for f; $D_\eta \subset \mathbb{R}^2$ is the disc of radius η around $0 \in \mathbb{R}^2$, $\epsilon \gg \eta > 0$. In fact, the same statement holds for $B_\epsilon \cap f^{-1}(D_\eta \setminus \{0\})$, which fibers over $D_\eta \setminus \{0\}$.

(ii)   *For every sufficiently small $\epsilon > 0$ one has a commutative diagram of fiber bundles,*

$$
\begin{array}{ccc}
B_\epsilon \setminus V & \xrightarrow{\;\Phi\;} & S^1 \\[2pt]
 & {\Psi}\searrow & \downarrow{\pi} \\[2pt]
 & & \mathbb{R}P^1
\end{array}
$$

   which restricts to a fiber bundle $\phi = \dfrac{f}{|f|} : S_\epsilon \setminus K_\epsilon \to S^1$.

(iii)   *The two fibrations above, one on the Milnor tube and one on the sphere, are equivalent.*

Proof.   Recall that we are assuming f has the Thom A_f property. Thus for $\epsilon \gg \delta > 0$ sufficiently small one has a *solid* Milnor tube,

$$SN(\epsilon, \delta) := B_\epsilon \cap f^{-1}(D_\delta \setminus \{0\}) \,,$$

and a fiber bundle $f : SN(\epsilon, \delta) \longrightarrow D_\delta \setminus \{0\}$.

The restriction of this locally trivial fibration to the boundary of D_δ gives the fibration in the first statement in Theorem 3.23.

Now define $\pi_1 : D_\delta \setminus \{0\} \to \mathbb{S}^1$ by $t \mapsto t/|t|$, let $\pi_2 : \mathbb{S}^1 \to \mathbb{R}P^1$ be the canonical projection, and set

$$\Psi := \pi_2 \circ \pi_1 \circ f : SN(\epsilon, \delta) \longrightarrow \mathbb{R}P^1 \,.$$

This is a fiber bundle with fibers $X_\theta \cap SN(\epsilon, \delta)$, and hence yields statement (ii) in Theorem 3.23 restricted to the (solid) Milnor tube $SN(\epsilon, \delta)$. We now use the vector field in Proposition 3.22 to complete the proof of the theorem. $\square$

Corollary 3.24. *Let* $f : (\mathbb{R}^n, 0) \to (\mathbb{R}^2, 0)$ *be a locally surjective real analytic map with an isolated critical value at* $0 \in \mathbb{R}^2$. *Suppose that* f *has the Thom* A_f *property at* 0 *and* $\dim V > 0$. *Then* f *is* d-*regular if and only if the map*

$$\frac{f}{|f|} : S_\epsilon^{n-1} \to S^1$$

is a smooth locally trivial fibration.

Proof. If f is d-regular, by Theorem 3.23(ii) $f/|f|$ is a smooth locally trivial fibration. Conversely, if $f/|f|$ is a smooth locally trivial fibration, then it is a submersion, and by Proposition 3.21(iv) $\Leftrightarrow$(i) f is d-regular. $\square$

Acknowledgements. The first author thanks the second author for inviting him to write the present article. He was partially supported by CONACYT J-49048-F, and DGAPA-UNAM: PAPIIT IN102208, Mexico.

We want to thank Prof. C. T. C. Wall and the referee for their comments and suggestions which greatly improved this article.

References

1. N. A'Campo, Le nombre de Lefschetz d'une monodromie, *Nederl. Akad. Wetensch. Proc. Ser. A* **76** = *Indag. Math.* **35** (1973) 113–118.
2. A. Bodin and A. Pichon, Meromorphic functions, bifurcation sets and fibred links, *Math. Res. Lett.* **14**(3) (2007) 413–422.
3. P. T. Church and K. Lamotke, Non-trivial polynomial isolated singularities, *Nederl. Akad. Wetensch. Proc. Ser. A* **78**=*Indag. Math.* **37** (1975) 149–154.
4. J. L. Cisneros-Molina, Join theorem for polar weighted homogeneous singularities, in *Singularities II: Geometric and Topological aspects*, *Proc. of the Conference in honour of Lê Dũng Tráng*, eds. J. P. Brasselet et al., (Cuernavaca, Mexico, 2007), *Contemporary Mathematics* **475** (2008) 43–59.
5. J. L. Cisneros-Molina, J. Seade, and J. Snoussi, Refinements of Milnor's Fibration Theorem for Complex Singularities, *Advances in Mathematics* **222** (2009) 937–970.
6. A. Durfee, Neighborhoods of algebraic sets, *Trans. Amer. Math. Soc.* **276**(2) (1983) 517–530.
7. T. Gaffney, Non isolated completed intersection singularities and the A_f–condition, in *Singularities I: Algebraic and Analytic aspects*, *Proc. of the Conference in honour of Lê Dũng Tráng*, eds. J. P. Brasselet et al., (Cuernavaca, Mexico, 2007), *Contemporary Mathematics* **474** (2008) 85–94.

8. H. Hamm, Lokale topologische Eigenschaften komplexer Räume, *Math Ann.* **191** (1971) 235–252.

9. H. Hamm and D. T. Lê, Un Théorème de Zariski du type de Lefschetz, *Ann. Sci. Éc. Norm. Sup.* **6(4)** (1973) 317–366.

10. H. Hironaka, Stratification and flatness, in *Real and complex singularities (Proc. Ninth Nordic Summer School/NAVF Sympos. Math., Oslo, (August 5–25, 1976)*, ed. P. Holm (1977) 199–265.

11. A. Jaquemard, Fibrations de Milnor pour des applications réelles, *Boll. Un. Mat. Ital.* **37(1**, 3-B (1989) 45–62.

12. A. Jacquemard, *Thèse 3ème cycle*, Université de Dijon, 1982.

13. D. T. Lê, Vanishing cycles on complex analytic sets, *Sûrikaisekikenkyûsho Kôkyûroku, Various problems in algebraic analysis (Proc. Sympos., Res. Inst. Math. Sci., Kyoto Univ., Kyoto, 1975)*, **266** (1976) 299–318.

14. D. T. Lê, Some remarks on relative monodromy, in *Real and complex singularities (Proc. Ninth Nordic Summer School/NAVF Sympos. Math., Oslo, (August 5–25, 1976)*, ed. P. Holm (1977) 397–403.

15. D. T. Lê, H. Maugendre, and C. Weber, Geometry of critical loci, *J. London Math. Soc.* (2) **63(3)** (2001) 533–552.

16. E. Looijenga, A note on polynomial isolated singularities, *Indag. Math.* **33** (1971) 418–421.

17. E. Looijenga, Isolated Singular Points on Complete Intersections, *London Math. Soc. Lect. Notes* **77**, 1984.

18. D. Massey, Real Analytic Milnor Fibrations and a Strong Lojasiewicz Inequality. This volume.

19. J. Mather. *Notes on Topological Stability*, Harvard University, July, 1970.

20. J. Milnor. On isolated singularities of hypersurfaces, *Preprint* June 1966. Unpublished.

21. J. Milnor, *Singular points of complex hypersurfaces*, Ann. of Math. Studies **61** Princeton University Press, 1968.

22. M. Oka, Topology of Polar Weighted Homogeneous Hypersurfaces, *Kodai Math. J.* **31(2)** (2008) 163–182.

23. M. Oka, *Non-degenerate complete intersection singularity*, Actualités Mathématiques. [Current Mathematical Topics]. Hermann, Paris, 1997.

24. M. Oka, Non-degenerate mixed functions, *Kodai Math. J.* **33(1)** (2010) 1–62.

25. A. Pichon and J. Seade, Fibered multilinks and singularities $f\bar{g}$, *Math. Ann.* **342** (2008) 487–514.

26. M. A. S. Ruas and R. Araújo dos Santos, Real Milnor Fibrations and (C)−regularity', *Manuscripta Math.* **117(2)** (2005) 207–218.

27. M. A. S. Ruas, J. Seade and A. Verjovsky, On Real Singularities with a Milnor Fibration, in *Trends in singularities*, eds. A. Libgober and M. Tibăr, *Trends Math.* 191–213. Birkhäuser, Basel, 2002.

28. R. N Araújo dos Santos, Uniform (m)-condition and strong Milnor fibrations, in *Singularities II: Geometric and Topological aspects, Proc. of the Conference in honour of Lê Dũng Tráng*, eds. J. P. Brasselet et al., Cuernavaca, Mexico, 2007, *Contemporary Mathematics* **475** (2008) 43–59.

29. R. N. Araújo dos Santos, 'Equivalence of real Milnor fibrations for quasi-homogeneous singularities', *to appear in the Rocky Mountain Journal Mathematics*.

30. R. N. Araújo dos Santos and M. Tibar, Real map germs and higher open books, *To appear in Geometriae Dedicata* doi: 10.1007/S10711-009-9449-z.

31. J. Seade, Open Book Decompositions Associated to Holomorphic Vector Fields, *Bol.Soc.Mat.Mexicana* **3(3)** (1997) 323–336.

32. J. Seade, *On the topology of isolated singularities in analytic spaces*, Progress in Mathematics, 241. Birkhäuser Verlag, Basel, 2006.

33. J. Seade, On Milnor's fibration theorem for real and complex singularities, in *Singularities in geometry and topology, World Sci. Publ., Hackensack, NJ* (2007) 127–158.

J. L. Cisneros-Molina
Instituto de Matemáticas, Unidad Cuernavaca
Universidad Nacional Autónoma de México
Av. Universidad s/n, Lomas de Chamilpa
62210 Cuernavaca, Morelos, A. P. 273-3
México
jlcm@matcuer.unam.mx

R. N. Araújo dos Santos
Instituto de Ciências Matemáticas e de Computação
Universidade de São Paulo
Av. Trabalhador São-Carlense, 400
13560-970 São Carlos
Brazil
rnonato@icmc.usp.br

7

Counting hypersurfaces invariant by one-dimensional complex foliations

MAURÍCIO CORRÊA JR. AND MÁRCIO G. SOARES

Abstract

We address the question of counting hypersurfaces which are invariant by a one-dimensional projective foliation. We obtain some bounds for the number of such hypersurfaces, of a given fixed degree k, in terms of k, the dimension of the projective space and the degree of the foliation. These generalize some already known bounds.

1. Introduction

In this note we address the question of counting hypersurfaces which are invariant by a one-dimensional projective foliation. The method adopted here stems from the work of J.V. Pereira [7], where the notion of *extactic* variety is exploited (for several references related to this notion see [2]). Let us briefly digress on extactic varieties and their main properties.

We start by recalling one-dimensional holomorphic foliations on $\mathbb{P}^n_{\mathbb{C}}$. These are given, in $\mathbb{C}^{n+1}$, by a homogeneous polynomial vector field $X = \sum_{i=0}^{n} P_i \frac{\partial}{\partial z_i}$, with $\deg P_i = d$ and $P_0, \ldots, P_n$ do not have a common factor, modulo addition of a vector field of the form $h.R$, where h is a homogeneous polynomial of degree $d - 1$ and $R = \sum_{i=0}^{n} z_i \frac{\partial}{\partial z_i}$ is the radial vector field. Such a vector field defines a field of directions on $\mathbb{P}^n_{\mathbb{C}}$ and hence a foliation, denoted $\mathcal{F}_X$. Equivalently, X can be seen as a holomorphic section $X : \mathbb{P}^n_{\mathbb{C}} \longrightarrow T\mathbb{P}^n_{\mathbb{C}} \otimes \mathcal{O}_{\mathbb{P}^n_{\mathbb{C}}}(d - 1)$.

2000 *Mathematics Subject Classification* 32S65.
Work partially supported by CAPES-DGU, FAPEMIG, CNPq and Pronex/FAPERJ-CNPq (Brasil).

The number d is called the *degree* of $\mathcal{F}_X$. We consider the vector field X as a derivation and put $X^0(F) = F$ and $X^m(F) = X(X^{m-1}(F))$, for $m > 0$.

We remark that to consider a finite dimensional linear system V on $\mathbb{P}^n_{\mathbb{C}}$ is the same as to consider a finite dimensional linear space of homogeneous polynomials V in the variable $z = (z_0, \dots, z_n)$. Suppose now that V is a finite dimensional linear system and let $v_1, \dots, v_\ell \in \mathbb{C}[z_0, \dots, z_n]$ be a basis of V. Consider the matrix

$$
E(V, X) = \begin{pmatrix}
v_1 & v_2 & \cdots & v_\ell \\
X(v_1) & X(v_2) & \cdots & X(v_\ell) \\
\vdots & \vdots & \ddots & \vdots \\
X^{\ell-1}(v_1) & X^{\ell-1}(v_2) & \cdots & X^{\ell-1}(v_\ell)
\end{pmatrix}. \tag{1.1}
$$

The *extactic of X associated to V* is $\mathcal{E}(V, X) = \det E(V, X)$, and the *extactic variety of X associated to V* is the variety $Z(\mathcal{E}(V, X))$. By introducing the notion of *extactic divisor* on a complex manifold, J.V. Pereira [7] obtained the following results, which elucidate the role of the extactic variety:

Proposition 1.1 ([7], Proposition 5). *Let $\mathcal{F}$ be a one-dimensional holomorphic foliation on a complex manifold M. If V is a finite dimensional linear system, then every $\mathcal{F}$-invariant hypersurface which is contained in the zero locus of some element of V must be contained in the zero locus of $\mathcal{E}(V, \mathcal{F})$.*

Suppose $M = \mathbb{P}^n_{\mathbb{C}}$, $\mathcal{F}_X$ and $\mathcal{E}(V, X)$ are as above. In this case proposition 1.1 is easy to exemplify. Let f be a defining equation for an irreducible $\mathcal{F}_X$-invariant hypersurface, which is in the zero locus of the linear system V. Change basis so that V is generated by $f, w_2, \dots, w_\ell$. Since $\{f = 0\}$ is invariant we have $X^j(f) = h_j f$, $1 \leq j \leq \ell - 1$, where h_j is a polynomial. Then $E(V, X)$ becomes

$$
E(V, X) = \begin{pmatrix}
f & w_2 & \cdots & w_\ell \\
h_1 f & X(w_2) & \cdots & X(w_\ell) \\
\vdots & \vdots & \ddots & \vdots \\
h_{\ell-1} f & X^{\ell-1}(w_2) & \cdots & X^{\ell-1}(w_\ell)
\end{pmatrix} \tag{1.2}
$$

and f factors $\mathcal{E}(V, X)$.

Now, if f is a defining equation for an irreducible $\mathcal{F}_X$-invariant hypersurface, which is also in the zero locus of V, its *multiplicity* is the largest integer m such that f^m divides $\mathcal{E}(V, X)$.

Before going any further, it's worth remarking that a generic one-dimensional singular holomorphic foliation $\mathcal{F}$ on a complex projective space leaves no algebraic set invariant, except for its singular locus. Here, generic means that $\mathcal{F}$ lies in an open and dense set of the space of such foliations. For results in this direction see [1], [6] and [3].

If $\mathcal{F}$ is a holomorphic one-dimensional foliation on a complex manifold M, then a *first integral* for $\mathcal{F}$ is a holomorphic map $\theta : M \longrightarrow N$, where N is a complex manifold, such that the fibers of θ are $\mathcal{F}$-invariant. Then we have:

Theorem 1.2 ([7], Theorem 3). *Let $\mathcal{F}$ be a one-dimensional holomorphic foliation on a complex manifold M. If V is a finite dimensional linear system such that $\mathcal{E}(V, \mathcal{F})$ vanishes identically, then there exits an open and dense set U where $\mathcal{F}_{|U}$ admits a first integral. Moreover, if M is a projective variety, then $\mathcal{F}$ admits a meromorphic first integral.*

Since we will be dealing with foliations by curves on projective spaces, by a meromorphic first integral we mean a rational map $\theta : \mathbb{P}^n_{\mathbb{C}} \dashrightarrow \mathbb{P}^1_{\mathbb{C}}$ whose fibers are $\mathcal{F}$-invariant. Although we could use Theorem 1.2 to invoke their existence in the proof of our main result, Theorem 2.3 below, we thought it better to give an explicit and more elementary proof of this fact, closely related to the proof of Theorem 4.3 of [2] (see Section 4).

2. Statement of results

Through the use of extactic varieties the following estimates were obtained in [7]:

Proposition 2.1 ([7] Proposition 2 and Corollary 1). *Let $\mathcal{F}_X$ be the foliation on $\mathbb{P}^2_{\mathbb{C}}$ induced by the homogeneous vector field X in $\mathbb{C}^3$, of degree d. Let $N_i(X)$ be the number of irreducible algebraic solutions of degree i of X, counting multiplicities. If X does not admit a first integral of degree $\leq k$ then*

$$\sum_{i=1}^{k} i \, N_i(X) \leq k \binom{2+k}{k} + (d-1)\binom{\binom{2+k}{k}}{2}. \qquad (2.1)$$

Also, the number of invariant irreducible curves of degree k, counting multiplicities, is at most

$$\binom{2+k}{k} + \frac{1}{k}\binom{\binom{2+k}{k}}{2}(d-1). \qquad (2.2)$$

In a similar way, we find in [4] the following result:

Theorem 2.2 ([4] Theorem 4). *Suppose the affine polynomial vector field* $Y = \sum_{i=1}^{n} Y_i \frac{\partial}{\partial z_i}$ *in* $\mathbb{C}^n$, $n \geq 2$, *with* $\deg Y_i = d$, $1 \leq i \leq n$, *has finitely many invariant hyperplanes. Then the following hold:*

(a) *The number of invariant hyperplanes of* Y, *counting multiplicities, is at most*

$$nd + \binom{n}{2}(d-1). \qquad (2.3)$$

(b) *The number of parallel invariant hyperplanes of* Y, *counting multiplicities, is at most* d.

(c) *The number of distinct invariant hyperplanes of* Y *through a single point, counting multiplicities, is at most*

$$(n-1)d + \binom{n-1}{2}(d-1) + 1. \qquad (2.4)$$

We now present counterparts of Proposition 2.1 and Theorem 2.2, items (a) and (c), for the case of hypersurfaces invariant by one-dimensional foliations on $\mathbb{P}^n_{\mathbb{C}}$. The proofs are given in section 4. More precisely,

Theorem 2.3. *Let* $\mathcal{F}_X$ *be a one-dimensional holomorphic foliation on* $\mathbb{P}^n_{\mathbb{C}}$ *with* $\deg \mathcal{F}_X = d$. *Suppose* $\mathcal{F}_X$ *does not admit a rational first integral. Then:*

(i) *Let* $N_i(X)$ *be the number of irreducible* $\mathcal{F}_X$-*invariant hypersurfaces of degree* i, *counting multiplicities. We have, for* $i \leq k$,

$$\sum_{i=1}^{k} i N_i(\mathcal{F}) \leq \sum_{i=0}^{K-1} [i(d-1) + k] = kK + (d-1)\binom{K}{2},$$

$$where\ K = \binom{n+k}{k}. \qquad (2.5)$$

(ii) *The number of* $\mathcal{F}_X$-*invariant irreducible hypersurfaces of degree* k, *counting multiplicities, is at most*

$$\binom{n+k}{k} + \frac{1}{k}\binom{\binom{n+k}{k}}{2}(d-1). \qquad (2.6)$$

(iii) *The number of* $\mathcal{F}_X$-*invariant irreducible hypersurfaces of degree* k *through a point, counting multiplicities, is at most*

$$\frac{1}{k}\binom{\binom{n+k}{k}-1}{2}(d-1) + \binom{n+k}{k} - 1. \qquad (2.7)$$

Remark. The hypothesis of Theorem 2.3 is stronger than that of Proposition 2.1. The reason for this comes from the fact that, in the two dimensional projective case, the fibers of a rational first integral have the same dimension as the leaves of the foliation, which allows for showing that if $\mathcal{E}(V^k, X) \equiv 0$, where $V^k \subset \mathbb{C}[z_0, \ldots, z_n]$ is the linear space of homogeneous polynomials of degree k, then X admits a rational first integral of degree $\leq k$ (see the proof of Theorem 4.3 in [2]). In the higher dimensional case, the fibers of a rational first integral have dimension $n - 1$ and, in this situation, we were only able to show that, if $\mathcal{E}(V^k, X) \equiv 0$, then X has a rational first integral (see section 4). However, in case X has a rational first integral θ of degree m, then $\mathcal{E}(V^m, X) \equiv 0$. This is because, writing $\theta = \dfrac{F}{G}$, where F and G share no common factor, the level hypersurfaces $F + \lambda G = 0$ are $\mathcal{F}_X$-invariant and hence they all do factor $\mathcal{E}(V^m, X)$. Since there are infinitely many of them, this extactic must vanish identically. From this it follows that $\mathcal{E}(V^\ell, X) \equiv 0$ for $\ell \geq m$.

Specializing (2.5) and (2.6) to $n = 2$ we obtain precisely (2.1) and (2.2). On the other hand, specializing (2.6) and (2.7) to $k = 1$, i.e., to the case of invariant hyperplanes, we obtain

$$(2.6)_{\{k=1\}} = n + 1 + \binom{n+1}{2}(d-1) = \left[nd + \binom{n}{2}(d-1)\right] + 1$$

$$= (2.3) + 1. \tag{2.8}$$

and

$$(2.7)_{\{k=1\}} = n + \binom{n}{2}(d-1) = (n-1)d + \binom{n-1}{2}(d-1) + 1 = (2.4).$$

$$\tag{2.9}$$

The reason why we have, in (2.8), a difference of 1 between $(2.6)_{\{k=1\}}$ and (2.3) comes from the fact that Theorem 2.2 (a) is affine, whereas Theorem 2.3 (iii) is projective, hence it contemplates the possibility of the hyperplane at infinity being $\mathcal{F}_X$-invariant.

Now, $(2.7)_{\{k=1\}} = (2.4)$ because both affirmatives are affine in scope.

Let us explore a little further item (iii) of Theorem 2.3 in the special case of hyperplanes. We have

$$(2.7)_{\{k=1\}} = n + \binom{n}{2}(d-1). \tag{2.10}$$

This bounds the number of invariant hyperplanes passing through a point p. However, if p is a regular point of the foliation $\mathcal{F}_X$, then this bound drops to (see corollary 2.4 below):

$$n - 1 + \binom{n-1}{2}(d-1).$$

This is because the direction defined by X at p is contained in every invariant hyperplane through p hence the line $\mathbb{L}^1$, passing through p and determined by the direction of $X(p)$, is $\mathcal{F}_X$-invariant. With this in mind we have

Corollary 2.4. *Let $\mathcal{F}_X$ be a one-dimensional singular holomorphic foliation on $\mathbb{P}^n_{\mathbb{C}}$ of $\deg \mathcal{F}_X = d$ and suppose $\mathcal{F}_X$ does not admit a rational first integral. Then, the number of $\mathcal{F}_X$-invariant hyperplanes which contain a fixed ℓ-plane, $0 \leq \ell \leq n - 1$, is bounded by*

$$n - \ell + \binom{n - \ell}{2}(d - 1). \tag{2.11}$$

Note that this corollary is of affine nature.

3. Examples

We now present examples which show that the bounds given in (2.8) and (2.11) are optimal.

Before doing this recall that, given a vector field X in $\mathbb{C}^n$, the induced foliation $\mathcal{F}_X$ on $\mathbb{P}^n_{\mathbb{C}}$, which we assume to have degree d, leaves the hyperplane at infinity, $\mathbb{P}^{n-1}_\infty$, invariant if, and only if, X is of the form

$$X = gR + \sum_{i=0}^{d} X_i \tag{3.1}$$

where $R = \sum_{i=1}^{n} z_i \dfrac{\partial}{\partial z_i}$ is the radial vector field, $g \in \mathbb{C}[z_1, \ldots, z_n]$ is homogeneous of degree d and $X_i = \sum_{j=1}^{n} X_{ij} \dfrac{\partial}{\partial z_j}$, with $X_{ij} \in \mathbb{C}[z_1, \ldots, z_n]$ homogeneous of degree i, $0 \leq i \leq d$. The variety $\{g = 0\} \subset \mathbb{P}^{n-1}_\infty$ is precisely the locus of tangencies of $\mathcal{F}_X$ with $\mathbb{P}^{n-1}_\infty$.

Consider the vector fields, defined in affine coordinates $z_0 = 1$, by

$$X_0 = \sum_{i=1}^{n} z_i \left(z_i^{d-1} - 1\right) \frac{\partial}{\partial z_i}. \tag{3.2}$$

$$X_1 = \frac{\partial}{\partial z_1} + \sum_{i=2}^{n} z_i \left(z_i^{d-1} - 1\right) \frac{\partial}{\partial z_i}. \tag{3.3}$$

$$X_\ell = \sum_{j=1}^{\ell} \left(z_1^d + \cdots + \widehat{z_j^d} + \cdots + z_\ell^d\right) \frac{\partial}{\partial z_j}$$

$$+ \sum_{j=\ell+1}^{n} z_j \left(z_j^{d-1} - 1\right) \frac{\partial}{\partial z_j}, \quad 2 \leq \ell \leq n - 1. \tag{3.4}$$

Remark that the foliations $\mathcal{F}_{X_\ell}$ on $\mathbb{P}^n_{\mathbb{C}}$ induced by $X_\ell, 0 \le \ell \le n-1$, do all leave the hyperplane at infinity invariant.

X_0 is an n-dimensional version of a member of the so-called "family of degree four" in $\mathbb{P}^2_{\mathbb{C}}$, one of the examples given by A. Lins Neto in [5]. A straightforward calculation shows that the $n + 1 + \binom{n+1}{2}(d-1)$ hyperplanes listed below are invariant by $\mathcal{F}_{X_0}$:

$$z_j = 0, \ 0 \le j \le n, \quad z_i^{d-1} - z_j^{d-1} = 0, \ 0 \le i < j \le n. \tag{3.5}$$

It's worth remarking that all the singularities of X_0 have the same analytic type and are determined precisely by the intersections of these hyperplanes. This shows the bound given in (2.8) is sharp.

Now, looking affinely, the vector field X_0 leaves invariant the $n + \binom{n}{2}(d-1)$ hyperplanes

$$z_j = 0, \ 1 \le j \le n, \quad z_i^{d-1} - z_j^{d-1} = 0, \ 1 \le i < j \le n. \tag{3.6}$$

X_1 leaves invariant the line $\mathbb{L}^1 = \{z_2 = \cdots = z_n = 0\}$, which is the base locus of the linear system $\sum_{j=2}^n \lambda_j z_j$. Moreover, the $n - 1 + \binom{n-1}{2}(d-1)$ hyperplanes listed below are X_1-invariant and contain $\mathbb{L}^1$:

$$z_j = 0, \ 2 \le j \le n, \quad z_i^{d-1} - z_j^{d-1} = 0, \ 2 \le i < j \le n. \tag{3.7}$$

As for $X_\ell, 2 \le \ell \le n-1$, the ℓ-plane $\mathbb{L}^\ell = \{z_{\ell+1} = \cdots = z_n = 0\}$ is left invariant, as are the $n - \ell + \binom{n-\ell}{2}(d-1)$ hyperplanes, which do all contain $\mathbb{L}^\ell$:

$$z_j = 0, \ \ell + 1 \le j \le n, \quad z_i^{d-1} - z_j^{d-1} = 0, \ \ell + 1 \le i < j \le n. \tag{3.8}$$

(3.6), (3.7) and (3.8) show that $(2.11)_\ell$ is sharp for $0 \le \ell \le n-1$.

4. Proofs

Lemma 4.1. Let $V = \langle v_1, \ldots, v_\ell \rangle_{\mathbb{C}}$ be a linear subspace of $\mathbb{C}[z_0, \ldots, z_n]$ where the v_j's are homogeneous polynomials of degree k. If $\mathcal{E}(V, X) \not\equiv 0$, then $\deg \mathcal{E}(V, X) = \sum_{i=0}^{\ell-1} \deg X^i(v_j)$, for any $1 \le j \le \ell$, where $X^0(v_j) = v_j$.

Proof. The elements in each row of matrix (1.1) are homogeneous of the same degree and hence $\mathcal{E}(V, X)(tz) = t^{\sum_{i=0}^{\ell-1} \deg X^i(v_j)} \mathcal{E}(V, X)(z)$. $\qquad \square$

Proof of Theorem 2.3. Consider the linear system $H^0(\mathbb{P}^n_\mathbb{C}, \mathcal{O}_{\mathbb{P}^n_\mathbb{C}}(k))$ or, equivalently, the linear subspace, or grading $V^k \subset \mathbb{C}[z_0, \ldots, z_n]$, of homogeneous polynomials of degree k, and let $\mathcal{F}_X$ be induced by $X \in H^0(\mathbb{P}^n_\mathbb{C}, T\mathbb{P}^n_\mathbb{C} \otimes \mathcal{O}_{\mathbb{P}^n_\mathbb{C}}(d-1))$. Let $w_1, \ldots, w_K$ be a basis of V^k, $K = \binom{n+k}{k}$, and recall that the extactic variety is independent of the chosen basis.

Now, the non-existence of a rational first integral for $\mathcal{F}_X$ assures that $\mathcal{E}(V^k, X) \not\equiv 0$. In fact, by Theorem 1.2, if $\mathcal{E}(V^k, X) \equiv 0$ then we would have a first integral for X. However, since we are working on $\mathbb{P}^n_\mathbb{C}$ we prefer, by the sake of completeness, to give explicit arguments to prove this. These follow closely to those given in the proof of Theorem 4.3 of [2]. To say $\mathcal{E}(V^k, X) \equiv 0$ means that the columns of the matrix

$$
E(V^k, X) = \begin{pmatrix} w_1 & w_2 & \cdots & w_K \\ X(w_1) & X(w_2) & \cdots & X(w_K) \\ \vdots & \vdots & \ddots & \vdots \\ X^{K-1}(w_1) & X^{K-1}(w_2) & \cdots & X^{K-1}(w_K) \end{pmatrix} \tag{4.1}
$$

are dependent. Hence, there are rational functions $\theta_1, \ldots \theta_K : \mathbb{P}^n_\mathbb{C} \dashrightarrow \mathbb{P}^1_\mathbb{C}$ such that

$$
M_i = \sum_{j=1}^{K} \theta_j X^i(w_j) = 0, \quad \text{for } 0 \le i \le K-1. \tag{4.2}
$$

Now, let s be the smallest integer with the property that there exist rational functions $\theta_1, \ldots, \theta_s : \mathbb{P}^n_\mathbb{C} \dashrightarrow \mathbb{P}^1_\mathbb{C}$ and $w_1, \ldots, w_s \in V^k$, linearly independent, such that (4.2) holds. We clearly have $1 < s \le K$ and we may assume $\theta_s = 1$.

Applying the derivation X to both sides of (4.2) we get

$$
X(M_i) = X(\theta_1) X^i(w_1) + \theta_1 X^{i+1}(w_1) + \cdots + \\ \underbrace{X(\theta_s)}_{=0} X^i(w_s) + \underbrace{\theta_s}_{=1} X^{i+1}(w_s) = 0 \tag{4.3}
$$

for $0 \le i \le s-2$. Subtract

$$
M_{i+1} = \sum_{j=1}^{s} \theta_j X^{i+1}(w_j) = 0, \quad 0 \le i \le s-2 \tag{4.4}
$$

from (4.3) to get

$$
X(M_i) - M_{i+1} = X(\theta_1) X^i(w_1) + \cdots + X(\theta_{s-1}) X^i(w_{s-1}) = 0, \tag{4.5}
$$

for $0 \leq i \leq s - 2$. By the minimality of s we must have $X(\theta_1) = \cdots = X(\theta_{s-1}) = 0$ and hence, provided these are not all constants, we have a first integral for X. This in fact occurs because, since M_0 reads

$$\theta_1 \, w_1 + \cdots + \theta_{s-1} \, w_{s-1} + w_s = 0, \tag{4.6}$$

we conclude that not all the θ_i's could be constant since $w_1, \ldots, w_s \in V^k$ are linearly independent.

Recall that, by Proposition 1.1, any $\mathcal{F}_X$-invariant hypersurface, which is contained in the zero locus of some element of $H^0(\mathbb{P}^n_{\mathbb{C}}, \mathcal{O}_{\mathbb{P}^n_{\mathbb{C}}}(k))$, is also contained in the zero locus of $\mathcal{E}(V^k, X)$. Hence, if we let $N_i(\mathcal{F}_X)$ denote the number of irreducible $\mathcal{F}_X$-invariant hypersurfaces of degree i then, for $i \leq k$, we have

$$\sum_{i=1}^{k} i \, N_i(\mathcal{F}_X) \leq \deg \mathcal{E}(V^k, X) = \sum_{i=0}^{K-1} \deg X^i(w_j) \tag{4.7}$$

by Lemma 4.1. It remains to evaluate $\deg \mathcal{E}(V^k, X)$. Since $\deg X = d$, it follows immediately that $\deg X^i(w_j) = i(d-1) + k$ and

$$\deg \mathcal{E}(V^k, X) = \sum_{i=0}^{K-1} \deg X^i(w_j) = \sum_{i=0}^{K-1} [i(d-1) + k] = kK + (d-1)\binom{K}{2}, \tag{4.8}$$

where $K = \binom{n+k}{k}$. This proves (i) of Theorem 2.3.

To prove (ii) we simply observe that

$$N_k(\mathcal{F}_X) \leq \frac{\deg \mathcal{E}(V^k, X)}{k} = \binom{n+k}{k} + \frac{(d-1)}{k} \binom{\binom{n+k}{k}}{2}. \tag{4.9}$$

To prove (iii) we remark that we may choose the point at our will, hence we choose it to be $(1 : 0 : \cdots : 0)$. Then we consider the linear space $V_0^k \subset \mathbb{C}[z_0, \ldots, z_n]$, of homogeneous polynomials of degree k in which the monomial z_0^k is absent. We have $\dim_{\mathbb{C}} V_0^k = \binom{n+k}{k} - 1$ and now repeat, verbatim, the proof of case (i). $\qquad \square$

Proof of Corollary 2.4. We may assume the ℓ-plane $\mathbb{L}^\ell$ is the base locus of the linear system $V_{n-\ell}$ generated by $z_{\ell+1}, \ldots, z_n$. Any hyperplane containing $\mathbb{L}^\ell$ belongs to $V_{n-\ell}$ and hence it's enough to consider the extactic

$\mathcal{E}(V_{n-\ell}, X) = \det E(V_{n-\ell}, X)$ where

$$E(V_{n-\ell}, X) = \begin{pmatrix} z_{\ell+1} & z_{\ell+2} & \cdots & z_n \\ X(z_{\ell+1}) & X(z_{\ell+2}) & \cdots & X(z_n) \\ \vdots & \vdots & \ddots & \vdots \\ X^{n-\ell-1}(z_{\ell+1}) & X^{n-\ell-1}(z_{\ell+2}) & \cdots & X^{n-\ell-1}(z_n) \end{pmatrix}.$$

(4.10)

Since the degree of $\mathcal{E}(V_{n-\ell}, X)$ is $n - \ell + \binom{n-\ell}{2}(d-1)$, the result follows by the arguments in the proof of theorem 2.3. □

Acknowledgements. The second author is grateful to ICTP for hospitality.

References

1. M. Brunella, Inexistence of invariant measures for generic rational differential equations in the complex domain, *Bol. Soc. Mat. Mexicana* (3) **12(1)** (2006) 43–49.
2. C. Christopher, J. L. Libre, J. V. Pereira, Multiplicity of invariant algebraic curves in polynomial vector fields, *Pacific J. Math.* **229(1)** (2007) 63–117.
3. S. C. Coutinho, J. V. Pereira, On the density of algebraic foliations without algebraic invariant sets, *J. Reine Angew. Math.* **594** (2006) 117–135.
4. J. L. Libre, J. C. Medrado, On the invariant hyperplanes for *d*-dimensional polynomial vector fields, *J. Phys. A: Math. Theor.* **40** (2007) 8385–8391.
5. A. Lins Neto, Some examples for the Poincaré and Painlevé problems, *Ann. Scient. Éc. Norm. Sup.* 4e série **35** (2002) 231–266.
6. A. Lins Neto, M. G. Soares, Algebraic solutions of one-dimensional foliations, *J. Differential Geom.* **43(3)** (1996) 652–673.
7. J. V. Pereira, Vector fields, invariant varieties and linear systems, *Ann. Inst. Fourier (Grenoble)* **51(5) (2001)** 1385–1405.

Maurício Corrêa Jr.
Dep. Matemática, UFV
Av. P. H. Rolfs s/n
36571-000 Viçosa
Brazil

Márcio G. Soares
Dep. Matemática – UFMG
Av. Antonio Carlos 6627
Brazil
msoares@mat.ufmg.br

8

A note on topological contact equivalence

J.C.F. COSTA

Abstract

In this work we discuss some properties and examples of the topological contact equivalence of smooth map germs. For example, we prove that all finitely C^0-$\mathcal{K}$-determined map germs $f : (\mathbb{R}^n, 0) \to (\mathbb{R}^p, 0)$, with $n < p$, are topologically contact equivalent.

1. Introduction

The contact equivalence (or $\mathcal{K}$-equivalence) was introduced by J. Mather [10] to reduce the problem of C^∞ classification of C^∞ stable map germs to the problem of isomorphic classification of $\mathbb{R}$-algebras. Many properties and invariants of $\mathcal{K}$-equivalence are well-known and appear in the classical literature of Singularity Theory (cf. for example, [10], [11], [12], [19]). However, there exist moduli for the $\mathcal{K}$-orbits. Hence it seems natural to investigate weaker versions of $\mathcal{K}$-equivalence. Some recent works in this direction are [13], [14], [1], [2], [3]. In this sense, the purpose of this article is to describe some results with respect to topological $\mathcal{K}$-equivalence (or C^0-$\mathcal{K}$-equivalence) obtained during the work of my Ph.D. thesis [4] realized at the ICMC-USP, São Carlos. For example, in this paper we show that finitely C^0-$\mathcal{K}$-determined map germs $f : (\mathbb{R}^n, 0) \to (\mathbb{R}^p, 0)$ with $n < p$ are all topologically $\mathcal{K}$-equivalent. For the case $n \geq p$, we obtain a particular result explaining the geometrical interpretation of topological $\mathcal{K}$-equivalence and we give conditions for the C^0-$\mathcal{K}$-triviality of families of map germs. The special case of functions germs was treated in [1].

2000 *Mathematics Subject Classification* 58C27 (primary) 32S15, 32S05 (secondary).
The author thanks the partial support from FAPESP grant no. 2007/01274-6 and PROCAD-CAPES grant no. 190/2007.

114

2. Topological contact equivalence

Definition 2.1. Two map germs $f, g : (\mathbb{R}^n, 0) \to (\mathbb{R}^p, 0)$ are said to be *topologically contact equivalent* (or *topologically $\mathcal{K}$-equivalent*) if there exist two germs of homeomorphisms

$$H : (\mathbb{R}^n \times \mathbb{R}^p, 0) \to (\mathbb{R}^n \times \mathbb{R}^p, 0) \quad \text{and} \quad h : (\mathbb{R}^n, 0) \to (\mathbb{R}^n, 0)$$

such that $H(\mathbb{R}^n \times \{0\}) = \mathbb{R}^n \times \{0\}$ and the following diagram is commutative:

$$
\begin{array}{ccccc}
(\mathbb{R}^n, 0) & \xrightarrow{(id, f)} & (\mathbb{R}^n \times \mathbb{R}^p, 0) & \xrightarrow{\pi_n} & (\mathbb{R}^n, 0) \\
h \downarrow & & H \downarrow & & h \downarrow \\
(\mathbb{R}^n, 0) & \xrightarrow{(id, g)} & (\mathbb{R}^n \times \mathbb{R}^p, 0) & \xrightarrow{\pi_n} & (\mathbb{R}^n, 0)
\end{array}
$$

where $id : (\mathbb{R}^n, 0) \to (\mathbb{R}^n, 0)$ is the identity mapping of $\mathbb{R}^n$ and $\pi_n : (\mathbb{R}^n \times \mathbb{R}^p, 0) \to (\mathbb{R}^n, 0)$ is the canonical projection germ.

Sometimes we use the notation "C^0-$\mathcal{K}$-equivalence" to indicate the topological contact equivalence. In general, it is usual to write "C^0-$\mathcal{G}$-equivalence" to describe the topological version of $\mathcal{G}$-equivalence, where $\mathcal{G}$ is one of classical Mather's group, i.e, $\mathcal{G} = \mathcal{R}, \mathcal{L}, \mathcal{A}, \mathcal{C}$ or $\mathcal{K}$.

Definition 2.2. A C^∞ map germ $f : (\mathbb{R}^n, 0) \to (\mathbb{R}^p, 0)$ is said to be *finitely C^0-$\mathcal{K}$-determined* if there exists a positive number k such that for any C^∞ map germ $g : (\mathbb{R}^n, 0) \to (\mathbb{R}^p, 0)$ with $j^k f(0) = j^k g(0)$, f is topologically $\mathcal{K}$-equivalent to g.

Several characterizations of C^0-$\mathcal{K}$-determinacy appear in the survey [19].

3. Properties and invariants

3.1. Case $n = p$

In the case $n = p$, T. Nishimura [14] has a beautiful result that shows a complete invariant to the topological $\mathcal{K}$-equivalence:

Theorem 3.1 ([14], p. 83). *Let $f, g : (\mathbb{R}^n, 0) \to (\mathbb{R}^n, 0)$ be two finitely C^0-$\mathcal{K}$-determined map germs and $n \neq 4$. Then, f and g are topologically $\mathcal{K}$-equivalent if and only if $|\deg(f)| = |\deg(g)|$.*

The property of C^0-$\mathcal{K}$-determinacy of a germ $f : (\mathbb{R}^n, 0) \to (\mathbb{R}^n, 0)$ guarantees that $f^{-1}(0) = \{0\}$ as germs and consequently the mapping degree $\deg(f)$ is defined. This degree can be algebraically calculated using the Eisenbud-Levine formula [5] when the map germ is finitely $\mathcal{K}$-determined. The restriction $n \neq 4$

that appears in the Theorem 3.1 is due to the Poincare's conjecture since it was not resolved at that time.

The key to the proof of Theorem 3.1 is the following lemma:

Lemma 3.2 ([14], p. 85). *Let $f, g : U \to \mathbb{R}^p$ be two continuous mappings, where U is a neighbourhood of the origin in $\mathbb{R}^n$. Suppose that there exists a family of continuous mappings $F_t : U \to \mathbb{R}^p, t \in [0, 1]$, such that the following conditions hold:*

 i) *$F_0 = f$ and $F_1 = g$ or $\bar{g} = (g_1, \ldots, g_{p-1}, -g_p)$;*
 ii) *$F_t^{-1}(0) = f^{-1}(0)$, for all $t \in [0, 1]$;*
 iii) *For any $t \in [0, 1]$, the vector $F_t(x)$ is not included in the set $\{\alpha F_0(x) \mid \alpha \in \mathbb{R}, \alpha < 0\}$, for any $x \in U - f^{-1}(0)$.*

Then F_t is topologically $\mathcal{K}$-equivalent to $F_{t'}$, for any $t, t' \in [0, 1]$. In particular, f and g are topologically $\mathcal{K}$-equivalent.

3.2. Case $n < p$

Let $f, g : (\mathbb{R}^n, 0) \to (\mathbb{R}^p, 0)$, $n < p$, be two finitely $\mathcal{C}^0$-$\mathcal{K}$-determined map germs. We are interested in describing when these germs are topologically $\mathcal{K}$-equivalent. In fact, we prove in Theorem 3.6 that the C^0-$\mathcal{K}$-equivalence does not distinguish finitely $\mathcal{C}^0$-$\mathcal{K}$-determined map germs. The next examples illustrate this fact and motivate it.

Example 3.3. The germs $f(x) = (x^2, 0)$ and $g(x) = (x, 0)$ are $\mathcal{C}^0$-$\mathcal{K}$-equivalent.

Initially, note that the germ $f(x) = (x^2, 0)$ is clearly $\mathcal{K}$-equivalent to the germ $\psi_1(x) = (x^2, x^3)$ (cf. Mather [11]) and the same happens with the germs $g(x) = (x, 0)$ and $\psi_2(x) = (x^2, x)$. Moreover, taking the pair of homeomorphisms $h(x) = x$ and $H(x, y_1, y_2) = (x, y_1, y_2^3)$, it follows from Definition 2.1 that the germs ψ_2 and ψ_1 are $\mathcal{C}^0$-$\mathcal{K}$-equivalent. Then, f and g are $\mathcal{C}^0$-$\mathcal{K}$-equivalent.

However, observe that the function germs x and x^2 are not $\mathcal{C}^0$-$\mathcal{K}$-equivalent by Theorem 3.1.

Remark. More generally, with the same arguments as in the Example 3.3, it is easy to prove that $f(x) = (x^l, 0)$ is $\mathcal{C}^0$-$\mathcal{K}$-equivalent to $g(x) = (x^s, 0)$, for any l, s positive integers.

It follows from the finite determinacy theory of contact equivalence (cf. [19], [10]) that if a germ $f : (\mathbb{R}, 0) \to (\mathbb{R}^2, 0)$ is finitely $\mathcal{K}$-determined, then f

is $\mathcal{K}$-equivalent to the germ $g(x) = (x^k, 0)$, for some integer k. Using this fact and the previous Remark we can prove the following result:

Proposition 3.4. *Let* $f, g : (\mathbb{R}, 0) \to (\mathbb{R}^2, 0)$ *be two finitely $\mathcal{K}$-determined map germs. Then, f and g are C^0-$\mathcal{K}$-equivalent.*

Example 3.5. The cross-cap $f(x, y) = (x, y^2, xy)$ is C^0-$\mathcal{K}$-equivalent to the $g(x, y) = (x, y, 0)$.

In fact, since $g \overset{\mathcal{K}}{\sim} (x, y^2, y)$ and $f \overset{\mathcal{K}}{\sim} (x, y^2, y^3)$, it is sufficient to show that $(x, y^2, y) \overset{C^0\text{-}\mathcal{K}}{\sim} (x, y^2, y^3)$, where the notations $\overset{\mathcal{K}}{\sim}$ and $\overset{C^0\text{-}\mathcal{K}}{\sim}$ denote the $\mathcal{K}$-equivalence and the C^0-$\mathcal{K}$-equivalence, respectively.

Taking

$$h(x, y) = (x^3, y^3) \quad \text{and} \quad H(x, y, z_1, z_2, z_3) = \left(x^3, y^3, z_1^3, z_2^3, z_3^9\right)$$

we get the required equivalence between f and g.

In general, if $n < p$ we can prove that any two finitely C^0-$\mathcal{K}$-determined map germs are C^0-$\mathcal{K}$-equivalent. The proof of this fact uses the same arguments and notations as in [14].

Theorem 3.6. *Let* $n < p$ *and* $f, g : (\mathbb{R}^n, 0) \to (\mathbb{R}^p, 0)$ *be two finitely $\mathcal{K}$-determined map germs. Then f and g are topologically $\mathcal{K}$-equivalent.*

Proof. By Fukuda's Cone Structure Theorem [6] we can assume that there exists a topologically cone-like germ $\tilde{f} : (\mathbb{R}^n, 0) \to (\mathbb{R}^p, 0)$ such that f is C^0-$\mathcal{K}$-equivalent to $\tilde{f}$ (resp. there exists $\tilde{g}$ such that g is C^0-$\mathcal{K}$-equivalent to $\tilde{g}$). A C^∞ map germ $f : (\mathbb{R}^n, 0) \to (\mathbb{R}^p, 0)$ is called *topologically cone-like* if there exists a positive number ε_0 such that for any number ε with $0 < \varepsilon \leq \varepsilon_0$ and any representative of f the following properties are satisfied:

i) the set $f^{-1}(S_\varepsilon^{p-1})$ is a smooth submanifold without boundary, which is homeomorphic to the standard sphere S_1^{n-1};

ii) the restriction mapping $f : f^{-1}(D_\varepsilon^p) \to D_\varepsilon^p$ is topologically right-left equivalent to the cone

$$c(f) : f^{-1}\left(S_\varepsilon^{p-1}\right) \times [0, \varepsilon)/f^{-1}(S_\varepsilon^{p-1}) \times \{0\} \to S_\varepsilon^{p-1} \times [0, \varepsilon)/S_\varepsilon^{p-1} \times \{0\}$$

of the restricted mapping $f : f^{-1}(S_\varepsilon^{p-1}) \to S_\varepsilon^{p-1}$ defined by $c(f) = (f(x), t)$.

It follows from condition i) that there exists a homeomorphism $\phi_f : S_{\varepsilon_0}^{n-1} \to \tilde{f}^{-1}(S_{\varepsilon_0}^{p-1})$ (resp. $\phi_g : S_{\varepsilon_0}^{n-1} \to \tilde{g}^{-1}(S_{\varepsilon_0}^{p-1})$). Define the mapping $F : S_{\varepsilon_0}^{n-1} \to S_{\varepsilon_0}^{p-1}$ by $F(x) = \tilde{f} \circ \phi_f(x)$ (resp. $G(x) = \tilde{g} \circ \phi_g$).

By hypothesis $n < p$, it follows that F and G are homotopic, i.e., there exists a homotopy $H_t : S_{\varepsilon_0}^{n-1} \to S_{\varepsilon_0}^{p-1}$, $t \in [0, 1]$ such that $H_0 = F$ and $H_1 = G$.

Take the cone

$$c(H_t) : S_{\varepsilon_0}^{n-1} \times [0, 1] \,/\, S_{\varepsilon_0}^{n-1} \times \{0\} \to S_{\varepsilon_0}^{p-1} \times [0, 1] \,/\, S_{\varepsilon_0}^{p-1} \times \{0\}$$

of H_t defined by $c(H_t)(x, s) = (H_t(x), s)$.

Then we have $c(H_0) = c(F)$, $c(H_1) = c(G)$ and $c(H_t)^{-1}(0) = \{0\}$ for any $t \in [0, 1]$. Moreover, as $c(H_t)$ is the cone of H_t, by compactness of $[0, 1]$ it follows that there exists a finite subset $\{t_0, \ldots, t_k\}$ of $[0, 1]$, $0 = t_0 < \cdots < t_k = 1$ and for any integer i, with $0 \le i \le k - 1$, and for any $t \in [t_i, t_{i+1}]$, the vector $c(H_t)(x) \notin \{\alpha c(H_{t_i}) \,|\, \alpha \in \mathbb{R}_-\}$ for any $x \in S_{\varepsilon_0}^{n-1} \times [0, 1] \,/\, S_{\varepsilon_0}^{n-1} \times \{0\} \approx D_\varepsilon^{n-1}$.

By Lemma 3.2, $c(F)$ and $c(G)$ are $\mathcal{C}^0$-$\mathcal{K}$-equivalent.

Hence

$$f \overset{\mathcal{C}^0\text{-}\mathcal{K}}{\sim} \tilde{f} \overset{\mathcal{C}^0\text{-}\mathcal{A}}{\sim} c(\tilde{f}) \overset{\mathcal{C}^0\text{-}\mathcal{R}}{\sim} c(F) \overset{\mathcal{C}^0\text{-}\mathcal{K}}{\sim} c(G) \overset{\mathcal{C}^0\text{-}\mathcal{R}}{\sim} c(\tilde{g}) \overset{\mathcal{C}^0\text{-}\mathcal{A}}{\sim} \tilde{g} \overset{\mathcal{C}^0\text{-}\mathcal{K}}{\sim} g.$$

Then f and g are $\mathcal{C}^0$-$\mathcal{K}$-equivalent. $\qquad\square$

3.3. Case $n > p$

First we consider the case $p = 1$. In [1], the author with S. Alvarez, L. Birbrair and A. Fernandes considers a classification problem, with respect to topological $\mathcal{K}$-equivalence of function germs definable in an o-minimal structure. In fact, we present some special classes of piecewise linear functions, so-called *tent-functions*. We prove that these tent-functions are models for equivalence classes. Namely, we prove that, for each definable germ f, there exists a tent-function g such that f and g are topologically $\mathcal{K}$-equivalent. The main technical tool for this is a definable version of Lemma 3.2. For analytic function germs of two variables, the equivalence classes of $\mathcal{K}$-decomposition defined by tent-functions can be described as equivalence classes of finite collections of elements -1, 0 or 1, by cyclic permutations. These classes are called $\mathcal{K}$-symbols. We prove that $\mathcal{K}$-symbol is a complete invariant of definable function germs, with respect to topological $\mathcal{K}$-equivalence. Moreover, we prove that any analytic $\mathcal{K}$-symbol admits a polynomial realization. We also present normal forms, for function germs of two variables which are finitely C^0-$\mathcal{K}$-determined. All results and definitions cited above can be found in [1].

In this paper, the case $p = 1$ is also considered. In fact, in the Subsection 3.3.1 we present a geometrical interpretation for the C^0-$\mathcal{K}$-equivalence when $p = 1$. For other cases, that is, when $n > p$, $p \ne 1$, we consider the problem of C^0-$\mathcal{K}$-triviality of families of map germs satisfying an integral closure condition (Subsection 3.3.2). We note that an integral closure condition is used here to guarantee the construction of appropriate integrable vector fields that provide the C^0-$\mathcal{K}$-triviality required. This type of argument is utilized by several authors (see for example, [7], [18]).

3.3.1. *Geometrical interpretation of* C^0-$\mathcal{K}$-*equivalence*

By definition two map germs $f, g : (\mathbb{R}^n, 0) \to (\mathbb{R}^p, 0)$ are $\mathcal{K}$-equivalent if there exists a C^∞-diffeomorphism in the product space $(\mathbb{R}^n \times \mathbb{R}^p, 0)$ which leaves $\mathbb{R}^n \times \{0\}$ invariant and maps $\operatorname{graph}(f)$ to $\operatorname{graph}(g)$ ([10]). J. Montaldi [12] interprets the contact equivalence by introducing a natural geometric notion, called *contact between submanifolds*. In fact, Montaldi clarifies the relationship between the $\mathcal{K}$-class of equivalence of maps and the contact type of submanifolds. The notion of contact type between two pairs of submanifolds germs at the origin in $\mathbb{R}^n$ is given by the existence of a diffeomorphism germ of $(\mathbb{R}^n, 0)$ taking one pair to the other. For the topological case, we adapt the definition of same topological contact type and we obtain an analogous result for the hypersurface case, with respect to C^0-$\mathcal{K}$-equivalence.

Definition 3.7. Let $\mathcal{X}_i, \mathcal{Y}_i, i = 1, 2$, be submanifolds of $\mathbb{R}^n$. We say that the pairs $(\mathcal{X}_1, \mathcal{Y}_1)$ and $(\mathcal{X}_2, \mathcal{Y}_2)$ have the *same topological contact type* at the origin in $\mathbb{R}^n$ if there exists a homeomorphism map germ $H : (\mathbb{R}^n, 0) \to (\mathbb{R}^n, 0)$ such that $H(\mathcal{X}_1) = \mathcal{X}_2$ and $H(\mathcal{Y}_1) = \mathcal{Y}_2$.

Notation:
$$C^0\text{–}K(\mathcal{X}_1, \mathcal{Y}_1, 0) = C^0\text{–}K(\mathcal{X}_2, \mathcal{Y}_2, 0).$$

Theorem 3.8. *Let* $\mathcal{X}_i, \mathcal{Y}_i$ *be submanifolds of* $\mathbb{R}^n$ *such that* dim $\mathcal{X}_i =$ dim $\mathcal{Y}_i = n - 1, i = 1, 2$. *Let* $g_i : (\mathcal{X}_i, 0) \to (\mathbb{R}^n, 0)$ *be immersion germs and* $f_i : (\mathbb{R}^n, 0) \to (\mathbb{R}, 0)$ *submersion germs, with* $f_i^{-1}(0) = \mathcal{Y}_i$, *such that* $f_i \circ g_i, i = 1, 2$, *are finitely* C^0-$\mathcal{K}$-*determined. Then*
$$C^0\text{-}K(\mathcal{X}_1, \mathcal{Y}_1, 0) = C^0\text{-}K(\mathcal{X}_2, \mathcal{Y}_2, 0) \Leftrightarrow f_1 \circ g_1 \quad and \quad f_2 \circ g_2 \quad are \quad C^0\text{-}K\text{-}$$
equivalent.

Proof. ($\Rightarrow$) Let $H : (\mathbb{R}^n, 0) \to (\mathbb{R}^n, 0)$ be a homeomorphism germ such that
$$H(\mathcal{X}_1) = \mathcal{X}_2 \quad \text{and} \quad H(\mathcal{Y}_1) = \mathcal{Y}_2.$$

Observe that $H(\mathcal{X}_1 \cap \mathcal{Y}_1) = \mathcal{X}_2 \cap \mathcal{Y}_2$, hence $\mathcal{X}_1 \cap \mathcal{Y}_1$ is homeomorphic to $\mathcal{X}_2 \cap \mathcal{Y}_2$. This homeomorphism will be indicated by $\overset{homeo}{\cong}$. Making appropriate identifications we can obtain
$$(f_1 \circ g_1)^{-1}(0) = g_1^{-1}(\mathcal{X}_1 \cap \mathcal{Y}_1) \overset{diffeo}{\cong} \mathcal{X}_1 \cap \mathcal{Y}_1 \overset{homeo}{\cong} \mathcal{X}_2 \cap \mathcal{Y}_2 \overset{diffeo}{\cong}$$
$$\overset{diffeo}{\cong} g_2^{-1}(\mathcal{X}_2 \cap \mathcal{Y}_2) = (f_2 \circ g_2)^{-1}(0),$$

where $\overset{diffeo}{\cong}$ indicates the diffeomorphism between the sets. Hence, the function germs $f_1 \circ g_1$ and $f_2 \circ g_2$ are C^0-$\mathcal{V}$-*equivalent*, that is, there exists a homeomorphism germ in $(\mathbb{R}^n, 0)$ which maps $(f_1 \circ g_1)^{-1}(0)$ to $(f_2 \circ g_2)^{-1}(0)$.

However, for function germs finitely C^0-$\mathcal{K}$-determined, Nishimura ([14]) showed that the C^0-$\mathcal{V}$-equivalence is equivalent to C^0-$\mathcal{K}$-equivalence. Then it follows that $f_1 \circ g_1$ and $f_2 \circ g_2$ are C^0-$\mathcal{K}$-equivalent.

($\Leftarrow$) By hypothesis we have dim $\mathcal{X}_i = $ dim $\mathcal{Y}_i = n - 1 = k$. The idea is to express $\mathcal{X}_i$ as the graph of $\phi_i : \mathbb{R}^k \to \mathbb{R}$ and $\mathcal{Y}_i$ as the graph of the null function of $\mathbb{R}^k$ to $\mathbb{R}$. For this, consider a coordinate system of $\mathbb{R}^n$ such that

$$f_1(x_1, \ldots, x_n) = x_n.$$

Thus, $\mathcal{Y}_1 = f_1^{-1}(0) = \mathbb{R}^k \times \{0\}$. Let V_1 be a 1-dimensional transverse subspace to $\mathcal{X}_1$ and $\mathcal{Y}_1$, such that $\mathbb{R}^n = \mathcal{Y}_1 \times V_1$.

Let $\pi : \mathbb{R}^n = \mathcal{Y}_1 \times V_1 \to \mathcal{Y}_1$ be the canonical projection into the first factor. Hence,

$$\pi|_{\mathcal{X}_1} : \mathcal{X}_1 \subset \mathbb{R}^n \to \mathcal{Y}_1$$

is a diffeomorphism that induces a coordinate system in $\mathcal{X}_1$.

With respect to this coordinate system, $\mathcal{Y}_1$ is the graph of the null function and $\mathcal{X}_1$ is the graph of $f_1 \circ g_1$, where f_1 is considered to be restricted to $f_1 : \mathcal{Y}_1 \times V_1 \longrightarrow V_1$.

By a similar method, we can do the same construction to $\mathcal{X}_2$ and $\mathcal{Y}_2$.

It follows from the C^0-$\mathcal{K}$-equivalence between $f_1 \circ g_1$ and $f_2 \circ g_2$ that there exists a homeomorphism germ $H : (\mathbb{R}^k \times \mathbb{R}, 0) \to (\mathbb{R}^k \times \mathbb{R}, 0)$ such that $H(\mathbb{R}^k \times \{0\}) = \mathbb{R}^k \times \{0\}$ and H takes the graph of $f_1 \circ g_1$ to the graph of $f_2 \circ g_2$. In other words, there exists a homeomorphism H such that $H(\mathcal{Y}_1) = \mathcal{Y}_2$ and $H(\mathcal{X}_1) = \mathcal{X}_2$.

Thus, $C^0 - K(\mathcal{X}_1, \mathcal{Y}_1, 0) = C^0 - K(\mathcal{X}_2, \mathcal{Y}_2, 0)$. $\qquad\square$

3.3.2. C^0-$\mathcal{K}$-*triviality*

There are several works interested in the study of the triviality of families of map germs with respect to some equivalence relations. For C^0-$\mathcal{K}$-triviality we can cite, for example, [15], [16], [17].

For a family of real map germs satisfying the condition of no coalescing, King [8] showed that the topological triviality of the zero-sets of this family implies the C^0-$\mathcal{R}$-triviality of this family and consequently its C^0-$\mathcal{K}$-triviality. In our result (Theorem 3.11) we put a condition so-called *good $\mathcal{K}$-deformation* to the family of map germs. We also demand an integral closure condition. The prerequisites for the study of the integral closure of ideals and modules can be found in the work of T. Gaffney [7].

Definition 3.9. An analytic deformation $F : (\mathbb{R}^n \times \mathbb{R}, 0) \to (\mathbb{R}^p, 0)$ of a germ $F_0 : (\mathbb{R}^n, 0) \to (\mathbb{R}^p, 0)$ is a *good $\mathcal{K}$-deformation* if there exists a neighbourhood of the origin $U \subset \mathbb{R}^n$ such that

$$U \cap F_t^{-1}(0) \cap \Sigma F_t - \{0\} = \emptyset,$$

for all $t \in \mathbb{R}$, where ΣF_t denotes the singular set of F_t.

Let A_n be the set of analytic function germs in n variables at the origin and $X \subset \mathbb{R}^n$ be an analytic set. We define $A_{X,x}^p$ the module formed by p-copies of

$$A_{X,x} = A_n/\langle \text{functions that define } X \text{ in a neighbourhood of } x \in X \rangle.$$

Definition 3.10. Suppose $\mathcal{M}$ is a submodule of $A_{X,x}^p$, X a real analytic set. Then the *real integral closure of* $\mathcal{M}$, denoted by $\overline{\mathcal{M}}$ in $A_{X,x}^p$ is the set of $h \in A_{X,x}^p$ such that for any analytic curve $\phi : (\mathbb{R}, 0) \to (X, x)$, we have $h \circ \phi \in (\phi^*(\mathcal{M}))A_1$.

Let $\mathcal{M}$ be a submodule of $A_{X,x}^p$. We denote by $[\mathcal{M}]$ the matrix of generators of $\mathcal{M}$ and by $J_k(\mathcal{M})$ the ideal generated by $k \times k$-minors of $[\mathcal{M}]$.

The main result of this subsection is the next theorem which gives a sufficient condition to $\mathcal{C}^0$-$\mathcal{K}$-triviality. The idea of Theorem 3.11 is to use a convenient condition of integral closure to construct appropriate integrable vector fields that establish the $\mathcal{C}^0$-$\mathcal{K}$-triviality required.

Theorem 3.11. *Let $F : (\mathbb{R}^n \times \mathbb{R}, 0) \to (\mathbb{R}^p, 0)$ be a good analytic $\mathcal{K}$-deformation of a germ $F_0 = f : (\mathbb{R}^n, 0) \to (\mathbb{R}^p, 0)$. Consider $F_t(x) = F(x, t)$. If*

$$\frac{\partial F}{\partial t} \in \overline{\mathrm{d}\, F_t(m_n \theta_{n+1}) + m_n F^\star(m_p)\theta_{F_t}},$$

then F is $\mathcal{C}^0$-$\mathcal{K}$-trivial.

Proof. For simplicity we denote $h = \dfrac{\partial F}{\partial t}$ and $\mathcal{M} = \mathrm{d}\, F_t(m_n \theta_{n+1}) + m_n F^\star(m_p)\theta_{F_t}$.

Consider $\{\alpha_1, \dots, \alpha_m\}$ the generators of $m_n \theta_{n+1}$ where each α_s is of the type $x_i \frac{\partial}{\partial x_j}$, for any $i = 1, \dots, n$ and for any $j = 1, \dots, n + 1$. Let $\{\mathrm{d}\, F_t(\alpha_1), \dots, \mathrm{d}F_t(\alpha_m)\}$ be the generators of submodule $\mathrm{d}\, F_t(m_n \theta_{n+1})$ and $\{x_l F_{t,i} e_j, l = 1, \dots, n; i = 1, \dots p; j = 1, \dots, p\}$ be the generators of submodule $m_n F^\star(m_p)\theta_{F_t}$, where $F_{t,i}$ are the components of F_t.

Let $\{\rho_1, \ldots, \rho_r\}$ be the generators of $J_p(\mathcal{M})$ and $\rho = \sum_{i=1}^{r} |\rho_i|^2$.

Since F is a good analytic $\mathcal{K}$-deformation, we have that $\rho(x) = 0$ if and only if $x = 0$. Observe that

$$\rho_k \cdot h \in J_p(\mathcal{M}).h.$$

Then, it follows from the real analytic version of Lemma 1.6, p. 303 of [7] that

$$\rho_k \cdot h \in \mathcal{M}.J_p((h, \mathcal{M})).$$

Hence we have

$$\rho_k \cdot h = \sum_{j}^{m} \mathrm{d}\, F_t(\alpha_j) a_{kj} + \sum_{l,i,j}^{n,p,p} x_l F_{t,i} e_j b_{ijk},$$

with $a_{kj}, b_{ijk} \in J_p((h, \mathcal{M}))$.

As ρ_k is a generator of $J_p(\mathcal{M})$, we have that ρ_k is a $p \times p$-minor of matrix $[\mathcal{M}]$ and $\rho = \sum_{i=1}^{r} |\rho_i|^2$.

Then it follows that

$$|\rho_k|^2 \cdot h = \sum_{j}^{m} \mathrm{d}\, F_t(\alpha_j) a_{kj} \rho_k + \sum_{l,i,j}^{n,p,p} x_l F_{t,i} e_j b_{ijk} \rho_k \implies$$

$$\rho \cdot h = \mathrm{d}F_t \left[\sum_{k,j}^{r,m} a_{kj} \rho_k \alpha_j \right] + \sum_{l,i,j,k}^{n,p,p,r} \rho_k b_{ijk} x_l e_j F_{t,i}.$$

By hypothesis $h \in \overline{\mathcal{M}}$ and by Gaffney's result on integral closure (see the real analytic version of Proposition 1.7, p. 304 of [7]) we have that

$$J_p((h, \mathcal{M})) \subseteq \overline{J_p(\mathcal{M})}.$$

Then it follows that $a_{kj}, b_{ijk} \in \overline{J_p(\mathcal{M})}$.

Consider the vector field $\varepsilon(x, t) = \sum_{k,j}^{r,m} \dfrac{a_{kj} \rho_k \alpha_j}{\rho}$. As $a_{kj} \in \overline{J_p(\mathcal{M})}$ we have

$$|a_{kj}| \leq c_1 \sup_i \{\text{generators of } J_p(\mathcal{M})\}, \quad \text{i.e.,}$$

$$|a_{kj}| \leq c_1 \sup_i \{|\rho_i|\}.$$

Hence, $|\varepsilon(x, t)| \leq c|x|$, and the vector field is integrable (cf. [9]).

Put

$$\eta(x, y, t) = \sum_{l,i,j,k}^{n,p,p,r} \frac{(\rho_k b_{ijk} x_l e_j) y_i}{\rho}.$$

Since $b_{ijk} \in \overline{J_p(\mathcal{M})}$, we have

$$|b_{ijk}| \leq c_2 \sup_i \{|\rho_i|\},$$

and then $|\eta(x, y, t)| \leq c|x||y|$. This means that η is also integrable.

The integrability of the vector fields ε and η provides the homeomorphisms giving the C^0-$\mathcal{K}$-triviality of the family F. □

Acknowledgements. I would like to thank Professor Maria Ruas for very helpful and interesting discussions and also I would like to congratulate her on the occasion of her 60th birthday.

The author is grateful to the referee for valuable suggestions.

References

1. S. Alvarez, L. Birbrair, J. C. F. Costa and A. Fernandes, Topological K- equivalence of analytic function germs (2009). Submitted for publication. Available from http://www.maths.manchester.ac.uk/raag/index.php?preprint=0120.

2. L. Birbrair, J. C. F. Costa, A. Fernandes and M. A. S. Ruas, $\mathcal{K}$-bi-Lipschitz equivalence of real function-germs, *Proc. Amer. Math. Soc.* **135(4)** (2007) 1089–1095.

3. L. Birbrair, J. C. F. Costa and A. Fernandes, Topological contact equivalence of map germs. *To appear in Hokkaido Math. Journal* (2009).

4. J. C. F. Costa, *Equivalências de contato topológica e bi-Lipschitz de germes de aplicações diferenciáveis*, Ph.D. Thesis (ICMC-USP, São Carlos (2005).

5. D. Eisenbud and H. I. Levine, An algebraic formula for the degree of a C^∞ mapgerm, *Ann. Math.* **106** (1977) 19–44.

6. T. Fukuda, Local topological properties of differentiable mappings I, *Inventiones Math.* **65(2)** (1981) 227–250.

7. T. Gaffney, 'Integral closure of modules and Whitney equisingularity', *Inventiones Math.* **107(2)** (1992) 301–322.

8. H. C. King, Topological type in families of germs, *Invent. Math.* **62** (1980) 1–13.

9. T. C. Kuo, On C^0-sufficiency of jets of potential functions, *Topology* **8** (1969) 167–171.

10. J. Mather, Stability of C^∞-mappings, III: finitely determined map-germs, *Publ. Math. I.H.E.S.* **35** (1969) 127–156.

11. J. Mather, Stability of C^∞-mappings, IV: classification of stable map-germs by $\mathbb{R}$-algebras, *Publ. Math. I.H.E.S.* **37** (1970) 223–248.

12. J. Montaldi, On contact between submanifolds, *Michigan Math. J.* **33** (1986) 195–199.

13. T. Nishimura, Topological types of finitely C^0-$\mathcal{K}$-determined map-germs, *Transactions of the Amer. Math. Soc.* **312(2)** (1989) 621–639.

14. T. Nishimura, Topological $\mathcal{K}$-equivalence of smooth map-germs, *Stratifications, Singularities and Differential Equations*, I (Marseille 1990, Honolulu, HI 1990), Travaux en Cours, **54** Hermann, Paris (1997) 82–93.

15. M. A. S. Ruas, C^l-*determinação finita e aplicações*, Ph.D. Thesis (ICMC-USP), São Carlos (1983).

16. M. A. S. Ruas, On the degree of C^l-determinacy, *Math. Scand.* **59** (1986) 59–70.

17. M. A. S. Ruas and M. J. Saia, C^l-determinacy of weighted homogeneous germs, *Hokkaido Math. J.* **26** (1997) 89–99.

18. M. A. S. Ruas and J. N. Tomazella, Topological triviality of families of functions on analytic varieties, *Nagoya Math. J.* **175** (2004) 39–50.

19. C. T. C. Wall, Finite determinacy of smooth map-germs, *Bull. London Math. Soc.* **13(6)** (1981) 481–539.

J. C. Ferreira Costa
Dep. Matemática
IBILCE-UNESP
R. C. Colombo, 2265
15054-000 S. J. Rio Preto
Brazil
jcosta@ibilce.unesp.br

9

Bi-Lipschitz equivalence, integral closure and invariants

TERENCE GAFFNEY

Abstract

In this note we relate several different ways of looking at an infinitesimal notion of bi-Lipschitz equisingularity. In the case of curves we show how invariants related to these notions can be used to tell if a family is bi-Lipschitz equisingular.

1. Introduction

The study of bi-Lipschitz equisingularity was started by Zariski [22], Pham and Teissier [18], [19], and was further developed by Lipman [13], Mostowski [14], [15], Parusinski [16], [17], Birbrair [2] and others.

In this note, we begin the study of bi-Lipschitz equisingularity from the perspective of our previous work on Whitney equisingularity. In the approach of that work, the study of the equisingularity condition is developed along two avenues. One direction is the study of the appropriate closure notion on modules and applying it to the Jacobian module of a singularity [4]. The other direction is through the study of analytic invariants which control the particular stratification condition [5]. The interaction between the two approaches is useful in understanding each approach.

In section two of this paper, we work on the first approach, looking to the hypersurface case, and to similar constructions for motivation. The construction which seems most promising, defining the saturation of I via the blow-up of X by I, is the analogue of a construction used to study the weak sub-integral closure of an ideal in [11], written with Marie Vitulli.

2000 *Mathematics Subject Classification* 32S15 (primary), 14B05, 13H15 (secondary).
Please refer to http://www.ams.org/msc/ for a list of codes

In section three, we apply the theory of analytic invariants to describe when a family of space curves is bi-Lipschitz equisingular.

This note was written while the author was the guest of USP-Sao Carlos. He thanks the Institute and the people of the singularities group there for their support during the period of this work. He also thanks the referee for a careful reading of this paper, and for helpful suggestions.

2. The Lipschitz saturation of an ideal or module

As we will see, the integral closure of an ideal or module can be related to the integral closure of a ring, hence of its normalization. So the starting point for us is the Lipschitz saturation of a space, as developed by Pham-Teissier ([19]). When we understand how to modify spaces, we can modify ideals and modules.

In the approach of Pham-Teissier, let A be a commutative local ring over $\mathbf{C}$, and $\bar{A}$ its normalization. (We can assume A is the local ring of an analytic space X at the origin in $\mathbf{C}^n$.) Let I be the kernel of the inclusion

$$\bar{A} \otimes_{\mathbf{C}} \bar{A} \to \bar{A} \otimes_A \bar{A}.$$

Then the Lipschitz saturation of A, denoted $\tilde{A}$, consists of all elements $h \in \bar{A}$ such that $h \otimes 1 - 1 \otimes h \in \bar{A} \otimes_{\mathbf{C}} \bar{A}$ is in the integral closure of I. The connection between this notion and that of Lipschitz functions is as follows. If we pick generators $(z_1, \ldots, z_n)$ of the maximal ideal of the local ring A, then $z_i \otimes 1 - 1 \otimes z_i \in \bar{A} \otimes_{\mathbf{C}} \bar{A}$ give a set of generators of I. Choosing z_i so that they are the restriction of coordinates on the ambient space, the integral closure condition is equivalent to

$$|h(z_1, \ldots, z_n) - h(z_1', \ldots, z_n')| \leq C \sup_i |z_i - z_i'|$$

holding on some neighborhood U, of $(0, 0)$ on $X \times X$. This last inequality is what is meant by the meromorphic function h being Lipschitz at the origin on X. (Note that the integral closure condition is equivalent to the inequality holding on a neighborhood U for some C for any set of generators of the maximal ideal of the local ring A. The constant C and the neighborhood U will depend on the choice.)

If X, x is normal, then passing to the saturation doesn't add any functions.

There is another interesting operation on spaces that fits between normalization and saturation. If, in the definition of the saturation, we ask only that $h \otimes 1 - 1 \otimes h$ be in the radical of I, then the set of such functions h defines a subring of $\bar{A}$ which we denote $\bar{A}_{SN}$. We can view elements of $\bar{A}$ as pullbacks of bounded meromorphic functions on X. The condition that $h \otimes 1 - 1 \otimes h$ is in the radical of I is equivalent to h being the pullback of a continuous meromorphic function on X. Then we define the semi-normalization of X as $\text{Spec}(\bar{A}_{SN})$. The normalization, semi-normalization and saturation are the largest spaces of their type, a notion which we now make precise.

Given an analytic germ X, x, which for simplicity we assume irreducible, we can consider the set $\mathcal{M}$, of bounded meromorphic functions on X. If $f : Y, y \to X, x$ is a bi-meromorphic, finite map-germ, then the local ring of Y, y can be identified with a subring of $\mathcal{M}$. The relative size of the subring is a measure of the size of the underlying space. If X_N, x' is a germ which is the normalization of X, x, with normalization map f_N, then the local ring of X_N, x' is $\mathcal{M}$, for the semi-normalization, the elements of the local ring are the continuous elements of $\mathcal{M}$, while the elements of the saturation are the continuous elements which satisfy the Lipschitz condition. If $f : Y, y \to X, x$ is a bi-meromorphic, finite map-germ, then the local ring of Y can be identified with a subring of X_N, x', and there exists a map $g : X_N \to Y$ which is a normalization of Y such that $f \circ g$ is a normalization of X. If the local ring of Y is a subring of the continuous elements of $\mathcal{M}$, then if f is an analytic homeomorphism, there exists a map $g : X_{SN} \to Y$ which is a semi-normalization of Y such that $f \circ g$ is a semi-normalization of X, and g is a homeomorphism. If the local ring of Y is a subring of the Lipschitz elements of $\mathcal{M}$, then similar statements hold; in particular there is a Lipschitz homeomorphism from the saturation of X to Y.

There are analogous closure operations for ideals in the first two cases. Given an ideal $I \subset A$, then h is integrally dependent on I, if h satisfies a monic polynomial relation of the form $h^k + \sum_{i=0}^{n-1} g_i h^i = 0$ where $g_i \in I^{k-i}$. The integral closure of I consists of the set of elements of A which are integrally dependent on I. Given an ideal $I \subset A$, then h is weakly subintegral over I, provided that there exist $q \in \mathbb{N}$ and $a_i \in I^i$, for ($1 \le i \le 2q + 1$), such that

$$h^n + \sum_{i=1}^{n} \binom{n}{i} a_i h^{n-i} = 0 \quad (q + 1 \le n \le 2q + 1). \tag{2.1}$$

The weak subintegral closure of I, denoted $^*I$, consists of the set of elements of A weakly subintegral over I.

At this point there are various possibilities for defining the Lipschitz closure of an ideal I. If $h \in \mathcal{O}_{X,x}$, denote by h_D the germ $h(z) - h(z')$, and by I_D the ideal generated by the $\{f_{iD}\}$ where the $\{f_i\}$ are a set of generators of I.

1) $I_{S_1} = \{h \in \mathcal{O}_{X,x} | h \in I\bar{A}\}$.
2) $I_{S_2} = \{h \in \mathcal{O}_{X,x} | h_D \in \overline{I_D}, h \in \sqrt{I}\}$.

There is an analogy which leads to a third definition. This analogy also ties the closure operations on ideals defined above to the operations on ringed spaces of normalization, seminormalization and saturation. Consider $B_I(X)$, the blow-up of X by I. If we pass to the normalization of the blow-up, then h is in $\bar{I}$ iff and only if the pull back of h to the normalization is in the ideal generated by the pullback of I [12]. If we pass to the seminormalization of the blow-up, then h is in $^*I$, iff the pullback of h to the seminormalization is in the ideal generated by the pullback of I. (For a proof of this and more details on the weak subintegral closure cf. [11]). Denote the saturation of the blow-up by $SB_I(X)$, and the map to X by π_S. Then the third candidate definition is:

3) $I_{S_3} = \{h \in \mathcal{O}_{X,x} | \pi_S^*(h) \in \pi_S^*(I)\}$.

We will call I_{S_3} the **Lipschitz saturation** of the ideal I, and will try to justify this below.

There are some easy observations regarding these possible definitions which we combine into a proposition.

Proposition 2.1. *We have*

1) $I_{S_1} = I$ *if* (X, x) *is normal or seminormal.*
2) $I_{S_2} \subset \bar{I}$.
3) $I_{S_3} \subset {}^*I \subset \bar{I}$.

Proof. 1) If X, x is normal or seminormal, then the Lipschitz functions are smooth.

2) Suppose $h \in I_{S_2}$; if $x \in V(I)$ then $|h(y) - h(x)| = |h(y) - 0| \leq C sup_i |f_i(y)|$, which implies $h \in \bar{I}$ at each point of $V(I)$ (Cf. [4]).

3) This follows because the maps from the seminormalization and the normalization of $B_I(X)$ factor through the saturation of $B_I(X)$. $\square$

Similar constructions can be made for submodules of a free $\mathcal{O}_{X,x}$ module. Given M a submodule of a free $\mathcal{O}_X$ module F of rank p, we can view M as an ideal $\mathcal{M}$ in the ring $\mathcal{O}_X[T_1, \ldots, T_p]$. We can then blow-up by the ideal sheaf induced on $X \times \mathbf{P}^{p-1}$. Starting from this space, we can form the saturation, seminormalization and normalization as before, using these spaces to define the Lipschitz saturation, weak sub-integral closure, and to form the integral closure.

Returning to the ideal case, we see that I_{S_1} is too restrictive, as when X is normal, no functions are added to I. To evaluate the other two definitions we turn to the applications we have in mind.

Our application is to equisingularity; assume we are interested in a pair of strata, the small stratum Y being embedded as an affine space; assume it is one dimensional, so we identify it with our field k. For both W and bi-Lipschitz equisingularity of families of real or complex analytic sets, the starting point is the desire to control the vector field obtained by taking the constant unit field on k, extending it trivially to the ambient space then projecting to the tangent space of the large stratum at every point using orthogonal projection. If we work on $\mathbf{R} \times \mathbf{R}^n$ in the case where the closure of the open stratum is a hypersurface with equation $f = 0$ then the vector field of interest in coordinates is:

$$\frac{\partial}{\partial y} - \sum_{i=1}^{n} \frac{f_y f_{x_i}}{f_{x_1}^2 + \cdots + f_{x_n}^2} \frac{\partial}{\partial x_i}.$$

We could ask that the vector field be Lipschitz. By work of Mostowski, we know that this cannot hold generically. Here the ideal is $J(f)$, the Jacobian ideal of f. If we lift the components of the field to $B_{J(f)}X$, by composing with p the projection to X, and work on U_i, the open set on which $(f_{x_i} \circ p)$ is a reduction of $p^*(J(f))$, we see that at a neighborhood of a point of U_1,

$$\left\| \frac{f_y f_{x_i}}{f_{x_1}^2 + \cdots + f_{x_n}^2} \right\| \leq C \left\| \frac{f_y}{f_{x_1}} \right\|.$$

The analogous inequality holds over $\mathbf{C}$.

Looking at the meromorphic function $\frac{f_y}{f_{x_1}}$ on the open subset U, suggests asking that $f_y \in J(f)_{S_3}$ is an interesting condition. If this condition holds we get inequalities of the form

$$\left\| \frac{f_y \circ p \circ \pi_1}{f_{x_i} \circ p \circ \pi_1} - \frac{f_y \circ p \circ \pi_2}{f_{x_i} \circ p \circ \pi_2} \right\| \leq C sup_{i,j}\{|y \circ p \circ \pi_1 - y \circ p \circ \pi_2\|,$$

$$\|x_i \circ p \circ \pi_1 - x_i \circ p \circ \pi_2\|, \|T_j/T_1 \circ \pi_1 - T_j/T_1 \circ \pi_2\|\}$$

where we are working at a point on the diagonal of $U_1 \times U_1$, over the origin, and π_i is the projection to the i-th factor, T_j/T_1 coordinates on projective space. If we work on a subset of U on which $\{\|T_j/T_1 \circ \pi_1 - T_j/T_1 \circ \pi_2\|\}$ are bounded by the difference of the coordinates, then the original field is Lipschitz on the corresponding horn neighborhood on X. This condition can hold at all points of the blow-up without contradicting the results of Mostowski on the failure of generic bi-Lipschitz equisingularity. For when the surface is most curved, the terms $\|T_j/T_1 \circ \pi_1 - T_j/T_1 \circ \pi_2\|$ can have relatively large changes compared with the change in the coordinates, so the Lipschitz inequality can still hold on B_IX though it fails on X. From this viewpoint, Mostowski introduces the polar varieties as new strata in order to deal with the points where the $\|T_j/T_1 \circ \pi_1 - T_j/T_1 \circ \pi_2\|$ cannot be bounded in terms of the change in the coordinates.

If we have a family of complex analytic hypersurface singularities then by work of Mostowski we know that the field we have been studying is Lipschitz on some horn neighborhood, hence the condition for $f_y \in J(f)_{S_3}$ is checked generically for a Z-open subset of $B_I(X)$. We conjecture that it holds over a

Z-open subset of the parameter space. The following result may be helpful in checking this. It gives a description of I_{S_3} that works on $X \times X$. Some notation is needed. Let $h \in \mathcal{O}_{X,x}$; then h_D is in $\mathcal{O}^2_{X \times X, x, x}$, defined by $(h \circ \pi_1, h \circ \pi_2)$. Let I be an ideal in $\mathcal{O}_{X,x}$; then I_D is the submodule of $\mathcal{O}^2_{X \times X, x, x}$ generated by the h_D where h is an element of I.

We first need a lemma describing the generators of I_D.

Lemma 2.2. *Suppose I is generated by $\{f_1, \ldots, f_d\}$. Then I_D is generated by the pairs $\{(f_i \circ \pi_1, f_i \circ \pi_2), ((x_j f_i \circ \pi_1, x_j f_i \circ \pi_2))\}$.*

Proof. It is clear that the elements listed are in I_D. They imply that I_D contains the ideals $I_\Delta(0, \pi_2^*(I))$ and $I_\Delta(\pi_1^*(I), 0)$, where I_Δ is the ideal generated by the double of the coordinates. It is also clear that I_D is generated by $\{(g \circ \pi_1, g \circ \pi_2)$ where $g \in I$. Now

$$(g \circ \pi_1, g \circ \pi_2) = \left(\sum (h_i f_i) \circ \pi_1, \sum (h_i f_i) \circ \pi_2 \right),$$

so it suffices to show that $((h_i f_i) \circ \pi_1, (h_i f_i) \circ \pi_2)$ is in the module generated by our list of generators. This is true if and only if $(0, (h_i \circ \pi_1 - h_i \circ \pi_2) f_i \circ \pi_2))$ is. Now, $h_i \circ \pi_1 - h_i \circ \pi_2 \in I_\Delta$ finishes the proof. $\square$

Example 2.3. If $I = (x^2, y^2)$, we see easily that the set of generators of I_D must include more than the doubles of x^2 and y^2.

Theorem 2.4. *Suppose (X, x) is a complex analytic set germ, $I \subset \mathcal{O}_{X,x}, h \in \bar{I}$. Then $h \in I_{S_3}$ if and only if $h_D \in \bar{I}_D$.*

Proof. We use a mixture of techniques here. If $\Phi = (\phi_1, \phi_2)$ is a curve, $\Phi : \mathbf{C}, 0 \to X \times X, (x, x)$, with image not contained in $V(I) \times V(I)$ then Φ has a unique lift $\tilde{\Phi}$ to $B_I X \times B_I X$. Consider the case where $\tilde{\Phi}(0)$ is not on the diagonal of $B_I X \times B_I X$. In this case, there is nothing to check if we want to check $h \in I_{S_3}$. For to check $h \in I_{S_3}$, we work on $B_I X$, at points of E the exceptional divisor, we choose f a local generator of $p^*(I)$ and check that $h \circ p / f \circ p$ is Lipschitz at points close to the point e of E where we are working. This involves working in a neighborhood of $e \times e$ and working with curves passing though $e \times e$.

We claim it is also trivial to check that $h_D \in \bar{I}_D$ for such a curve. Because $\tilde{\Phi}(0)$ is not on the diagonal of E, the matrix of generators of $\tilde{\Phi}^* p^*(I_D)$ takes the form of

$$\langle f_i \circ \phi_1, f_j \circ \phi_2 \rangle M$$

where M contains an invertible submatrix and $f_i \circ p, f_j \circ p$ are local generators for $p^* I$ at $\tilde{\phi}_1(0), \tilde{\phi}_2(0)$ respectively. In turn this implies that $\tilde{\Phi}^* p^*(I_D)$ is the direct sum of $\phi_1^*(I)$ and $\phi_2^*(I)$, and this contains $(h \circ \phi_1, h \circ \phi_2)$.

So we may assume we are working in a neighborhood of a point (e, e) on the diagonal of $E \times E$. Label the generators of I so that f_1 is a local generator of $p^*(I)$ on a neighborhood of e. Then the condition that $h \in I_{S_3}$ is checked at (e, e) if the function h/f_1 is Lipschitz at e which means that

$$|h/f_1 \circ p \circ \pi_1(z) - h/f_1 \circ p \circ \pi_2(z)| \leq C \sum_i |(x_i \circ p)_D(z)|$$
$$+ \sum_j |(T_j/T_1)_D(z)|.$$

For the module condition, it suffices to show $(h \circ p \circ \pi_1, h \circ p \circ \pi_2)$ is in the integral closure of the pullback of I_D to the product $B_I X \times B_I X$ in a neighborhood of the point (e, e).

For this it is convenient to use the determinantal criterion of [4]. The integral closure condition holds if and only if the minors formed by adding $(h \circ p \circ \pi_1, h \circ p \circ \pi_2)$ as a column vector to the matrix of generators of the pullback of I_D and taking 2x2 minors is in the integral closure of the minors of the matrix of generators.

Using the previous lemma, we get that the minors of the matrix of generators have the form

$$\{(f_1 \circ p \circ \pi_1 f_1 \circ p \circ \pi_2)(T_j/T_1 \circ \pi_1 - T_j/T_1 \circ \pi_2),$$
$$(f_1 \circ p \circ \pi_1 f_1 \circ p \circ \pi_2)(x_i \circ p)_D(T_j/T_1) \circ \pi_2\}.$$

Meanwhile the additional minors that we get by adding $(h \circ p \circ \pi_1, h \circ p \circ \pi_2)$ as a column vector are the product of $(f_1 \circ p \circ \pi_1 f_1 \circ p \circ \pi_2)$ with the collection:

$$\{(h/f_1 \circ p \circ \pi_1 - h/f_1 \circ p \circ \pi_2), (h/f_1 \circ p \circ \pi_1)(T_j/T_1 \circ \pi_1 - T_j/T_1 \circ \pi_2),$$
$$(h/f_1 \circ p \circ \pi_1)((x_i \circ p_D)(T_j/T_1) \circ \pi_2\}.$$

Of these new minors all are in the integral closure of the original collection since $h \in \bar{I}$ with the possible exception of the first one.

So we need

$$|h/f_1 \circ p \circ \pi_1(z) - h/f_1 \circ p \circ \pi_2(z)| \leq C' \sum_i |(x_i \circ p)_D(z)(T_j/T_1) \circ \pi_2(z)|$$
$$+ \sum_j |(T_j/T_1)_D(z)|.$$

Since the $(T_j/T_1) \circ \pi_2(z)$ are locally bounded, this becomes

$$|h/f_1 \circ p \circ \pi_1(z) - h/f_1 \circ p \circ \pi_2(z)| \leq C \sum_i |(x_i \circ p)_D(z)(z)|$$
$$+ \sum_j |(T_j/T_1)_D(z)|,$$

which is the same as the other condition. □

The advantage of the previous result is that it allows us to work on $X \times X$ instead of the blow-up. Assuming that we can prove that this module condition is generic we can then use such tools as the principle of specialization of integral dependence to relate the generic Lipschitz condition to parameter values. This would show that invariants related to the modules I_D play a role in the bi-Lipschitz equisingularity of families of hypersurfaces. In the next section we take a different approach in the development of a theory of analytic invariants controlling bi-Lipschitz regularity.

3. Bi-Lipschitz equisingularity of families of curves

In the last section we developed a theory in terms of defining equations using Jacobian ideals and modules. In this section, we develop an infinitesimal theory using parameterizations of our sets and relate this theory to analytic invariants.

Let $X \subset \mathbf{C} \times \mathbf{C}^n$ be a Whitney equisingular family of irreducible curves with $\mathbf{C} \times 0 \subset X$ as parameter space. Suppose X is defined by a family of maps $\bar{f}(y, z) = f_y(z)$ with y coordinates on the parameter space $\mathbf{C}$ and z coordinates on $\mathbf{C}^n$. Then the hypothesis of Whitney equisingularity implies that the family of curves has a simultaneous smooth resolution, so we can assume there exists a family of maps $F(y, t) \colon \mathbf{C} \times \mathbf{C} \to \mathbf{C} \times \mathbf{C}^n$, $F(y, t) = (y, F_y(t))$, such that F is a homeomorphism, and is an equivalence off the parameter space. Here we assume that the parameter space is one dimensional only for simplicity of notation and argument.

The existence of F puts us into the setting of [6]. Then, not only can we ask that X be bi-Lipschitz equisingular, but also that the family of maps F be bi-Lipschitz left trivial. This question has the advantage that it is true iff a certain canonical vectorfield is Lipschitz. We know that

$$f \circ F = 0$$

This implies that

$$\frac{\partial}{\partial y} + \sum_i \frac{\partial F_y}{\partial y} \frac{\partial}{\partial z_i}$$

is tangent to X at all points where X is smooth, and is tangent to the y axis. Note that this vector field is only meromorphic on X. Though the $\frac{\partial F_y}{\partial t}$ depend holomorphically on (y, t), (y, t) depend meromorphically on (y, z).

In order to show that these are Lipschitz on X, in contrast to the last section, we work directly with the Lipschitzian saturation of the ring of X, showing that the components of the field are in the Lipschitzian saturation. Here we follow [19], but work in local coordinates. Some needed notation which **diverges somewhat from the notation of the previous section**. Given $h \in \mathcal{O}_{X,x}$, $X \subset \mathbf{C}^n$, the double of h denoted h_D is the element $h(z) - h(z')$ of $\mathcal{O}_{X \times X,(x,x)}$. Given a map $(f_1, \ldots, f_p)$ if we are interested in the double points of the map we form the ideal $(f_{1,D}, \ldots, f_{p,D})$. Denote this ideal by $I_D(f)$.

Given the local ring of an analytic set $\mathcal{O}_{X,x}$, let $(z_1, \ldots z_n)$ be the generators of the maximal ideal; let $N\mathcal{O}_{X,x}$ denote the normalization of $\mathcal{O}_{X,x}$. Then $h \in N\mathcal{O}_{X,x}$ is in the Lipschitizian saturation of $\mathcal{O}_{X,x}$, if h_D is in $\overline{I_D(z_1, \ldots z_n)}$, where the integral closure is taken in the local ring of the product of the normalizations.

One of the goals of this note to find a set of invariants depending only on the members of the family whose independence of parameter is necessary and sufficient for the bi-Lipschitz equisingularity of the family.

A key role is played in this by the second Segre number of $I_D(F_y)$, denoted $s_2(I_D(F_y))$, and the multiplicity of the pair $(I_D(F_y), I_\Delta)$, denoted $e((I_D(F_y), I_\Delta))$, where I_Δ is the ideal that defines the diagonal in $\mathbf{C} \times \mathbf{C}$. The Segre numbers were defined in [10], and used in [6] in a similar context. For information about the multiplicity of a pair of ideals or modules see [7], [8], [9]. Both numbers have a geometric interpretation in the curve case which we now describe.

Proposition 3.1. *Suppose* $f : \mathbf{C}, 0 \to \mathbf{C}^n, 0$ *is the normalization of its image. Suppose* $\pi : \mathbf{C}^n \to \mathbf{C}^2$ *is a generic projection, and the image of* $\pi \circ f$ *is* X. *Then* $s_2(I_D(f)) = \mu(X) + m(X) - 1$, $e(I_D(f), I_\Delta) = \mu(X)$.

Proof. For both invariants we are working in the ring $\mathcal{O}_2$. Then there exists a Z-open set of 2 linear combinations of our generators g_1 and g_2 which generate a reduction of $I_D(f)$, ie. a subideal of $I_D(f)$ with the same integral closure. Choosing two linear combinations of our generators $(f_i(t) - f_i(t'))$, $1 \le i \le n$, amounts to choosing a linear projection π_2 from $\mathbf{C}^n \to \mathbf{C}^2$, and the ideal so generated is just $I_D(\pi \circ f)$. So $s_2(I_D(f)) = s_2(I_D(\pi_2 \circ f))$; in [6] it is proved that this is $\mu(X) + m(X) - 1$, by showing that if we deform f so that the image of $\pi_2 \circ f$ has only nodes, then $s_2(I_D(\pi_2 \circ f))$ breaks up into a sum in which each node counts twice, as there are two double points in the source, and the ramification locus of $\pi_1 \circ f$ breaks into $m(X) - 1$ points, where π_1 is a generic projection to $\mathbf{C}$. (See Cor. 3.3 of [6] for a discussion.)

Again $e(I_D(f), I_\Delta) = e(I_D(\pi_2 \circ f), I_\Delta)$ since $I_D(\pi_2 \circ f)$ is a reduction of $I_D(f)$. Then again varying f so that the image of $\pi_2 \circ f$ has only nodes, the multiplicity polar theorem shows that each double point of $\pi_2 \circ f_t$ counts once toward $e(I_D(\pi_2 \circ f), I_\Delta)$. (Neither ideal has a polar curve as both have only two generators.) So, $e(I_D(f), I_\Delta) = 2\delta = \mu$. $\square$

Remark. Artin-Nagata have shown that the number of double points is the dimension over $\mathbf{C}$ of the quotient of I_Δ by $I_D(\pi_2 \circ f)$, so this gives another way of computing $e(I_D(f), I_\Delta)$. See [6] discussion before Cor. 3.5.

The fact that choosing two generic elements of $I_D(f)$ amounts to choosing a generic projection to $\mathbf{C}^2$ may give some insight into Mostowski's construction of Lipschitz stratifications. For example, looking at the double point curve of a generic linear projection from $X^2, 0 \subset \mathbf{C}^n$ to $\mathbf{C}^3$, is related to looking at the polar curve of the ideal $I_D(z_1, \ldots, z_n)$, where $(z_1, \ldots, z_n)$ are generators of the maximal ideal of $X, 0$.

Proposition 3.2. *Suppose $F(y, t): \mathbf{C} \times \mathbf{C} \to \mathbf{C} \times \mathbf{C}^n$ is a homeomorphism onto its image, defining a family of irreducible curves with singular locus $\mathbf{C} \times 0$. Then the following are equivalent:*

1) *There exists a set U of generic projections such that $(y, \pi \circ F_y)$ defines an equisingular family of plane curves.*
2) *$e(I_D(F_y), I_\Delta)$ is independent of y.*
3) *$s_2(I_D(F_y))$ is independent of y.*

Proof. By the discussion above 1) clearly implies 2) and 3) since the curves are equisingular, μ and m are independent of parameter.

We can choose a projection to $\mathbf{C}^2$ such that for a Z-open set U of parameters the ideal $I_D(\pi_2 \circ F_y)$ is a reduction of $I_D(F_y)$ as ideals on $U \times \mathbf{C}$. We can also ask that it be a reduction when restricted to $0 \times \mathbf{C}$. In general, this is not enough to say that $I_D(\pi_2 \circ F_y)$ is a reduction of $I_D(F_y)$ as ideals on $\mathbf{C} \times \mathbf{C}$, but either of the hypotheses of 2) or 3) imply it is. (For 2) see [10]. For 3) the argument uses the multiplicity polar theorem again; the numerical condition implies that the ideal has no polar curve, hence there is no vertical component in the exceptional divisor in the blowup of the ideal.)

In particular $\mathbf{C} \times \Delta_{\mathbf{C}^2} = V(I_D(F_y)) = V(I_D(\pi_2 \circ F_y))$, so $(y, \pi_2 \circ F_y)$ is homeomorphic onto its image, which has constant Milnor number, hence is an equisingular family. $\square$

The next step is to show that 1) implies the Lipschitz equisingularity of the family defined by F_t. This has already been proved by Mostowski, as his work shows that with this hypothesis, a Lipschitz stratification of X is given by the y-axis and its complement. Here we use integral closure methods to achieve the same result.

Proposition 3.3. *Suppose $F(y, t): \mathbf{C} \times \mathbf{C} \to \mathbf{C} \times \mathbf{C}^n$ is a homeomorphism onto its image, defining a family of irreducible curves with singular locus*

$\mathbf{C} \times 0$, *and any of the conditions of proposition 2 hold. Then the vector field*

$$\frac{\partial}{\partial y} + \sum_i \frac{\partial F_y}{\partial y} \frac{\partial}{\partial z_i}$$

is Lipschitz.

Proof. We may assume $F(y, t)$ has the form $(y, f_0(t) + \sum h_i(t)y^i)$ where f_0, and h_i are n-tuples of functions. Note that $\frac{\partial F_y}{\partial y}$ is $\sum i h_i(t)y^{i-1}$, and its double is

$$\sum i(h_i(t)y^{i-1} - h_i(t')y'^{i-1})$$

We want to show that this n-tuple of elements is in $\overline{I_D(F(y, t))}$, which is the integral closure of the ideal generated by $(y - y', f_0(t) - f_0(t') + \sum h_i(t)y^i - h_i(t')y'^i)$. Using the $(y - y')$ term, we see that this is the same as showing:

$$\sum i(h_i(t)y^{i-1} - h_i(t')y^{i-1}) = \sum i(h_i(t) - h_i(t'))y^{i-1}$$

is in the integral closure of the ideal generated by $(y - y', f_0(t) - f_0(t') + \sum(h_i(t) - h_i(t'))y^i)$

We will show that

$$\sum i(h_i(t) - h_i(t'))y^{i-1} \subset \overline{(f_0(t) - f_0(t') + \sum(h_i(t) - h_i(t'))y^i))}$$

along (y, t, t'). Consider the family of ideals on $y \times \mathbf{C}^2$ as y varies, gotten by taking the underlying ideal on the right hand side and restricting to $y \times \mathbf{C}^2$. Using our hypothesis, by [3], we know that the family of curves is equisaturated. Further by Lemme IV.1.3 of [3], we know that the saturation of the ring of each fiber contains all monomials in t that appear in the parameterization of the curve. Since generically all of the monomials in all of the h_i and f_0 appear, all of these monomials have their doubles in the integral closure of $(f_0(t) - f_0(t') + \sum(h_i(t) - h_i(t'))y^i))$. Since the Segre numbers of these ideals are independent of y, it follows by the principle of specialization of integral dependence ([10]) that all of these monomials are actually in $\overline{(f_0(t) - f_0(t') + \sum(h_i(t) - h_i(t'))y^i))}$ along (y, t, t'). This implies the desired inclusion for $\sum i(h_i(t) - h_i(t'))y^{i-1}$. This finishes the proof. $\square$

It is not surprising that the results of this section are related to others appearing in the literature, as space curves have been much studied. Teissier, in [20], studied the ideal of the diagonal $I_D(f)$ in the case of plane curves, showing that the ideal is the product of the ideal defining the diagonal Δ in IC^2 with a residual ideal (p. 118 [20]); it is not hard to see that the multiplicity of this residual ideal is the multiplicity of the pair of ideals $I_D(f), I_\Delta$. Teissier also mentions (p. 120 [20]) that, in general, the multiplicity of this residual ideal is related to the δ invariant of the image of the curve under a generic projection to the plane. Other results, related to our proposition 3.3 are proved in [21] p357–361 and use the principle of specialization of integral dependence for ideals of finite colength. The value of our approach lies in the possibility of extending the theory beyond the curve case, as the machinery of Segre numbers, the multiplicity of pairs of ideals and the multiplicity polar theorem are not limited to ideals of finite colength.

References

1. Adkins, William A. Weak normality and Lipschitz saturation for ordinary singularities. *Compositio Math.* **51(2)** (1984) 149–157.
2. L. Birbrair, Local bi-Lipschitz classification of 2-dimensional semialgebraic sets. *Houston J. Math.* **25(3)** (1999) 453–472.
3. J. Briançon, A. Galligo, M. Granger, Déformations équisingulières des germes de courbes gauches réduites *Mémoires de la Société Mathématique de France* Sér. **2(1)** (1980) 1–60.
4. T. Gaffney, Integral closure of modules and Whitney equisingularity, *Invent. Math.* **107** (1992) 301–22.
5. T. Gaffney, Polar multiplicities and equisingularity of map germs. *Topology* **32(1)** (1993) 185–223.
6. T. Gaffney, L^0-equivalence of maps, Math. Proc. Cambridge Philos. Soc. **128(3)** (2000) 479–496.
7. T. Gaffney, Polar methods, invariants of pairs of modules and equisingularity, Real and Complex Singularities- Sao Carlos (2002), Ed. T. Gaffney and M.Ruas, *Contemp. Math.* **354** Amer. Math. Soc., Providence, RI, June (2004) 113–136.
8. T. Gaffney, *The multiplicity of pairs of modules and hypersurface singularities*, Real and Complex Singulartities – Sao Carlos, *Trends in Mathematics*, Birkhauser Verlag Basel/Switzerland (2004) 143–168.
9. T. Gaffney, The Multiplicity Polar Theorem and Isolated Singularities, *J. Alg. Geom.* math.AG/0509285. To appear.
10. T. Gaffney, and R. Gassler, Segre numbers and hypersurface singularities, *J. Alg. Geom.* **8** (1999) 695–736.
11. T. Gaffney, and M. Vitulli, *Weak subintegral closure of ideals.* Math arXiv:0708.3105 25 pages.
12. M. Lejeune-Jalabert and B. Teissier, Clôture intégrale des idéaux et equisingularité, Séminaire Lejeune-Teissier, Centre de Mathématiques École Polytechnique, (1974) Publ. Inst. Fourier St. Martin d'Heres, F–38402 (1975).
13. J. Lipman, Relative Lipschitz-saturation. *Amer. J. Math.* **97(3)** (1975) 791–813.
14. T. Mostowski, A criterion for Lipschitz equisingularity. *Bull. Polish Acad. Sci. Math.* **37(1–6)** (1989), 109–116 (1990).
15. T. Mostowski, Tangent cones and Lipschitz stratifications. Singularities (Warsaw, 1985), 303–322, Banach Center Publ. **20** PWN, Warsaw, 1988.

16. A. Parusinski, Lipschitz stratification of real analytic sets. Singularities (Warsaw, 1985), 323–333, Banach Center Publ. **20** PWN, Warsaw, 1988.

17. A. Parusinski, Lipschitz properties of semi-analytic sets. *Ann. Inst. Fourier* (Grenoble) **38(4)** (1988) 189–213.

18. F. Pham, *Fractions lipschitziennes et saturation de Zariski des algèbres analytiques complexes.* Exposé d'un travail fait avec Bernard Teisser. Fractions lipschitziennes d'une algèbre analytique complexe et saturation de Zariski, Centre Math. lÉcole Polytech., Paris, 1969. Actes du Congrès International des Mathématiciens (Nice, 1970), Tome 2, pp. 649–654. Gauthier-Villars, Paris, 1971.

19. F. Pham and B. Teissier, *Fractions lipschitziennes d'une algébre analytique complexe et saturation de Zariski*, Centre de Mathématiques de l'Ecole Polytechnique (Paris), http://people.math.jussieu.fr/ teissier/old-papers.html, June 1969.

20. B. Teissier, *II-Resolution Simultanee et Cycles Evanescents*, in Séminaire sur les Singularités des Surfaces. Centre de Mathématiques de l'École Polytechnique, Palaiseau, 1976–1977. Edited by Michel Demazure, Henry Charles Pinkham and Bernard Teissier. Lecture Notes in Mathematics **777** Springer, Berlin, 1980.

21. B. Teissier, Variétés polaires. II. Multiplicités polaires, sections planes, et conditions de Whitney, in Algebraic Geometry (La Rábida, 1981), 314–491, *Lecture Notes in Math.* **961** Springer, Berlin, 1982.

22. O. Zariski, General theory of saturation and of saturated local rings. II. Saturated local rings of dimension 1. *Amer. J. Math.* **93** (1971) 872–964.

Terence Gaffney
Department of Mathematics
Northeastern University
Boston, MA 02115
USA
gaff@neu.edu

10

Solutions to PDEs and stratification conditions

TERENCE GAFFNEY

Abstract

In this note we show the connection between Verdier's condition W and the smoothness of solutions of PDEs obtained by the method of characteristics in the two stratum case.

1. Introduction

If we consider the behavior of a partial differential equation (PDE) on a singular space, a natural question to ask is "when is the PDE and its solutions similar along a singular stratum?" A solution to this problem will involve stratifying a space relative to the PDE, as well as finding a good definition of "similar". In this note we consider the simplest case–those PDEs for which the method of characteristics applies.

We show that there is an intimate connection between the applicability of this method and Verdier's condition W.

2. The method of characteristics and W stratifications

The method of characteristics is a tool for reducing the solution of a PDE to the solution of an ODE (ordinary differential equation). To recall how this method works consider the first order linear homogeneous PDE:

$$\sum\nolimits_{i=1}^{n} h_i(x)\frac{\partial u}{\partial x_i} = 0 \qquad *$$

2000 Mathematics Subject Classification 00000.

Here we are working on an open subset U of $\mathbf{K}^n$, where $\mathbf{K} = \mathbf{C}$ or $\mathbf{R}$, and h_i are analytic functions. To apply the method of characteristics, consider the vector field

$$V = \sum_{i=1}^{n} h_i(x) \frac{\partial}{\partial x_i}.$$

The integral curves of V are called characteristic curves. Note that if $u(x)$ is a solution of $*$, then $D(u \circ \phi) = 0$, where ϕ is any integral curve of V, hence u is constant on ϕ. If we can find a smooth hypersurface H transverse to V at every point, and we can specify the value of u on H, then we can find the value of u at an point on U by flowing back to H using the characteristic curves and evaluating u on H. H is a called a non-characteristic since it is nowhere tangent to V.

We illustrate these ideas with an example adapted from [4]. The transport equation is the initial value problem

$$u_t + \vec{v} \cdot \nabla u = 0$$
$$u(0, \vec{x}) = h(\vec{x}).$$

These equations model the flow of a fluid along a domain in $\mathbf{R}^n$ with constant velocity $\vec{v} = \langle v_1, \ldots, v_n \rangle$, with $u(t, \vec{x})$ the density at time t and position $\vec{x}$ of a contaminant.

In this problem the associated vector field is

$$\vec{V} = \frac{\partial}{\partial t} + \sum_{i=1}^{n} v_i \frac{\partial}{\partial x_i}$$

which is clearly a well defined vector field on $\mathbf{R}^{n+1}$.

The characteristic curves are just the parallel lines $\vec{r}(s) = (0, \vec{x}) + s\langle 1, v_1, \ldots, v_n \rangle$, and the non-characteristic is the hyperplane $t = 0$. If we start at $(t, \vec{x})$ and flow back to the non-characteristic, we end at $(0, \vec{x} - t\vec{v})$. For the initial condition $h(\vec{x})$, the solution becomes $u(t, x) = h(\vec{x} - t\vec{v})$. As the flow is analytic, the solution has the same smoothness as h.

(For further information about the method of characteristics see [4], or [3].)

Now we describe how this transfers to PDEs on singular spaces.

Set-up: Let $Y = \mathbf{K}^k \times 0 \subset X^{d+k} \subset \mathbf{K}^k \times \mathbf{K}^n$, X the germ of an analytic set, $X_0 := X - Y$ smooth. Assume y are coordinates on Y and z are coordinates on $\mathbf{K}^n$, m_n the ideal generated by the z coordinates. If the ideal $I(X)$ defining X is generated by $\{f_1, \ldots, f_p\}$, denote by f the mapgerm with components $(f_1, \ldots, f_p)$. Denote by $J M_z(f)$ the module generated by the partial derivatives of f with respect to the z coordinates , and by $\overline{J M_z(f)}$ its integral closure, if working over $\mathbf{C}$ or its real integral closure if working over $\mathbf{R}$. (Cf. [2] for information on these two closure notions.)

Given a linear PDE as above, if $x \in S$ where $S \in \{X_0, Y\}$, we ask that V, the associated vector field, is tangent to S at every x.

In applying the method of characteristics, we need a non-characteristic hypersurface, and we can ask about the smoothness of any solution. If the method of characteristics applies to our set-up, then the solutions of the PDE along the singular stratum Y are similar, as we can use the flow to "push" H along.

In this note we use Verdier's condition W as our main stratification condition. Recall how this is defined.

Suppose A, B are linear subspaces at the origin in $\mathbf{K}^N$, then define the distance from A to B as:

$$\text{dist}(A, B) = \sup_{\substack{u \in B^{\perp} - \{0\} \\ v \in A - \{0\}}} \frac{|(u, v)|}{\|u\| \, \|v\|}.$$

In the applications B is the "big" space and A the "small" space. (Note that $\text{dist}(A, B)$ is not in general the same as $\text{dist}(B, A)$.) Suppose $Y \subset \bar{X}$, where X, Y are strata in a stratification of an analytic space, and $\text{dist}(TY_0, TX_x) \leq C\text{dist}(x, Y)$ for all x close to Y. Then the pair (X, Y) satisfies Verdier's condition W at $0 \in Y$.

This notion is useful for producing rugose vectorfields as extensions of constant vectorfields on Y. A statified vector field v is called **rugose** near $y \in S_\alpha$, where S_α is a stratum of X, when there exists a neighbourhood W_y of y and a constant $K > 0$, such that

$$\|v(y') - v(x)\| \leq K\|y' - x\|$$

for every $y' \in W_y \cap S_\alpha$ and every $x \in W_y \cap S_\beta$, with $S_\alpha \subset \overline{S_\beta}$.

In [2], an integral closure formulation of this condition is given, namely that the pair (X_0, Y) satisfies condition W at the origin if and only

$$\frac{\partial f}{\partial y_i} \in \overline{m_n J M_z(f)}$$

for all i, $1 \leq i \leq k$.

Our first goal is to construct a PDE on X which is solvable by the method of characteristics. This will require more notation. Let $I = (i_1, \ldots, i_{n-d})$, $J = (j_1, \ldots, j_{n-d})$ be collections of increasing integers $1 \leq i_s \leq n$, $1 \leq j_s \leq p$. Let $D_z f$ denote the matrix of partial derivatives of f with respect to the z coordinates. Let $j_{I,J}(f)$ denote the minor of size $n - d$ of $D_z f$ formed using the columns indexed by I and rows indexed by J. Let $J_z(f, n - d)$ denote the ideal generated by these minors using all valid indexing sets. If in forming $j_{I,J}(f)$ we replace the i_s column of $D_z f$ by $\frac{\partial f}{\partial y_i}$ denote the resulting minor by $j_{I,s,J}(f)$

Notice that we have the

Proposition 2.1. *In the above set-up we have the relations*

$$\| j_{I,J}(f) \|^2 \frac{\partial f}{\partial y_i} = \sum_{i_s \in I} \overline{j_{I,J}(f)} j_{I,s,J}(f) \frac{\partial f}{\partial z_{i_s}}.$$

Proof. If $j_{I,J}(f)$ is identically 0 on a component there is nothing to prove for that component; so assume it's not. Then the result follows if we can show

$$j_{I,J}(f) \frac{\partial f}{\partial y_i} = \sum_{i_s \in I} j_{I,s,J}(f) \frac{\partial f}{\partial z_{i_s}}.$$

Work on the Z-open set U of points where $j_{I,J}(f)$ is not zero. We know the matrix obtained by adding the column $j_{I,J}(f) \frac{\partial f}{\partial y_i} - \sum_{i_s \in I} j_{I,s,J}(f) \frac{\partial f}{\partial z_{i_s}}$ to the matrix $D_z(f)$ still has the same rank $n - d$ as $D_z(f)$ has. But this is a contradiction unless the added column vector is zero on U, for the new matrix contains the submatrix of rank $n - d$ defined by I and J, and by Cramer's rule the entries of the added vector in the rows indexed by J are 0. So if any entries of the added row are non-zero, the rank of the new matrix must be at least $n - d + 1$. □

Now fix i and add together the resulting equations to obtain:

$$\sum_{I,J} \| j_{I,J}(f) \|^2 \frac{\partial f}{\partial y_i} = \sum_{I,J,i_s \in I} \overline{j_{I,J}(f)} j_{I,s,J}(f) \frac{\partial f}{\partial z_{i_s}}.$$

Dividing by the coefficient of $\frac{\partial f}{\partial y_i}$ we obtain:

$$\frac{\partial f}{\partial y_i} - \sum_{I,J,i_s \in I} \frac{\overline{j_{I,J}(f)} j_{I,s,J}(f)}{\sum_{I,J} \| j_{I,J}(f) \|^2} \frac{\partial f}{\partial z_{i_s}} = 0.$$

This condition implies that the vector field

$$V_i = \frac{\partial}{\partial y_i} - \sum_{I,J,i_s \in I} \frac{\overline{j_{I,J}(f)} j_{I,s,J}(f)}{\sum_{I,J} \| j_{I,J}(f) \|^2} \frac{\partial}{\partial z_{i_s}}$$

is tangent to X_0. Note that the field is well defined on X_0 as the denominator $\sum_{I,J} \| j_{I,J}(f) \|^2$ is only 0 on the singular locus of X which is Y by the set-up. We define the field on all of X by setting $V_i | Y = \frac{\partial}{\partial y_i}$.

Given the field we can write the PDE it is linked to in two different forms, both of which have the same solutions.

$$\sum_{I,J} \|j_{I,J}(f)\|^2 \frac{\partial u}{\partial y_i} - \sum_{I,J,i_s \in I} \overline{j_{I,J}(f)} j_{I,s,J}(f) \frac{\partial u}{\partial z_{i_s}} = 0. \qquad (P_i)$$

$$\frac{\partial u}{\partial y_i} - \sum_{I,J,i_s \in I} \frac{\overline{j_{I,J}(f)} j_{I,s,J}(f)}{\sum_{I,J} \|j_{I,J}(f)\|^2} \frac{\partial u}{\partial z_{i_s}} = 0. \qquad (P_i')$$

The first equation has analytic coefficients, while the second, though it only has meromorphic coefficients, defines the vector field we integrate to get the characteristics. Each of the components of f are solutions to both equations, corresponding to the initial condition $u(0, z) = 0$. We will see shortly that if the pair (X_0, Y) satisfies W then, because the PDE is linear, the solution is determined by its values on $X \cap 0 \times \mathbf{K}^n$.

Theorem 2.2. *Suppose (X_0, Y) satisfies W, then $X \cap (0 \times \mathbf{C}^n)$ is non-characteristic for each P_i' and the vector field associated to each P_i' is rugose. If $h(0, z)$ is the initial condition for P_i', then $u = h \circ \Phi^{-1}$ is the unique solution to the PDE P' on X, where Φ is the flow defined by the rugose field.*

Proof. From the definition of V_i, it is clear that V_i is not tangent to $X \cap (0 \times \mathbf{C}^n)$ on some neighborhood of the origin, hence $X \cap (0 \times \mathbf{C}^n)$ is non-characteristic. We show that V_i is rugose. It is clearly real analytic on X_0 and on Y and tangent to both by construction. It suffices to show that

$$\left\| \frac{\overline{j_{I,J}(f)} j_{I,s,J}(f)}{\sum_{I,J} \|j_{I,J}(f)\|^2}(y, z) \right\| \le C \|(z_1, \dots, z_n)\|$$

for each I, J. In turn to show this, it suffices to show

$$\|\overline{j_{I,J}(f)} j_{I,s,J}(f)(y, z)\| \le C \|(z_1, \dots, z_n)\| \sup_{I,J} \|j_{I,J}(f)(y, z)\|^2$$

$$\|j_{I,s,J}(f)(y, z)\| \le C \|(z_1, \dots, z_n)\| \sup_{I,J} \|j_{I,J}(f)(y, z)\|$$

The condition that W holds implies

$$\frac{\partial f}{\partial y_i} \in \overline{m_n J M_z(f)}.$$

By the determinantal criterion of [2] this implies the ideal generated by the determinants of size $n - d$ formed by taking minors of a matrix of generators of $m_n J M_z(f)$ contains those minors obtained by replacing one of the columns of the matrix of generators $m_n J M_z(f)$ by $\frac{\partial f}{\partial y_i}$.

Applying the curve criterion, we see that this is equivalent to $j_{I,s,J}(f)$ in the integral closure of $m_n J_z(f, n - d)$; the other factors of z_i canceling off. In turn this implies

$$\|j_{I,s,J}(f)(y,z)\| \le C\|(z_1,\dots,z_n)\| \sup_{I,J} \|j_{I,J}(f)(y,z)\|$$

which shows V_i are rugose.

Since V_i is rugose it can be integrated to give a flow Φ ([5]), which is real analytic on X_0 and continuous on X.

Then $h \circ \Phi^{-1}$ is a solution to P' which satisfies the initial condition. If u is another continuous solution which also satisfies the initial condition, then $h \circ \Phi^{-1} - u$ is a solution by linearity and is 0 on $X \cap 0 \times \mathbf{K}^n$. Since $h \circ \Phi^{-1} - u$ is a solution, it is constant on characteristics hence is identically 0 on X_0, hence 0 on X. $\qquad\square$

Note that in general the solution to P_i or P_i' is as smooth as Φ^{-1} is. To see this let $h(0,z) = z_j$; then the solution with this initial condition is just the j-th component of Φ^{-1}. So the geometry of X puts restrictions on the smoothness in general of the solutions of the PDE P_i.

There is a converse to the previous theorem.

Proposition 2.3. *Suppose X is an analytic set as in the set-up for this section. Suppose that the vector field associated to each P_i' is rugose; then (X_0, Y) satisfies W.*

Proof. From the last proof, we can see that the condition that the fields V_i are rugose implies

$$\frac{\partial f}{\partial y_i} \in \overline{m_n J M_z(f)}.$$

In turn, this implies (X_0, Y) satisfies W. $\qquad\square$

The last proposition has an extension.

Theorem 2.4. *Suppose X is an analytic set as in the set-up for this section. Suppose for each tangent vector to Y, $\frac{\partial}{\partial y_i}$, there exists a linear homogenous first order PDE, with associated rugose vector field whose restriction to Y is $\frac{\partial}{\partial y_i}$; then W holds for the pair (X_0, Y).*

Proof. This proof is similar to the first half of the proof of Theorem 2.5 of [2], but easier. In the complex case we assume our vector fields are complex valued. (The P_i' are good examples in the complex case.) The hypothesis implies that on a neighborhood of the origin we can find k linearly independent fields, which restrict to a basis of TY_y, which are tangent to X_0 thereby defining a k-dimensional subspace $T_{k,x}$ of TX_x at $x \in X_0$. Further, we can find linear combinations of these fields so that we can assume they have the form:

$$v_i = \frac{\partial}{\partial y_i} + \sum_j a_{i,j} \frac{\partial}{\partial z_j}$$

Now we are going to compute the distance between TY_y and $TX_{y,z}$.   □

A basis for the vectors which are orthogonal to $T_{k,x}$ is given by

$$V^\perp = \left\{ \frac{\partial}{\partial z_j} - \sum_i \overline{a_{i,j}} \frac{\partial}{\partial y_i} \right\}.$$

Now

$$\text{dist}(TX_{y,z}, TY_y) \le \text{dist}(T_{k,x}, TY_y) \le C\sup_{i,j} \|a_{i,j}\| \le C'\|z\|.$$

Together Theorems 2.2 and 2.4 offer another characterization of condition W in the two strata case.

This result shows that if Y has dimension 1, and W fails at the origin, then the method of characteristics does not apply if we insist that the associated field be rugose.

So far in Theorem 2.2, in the case where (X_0, Y) satisfies W, we have exhibited a linear homogeneous PDE which can be solved by the method of characteristics. In Theorem 2.4 we showed W holds for the pair (X_0, Y), if there exists a linear homogenous first order PDE, with associated rugose vector field whose restriction to Y is $\frac{\partial}{\partial y_i}$. It is natural to ask that if (X_0, Y) satisfies W, are there other examples of PDEs of the type described in 2.4 other than those in 2.2? We give a set of examples indicating these should be plentiful, and giving some idea of how to find them.

The context of our examples is the paper [1]. This deals with families of hypersurfaces defined by functions $F(y, z)$, $F: \mathbf{K}^k \times \mathbf{K}^n \to \mathbf{K}$ where the function can be either real or complex analytic. We denote $F(0, z)$ by $f_0(z)$ which we assume has an algebraically isolated singularity at the origin. We can associate to f_0 its Newton polygon, giving rise to the Newton filtration of C_z of germs of analytic functions at the origin. We denote by $\mathcal{A}_l$ all germs of filtration l or greater. We assume further that f_0 is *fit* (ie. its Newton polyogon intersects each of the coordinate axes), and is *Newton non-degenerate*. We also assume that $F(y, z)$ is a deformation of $f_0(z)$ by terms above or on the Newton boundary. In this set-up it is shown in [1] that the pair $(X - Y, Y)$ satisfies W at the origin, where X is the hypersurface defined by F and Y, as usual, is the parameter space $\mathbf{K}^k \times 0$.

We say the level $\mathcal{A}_l$ is fit, if all of its vertices are lattice points of $\mathbf{R}^n$. As discussed on p 339 of [1] if $\mathcal{A}_l$ is fit, so is $\mathcal{A}_{rl}$, for all integers $r \ge 1$. Denote the set of monomials corresponding to the vertices of $\mathcal{A}_l$ by $\text{ver}(\mathcal{A}_l)$. Further, the filtration on C_z extends to a filtration on $C_{z,y}$; we denote germs of filtration l by $\mathcal{A}_{l,z,y}$. In the current set-up, it is shown in [1], that the condition of theorem 2 on p340 holds for some positive integer l and for all j. This condition is:

$$\mathrm{ver}(\mathcal{A}_{lm}) \cdot \frac{\partial F}{\partial y_j} \subset (\mathcal{A}_{lm,z,y}) \cdot \left(z_i \frac{\partial F}{\partial z_i} \right)$$

where m is the filtration level of f_0, and $(z_i \frac{\partial F}{\partial z_i})$ is the ideal generated by the germs $z_i \frac{\partial F}{\partial z_i}$. Let $\vec{\mathbf{V}}_{lm}$ be the tuple of monomials made up of the elements of $\mathrm{ver}(\mathcal{A}_{lm})$, and let ρ_{lm} denote $\| \vec{\mathbf{V}}_{lm} \|^2$. The above filtration condition then implies that there exists a system of PDEs on X of the form

$$\frac{\partial u}{\partial y_j} - \sum_i \frac{h_{j,i}}{\rho_{lm}} z_i \frac{\partial u}{\partial z_i} = 0, 1 \leq j \leq k$$

with the filtration of $h_{j,i}$ greater than or equal to the filtration of ρ_{lm}. This ensures that the associated vector field is rugose, which finishes the example.

A partial differential operator of the form $h \frac{\partial}{\partial y} + \sum_i h_i \frac{\partial}{\partial z_i}$ must satisfy the equation $h \frac{\partial f}{\partial y} + \sum_i h_i \frac{\partial f}{\partial z_i} = 0$ for the operator to be well defined on the cotangent bundle of X in our set-up, where f is any element of the ideal of X. Thus, $h \frac{\partial f}{\partial y}$ lies in the module generated by the partials of F with respect to the z variables. Looking back over the above set of examples, shows that being able to prove that the method of characteristics applies yielding rugose solutions involves knowing where $h \frac{\partial f}{\partial y}$ sits in this module.

References

1. J. Damon and T. Gaffney, *Topological Triviality of Deformations of Functions and Newton Filtrations*, Inventiones **72** (1983) 335–358.
2. T. Gaffney, *Integral closure of modules and Whitney equisingularity*, Invent. Math. **107** (1992) 301–22.
3. L. Hörmander, *The analysis of linear partial differential operators. I. Distribution theory and Fourier analysis.* Grundlehren der Mathematischen Wissenschaften [Fundamental Principles of Mathematical Sciences], **256**. Springer-Verlag, Berlin, 1983.
4. R. C. McOwen, *Partial Differential Equations: Methods and Applications*, Pearson Education Inc. New Jersey, 2003.
5. J.-L. Verdier, *Stratifications de Whitney et thorme de Bertini-Sard.* Invent. Math. **36** (1976) 295–312.

Terence Gaffney
Department of Mathematics
Northeastern University
Boston, MA 02115
USA
gaff@neu.edu

11

Real integral closure and Milnor fibrations

TERENCE T. GAFFNEY AND RAIMUNDO ARAÚJO
DOS SANTOS

Abstract

We give a condition to guarantee the existence of a Milnor fibration for real map germs of corank 1, which include cases that are not L-maps in the sense of Massey. Our approach exploits the structure of a family of functions.

1. Introduction

The existence of Milnor fibrations for non-isolated singularities has been studied by many mathematicians using different approaches, for example, in [PS] the authors studied the existence of real Milnor fibrations in the Milnor tube and in spheres where the projection map is given by $\dfrac{f}{\|f\|}$, or the **strong Milnor fibration** (for further details see [AT], [AR], [AR1], [RSV] and [Se1]). In [Se2] Seade presented a beautiful survey about the existence of real Milnor fibrations for non-isolated singularities as well as interesting new results. In [AT] the authors studied the existence of strong Milnor fibrations for non-isolated singularities using the idea of open book structures in higher dimensions.

In [DM], Massey studies the existence of real Milnor fibrations on the Milnor tube involving the singular zero level of map, as approached by Lê D. Tráng in [LD] and H. Hamm and Lê in [HL].

2000 *Mathematics Subject Classification* Primary 32S55, 32C18, 14P25; Secondary 32S05, 32S10, 14P05.
The second author would like to thank the Brazilian agencies FAPESP/São Paulo and the CNPq, grant PDE, number: 200643/2007-0. The former for supporting his first four month stay in U.S. in 2006, when the post-doc project started, and the latter for supporting his stay in U.S. from October/2007 to July/2008 during the post-doc. Also the CNPq PQ-grant number 305183/2009-5 Thanks !

He considers two conditions called **Milnor condition (a)** and **Milnor condition (b)**, which are sufficient to guarantee the existence of Milnor fibrations for real analytic map germs $f : \mathbb{R}^n, 0 \to \mathbb{R}^k, 0$ for $n \geq k \geq 2$. In what follows we will give a short description of these conditions:

Let $f : \mathbb{R}^n, 0 \to \mathbb{R}^k, 0, f(x) = (f_1(x), f_2(x), \ldots, f_k(x))$ a real analytic map germ. Define the following analytic sets:

The variety of f, denoted $V(f)$, and defined as $V(f) := \{x \in \mathbb{R}^n, 0 : f_1(x) = f_2(x) = \cdots = f_k(x) = 0\}$. Let $\sum(f)$ denote the critical set of f, the points $\{x \in \mathbb{R}^n, 0 : \nabla f_1(x), \ldots, \nabla f_k(x),$ are linearly dependent$\}$.

Now let r be the function given by the square of distance from the origin and define $\mathcal{B} = \{x \in \mathbb{R}^n, 0 : \nabla f_1(x), \ldots, \nabla f_k(x), \nabla r(x),$ are linearly dependent$\}$. It is clear that $\sum(f) \subseteq \mathcal{B}$. Following Massey, we make the following definitions.

Definition 1.1. We say that a map germ f satisfies **Milnor's condition (a) at 0** if and only if $0 \notin \overline{\sum(f) - V(f)}$, i.e., $\sum(f) \subseteq V(f)$.

Definition 1.2. We say that the map germ f satisfies **Milnor's condition (b) at 0**, if and only if, **0** is an isolated point of $V(f) \cap \overline{\mathcal{B} - V(f)}$.

We say that $\epsilon > 0$ is a **Milnor radius for** f **at origin 0**, provided that $\overline{B_\epsilon} \cap (\overline{\sum(f) - V(f)}) = \phi$ (empty) and $\overline{B_\epsilon} \cap V(f) \cap \overline{\mathcal{B} - V(f)} \subseteq \{0\}$, where B_ϵ denotes the open ball in $\mathbb{R}^n$ with radius ϵ and $\overline{B_\epsilon}$ its topological closure in $\mathbb{R}^n$.

Theorem 1.3 [DM]. *Suppose that f satisfies Milnor's conditions (a) and (b) at* **0***, and let $\epsilon_0 > 0$ be a Milnor radius for f at* **0**. *Then, for all $0 < \epsilon \leq \epsilon_0$, there exists $\delta > 0$, $0 < \delta \ll \epsilon$, such that $f_| : \overline{B_\epsilon} \cap f^{-1}(B_\delta^*) \to B_\delta^*$ is a surjective, smooth, proper, stratified submersion and, hence, a locally-trivial fibration.*

Proof (Idea of proof). Since f has a Milnor radius $\epsilon_0 > 0$, we have that $\sum(f) \cap \overline{B_{\epsilon_0}} \subset V(f) \cap \overline{B_{\epsilon_0}}$. It means that, for all $0 < \epsilon \leq \epsilon_0$ the map $f_| : B_\epsilon \setminus V(f) \to R^k$ is a smooth submersion. Now from the Milnor condition (b), and the remark above, it follows that: for each ϵ there exists δ, $0 < \delta \ll \epsilon$, such that

$$f_| : S_\epsilon^{n-1} \cap f^{-1}(B_\delta - \{0\}) \to B_\delta - \{0\}$$

is a submersion on the boundary S_ϵ^{n-1} of the closed ball $\overline{B_\epsilon}$. Now, combining these two conditions we have that, for each ϵ, we can choose δ such that

$$f_| : \overline{B_\epsilon} \cap f^{-1}(B_\delta - \{0\}) \to B_\delta - \{0\} \quad (1)$$

is a proper smooth submersion. Applying the version of Ehresmann Fibration Theorem for the manifold with boundary $\overline{B_\epsilon}$, we get that it is a smooth locally trivial fibration.   □

Remark 1. *Instead of using the Ehresmann theorem for manifold with boundary, we also can apply Thom's 1st isotopy lemma by considering the map f_1 as a proper stratified submersion.*

Definition 1.4. We say that the map germ f satisfies the **strong Lojasiewicz inequality at the origin 0** or is an **L-map** if, and only if, there exists an open neighborhood $0 \in U$, and constants $c > 0, 0 < \theta < 1$, such that for all $x \in U$,

$$|f(x)|^\theta \le c \cdot \min_{\|(a_1,\ldots,a_k)\|=1} \|a_1 \nabla f_1(x) + \cdots + a_k \nabla f_k(x)\|$$

In [DM], the author proved that if f is an L-map at origin then Milnor's conditions (a) and (b) is satisfied.

It is easy to see that pairs of functions which come from a holomorphic function or which come from mixing holomorphic and anti-holomorphic functions are natural candidates to satisfy the L-map properties. (Cf. example 3.15 of [DM].)

The problem is that the class of L-maps may not behave very well for truly real analytic maps, as we now show.

Example 1.5. Consider the maps germs $G(x, y, z) = (x, y(x^2 + y^2 + z^2))$, $G : \mathbb{R}^3, 0 \to \mathbb{R}^2, 0$ and $H(x, y) = (x, y^3 + x^2 y)$, $H : \mathbb{R}^2, 0 \to \mathbb{R}^2, 0$. It is easy to see that $V(G) = Oz-$axis, $\Sigma(G) = \{(0, 0, 0)\}$ and $\Sigma(H) = \{(0, 0)\}$, but $\Sigma(H \circ G) = Oz-$axis. The map G is not an L-map at the origin; if it were, the defining inequality would hold along every curve. However, restricting to the curve $\phi(t) = (t, 0, 0)$, and using $a_1 = 0, a_2 = 1$, if the defining inequality holds, it implies that

$$|t|^\theta \le c \cdot |(\nabla y(x^2 + y^2 + z^2)) \circ \phi(t)| = c \cdot |t^2|$$

which is impossible. The map G is an L-map at other points of the $Oz-$axis, because G is a submersion there.

Example 1.6. Consider now the composed map $H \circ G = (x, (y(x^2 + y^2 + z^2))^3 + x^2(y(x^2 + y^2 + z^2)))$. The map $H \circ G$ is not L analytic along $Oz-$axis. Again, this is easily seen using curves. Suppose $(0, 0, z_0)$ a point on the $Oz-$axis different from the origin. Consider the curve $\phi(t) = (t, 0, z_0)$. So, $x \circ \phi(t) = t$, while $\nabla((y(x^2 + y^2 + z^2))^3 + x^2(y(x^2 + y^2 + z^2))) \circ \phi(t) = (0, z_0^2 t^2, 0)$ module t^3. Then, the inequality which defines the L analytic condition fails along this curve.

The problem in both cases is that the function x is too large in norm compared with the norm of the gradient of the second component function.

We note that the map germ G has an isolated critical point at the origin, so by Milnor's result [Mi] the Milnor fibration (1) exists in the "full Milnor tube" involving(but without) the zero level, but this map germ does not satisfy the strong Lojasiewicz inequality. Furthermore, the map germ H is an analytic homeomorphism; since G is an L-map on points of $V(G)$ different from the origin, you might hope that the composition $H \circ G$ would continue to be an L-map at such points, but, as the example shows, this is false.

Note that these examples can be thought of as families of functions parameterized by x. These examples suggest that better results can be obtained by studying the maps as families of functions.

In what follows we will show how to use the real integral closure tools for modules, as defined by the first author in [G], to prove that under certain natural conditions these kind of map germs, some with non-isolated singular set, have Thom's A_f−condition and consequently, have a Milnor's fibration.

We also introduce the notion of the uniform Lojasiewicz inequality after proposition 3.2. This inequality plays the role of Massey's strong Lojasiewicz inequality in this paper. It is essentially a Lojasiewicz inequality, but at the level of a family of functions.

The existence of Thom's A_f-condition for non isolated complex complete intersection singularities have been studied by the first author in [G1] as well. Here we are concerned with the real case.

2. Notations and setup

Denote by $(\mathcal{A}_n, \mathfrak{m}_n)$ the local ring of real analytic function germs at the origin in $\mathbb{R}^n$ and by $\mathcal{A}_n^p$, the $\mathcal{A}_n$ free module of rank p. If (X, x) is the germ of a real analytic set at x, denote by $\mathcal{A}_{X,x}$ the local ring of real analytic function germs on (X, x), and by $\mathcal{A}_{X,x}^p$ the corresponding free module of rank p. If f is a map-germ at $\mathbb{R}^n$, 0, let $I(f)$ denote the ideal in $\mathcal{A}_n$ generated by the component functions of f. Then, as usual, $I(f)\mathcal{A}_n^p$ is the submodule of $\mathcal{A}_n^p$ made up of p-tuples of elements of $I(f)$.

Definition 2.1 [G], [GTW]. Suppose $(X, 0)$ is a real analytic set germ in $\mathbb{R}^n$, M a submodule of $\mathcal{A}_{X,x}^p$. Then:

1) $h \in \mathcal{A}_{X,x}^p$ is in the real integral closure of M, denoted by $\overline{M}$, iff for all analytic paths $\phi : (\mathbb{R}, 0) \to (X, 0), h \circ \phi \in (\phi^* M)\mathcal{A}_1$;
2) $h \in \mathcal{A}_{X,x}^p$ is strictly dependent on M, iff for all analytic paths $\phi : (\mathbb{R}, 0) \to (X, 0), h \circ \phi \in \mathfrak{m}_1(\phi^* M)\mathcal{A}_1$. We denote by $M^\dagger$ the set of elements strictly dependent on M.

The definition of the real integral closure of a module is equivalent to the following formulation using analytic inequalities.

Proposition 2.2 ([G], page 318). *Suppose $h \in \mathcal{A}_{X,x}^p$, M a submodule of $\mathcal{A}_{X,x}^p$. Then $h \in \overline{M}$ if, and only if, for each choice of generators $\{s_i\}$ of M, there exists a constant $C > 0$ and a neighborhood U of x such that for all $\psi \in \Gamma(Hom(\mathbb{R}^p, \mathbb{R}))$,*

$$||\psi(z) \cdot h(z)|| \leq C ||\psi(z) \cdot s_i(z)||$$

for all $z \in U$.

Consider $f : \mathbb{R}^n, 0 \to \mathbb{R}^2, 0$ and $I(f)$ as above. If you alter the form of Massey's inequality allowing $\theta = 1$, then it is equivalent to asking that $I(f)\mathcal{A}_n^2$ is in the integral closure of the jacobian module of f, which is the submodule of $\mathcal{A}_n^2$ generated by the partial derivatives of f. The Lojasiewicz inequality in the complex analytic case is equivalent to asking that f is strictly dependent on the jacobian ideal at all points where the jacobian ideal is zero. We conjecture that Massey's inequality is equivalent to $I(f)\mathcal{A}_n^2$ being strictly dependent on the jacobian module of f at all critical points of f. If Massey's inequality holds at the origin on some neighborhood U, then $I(f)\mathcal{A}_n^2$ is strictly dependent on the Jacobian module of f at all critical points of f on U.

If we assume θ is a rational number p/q then more can be said. In this case, Massey's inequality takes the form

$$|f(x)|^p \leq c \cdot \min_{||(a_1,\ldots,a_k)||=1} ||(a_1 \nabla f_1(x) + \cdots + a_k \nabla f_k(x))||^q.$$

If f is a function germ then this amounts to saying that $I(f)^p$ is in the real integral closure of $J^q(f)$. The module case works as follows. Given the jacobian module of f, $JM(f) \subset \mathcal{A}_n^2$, we can view $JM(f)$ and $I(f)\mathcal{A}_n^2$ as ideals $\mathcal{M}(f)$ and $\mathcal{I}(f)$ in the ring $A_n[T_1, T_2]$, then the inequality says that the real integral closure of $(\mathcal{M}(f))^q$ contains $(\mathcal{I}(f))^p$.

3. Some results

Consider $f : \mathbb{R}^n, 0 \to \mathbb{R}^2, 0$, $f(x, y_1, \ldots, y_{n-1}) = (x, g(x, y_1, \ldots, y_{n-1}))$, where sometimes we let y denote $(y_1, \ldots, y_{n-1})$. We assume $f(\sum(f)) = 0$ or equivalently $\sum(f) \subseteq V(f)$ (or Milnor condition (a)), where $V(f)$ is the variety of f, and let $X_k := V(g - s^k) \subset \mathbb{R}^{n+1}$, $F_k := (s, x)|_{X_k}$, where now we are considering $g - s^k : \mathbb{R}^{n+1} \to \mathbb{R}$, $(s, x, y) \mapsto g(x, y) - s^k$ and $F : \mathbb{R}^{n+1} \to \mathbb{R}^2$, $F(s, x, y) = (s, x)$. So, $F_k = F|_{X_k}$. Also consider $\Pi : \mathbb{R}^{n+1} \to \mathbb{R}$, $\Pi(s, x, y) = x$ and $\Pi_k := \Pi|_{X_k}$.

The first result is:

Lemma 3.1. *In the setup above we have* $\sum(F_k) = \sum(\Pi|_{X_k}) = \sum(f) \times \{0\}$.

Proof. Consider the jacobian matrices of the respective maps.

$$J(F_k) = \begin{pmatrix} 1 & 0 & 0 & \cdots & 0 \\ 0 & 1 & 0 & \cdots & 0 \\ -ks^{k-1} & g_x & g_{y_1} & \cdots & g_{y_{n-1}} \end{pmatrix},$$

$$J(\Pi_k) = \begin{pmatrix} 0 & 1 & 0 & \cdots & 0 \\ -ks^{k-1} & g_x & g_{y_1} & \cdots & g_{y_{n-1}} \end{pmatrix}$$

and

$$J(f) = \begin{pmatrix} 1 & 0 & 0 & \cdots & 0 \\ g_x & g_{y_1} & g_{y_2} & \cdots & g_{y_{n-1}} \end{pmatrix}.$$

It is easy to see that the critical locus of f is given by $\sum(f) = V(g_{y_1}, \ldots, g_{y_{n-1}}) \subset \mathbb{R}^n \times \{0\}$. In the other two cases $(g_{y_1}, \ldots, g_{y_{n-1}})$ also vanish on the respective singular sets, and since we assume Milnor condition (a), this implies x and g are zero as well. In turn, on X_k, this implies $s = 0$, which shows that the singular locus of Π_k is the same as F_k. $\square$

Note that the singular locus of X_k is given by $V(J(g(x, y))) \cap \{s = 0\}$, and this is contained in $\sum(f) \times \{0\}$. Assume now that we have a stratification of X_k which satisfies A_{Π_k}–condition. Since, by the Milnor condition (a), $\sum(f) \times \{0\}$ lies in $x = 0$, it follows that the strata of $\sum(f) \times \{0\}$ satisfy the Whitney A condition. Assume also that there exists a neighborhood $0 \in U \subset \mathbb{R}^n$, such that

$$|g(x, y)|^\theta \leq c \sup_{i=1,\ldots,n-1} \{\|g_{y_i}(x, y)\|\}, \quad 0 < \theta < 1$$

holds for some neighborhood V_z of any $z = (x, y) \in U \setminus \{0\}$, where the constant c may depend on z, but the constant θ does not. This inequality plays the role in this paper that Massey's inequality does in the definition of L-maps. If we think of $g(x, y)$ as a family of functions parameterized by x, then our inequality is a **uniform Lojasiewicz inequality** and we denote it by such in the following, as it relates the values of g to the values of the partials of g in the y variables, which are the "state space" variables.

Notice that the above inequality implies that $\sum(f) \subset V(g)$, so if we assume this inequality we need only assume $\sum(f) \subset V(x)$, and this coupled with the inequality implies Milnor condition (a). Thinking of f as a family of functions, the Milnor condition (a) means that $\sum(g_x)$ is non-empty only for $x = 0$.

Proposition 3.2. *Suppose that in the setup above we have* $\theta < \dfrac{k-1}{k}$. *Then the* A_{Π_k} *stratification of* X_k *is also an* A_{F_k} *stratification, perhaps adding the origin as a stratum.*

Proof. We have the singular set of F_k is $\sum(f) \times \{0\}$, and this lies in $F_k = 0$, and the stratification of $\sum(f) \times \{0\}$ satisfies Whitney A condition. So it suffices to consider the pairs $(X_k \setminus \sum(F_k), S_{(p,0)})$, $(p,0) \in \sum(F_k)$, $p \neq 0$, $S_{(p,0)}$ is the stratum of $\sum(F_k)$ which contains the point $(p,0)$.

Suppose we have such a point $(p,0) \in \sum(F_k)$, $p \neq 0$, where the A_{F_k}–condition fails for the pair $(X_k \setminus \sum(F_k), S_{(p,0)})$. By assumption we know that the A_{Π_k}–condition holds at $(p,0)$, and we know that the tangent vectors to the stratum $S_{(p,0)}$ at $(p,0)$ are in $\ker(F(p,0))$. The last follows because $\sum(F_k) = \sum(f) \times 0 \subset V(x) \times 0$.

Now the A_{Π_k}–condition holds iff the column vector $J(\Pi_k) \cdot v$ is strictly dependent on the module generated by the partial derivatives with respect to variables s, y at $(p,0)$, where v is any tangent vector to $S_{(p,0)}$ at $(p,0)$. This follows because it is known that the strict dependence condition must hold for the module generated by all the partial derivatives; but since $v \in V(x,s)$, $J(\Pi_k) \cdot v$ lies in $(0, I)$ where I is the ideal generated by $(g_{y_1}, \ldots, g_{y_{n-1}})$, it follows that only the partials with respect to s, y can be used, as only they give elements with a zero as first entry.

The A_{F_k}–condition holds iff $J(s, x, g(x,y) - s^k) \cdot v$ is strictly dependent on the module generated by the partial derivatives with respect to y by the reasoning of the previous paragraph.

If A_{F_k} fails and A_{Π_k} holds we have that there exists a curve $\varphi : \mathbb{R}, 0 \to X_k, (p,0)$ such that while $\nabla_y g(x,y).v \in (s^{k-1}, g_{y_1}, \ldots, g_{y_{n-1}})^\dagger$, for all v tangent to $S_{(p,0)}$, $(\nabla_y g \cdot v \circ \varphi) \notin m_1 \varphi^*(g_{y_1}, \ldots, g_{y_{n-1}})$, for some v, where m_1 is the maximal ideal in the ring of germs of analytic functions of one variable.

Claim: These two facts imply that $(s^{k-1}) \notin \overline{(g_{y_1}, \ldots, g_{y_{n-1}})}$. Further, if $s^{k-1} \in \overline{(g_{y_1}, \ldots, g_{y_{n-1}})}$, then the A_{F_k}–condition holds.

Proof of claim. If $(s^{k-1}) \in \overline{(g_{y_1}, \ldots, g_{y_{n-1}})}$, then

$$(s^{k-1}, g_{y_1}, \ldots, g_{y_{n-1}})^\dagger = (g_{y_1}, \ldots, g_{y_{n-1}})^\dagger.$$

However, the first fact implies that $\nabla_y g \cdot v$ is in the left hand side of the above equation, while the second fact implies it is not in the right, so $(s^{k-1}) \notin \overline{(g_{y_1}, \ldots, g_{y_{n-1}})}$.

If $(s^{k-1}) \in \overline{(g_{y_1}, \ldots, g_{y_{n-1}})}$, then the A_{F_k}–condition holds because then $(g_{y_1}, \ldots, g_{y_{n-1}})^\dagger$ will contain $\nabla_y g(x,y) \cdot v$, for all v tangent to $S_{(p,0)}$.

So we may suppose $(s^{k-1}) \notin \overline{(g_{y_1}, \ldots, g_{y_{n-1}})}$. This implies there exists a curve $\varphi : \mathbb{R}, 0 \to X_k, (p, 0)$ such that the $ord(s^{k-1} \circ \varphi) < \min\{ord(g_{y_i} \circ \varphi)\}$.
(*)

Now as $s^{k-1} = (s^k)^{\frac{k-1}{k}}$ and using the fact that under X_k we have $g(x, y) = s^k$, so we have $(s^k)^{\frac{k-1}{k}} \circ \varphi = g^{\frac{k-1}{k}} \circ \varphi$.

Now $|g^{\frac{k-1}{k}} \circ \varphi(t)| < C \min_{i=1,\ldots,n-1}\{|g_{y_i} \circ \varphi(t)|\}$, for all t small enough by the uniform Lojasiewicz inequality. This implies that $ord(g^{\frac{k-1}{k}} \circ \varphi) > \min\{ord(g_{y_i} \circ \varphi)\}$. Then, $ord(s^{k-1} \circ \varphi) > \min\{ord(g_{y_i} \circ \varphi)\}$. This contradicts (*) above. $\qquad\square$

Lemma 3.3. *Suppose we have an A_f stratification on $\mathbb{R}^n$, then for all k we have an A_F stratification on X_k(or A_{F_k}−condition holds), in which the strata on $X_k \cap (\mathbb{R}^n \times \{0\})$ are the same as for the A_f stratification, and the union of the open strata is the complement of $\sum(F_k) = \sum(f) \times 0$.*

Proof. Note that the A_f−condition implies the Whitney A condition on strata of $X_k \cap (\mathbb{R}^n \times \{0\}) = V(g)$. So, we only have to check for the open strata. Suppose we have a curve φ such that $ord((D(g) \cdot v) \circ \varphi)$ fails to be greater than the $ord(J(g) \circ \varphi)$ where v is a tangent vector to a stratum of $X_k \cap (\mathbb{R}^n \times \{0\})$ at $\varphi(0)$. Then using $\Pi_{\mathbb{R}^n} \circ \varphi$ (where $\Pi_{\mathbb{R}^n}$ stands for the projection of $\mathbb{R}^{n+1} \to \mathbb{R}^n$), we get a curve where their order inequality again holds. This implies A_f−condition fails. $\qquad\square$

Lemma 3.4. *Suppose S is an A_{F_k} stratification of the X_k, where $\sum(F_k) = \sum(f) \times 0$ is a union of strata, and the complement of $\sum(F_k)$ is a union of open strata. Then, this induces an A_f stratification of $\mathbb{R}^n$.*

Proof. Again we know that the Whitney A condition holds between the strata of $\sum(f)$, so we just need to consider the open strata. Suppose A_f fails at some point p, where A_F−condition holds for X_k at $(p, 0)$, then there exists a curve φ, and functions $\psi_1(t)$, $\psi_2(t)$ such that

$\lim_{t \to 0} \frac{1}{t^p}[(\psi_1(t), \psi_2(t)) \cdot J(f)(\varphi(t))]$, which is a limiting conormal vector to the fibers of f, fails to contain $T_p(S_p)$. Now we re-parameterize φ. Suppose $g(\varphi(t))$ has order r, and use $t = \tau^k$. So, $g(\varphi(\tau^k))$ has order kr. This implies that $g \circ \varphi(\tau^k) = \tau^{kr}h$, for some h, with $h(0) \neq 0$. Consider $g \circ \varphi(\tau^k) = [\tau^r h^{\frac{1}{k}}]^k$.

Let $s \circ \widehat{\varphi} = \tau^r h^{\frac{1}{k}}$ define an extension of φ, then $\widehat{\varphi}(\tau) = (\varphi(\tau^k), \tau^r h^{\frac{1}{k}})$ lies on X_k over p.

Now set $\psi_0 = -\psi_2(\tau^k)s^{k-1} \circ \widehat{\varphi}(\tau)$, then

$$\lim_{\tau \to 0} \frac{1}{\tau^{pk}} \langle \psi_0(\tau), \psi_1(\tau), \psi_2(\tau) \rangle \cdot \begin{pmatrix} 1 & 0 & \dots 0 \\ 0 & 1 & \dots 0 \\ s^{k-1} & g_x & g_y \end{pmatrix} \circ \varphi(\tau^k) =$$

$$= \lim_{\tau \to 0} \frac{1}{\tau^{pk}} \langle \psi_1(\tau), \psi_2(\tau) \rangle \cdot \begin{pmatrix} 1 & \dots 0 \\ g_x & g_y \end{pmatrix} \circ \varphi(\tau^k), \text{ so this limit is the same}$$

and A_F−condition fails. $\qquad\square$

Theorem 3.5. *Suppose f as in our setup, and the uniform Lojasiewicz inequality holds, then there exists an A_f stratification of $\mathbb{R}^n$ in which $\sum(f)$ is a union of strata, and the open strata are the complement of $\sum(f)$.*

Proof. Choose k as in the proof of proposition 3.2, then we know that since Π_k is an analytic function, there exists an A_{Π_k} stratification. (For the complex analytic case this is proved in [H], and the complex stratification can be used to give a refinement of a real Whitney stratification of X_k in which $\sum(f) \times \{0\}$ is a stratified set, which is an A_{Π_k} stratification.) Since $\sum(\Pi_k) = \sum(f) \times 0$ by proposition 3.1, $\sum(f) \times 0$ is a union of strata, and its complement is a union of open strata. Now by proposition 3.2, this stratification is also an A_{F_k} stratification, and by lemma 3.3, induces the desired A_f stratification. $\qquad\square$

Example 3.6. It is useful to see how the above proof breaks down for the map $f = (x, xy)$ which is the basic example of a map which does not have an A_f stratification. Here the singular locus is the Oy-axis, so the map satisfies Milnor condition (a). Further, X_k has equation $xy - s^k = 0$, and a A_{Π_k} stratification is given by $\{\{0\}, \{Oy - 0\}, \{X_k - Oy\}$. However, no uniform Lojasiewicz inequality holds, for in a neighborhood of points $(0, y)$, $y \neq 0$, $C|xy| > |g_y| = |x|$ for $C = 2/|y|$.

Example 3.7. We wish to show that the key inequality applies to the examples 1 and 2. We first consider $G(x, y, z) = (x, y(x^2 + y^2 + z^2)) = (x, g_1(x, y, z))$, and show

$$|g_1(x, y, z)|^{2/3} \le c \sup\{|g_{1,y}(x, y, z)|, |g_{1,z}(x, y, z)|\}.$$

(Here the sup is taken over the values of the partial derivatives $g_{1,y}(x, y, z)$, $g_{1,z}(x, y, z)$ at each point in a neighborhood of the origin, and compared with the value of $|g_1(x, y, z)|^{2/3}$ at that point.)

This will follow if we show that

$$|y(x^2 + y^2 + z^2)|^2 \le c|x^2 + 3y^2 + z^2|^3.$$

But, this is obvious.

Now consider $H \circ G = (x, g(x, y, z))$, where $H(x, y) = (x, y^3 + x^2 y) = (x, h(x, y))$.

As before we know that

$$|h(x, y)|^2 \le C |h_y(x, y)|^3$$

at all points. Now we want to show that the composition $H \circ G$ also satisfies a uniform Lojasiewicz inequality.

We have $h^p \circ G = g_1^p (g_1^2 + x^2)^p$, while $g_y^{p+1} = (h_y(G) g_{1,y})^{p+1}$. Now $y^2 + x^2 \le 3y^2 + x^2 = h_y(x, y)$, so

$$|h^p \circ G(x, y, z)| \le |g_1^p(x, y, z)||h_y^p(G(x, y, z))|.$$

We want to show

$$|h^p \circ G(x, y, z)| \le |h_y(G(x, y, z))|^{p+1} |g_{1,y}(x, y, z)|^{p+1}.$$

It suffices to show that

$$\left| g_1^p(x, y, z) \right| \le |h_y(G(x, y, z))||g_{1,y}(x, y, z)|^{p+1}.$$

From the form of g_1 we see that

$$\left| g_1^p(x, y, z) \right| = |y^{p-2}|(|y^2||x^2 + y^2 + z^2|^p).$$

Now $|y^2||x^2 + y^2 + z^2|^p \le |g_{1,y}(x, y, z)|^{p+1}$, so it suffices to prove that there exists p such that $|y^{p-2}| \le |h_y(G(x, y, z))|$. Looking at the form of $h_y(G(x, y, z))$ we see that $p = 6$ suffices.

For our final result we show that the property of satisfying an uniform Lojasiewicz inequality is an analytic invariant of a family. In general, it remains open as to whether or not the families obtained by the composition of two maps whose associated families satisfy an uniform Lojasiewicz inequality also satisfy an uniform Lojasiewicz inequality.

Proposition 3.8. *Suppose* $R(x, y_1, \ldots, y_{n-1}) = (x, r(x, y_1, \ldots, y_{n-1}))$ *and* $L(x, Y) = (x, l(x, Y))$ *are bi-analytic map germs at the origin of* $\mathbb{R}^n$ *and* $\mathbb{R}^2$ *respectively,* $F(x, y) = (x, f(x, y_1, \ldots, y_{n-1}))$ *as analytic germ at the origin,* F *satisfies an uniform Lojasiewicz inequality with rational* θ. *Assume* L *preserves the y-axis. Then,* $L \circ F \circ R$ *satisfies an uniform Lojasiewicz inequality with the same* θ *as for* F.

Proof. Notice that it suffices to prove the proposition for R and L separately. Consider $L \circ F$. We know that

$$|f(x, y_1, \ldots, y_{n-1}))|^{\theta} \le C \sup_{i=1,\ldots,n-1} \{\| f_{y_i}(x, y_1, \ldots, y_{n-1})\|\}$$

Since

$$l(x, y) = y(l_0(x, y) \text{ with } l_0(0, 0) \neq 0$$

and

$$(l \circ F)_{y_i}(x, y) = l_y(x, f(x, y)) f_{y_i}(x, y), \text{ with } l_y(0, f(0, 0)) \neq 0,$$

it follows that

$$|l(f(x, y_1, \ldots, y_{n-1}))|^\theta \leq C' \sup_{i=1,\ldots,n-1} \{|(l \circ F)y_i(x, y_1, \ldots, y_{n-1}))|\}.$$

For the case $F \circ R$ the result follows from the fact that the set $\{(f \circ R)_{y_i}\}$ generate the same ideal as $\{R^*(f_{y_i})\}$, so that the existing inequality pulls back under R to give the desired new inequality, with perhaps a different constant C.   □

Acknowledgements. The authors thank David Massey for sharing earlier drafts of [DM] with them, and acknowledge the inspiration received from this paper.

The second author thanks the first author, Terence Gaffney, for all his support and attention given during the development of the post-doc at NEU/USA under his supervision. He thanks Mrs. Mary Gaffney for all the generosity, kindness and assistance in all moments of his stay and to all the NEU staff for making the visiting period pleasant and productive.

Thanks a lot!

References

AR. R. Araújo dos Santos, *Uniform (m)-condition and strong Milnor fibrations,* in *Singularities II: Geometric and Topological aspects, Proc. of the Conference in honour of Lê Dũng Tráng,* eds. J. P. Brasselet et al., (Cuernavaca, Mexico, 2007), *Contemporary Mathematics,* **475** (2008) 43 59.

AR1. R. Araújo dos Santos, *Equivalence of real Milnor fibrations for quasi-homogeneous singularities.* Accepted for publication in The Rocky Mountain Journal Mathematics.

AT. R. Araújo dos Santos; M. Tibar, *Real map germs and higher open book structures.* To appear in *Geometriae Dedicata,* doi 10.1007/S10711-009-9449-z.

G.   T. Gaffney, *Integral closure of modules and Whitney equisingularity,* Invent. Math. **107(2)** 1992 301–322.

G1. T. Gaffney, *Non-isolated Complete Intersection Singularities and the A_f Condition, Singularities I: Algebraic and Analytic aspects, Proc. of the Conference in honour of Lê Dũng Tráng,* eds. J. P. Brasselet et al., (Cuernavaca, Mexico, 2007), *Contemporary Mathematics,* **474** (2008), 85–94.

G2. T. Gaffney, *The multiplicity of pairs of modules and hypersurface singularities*, Real and complex singularities, 143–168, Trends Math., Birkhäuser, Basel, 2007.

GTW. T. Gaffney, D. Trotman and L. Wilson, *Equisingularity of sections, (t^r) condition, and the integral closure of modules*, to appear Journal of Algebraic Geometry.

HL. H. A. Hamm; Lê D. Tráng *Un théorème de Zariski du type de Lefschetz*. Ann. Sci. École Norm. Sup. **4(6)** (1973) 317–355.

H. H. Hironaka, *Stratification and flatness,* Real and complex singularities (Proc. Ninth Nordic Summer School/NAVF Sympos. Math., Oslo, 1976), pp. 199–265. Sijthoff and Noordhoff, Alphen aan den Rijn, 1977.

Ja. A.Jaquemard, *Fibrations de Milnor pour des applications réelles,* Boll. Un. Mat. Ital., **37(1)** 3-B (1989) 45–62.

LD. Lê D. Tráng, *Some remarks on relative monodromy*. Real and complex singularities (Proc. Ninth Nordic Summer School/NAVF Sympos. Math., Oslo, 1976), pp. 397–403. Sijthoff and Noordhoff, Alphen aan den Rijn, 1977.

DM. D. Massey, *Real Analytic Milnor Fibrations and a Strong Lojasiewicz Inequality,* available on arXiv.org, math/0703613.

Mi. J. Milnor, *Singular points of complex hypersurfaces*, Ann. of Math. Studies **61**, Princeton 1968.

PS. A. Pichon and J. Seade, *Fibred Multilinks and singularities $f\bar{g}$,* Mathematische Annalen, **342(3)** (2008) 487–514.

RSV. M. A. Ruas; J. Seade; A. Verjovsky, *On real singularities with a Milnor fibration.* Trends in singularities, 191–213, Trends Math., Birkhäuser, Basel, 2002.

Se1. J. Seade, *On the topology of isolated singularities in analytic spaces*. Progress in Mathematics, 241. Birkhäuser Verlag, Basel, 2006. xiv+238 pp. ISBN: 978-3-7643-7322-1; 3-7643-7322-9.

Se2. J. Seade, *On Milnor's fibration theorem for real and complex singularities.* Singularities in geometry and topology, 127–158, World Sci. Publ., Hackensack, NJ, 2007.

Terence Gaffney
Department of Mathematics
Northeastern University
Boston, MA, 02115
USA
gaff@neu.edu

Raimundo Araújo dos Santos
ICMC
Universidade de São Paulo,
Av. Trabalhador São-Carlense, 400
13.560-970 São Carlos
Brazil
rnonato@icmc.usp.br

12

Surfaces around closed principal curvature lines, an inverse problem

R. GARCIA, L. F. MELLO AND J. SOTOMAYOR

Abstract

Given a non circular spacial closed curve whose total torsion is an integer multiple of 2π, we construct a germ of a smooth surface that contains it as a hyperbolic principal cycle.

1. Introduction

Let $\alpha : \mathbb{M} \to \mathbb{R}^3$ be a C^r immersion of a smooth, compact and oriented, two–dimensional manifold $\mathbb{M}$ into space $\mathbb{R}^3$ endowed with the canonical inner product $\langle ., . \rangle$. It will be assumed that $r \geq 4$.

The *Fundamental Forms* of α at a point p of $\mathbb{M}$ are the symmetric bilinear forms on $\mathbb{T}_p\mathbb{M}$ defined as follows [10], [11]:

$$I_\alpha(p; v, w) = \langle D\alpha(p; v), D\alpha(p; w) \rangle,$$

$$II_\alpha(p; v, w) = \langle -DN_\alpha(p; v), D\alpha(p; w) \rangle.$$

Here, N_α is the positive normal of the immersion α.

The first fundamental form in a local chart (u, v) is defined by $I_\alpha = Edu^2 + 2Fdudv + Gdv^2$, where $E = \langle \alpha_u, \alpha_u \rangle$, $F = \langle \alpha_u, \alpha_v \rangle$ and $G = \langle \alpha_v, \alpha_v \rangle$.

The second fundamental form relative to the unitary normal vector $N_\alpha = (\alpha_u \wedge \alpha_v)/|\alpha_u \wedge \alpha_v|$ is given by $II_\alpha = edu^2 + 2fdudv + gdv^2$, where

$$e = \frac{\det[\alpha_u, \alpha_v, \alpha_{uu}]}{\sqrt{EG - F^2}}, \quad f = \frac{\det[\alpha_u, \alpha_v, \alpha_{uv}]}{\sqrt{EG - F^2}}, \quad g = \frac{\det[\alpha_u, \alpha_v, \alpha_{vv}]}{\sqrt{EG - F^2}}.$$

2000 *Mathematics Subject Classification* 53A04, 34A09, 53A05, 57R30.
Acknowledgements: This paper was done under the project CNPq 473747/2006-5. The first author had the support of FUNAPE/UFG.

158

In a local chart (u, v) the principal directions of an immersion α are defined by the implicit differential equation

$$(Fg - Gf)dv^2 + (Eg - Ge)dudv + (Ef - Fe)du^2 = 0. \tag{1.1}$$

The *umbilic set* of α, denoted by $\mathcal{U}_\alpha$, consists on the points where the three coefficients of equation (1.1) vanish simultaneously.

The regular integral curves of equation (1.1) are called *principal curvature lines*. This means curves $c(t) = (u(t), v(t))$, differentiable on an interval, say J, with non–vanishing tangent vector there, such that, for every $t \in J$, it holds that

$$(Fg - Gf)(u(t), v(t)) \left(\frac{dv(t)}{dt} \right)^2 + (Eg - Ge)(u(t), v(t)) \frac{du(t)}{dt} \frac{dv(t)}{dt}$$

$$+ (Ef - Fe)(u(t), v(t)) \left(\frac{du(t)}{dt} \right)^2 = 0$$

and $c(J) \cap \mathcal{U}_\alpha = \emptyset$.

When the surface $\mathbb{M}$ is oriented, the principal curvature lines on $\mathbb{M} \setminus \mathcal{U}_\alpha$ can be assembled in two one–dimensional orthogonal foliations which will be denoted by $\mathcal{F}_1(\alpha)$ and $\mathcal{F}_2(\alpha)$. Along the first (resp. second), the normal curvature $II_\alpha(p)$ attains its minimum $k_1(p)$, denominated the *minimal principal curvature at* p, (resp. maximum $k_2(p)$, denominated the *maximal principal curvature at* p).

The triple $\mathcal{P}_\alpha = \{\mathcal{F}_1(\alpha), \mathcal{F}_2(\alpha), \mathcal{U}_\alpha\}$ is called the *principal configuration* of the immersion α, [5], [6]. For a survey about the qualitative theory of principal curvature lines see [2].

A closed principal curvature line, also called a *principal cycle*, is called *hyperbolic* if the first derivative of the Poincaré return map associated to it is different from one.

In [9] and [8] it was proved that a regular closed line of curvature on a surface has as total torsion a multiple of 2π. In this paper we consider the following inverse problem.

Problem. Given a simple closed Frenet curve, that is a smooth regular curve of $\mathbb{R}^3$ with non zero curvature, is there an oriented embedded surface that contains it as a hyperbolic principal cycle?

It will be shown that this Problem has a positive answer in the case that the curve is a Frenet, non circular, curve such that its total torsion is an integer multiple of 2π.

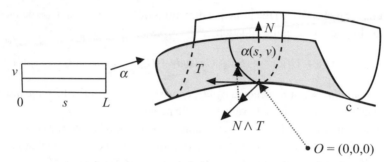

Figure 12.1. Germ of a parametrized surface $\alpha(s, v)$ near a curve $\mathbf{c}$.

The interest of hyperbolic principal cycles is that the asymptotic behavior of the principal foliation around them is determined. The first examples of hyperbolic principal cycles on surfaces were considered by Gutierrez and Sotomayor in [5], where their genericity and structural stability were also established.

2. Preliminary results

Let $c : [0, L] \to \mathbb{R}^3$ be a smooth *simple, closed, regular curve* in $\mathbb{R}^3$ with positive curvature k and of length $L > 0$, i.e., a Frenet curve. Let also $\mathbf{c} = c([0, L])$. Consider the Frenet frame $\{t, n, b\}$ along $\mathbf{c}$ satisfying the equations

$$t'(s) = k(s)n(s),$$

$$n'(s) = -k(s)t(s) + \tau(s)b(s), \qquad (2.1)$$

$$b'(s) = -\tau(s)n(s).$$

Here $k > 0$ is the curvature and τ is the torsion of $\mathbf{c}$.

Consider the parametrized surface of class C^r, $r \geq 4$, defined by the equation

$$\alpha(s, v) = c(s) + [\cos \theta(s)n(s) + \sin \theta(s)b(s)] v$$

$$+ [\cos \theta(s)b(s) - \sin \theta(s)n(s)] \left[\frac{1}{2}A(s)v^2 + \frac{1}{6}B(s)v^3 + v^4C(s, v) \right]$$

$$= c(s) + v(N \wedge T)(s) + \left[\frac{1}{2}A(s)v^2 + \frac{1}{6}B(s)v^3 + v^4C(s, v) \right] N(s).$$

$$(2.2)$$

For an illustration see Fig. 12.1 and [4], [5].

Here, $\mathbf{c}'(s) = t(s) = T(s)$, $\theta(s) = \theta(s + L)$, $A(s) = A(s + L)$, $B(s) = B(s + L)$, $C(s, v) = C(s + L, v)$, $C(s, 0) = 0$, are smooth L–periodic functions with respect to s and v is small.

Proposition 2.1. *The curve* $\mathbf{c}$ *is the union of principal curvature lines of* α *if and only if*

$$\tau(s) + \theta'(s) = 0, \ \theta(0) = \theta_0, \ \int_0^L \tau(s)ds = 2m\pi, \ m \in \mathbb{Z}. \tag{2.3}$$

Moreover, for any solution $\theta(s)$ *of equation (2.3) the parametric surface defined by equation (2.2) is a regular, oriented and embedded surface in a neighborhood of* $\mathbf{c}$. *The umbilic set* $\mathcal{U}_\alpha \cap \mathbf{c}$ *is defined by the equation*

$$A(s) + k(s)\sin\theta(s) = 0. \tag{2.4}$$

Proof. We have that $N(s) = N_\alpha(s, 0) = \cos\theta(s)b(s) - \sin\theta(s)n(s)$. By Rodrigues formula it follows that $\mathbf{c}$ is a principal curvature line (union of maximal and minimal principal lines) if and only if $N'(s) + \lambda(s)t(s) = 0$. Here λ is a principal curvature (maximal or minimal).

Differentiating N leads to

$$N'(s) = \sin\theta(s)k(s)t(s) - [\tau(s) + \theta'(s)][\cos\theta(s)n(s) + \sin\theta(s)b(s)].$$

Therefore $\mathbf{c}$ is a principal line (union of maximal and minimal principal lines and umbilic points) if and only if $\tau(s) + \theta'(s) = 0$.

By the definition of α it follows that $\alpha_s(s, 0) = t(s)$ and $\alpha_v(s, 0) = \cos\theta(s)n(s) + \sin\theta(s)b(s)$ are linearly independent and so by the local form of immersions it follows that α is locally a regular surface in a neighborhood of $\mathbf{c}$.

Since the total torsion is an integer multiple of 2π and $\tau(s) + \theta'(s) = 0$ it follows that for any initial condition $\theta(0) = \theta_0$ equation (2.2) defines an oriented and embedded surface containing $\mathbf{c}$ and having it as the union of principal lines and umbilic points.

Supposing that $\tau(s) + \theta'(s) = 0$ it follows that the coefficients of the first and second fundamental forms of α are given by

$$E(s, v) = 1 - 2k(s)\cos\theta(s)v$$
$$+ \left[\frac{1}{2}k(s)^2(1 + \cos 2\theta(s)) + k(s)A(s)\sin\theta(s)\right]v^2 + O(v^3),$$

$$F(s, v) = \frac{1}{2}A'(s)A(s)v^3 + O(v^4),$$

$$G(s, v) = 1 + A(s)^2v^2 + O(v^3), \tag{2.5}$$

$$e(s, v) = -k(s)\sin\theta(s) + k(s)\cos\theta(s)(\sin\theta(s) - A(s))v + O(v^2),$$

$$f(s, v) = A'(s)v + O(v^2),$$

$$g(s, v) = A(s) + B(s)v + O(v^2).$$

By equations (1.1) and (2.5) it follows that the coefficients of the differential equation of principal curvature lines are given by

$$L(s, v) = (Fg - Gf)(s, v) = -A'(s)v + O(v^2),$$

$$M(s, v) = (Eg - Ge)(s, v) = A(s) + k(s)\sin\theta(s)$$

$$+ \left[B(s) - k(s)A(s)\cos\theta(s) - \frac{1}{2}k(s)^2\sin 2\theta(s)\right]v + O(v^2),$$

$$N(s, v) = (Ef - Fe)(s, v) = A'(s)v + O(v^2). \tag{2.6}$$

The umbilic points along $\mathbf{c}$ are given by the solutions of $M(s, 0) = A(s) + k(s)\sin\theta(s) = 0$ which corresponds to the equality between the principal curvatures $k_1(s) = -k(s)\sin\theta(s)$ and $k_2(s) = A(s)$. $\qquad\square$

Remark 1. The one parameter family of surfaces $M(\theta_0) = \alpha_{\theta_0}([0, L] \times (-\epsilon, \epsilon)) \setminus \mathbf{c}$ defined by equations (2.2) and (2.3) is a foliation of a neighborhood of $\mathbf{c}$ after $\mathbf{c}$ is removed. For all θ_0 the curve $\mathbf{c}$ is a principal cycle of α_{θ_0}. This follows from the theorem of existence and uniqueness of ordinary differential equations and smooth dependence with initial conditions of α_{θ_0} and boundary conditions given by equation (2.3).

3. Hyperbolic principal cycles

In this section it will be given a solution to the problem formulated in the Introduction.

Let α_{θ_0} be the surface defined by equation (2.2) and associated to the Cauchy problem given by equation (2.3).

Theorem 3.1. *Consider the oriented parametric surface α_{θ_0} of class C^r, $r \geq 4$, defined by equations (2.2) and (2.3) such that $\mathcal{U}_{\alpha_{\theta_0}} \cap \mathbf{c} = \emptyset$. Then $\mathbf{c}$ is a hyperbolic principal cycle of α_{θ_0} if and only if*

$$\Lambda(\theta_0) = \int_0^L \frac{A'(s)}{A(s) + k(s)\sin\theta(s)} ds \neq 0. \tag{3.1}$$

The coefficient $\Lambda(\theta_0)$ *is called the characteristic exponent of the Poincaré map associated to* **c**.

Proof. The principal curvatures are given by

$$k_1(s, v) = -k(s)\sin\theta(s) + (k(s)\sin\theta(s) + A(s))k(s)\cos\theta(s)v + O(v^2),$$
$$k_2(s, v) = A(s) + B(s)v + O(v^2).$$

The first return map $\pi : \{s = 0\} \to \{s = L\}$ defined by $\pi(v_0) = v(L, v_0)$, with $v(0, v_0) = v_0$, satisfies the variational equation

$$M(s, 0)v_{sv_0}(s) + N_v(s, 0)v_{v_0}(s) = 0.$$

By equation (2.6) it follows that

$$-\frac{N_v}{M}(s, 0) = -\frac{A'(s)}{A(s) + k(s)\sin\theta(s)}.$$

Integration of the above equation leads to the result.   □

Remark 2. The criterium of hyperbolicity of a principal cycle was established by Gutierrez and Sotomayor in [5], [6]. They proved that a principal cycle **c** is hyperbolic if and only if

$$\int_{\mathbf{c}} \frac{dk_1}{k_2 - k_1} = \int_{\mathbf{c}} \frac{dk_2}{k_2 - k_1} = \frac{1}{2}\int_{\mathbf{c}} \frac{d\mathcal{H}}{\sqrt{\mathcal{H}^2 - \mathcal{K}}} \neq 0.$$

Here $\mathcal{H} = (k_1 + k_2)/2$ and $\mathcal{K} = k_1 k_2$ are respectively the *arithmetic mean* and the *Gauss* curvatures of the surface.

Proposition 3.2. *Consider the family of oriented parametric surfaces* α_{θ_0} *defined by equations (2.2) and (2.3) such that* $\mathcal{U}_{\alpha_{\theta_0}} \cap \mathbf{c} = \emptyset$ *for all* θ_0. *Then the following holds*

$$\Lambda'(\theta_0) = \int_0^L \frac{k(s)A'(s)\cos(\theta_0 - \int_0^s \tau(s)ds)}{[k(s)\sin(\theta_0 - \int_0^s \tau(s)ds) + A(s)]^2}ds. \tag{3.2}$$

Proof. Direct differentation of equation (3.1).   □

Theorem 3.3. *Let* **c** *be a smooth curve, that is a closed Frenet curve of length* L *in* $\mathbb{R}^3$ *such that* τ *is not identically zero and* $\int_0^L \tau(s)ds = 2m\pi$, $m \in \mathbb{Z}$. *Then there exists a germ of an oriented surface of class* C^r, $r \geq 4$, *containing* **c** *and having it as a hyperbolic principal cycle.*

Proof. Consider the parametric surface defined by equation (2.2). By Proposition 2.1, **c** is a principal cycle when $\theta'(s) = -\tau(s)$, $\theta(0) = \theta_0$, $\int_0^L \tau(s)ds = 2m\pi$. Taking $A(s) = (1 - \sin\theta(s))k(s)$ it follows that $M(s, 0) = k(s) > 0$ and so $\mathcal{U}_\alpha \cap \mathbf{c} = \emptyset$. So **c** is a closed principal line of the parametric surface α.

By Theorem 3.1 it follows that $\mathbf{c}$ is hyperbolic if and only if

$$\Lambda = \ln(\pi'(0)) = \int_0^L \frac{A'(s)}{A(s) + k(s)\sin\theta(s)}ds = -\int_0^L \frac{[k(s)\sin\theta(s)]'}{k(s)}ds \neq 0.$$

By assumption the function $k(s)\sin\theta(s)$ is not constant. In fact, $k(s) > 0$ and as τ is not identically equal to zero it follows that $\sin\theta(s) = \sin(\theta_0 - \int_0^s \tau(s)ds)$ is not constant.

If $\Lambda \neq 0$ it follows that $\mathbf{c}$ is hyperbolic and this ends the proof.

In the case when $k(s)\sin\theta(s)$ is not constant and $\int_0^L \frac{[k(s)\sin\theta(s)]'}{k(s)}ds = 0$, consider the deformation of α given by

$$\alpha_\epsilon(s, v) = \alpha(s, v) + \epsilon \frac{a(s)}{2}v^2[\cos\theta(s)b(s) - \sin\theta(s)n(s)],$$

$$a(s) = [k(s)\sin\theta(s)]'.$$

Then $\mathbf{c}$ is a principal cycle of α_ϵ and the principal curvatures are given by

$$k_1(s, 0, \epsilon) = -k(s)\sin\theta(s),$$

$$k_2(s, 0, \epsilon) = A(s) + \epsilon[k(s)\sin\theta(s)]' = (1 - \sin\theta(s))k(s) + \epsilon[k(s)\sin\theta(s)]'.$$

Therefore by Theorem 3.1 and Remark 2 it follows that

$$\Lambda(\epsilon) = \ln(\pi'_\epsilon(0)) = -\int_0^L \frac{k_1'(s, 0, \epsilon)}{k_2(s, 0, \epsilon) - k_1(s, 0, \epsilon)}ds = \int_0^L \frac{[k(s)\sin\theta(s)]'}{k(s) + \epsilon[k(s)\sin\theta(s)]'}ds.$$

Differentiating the above equation with respect to ϵ and evaluating in $\epsilon = 0$ it follows that

$$\Lambda'(0) = \frac{d}{d\epsilon}\big(\ln(\pi'_\epsilon(0))\big)|_{\epsilon=0} = \int_0^L \left[\frac{[k(s)\sin\theta(s)]'}{k(s)}\right]^2 ds \neq 0.$$

This ends the proof.   □

Remark 3. When the curve $\mathbf{c}$ is such that $\int_0^L \tau(s)ds = 2m\pi$, $m \in \mathbb{Z} \setminus \{0\}$ there are no ruled surfaces as given by equation (2.2) containing $\mathbf{c}$ and having it as a closed principal curvature line. In this situation we have always umbilic points along $\mathbf{c}$. In fact, in this case $k_2(s) = A(s) = 0$ and $m \neq 0$ implies that $sin\theta(s)$ always vanishes. These points, at which $k_1(s)$ also vanishes, happen to be the umbilic points. See Proposition 2.1.

Corollary 3.4. *Let $\mathbf{c}$ be a closed planar or spherical Frenet curve of length L in $\mathbb{R}^3$. Then there exists a germ of an oriented surface containing $\mathbf{c}$ and having it as a hyperbolic principal cycle if and only if $\mathbf{c}$ is not a circle.*

Proof. In the case of a planar curve, let $\mathbf{c}(s) = (x(s), y(s), 0)$ with curvature k and consider the Frenet frame $\{t, n, \mathbf{z}\}$, $\mathbf{z} = (0, 0, 1)$ associated to $\mathbf{c}$. Any parametrized surface α containing $\mathbf{c}$ as a principal curvature line has the normal vector equal to $N = \cos\theta(s)n(s) + \sin\theta(s)\mathbf{z}$. Therefore,

$$N' = -k(s)\sin\theta(s)t(s) + \theta'[-\sin\theta(s)\mathbf{z} + \cos\theta(s)n(s)].$$

By Rodrigues formula $N' = -\lambda(s)t(s)$ if and only if $\theta(s) = \theta_0 = cte$. One principal curvature is equal to $k_1(s) = k(s)\sin\theta_0$. By the criterium of hyperbolicity of a principal cycle, see Remark 2 and Theorem 3.1, the principal curvatures can not be constant along a principal cycle. A construction of the germ of surface containing $\mathbf{c}$ as a hyperbolic principal cycle can be done as in Proposition 2.1 and Theorem 3.3. In the case of spherical curves, any closed curve has total torsion equal to zero and the proof follows the same steps of the planar case. This ends the proof. $\qquad\qquad\square$

4. Concluding remarks

The study of principal lines goes back to the works of Monge, see [11, page 95], Darboux [1] and many others. In particular, the local behavior of principal lines near umbilic points is a classical subject of research, see [7] for a survey. The structural stability theory and dynamics of principal curvature lines on surfaces was initiated by Gutierrez and Sotomayor [5], [6] and also has been the subject of recent investigation [2].

The possibility of a Frenet (biregular) closed curve in the space to be a principal line of a surface along it depends only on its total torsion to be an integer multiple of 2π.

The presence of umbilic points on such surface depends on function $A(s)$ as well as on $k(s)$ and θ, which in turn depends on τ. In fact $\theta = -\int_0^s \tau(s)ds + \theta_0$, depending on a free parameter θ_0.

In fact the location of the umbilic points given by equation (2.4) change with the parameter θ_0.

In this paper it has been shown that a non circular closed Frenet curve $\mathbf{c}$ in $\mathbb{R}^3$ can be a principal line of a germ of surface provided its total torsion is $2m\pi$, $m \in \mathbb{Z}$. In Theorem 3.3 the germ of the surface has been constructed in such a way that $\mathbf{c}$ is a hyperbolic principal line. As it is well known that the total torsion of a closed curve can be any real number, the results of this paper show that closed principal lines (principal cycles) are special curves of $\mathbb{R}^3$. This completes the results established in [8] and [9].

The generic behavior of principal curvature lines near a regular curve of umbilics was studied in [3].

References

1. G. Darboux, Leçons sur la Théorie des Surfaces, vols. I, II, III, IV Paris: Gauthier Villars, (1896).
2. R. Garcia and J. Sotomayor, *Lines of Curvature on Surfaces, Historical Comments and Recent Developments*, São Paulo Journal of Mathematical Sciences, **02** (2008) 99–143.
3. R. Garcia and J. Sotomayor, *On the patterns of principal curvature lines around a curve of umbilic points*, Anais Acad. Bras. de Ciências, **77** (2005) 13–24.
4. R. Garcia and J. Sotomayor, Differential Equations of Classical Geometry, a Qualitative Theory, 27th Brazilian Math. Colloquium, Rio de Janeiro, IMPA, (2009).
5. C. Gutierrez and J. Sotomayor, *Structural Stable Configurations of Lines of Principal Curvature*, Asterisque, **98–99** (1982) 185–215.
6. C. Gutierrez and J. Sotomayor, Lines of Curvature and Umbilic Points on Surfaces, 18th Brazilian Math. Colloquium, Rio de Janeiro, IMPA, 1991. Reprinted as Structurally Stable Configurations of Lines of Curvature and Umbilic Points on Surfaces, Lima, Monografias del IMCA, (1998).
7. C. Gutierrez and J. Sotomayor, *Lines of curvature, umbilic points and Carathéodory conjecture*, Resenhas IME–USP, **03** (1998) 291–322.
8. C. C. Pansonato and S. I. R. Costa, *Total torsion of curves in three–dimensional manifolds*, Geom. Dedicata, **136** (2008) 111–121.
9. Yong–An Qin and Shi–Jie Li, *Total torsion of closed lines of curvature*, Bull. Austral. Math. Society, **65** (2002) 73–78.
10. M. Spivak, Introduction to Comprehensive Differential Geometry, Vol. III, Berkeley, Publish or Perish (1999).
11. D. Struik, Lectures on Classical Differential Geometry, Addison Wesley Pub. Co., Reprinted by Dover Publications, Inc. (1988).

Ronaldo A. Garcia
Instituto de Matemática e Estatística
Universidade Federal de Goiás,
Caixa Postal 131
74001-970 Goiânia
Brazil
ragarcia@mat.ufg.br

Jorge Sotomayor
Instituto de Matemática e Estatística
Universidade de São Paulo
05508-090 São Paulo
Brazil
sotp@ime.usp.br

Luis F. Mello
Instituto de Ciências Exatas
Univ. Federal de Itajubá
37500-903 Itajubá
Brazil
lfmelo@unifei.edu.br

13

Euler characteristics and atypical values

Abstract

The theorem of Hà and Lê says that one can check using the Euler characteristic of the fibres whether a polynomial mapping $\mathbb{C}^2 \to \mathbb{C}$ is locally trivial in the sense that it defines a C^∞ fibre bundle. This theorem will be generalized to the case of a polynomial mapping $g : Z \to \mathbb{C}$, where Z is a smooth closed algebraic subvariety of some $\mathbb{C}^N$, not necessarily of dimension 2. It is well-known that even in the case $Z = \mathbb{C}^n$, $n \geq 3$, it is no longer enough to look at the Euler characteristic of the fibre of g alone without serious additional assumptions. In this paper we will use the Euler characteristic of other spaces for this purpose in order to avoid explicit reference to some compactification.

The method of proof is the following: first it is shown that there is no vanishing cycle at infinity (with respect to a suitable compactification) and then that one can construct vector fields which lead to a local trivialization.

Introduction. Let $g : \mathbb{C}^n \to \mathbb{C}$ be a polynomial map. It is well-known that even if g has no critical points it may happen that g does not define a fibre bundle which is locally (and therefore globally) C^∞ trivial. This is due to the circumstance that g is not proper as soon as $n > 1$, so we cannot apply Ehresmann's theorem. There may be fibres which are of a different kind from the general ones. It is customary to call t_0 an atypical value for g if g does not define a topological fibre bundle over any neigbourhood ot t_0. The most famous example of an atypical value has been presented by Broughton [2].

2000 *Mathematics Subject Classification* 32S20 (primary), 14D05, 32S15, 58K15 (secondary).

Here we are first interested in the cohomological counterpart. More generally, let $g: Z \to V$ be an arbitrary holomorphic mapping between complex spaces and $k \geq 0$. Then we consider the groups $H^k(g^{-1}(t); \mathbb{C})$ for $t \in V$. If g defines a locally trivial topological fibre bundle these groups are isomorphic provided that V is connected. In fact we can say more: they are the stalks of the locally constant sheaf of $\mathbb{C}$-vector spaces $R^k g_* \mathbb{C}_Z$, where $\mathbb{C}_Z$ is the constant sheaf of complex numbers on Z and $R^k g_* \mathbb{C}_Z$ is associated to the presheaf $W \mapsto H^k(g^{-1}(W), \mathbb{C}_Z) = H^k(g^{-1}(W); \mathbb{C})$.

We will concentrate on the case where V is an open subset of $\mathbb{C}$ but will admit a constructible complex $\mathbb{F}$ of sheaves instead of the constant sheaf of complex numbers. The question will be under which condition the sheaf $\mathbb{R}^k g_* \mathbb{F}$ is (locally) constant and its stalk over t is isomorphic to $\mathbb{H}^k(g^{-1}(t), \mathbb{F})$. (Here $\mathbb{H}$ denotes hypercohomology and $\mathbb{R}^k g_*$ the k-th hyperdirect image.) It may happen that this is the case for some but not all k. If there is some k such that it is not true in any neighbourhood of t_0 we say that t_0 is a cohomologically atypical value with respect to $\mathbb{F}$.

Finally we will return to the original question under which condition we have a topological fibre bundle. In particular, we will prove the following theorem:

Theorem 0.1. *Let* $g: \mathbb{C}^n \to \mathbb{C}$ *be a polynomial map. Suppose that the fibre* $g^{-1}(\{0\})$ *is reduced and that for all* $0 \leq k \leq n - 2, 0 < j_1 < \ldots < j_k \leq n$ *and almost all* $(z_{j_1}^*, \ldots, z_{j_k}^*) \in \mathbb{C}^k$ *the following is true: The Euler characteristic of* $\{z \in \mathbb{C}^n \mid z_{j_1} = z_{j_1}^*, \ldots, z_{j_k} = z_{j_k}^*, g(z) = t\}$ *is independent of* t, $|t| \ll 1$. *Then* g *defines a* C^∞ *fibre bundle over some neighbourhood of* 0, *so* 0 *is not an atypical value of* g.

This is a generalization of a theorem of Hà-Lê [6] which concerns the case $n = 2$. For other conditions see e.g. [13].

The author would like to thank Lê Dũng Tráng, Mihai Tibăr and Jörg Schürmann for valuable discussions and the Deutsche Forschungsgemeinschaft for financial support.

1. Cohomological results

1.1. Let Z be a complex space, V an open subset of $\mathbb{C}$ and $g: Z \to V$ a holomorphic function. Let us suppose that we have a compactification of g, i.e. a proper holomorphic extension $\overline{g}: \overline{Z} \to V$, where $\overline{Z}$ is a complex space which contains Z as a Zariski open dense subset. Let Z_∞ be the space $\overline{Z} \setminus Z$. We suppose that the inclusion $j: Z \to \overline{Z}$ is Stein, i.e. that every $z \in Z_\infty$ has a fundamental system of open Stein neighbourhoods U of z in $\overline{Z}$ such that $U \cap Z$

is Stein. In particular, this hypothesis is fulfilled as soon as Z_∞ is locally defined in $\overline{Z}$ by one holomorphic function.

Let R be a principal ideal domain, e.g. $\mathbb{Z}$ or a field like $\mathbb{Q}$ or $\mathbb{R}$. Let $\mathbb{F}$ be a bounded complex of sheaves of R-modules on Z. We will consider this complex as an object of the derived category. Let us assume that $\mathbb{F}$ is constructible, which means that that the stalks of the cohomology sheaves are of finite type over R and that there is a stratification (i.e. a complex analytic partition) of Z such that every cohomology sheaf is locally constant along each stratum. Let $\mathbb{R}j_*\mathbb{F}$ be the direct image with respect to the right derived direct image functor; for simplicity we will omit the indication of derived functors and write $j_*\mathbb{F}$ instead. In fact we suppose that the partition above may be extended to an analytic partition of $\overline{Z}$: then $j_*\mathbb{F}$ is constructible, too. For the theory of constructible sheaves we refer to [10].

Let us fix an integer m. We suppose that:

a) $\dim\{z \in Z \mid h^k(\mathbb{F}_z) \neq 0\} \leq m - k$ for all k (support condition),
b) $\dim\{z \in Z \mid \lim_{\rightarrow} \mathbb{H}^k(U, U \setminus \{z\}; \mathbb{F}) \neq 0$ or $\lim_{\rightarrow} \mathbb{H}^{k+1}(U, U \setminus \{z\}; \mathbb{F})$ is not free$\} \leq k - m$ for all k (cosupport condition, cf. [5]).

Here U runs through the system of all open neighbourhoods of z in Z and $\mathbb{H}^k(U, U \setminus \{z\}; \mathbb{F})$ is defined similarly to algebraic topology.

We will then write $\mathbb{F} \in D^m(Z)$.

In the case where R is a field this means – up to shift – that $\mathbb{F}$ is perverse with respect to the middle perversity, and the conditions a) and b) are written as $\mathbb{F} \in D^{\leq m}(Z)$ resp. $\mathbb{F} \in D^{\geq m}(Z)$.

Example 1.1. If Z is smooth and everywhere of dimension m we have that R_Z as well as every locally free R_Z-module of finite rank is in $D^m(Z)$.

The same is true if Z is locally a complete intersection of dimension m.

In general, if $\dim Z = m$ the sheaf R_Z satisfies the support condition a) above, of course. If R_Z satisfies the cosupport condition b), i.e. $\dim\{z \in Z \mid \lim_{\leftarrow} H_k(U, U \setminus \{z\}; R) \neq 0\} \leq k - m$ for all k, we say that $rHd_R(Z) \geq m$ ($rHd_R = $ rectified homological R-depth); cf. [5] Déf. 2.4.2.

Now let us consider the situation near a point $t_0 \in V$. Without loss of generality we may suppose that $t_0 = 0$. Remember that we have the notion of the sheaf complex of vanishing cycles of $j_*\mathbb{F}$ with respect to $\overline{g}$: $\Phi_{\overline{g}}j_*\mathbb{F}$. This is a complex on $\overline{g}^{-1}(\{0\})$, the cohomology sheaves will be denoted by $\Phi^k_{\overline{g}}j_*\mathbb{F}$. We will not repeat the precise definition but recall that $(\Phi^k_{\overline{g}}j_*\mathbb{F})_z \simeq \mathbb{H}^k(U, U \cap \overline{g}^{-1}(t); j_*\mathbb{F}) = \mathbb{H}^k(U \cap Z, U \cap g^{-1}(t); \mathbb{F})$ where U is a suitable open neighbourhood of z in $\overline{Z}$ which is sufficiently small and $t \neq 0$ is sufficiently small compared with U. Notice our convention which may differ from the literature by a shift!

In contrast to [3] where cohomological methods have been already used in the context of compactifiable maps we will really exploit the central role of the sheaf complex of vanishing cycles here.

Of course, $\mathbb{R}^k g_* \mathbb{F} = \mathbb{R}^k \overline{g}_* (j_* \mathbb{F})$. We choose a Whitney regular stratification of $(\overline{Z}, Z_\infty)$ which is adapted to $j_* \mathbb{F}$, i.e. such that the cohomology sheaves of $j_* \mathbb{F}$ are locally constant along each stratum (this is the case as soon as the stratification of Z is adapted to $\mathbb{F}$, see [10] Prop. 4.0.2, p. 215). Note that $\overline{g}$ is submersive over $V \setminus \{0\}$ in the stratified sense after shrinking V if necessary. From now on let us suppose that V is chosen in such a way.

Using Thom's first isotopy lemma we obtain:

Lemma 1.2.
a) $(\mathbb{R}^k g_* \mathbb{F})_t \simeq \mathbb{H}^k(g^{-1}(\{t\}), \mathbb{F})$, $t \in V \setminus \{0\}$.
b) *Suppose that $\overline{g}$ is submersive at $z \in \overline{g}^{-1}(\{0\})$ in the stratified sense. Then* $(\Phi_{\overline{g}}^k j_* \mathbb{F})_z = 0$ *for all k.*
c) *If $\overline{g}$ is submersive in the stratified sense we have for every k: $\mathbb{R}^k g_* \mathbb{F}$ is locally constant, and $(\mathbb{R}^k g_* \mathbb{F})_t \simeq \mathbb{H}^k(g^{-1}(\{t\}), \mathbb{F})$, $t \in V$.*

Furthermore, let us notice that c) follows from b):

Lemma 1.3.
a) *Suppose that the complex $\Phi_{\overline{g}} j_* \mathbb{F}$ is acyclic, i.e. its cohomology sheaves are zero. Then we have for every k: $\mathbb{R}^k g_* \mathbb{F}$ is locally constant.*
b) *Suppose that $\Phi_{\overline{g}} j_* \mathbb{F}$ is acyclic near Z_∞. Then we have for every k:* $(\mathbb{R}^k g_* \mathbb{F})_t \simeq \mathbb{H}^k(g^{-1}(\{t\}), \mathbb{F})$, $t \in V$.

Proof. a) Let g_0 and $\overline{g}_0$ be the mappings of $g^{-1}(\{0\})$ resp. $\overline{g}^{-1}(\{0\})$ onto $\{0\}$. Note that $(\overline{g}_0)_* \Phi_{\overline{g}} j_* \mathbb{F}$ is acyclic. Since $\overline{g}$ is proper we get by base change:

$$(\overline{g}_0)_* \Phi_{\overline{g}} j_* \mathbb{F} = \Phi_{id} \overline{g}_* j_* \mathbb{F} = \Phi_{id} g_* \mathbb{F}.$$

Now $\Phi_{id}^k g_* \mathbb{F} = \mathbb{H}^k(g^{-1}(W), g^{-1}(t); \mathbb{F})$ if W is a suitable neighbourhood of 0 in V and $t \in W \setminus \{0\}$. So for all k: $\mathbb{H}^k(g^{-1}(W), g^{-1}(t); \mathbb{F}) = 0$, i.e. $\mathbb{H}^k(g^{-1}(W), \mathbb{F}) \longrightarrow (\mathbb{R}^k g_* \mathbb{F})_t$ is bijective. Of course, the latter is true also for $t = 0$.

b) Because of Lemma 1.2 it is sufficient to take $t = 0$.
Let us consider the following commutative diagram:

$$
\begin{array}{ccccc}
g^{-1}(\{0\}) & \overset{j_0}{\hookrightarrow} & \overline{g}^{-1}(\{0\}) & \overset{i_0}{\hookleftarrow} & \overline{g}^{-1}(\{0\}) \cap Z_\infty \\
k \downarrow & & \overline{k} \downarrow & & k_\infty \downarrow \\
Z & \overset{j}{\hookrightarrow} & \overline{Z} & \overset{i}{\hookleftarrow} & Z_\infty
\end{array}
$$

Now it is sufficient to show that $\mathbb{H}^k(g^{-1}(W), g^{-1}(0); \mathbb{F}) = 0$ for all k. Let l be the inclusion of $\{0\}$ in V. Then: $l^* g_* \mathbb{F} = l^* \overline{g}_* j_* \mathbb{F} = (\overline{g}_0)_* \overline{k}^* j_* \mathbb{F}$ and $(g_0)_* k^* \mathbb{F} = (g_0)_* k^* j^* j_* \mathbb{F} = (\overline{g}_0)_* (j_0)_* j_0^* \overline{k}^* j_* \mathbb{F}$. So the mapping $\mathbb{H}^k(g^{-1}(W), \mathbb{F}) \longrightarrow \mathbb{H}^k(g^{-1}(0); \mathbb{F})$ coincides with the map $\mathbb{R}^k(\overline{g}_0)_* (\overline{k}^* j_* \mathbb{F}) \longrightarrow \mathbb{R}^k(\overline{g}_0)_*((j_0)_* j_0^* \overline{k}^* j_* \mathbb{F})$. We have a distinguished triangle:

$$\longrightarrow (\overline{g}_0)_*(i_0)_* i_0^! \overline{k}^* j_* \mathbb{F} \longrightarrow (\overline{g}_0)_* \overline{k}^* j_* \mathbb{F} \longrightarrow (\overline{g}_0)_*(j_0)_* j_0^* \overline{k}^* j_* \mathbb{F} \overset{+1}{\longrightarrow}$$

Therefore we must show that the first complex is acyclic. But we have two further exact triangles:

$$\longrightarrow (\overline{g}_0)_*(i_0)_* i_0^! \overline{k}^* j_* \mathbb{F} \longrightarrow (\overline{g}_0)_*(i_0)_* i_0^! \Psi_{\overline{g}} j_* \mathbb{F} \longrightarrow (\overline{g}_0)_*(i_0)_* i_0^! \Phi_{\overline{g}} j_* \mathbb{F}[+1] \overset{+1}{\longrightarrow}$$

$$\longrightarrow (\overline{g}_0)_*(i_0)_* i_0^! \overline{k}^! j_* \mathbb{F} \longrightarrow (\overline{g}_0)_*(i_0)_* i_0^! \Phi_{\overline{g}} j_* \mathbb{F} \longrightarrow (\overline{g}_0)_*(i_0)_* i_0^! \Psi_{\overline{g}} j_* \mathbb{F}[-1] \overset{+1}{\longrightarrow}$$

Since $i_0^! \overline{k}^! j_* = k_\infty^! i^! j_* = 0$ and $\Phi_{\overline{g}} j_* \mathbb{F}$ is acyclic near Z_∞ we obtain the desired result. $\qquad \square$

1.2. In the remainder of this section we make the following

Assumption. The cohomology sheaves of the complex $\Phi_{\overline{g}} j_* \mathbb{F}$ are concentrated on a finite set S.

Lemma 1.4.

a) $\Phi_{\overline{g}} j_* \mathbb{F} \in D^m(\overline{g}^{-1}(0))$.
b) *For all* $z \in S$: $(\Phi_{\overline{g}}^k j_* \mathbb{F})_z = 0$ *for* $k \neq m$, $(\Phi_{\overline{g}}^m j_* \mathbb{F})_z$ *is free of finite rank.*

Proof. a) We have that $j_* \mathbb{F} \in D^m(\overline{Z})$: the support condition follows from [4] because the mapping j is Stein, the cosupport condition is trivial. According to a result of Schürmann [10] Theorem 6.0.2 we can conclude that $\Phi_{\overline{g}} j_* \mathbb{F} \in D^m(\overline{g}^{-1}(0))$.
b) follows from a) because S is finite. $\qquad \square$

Theorem 1.5.

a) $\mathbb{R}^k g_* \mathbb{F}$ *is locally constant,* $k \neq m - 1, m$,
b) *the mapping* $t \mapsto rk\,(\mathbb{R}^k g_* \mathbb{F})_t$ *is lower semicontinuous for* $k = m - 1$ *and upper semicontinuous for* $k = m$.
c) $(\mathbb{R}^k g_* \mathbb{F})_0 = \mathbb{H}^k(g^{-1}(\{0\}), \mathbb{F})$, $k \neq m - 2, m - 1, m$,
d) *the mapping* $t \mapsto rk\,\mathbb{H}^k(g^{-1}(\{t\}), \mathbb{F})$ *is lower semicontinuous for* $k = m - 1$ *and upper semicontinuous for* $k = m - 2, m$.

Proof. Let us take up the notation of the proof of Lemma 1.3. According to Lemma 1.4, $\Phi_{\overline{g}} j_* \mathbb{F} \in D^m(\overline{g}^{-1}(0))$.

By hypothesis the cohomology of this complex is concentrated on the finite set S. Therefore $(\overline{g}_0)_* \Phi_{\overline{g}} j_* \mathbb{F} \in D^m(0)$. Since $\overline{g}$ is proper we get by base change:

$$(\overline{g}_0)_* \Phi_{\overline{g}} j_* \mathbb{F} = \Phi_{id} \overline{g}_* j_* \mathbb{F} = \Phi_{id} g_* \mathbb{F}.$$

Now $\Phi_{id}^k g_* \mathbb{F} = \mathbb{H}^k(g^{-1}(W), g^{-1}(t); \mathbb{F})$ if W is a suitable neighbourhood of 0 in V and $t \in W \setminus \{0\}$. So $\mathbb{H}^k(g^{-1}(W), g^{-1}(t); \mathbb{F}) = 0$ for $k \neq m$, $\mathbb{H}^m(g^{-1}(W), g^{-1}(t); \mathbb{F})$ is free. (*)

This implies a) and b).

Now let us compare $\mathbb{H}^k(g^{-1}(0), \mathbb{F})$ and $\mathbb{H}^k(g^{-1}(t), \mathbb{F})$, $t \neq 0$: First, $\mathbb{H}^k(g^{-1}(t), \mathbb{F}) \simeq \Psi_{id}^k(g_* \mathbb{F}) = \Psi_{id}^k(\overline{g}_* j_* \mathbb{F}) = \mathbb{R}^k(\overline{g}_0)_* \Psi_{\overline{g}}(j_* \mathbb{F})$. Note that $(\overline{g}_0)_*(j_0)_* = (g_0)_*$ and $j_0^* \overline{k}^* = k^* j^*$, $j^* j_* = id$, so $(\overline{g}_0)_*(j_0)_* j_0^* \overline{k}^* j_* \mathbb{F} = (g_0)_* k^* j^* j_* \mathbb{F} = (g_0)_* k^* \mathbb{F}$. So, $\mathbb{H}^k(g^{-1}(0), \mathbb{F}) \simeq \mathbb{R}^k(g_0)_* k^* \mathbb{F}$.

Furthermore, $j_0^* \Psi_{\overline{g}} = \Psi_g j^*$, so $j_0^* \Psi_{\overline{g}} j_* = \Psi_g j^* j_* = \Psi_g$. Therefore we have a commutative diagram: (**)

$$\begin{array}{ccccc}
(\overline{g}_0)_* \overline{k}^* j_* \mathbb{F} & \longrightarrow & (\overline{g}_0)_*(j_0)_* j_0^* \overline{k}^* j_* \mathbb{F} & = & (g_0)_* k^* \mathbb{F} \\
\downarrow & & \downarrow & & \downarrow \\
(\overline{g}_0)_* \Psi_{\overline{g}} j_* \mathbb{F} & \longrightarrow & (\overline{g}_0)_*(j_0)_* j_0^* \Psi_{\overline{g}} j_* \mathbb{F} & = & (g_0)_* \Psi_g \mathbb{F}
\end{array}$$

We want to give a more geometric interpretation: Suppose that Z_∞ is defined in $\overline{Z}$ by $\phi = 0$, where $\phi : \overline{Z} \to [0, \infty[$ is a continuous subanalytic function. Let $Z_{\leq r} := \{z \in Z \mid \phi(z) \geq 1/r\}$. Choose $r \gg 0$, then W and $t \in W \setminus \{0\}$ small compared with $1/r$. Then the diagram (**) above yields, after applying $\mathbb{R}^k \ldots$, the following commutative diagram:

$$\begin{array}{ccc}
\mathbb{H}^k(g^{-1}(W), \mathbb{F}) & \longrightarrow & \mathbb{H}^k(g^{-1}(0), \mathbb{F}) \\
\downarrow & & \downarrow \\
\mathbb{H}^k(g^{-1}(t), \mathbb{F}) & \longrightarrow & \mathbb{H}^k(g^{-1}(t) \cap Z_{\leq r}, \mathbb{F})
\end{array}$$

Here, the right vertical is defined as composition:

$$\mathbb{H}^k(g^{-1}(0), \mathbb{F}) \longrightarrow \mathbb{H}^k(g^{-1}(0) \cap Z_{\leq r}, \mathbb{F}) \longleftarrow \mathbb{H}^k(g^{-1}(W) \cap Z_{\leq r}, \mathbb{F})$$
$$\longrightarrow \mathbb{H}^k(g^{-1}(t) \cap Z_{\leq r}, \mathbb{F}),$$

where the first two arrows are isomorphisms.

Now let us return to the diagram (**) of sheaf complexes before. The lower horizontal can be completed to a distinguished triangle:

$$\longrightarrow (\overline{g}_0)_*(i_0)_* i_0^! \Psi_{\overline{g}} j_* \mathbb{F} \longrightarrow (\overline{g}_0)_* \Psi_{\overline{g}} j_* \mathbb{F} \longrightarrow (\overline{g}_0)_*(j_0)_* j_0^* \Psi_{\overline{g}} j_* \mathbb{F} \xrightarrow{+1}$$

Let us consider the first term. We have another distinguished triangle

$$\longrightarrow (\overline{g}_0)_*(i_0)_* i_0^! \overline{k}^! j_* \mathbb{F} \longrightarrow (\overline{g}_0)_*(i_0)_* i_0^! \Phi_{\overline{g}} j_* \mathbb{F} \longrightarrow (\overline{g}_0)_*(i_0)_* i_0^! \Psi_{\overline{g}} j_* \mathbb{F}[-1] \xrightarrow{+1}$$

Since $i_0^! \overline{k}^! j_* = k_\infty^! i^! j_* = 0$ we get a quasiisomorphism $(\overline{g}_0)_* (i_0)_* i_0^! \Phi_{\overline{g}} j_* \mathbb{F} \longrightarrow$ $(\overline{g}_0)_* (i_0)_* i_0^! \Psi_{\overline{g}} j_* \mathbb{F}[-1]$, see end of proof of Lemma 1.3.

Since $\Phi_{\overline{g}} j_* \mathbb{F} \in D^m(\overline{g}^{-1}(\{0\})$ is concentrated on a finite set we get that $(\overline{g}_0)_* (i_0)_* i_0^! \Phi_{\overline{g}} j_* \mathbb{F} \in D^m(0)$.

Altogether this implies that the lower horizontal map above in (**) induces a mapping $\mathbb{R}^k (\overline{g}_0)_* \Psi_{\overline{g}} j_* \mathbb{F} \longrightarrow \mathbb{R}^k (\overline{g}_0)_* (j_0)_* j_0^* \Psi_{\overline{g}} j_* \mathbb{F}$, i.e. $\mathbb{H}^k(g^{-1}(t), \mathbb{F}) \longrightarrow$ $\mathbb{H}^k(g^{-1}(t) \cap Z_{\leq r}, \mathbb{F})$, which is a monomorphism for $k = m - 2$, an epimorphism for $k = m - 1$ and an isomorphism otherwise.

As for the right vertical of (**), we have a distinguished triangle

$$\longrightarrow (g_0)_* k^* \mathbb{F} \longrightarrow (g_0)_* \Psi_g \mathbb{F} \longrightarrow (g_0)_* \Phi_g \mathbb{F}[+1] \overset{+1}{\longrightarrow}$$

Now the cohomology of $\Phi_g \mathbb{F} = j_0^* \Phi_{\overline{g}} j_* \mathbb{F} \in D^m(g^{-1}(0))$ is concentrated on the finite set $S \cap Z$, so $(g_0)_* \Phi_g \mathbb{F} = (\overline{g}_0)_* (j_0)_* \Phi_g \mathbb{F} \in D^m(0)$. Thus we obtain that the mapping $\mathbb{R}^k (g_0)_* k^* \mathbb{F} \longrightarrow \mathbb{R}^k (g_0)_* \Psi_g \mathbb{F}$, i.e. $\mathbb{H}^k(g^{-1}(0), \mathbb{F}) \simeq \mathbb{H}^k(g^{-1}(W) \cap$ $Z_{\leq r}, \mathbb{F}) \longrightarrow \mathbb{H}^k(g^{-1}(t) \cap Z_{\leq r}, \mathbb{F})$, is a monomorphism for $k = m - 1$, an epimorphism for $k = m$ and an isomorphism otherwise.

Altogether we obtain: (***)

$$\begin{aligned} rk\, \mathbb{H}^k(g^{-1}(0), \mathbb{F}) &\leq rk\, \mathbb{H}^k(g^{-1}(t), \mathbb{F}), & k = m - 1 \\ &\geq & k = m - 2, m \\ &= & \text{otherwise} \end{aligned}$$

This implies d).

Note that we may argue for the left vertical of the first commutative diagram (**) similarly as for the right one. Altogether we obtain that the upper horizontal of (**) induces an isomorphism $\mathbb{R}^k (\overline{g}_0)_* \overline{k}^* j_* \mathbb{F} \longrightarrow \mathbb{R}^k (g_0)_* k^* \mathbb{F}$ for $k \neq m - 2, m - 1, m$, which implies c). $\qquad \square$

Furthermore let us look at the Euler characteristic
$\chi_t := \chi(g^{-1}(\{t\}), \mathbb{F}) := \sum_k (-1)^k rk\, \mathbb{H}^k(g^{-1}(\{t\}), \mathbb{F})$.
Define $\mu_z := rk(\Phi_{\overline{g}}^m j_* \mathbb{F})_z$, $z \in \overline{g}^{-1}(\{0\})$.
Of course, we have $\chi_t = \chi((g_* \mathbb{F})_t) := \sum_k (-1)^k rk(\mathbb{R}^k g_* \mathbb{F})_t$ for $t \neq 0$.

Theorem 1.6.

a) $\chi_0 = \chi((g_* \mathbb{F})_0)$,

b) $\chi_0 - \chi_t = (-1)^m \sum_z \mu_z$.

Proof. We refer to the proof of Theorem 1.5. The considerations there imply the following equations:

$\chi_0 - \chi((g_0)_* \Psi_g \mathbb{F}) = (-1)^m \sum_{z \notin Z_\infty} \mu_z$,

$\chi((g_0)_* \Psi_g \mathbb{F}) - \chi_t = (-1)^m \sum_{z \in Z_\infty} \mu_z$.

For the first equation, see the discussion of the right vertical in (**).

The lower horizontal in (**) shows that $\chi((g_0)_*\Psi_g\mathbb{F}) - \chi_t = -\chi((\bar{g}_0)_*(i_0)_*i_0^!\Psi_{\bar{g}}j_*\mathbb{F}) = \chi((\bar{g}_0)_*(i_0)_*i_0^!\Phi_{\bar{g}}j_*\mathbb{F})$.

The distinguished triangle

$$\longrightarrow (\bar{g}_0)_*(i_0)_*i_0^!\Phi_{\bar{g}}j_*\mathbb{F} \longrightarrow (\bar{g}_0)_*\Phi_{\bar{g}}j_*\mathbb{F} \longrightarrow (\bar{g}_0)_*(j_0)_*j_0^*\Phi_{\bar{g}}j_*\mathbb{F} \xrightarrow{+1}$$

shows that $\chi((\bar{g}_0)_*(i_0)_*i_0^!\Phi_{\bar{g}}j_*\mathbb{F}) = (-1)^m \sum_z \mu_z - (-1)^m \sum_{z \notin Z_\infty} \mu_z = (-1)^m \sum_{z \in Z_\infty} \mu_z$. So we get the second equation, too.

Therefore

$$\chi_0 - \chi_t = (-1)^m \sum_z \mu_z$$

which gives b).

Similarly, because of (*):

$$\chi((g_*\mathbb{F})_0) - \chi((g_*\mathbb{F})_t) = (-1)^m \sum_z \mu_z$$

A comparison between these formulas yields a), because $\chi_t = \chi((g_*\mathbb{F})_t)$. □

Theorem 1.7. *Suppose that the cohomology sheaves of the complex $\Phi_{\bar{g}}j_*\mathbb{F}$ are concentrated on a finite set S and that the mapping $t \mapsto \chi_t$ is locally constant. Then we have for all k:*

a) $\mathbb{R}^k g_*\mathbb{F}$ *is locally constant,*
b) $(\mathbb{R}^k g_*\mathbb{F})_0 = \mathbb{H}^k(g^{-1}(\{0\}), \mathbb{F})$,
c) $\Phi_{\bar{g}}^k j_*\mathbb{F} = 0$, *i.e. we may take $S = \emptyset$.*

Proof. Because of Theorem 1.6 b) we have that $\Phi_{\bar{g}}j_*\mathbb{F}$ is acyclic, i.e. c). By Lemma 1.3 we can derive a) and b). □

2. Absence of vanishing cycles

2.1. Now we return to the situation in section 1.1 and want to study the question under which condition the complex $\Phi_{\bar{g}}j_*\mathbb{F}$ is acyclic, avoiding the restriction made in Theorem 1.7. We want to present a condition which only uses some kind of Euler characteristics, putting $r = 1$ below.

Let g and $\bar{g}$ be as before (section 1.1). We assume that there are holomorphic mappings $g_1, \ldots, g_{n-1} : Z \to \mathbb{C}$ which extend to holomorphic mappings $\bar{g}_1, \ldots, \bar{g}_{n-1} : \bar{Z} \to \hat{\mathbb{C}} := \mathbb{C} \cup \{\infty\}$ such that $(\bar{g}_1, \ldots, \bar{g}_n) : \bar{Z} \to \hat{\mathbb{C}}^{n-1} \times V$ is finite, where $g_n := g$ and $\overline{g}_n := \overline{g}$. As before let us assume that the inclusion $j: Z \to \overline{Z}$ is Stein. Let $\mathbb{F}$ be a complex of sheaves of R-modules which belongs to $D^m(Z)$ and $r \geq 0$, where R is a principal ideal domain.

Theorem 2.1. *Assume that the cohomology sheaves of $\Phi_{\overline{g}_n} j_* \mathbb{F}$ vanish outside some analytic subset of dimension $\leq l$. Then the following conditions are equivalent:*

a) *For all $r \leq k \leq l$, $0 < j_1 < \ldots < j_k < n$ there is a dense subset S_j of $\mathbb{C}^k$ such that for all $(s^*_{j_1}, \ldots, s^*_{j_k}) \in S_j$ the following holds: The number $\chi(\{g_{j_1} = s^*_{j_1}, \ldots, g_{j_k} = s^*_{j_k}, g_n = t\}, \mathbb{F})$ is independent of t, $|t| \ll 1$.*
b) *The cohomology sheaves of $\Phi_{\overline{g}_n} j_* \mathbb{F}$ vanish outside some analytic subset of dimension $\leq r - 1$.*

Proof. a) $\Rightarrow$ b): Let us choose some Whitney statification of $(\overline{Z}, \overline{g}^{-1}(\{0\}))$ adapted to $\Phi_{\overline{g}_n} j_* \mathbb{F}$ which satisfies Thom's $a_{\overline{g}_n}$-condition (the last property is automatic by [1]). Let z be a point of some stratum $Z_i \subset \overline{g}_n^{-1}(\{0\})$ of dimension $k \geq r$. Assume that the cohomology sheaves of $\Phi_{\overline{g}_n} j_* \mathbb{F}$ do not all vanish at z; we can suppose that k is maximal with respect to this property. Of course, $k \leq l$. We must derive a contradiction. Put $\overline{h} := (\overline{g}_1, \ldots, \overline{g}_{n-1})$. Then $\overline{h}(Z_i)$ is analytic of dimension k. Take a smooth point $s^* = (s^*_1, \ldots, s^*_{n-1})$ of this set. After permutation of the components of $\overline{h}$ we may assume that the space $\{s_1 = s^*_1, \ldots, s_k = s^*_k\}$ is transversal to $\overline{h}(Z_i)$ at this point, $s^*_1, \ldots, s^*_k \neq \infty$. We may assume that $\overline{g}_j(z) = s^*_j$, $j = 1, \ldots, n-1$, then $\{\overline{g}_1 = s^*_1, \ldots, \overline{g}_k = s^*_k\}$ is transversal to Z_i at z. Let $S' := S_{1,\ldots,k}$ be as above, then we may assume that $(s^*_1, \ldots, s^*_k) \in S'$. Furthermore we may assume that $\{\overline{g}_1 = s^*_1, \ldots, \overline{g}_k = s^*_k\}$ is transversal to the whole stratification. Then the support of the cohomology of $\Phi_{\overline{g}_n} j_* \mathbb{F}$ intersects this set only in some set S which consists of finitely many points. Now let $\overline{Z}' := \overline{Z} \cap \{\overline{g}_1 = s^*_1, \ldots, \overline{g}_k = s^*_k\}$, $Z' := Z \cap \overline{Z}'$, $\overline{g}' := \overline{g}|\overline{Z}'$, j' the inclusion of Z' in $\overline{Z}'$, $\mathbb{F}' := \mathbb{F}|Z'$. Because of the transversality condition we have for all $z \in \overline{g}'^{-1}(\{0\})$: $(\Phi_{\overline{g}} j_* \mathbb{F})_z \simeq (\Phi_{\overline{g}'} j'_* \mathbb{F}')_z$. Note that we have base change for j_* because of [10] Prop. 4.3.1, p. 261, and for $\Phi_{\overline{g}}$ because of [10] Lemma 4.3.4, p. 265. Furthermore, $\mathbb{F}' \in D^m(Z')$. So we can apply Theorem 1.7 and conclude that the finite set S is empty, which is a contradiction to our assumption $z \in S$ before.

b) $\Rightarrow$ a): Without loss of generality we may assume $j_1 = 1, \ldots, j_k = k$. Let $(s^*_1, \ldots, s^*_k)$ be chosen generically, $\overline{Z}'$ etc. as above, $z \in \overline{g}^{-1}(\{0\})$. Then the cohomology of $(\Phi_{\overline{g}'} j'_* \mathbb{F}')_z \simeq (\Phi_{\overline{g}} j_* \mathbb{F})_z$ vanishes, because $r - 1 - k < 0$. This implies that the cohomology of $(\overline{g}'_0)_* \Phi_{\overline{g}'} j'_* \mathbb{F}') \simeq \Phi_{id}(\overline{g}')_* j'_* \mathbb{F}' \simeq \Phi_{id}(g')_* \mathbb{F}'$ vanishes, which yields the desired result. $\square$

Note that in the case $r = l = 0$ a) $\Rightarrow$ b) is just a consequence of Theorem 1.7.

Remark 1. The hypothesis that the cohomology sheaves of $\Phi_{\overline{g}_n} j_* \mathbb{F}$ vanish outside some analytic subset of dimension $\leq l$ is fulfilled if we take $l = \dim(supp\, \Phi_g \mathbb{F}) \cup (\overline{g}^{-1}(\{0\}) \cap Z_\infty)$.

Example 2.2. Let $\overline{Z}$ be a closed analytic subset of $\hat{\mathbb{C}}^N \times V$, $\overline{g}_j(z) := z_j$, $j = 1, \ldots, N+1$, $Z := (\mathbb{C}^N \times V) \cap \overline{Z}$, $g_j := \overline{g}_j | Z$, $\overline{g} := \overline{g}_{N+1}$, $g := g_{N+1}$. Then $j : Z \to \overline{Z}$ is Stein, and $(\overline{g}_1, \ldots, \overline{g}_{N+1}) : \overline{Z} \to \hat{\mathbb{C}}^N \times V$ is the inclusion of a closed subspace and therefore finite. So we can apply Theorem 2.1.

2.2. The most important case is the following: Assume $Z \subset \mathbb{C}^N$ Zariski-closed, $g : Z \to \mathbb{C}$ a polynomial mapping, $\overline{Z} :=$ closure of the graph of g in $\hat{\mathbb{C}}^N \times \mathbb{C}$, $\overline{g}(z) := z_{N+1}$. Of course we may identify Z with the graph of g. Put $\overline{g}_j(z) := z_j$, $j = 1, \ldots, N$. Then we can apply Theorem 2.1 (see Example 2.2) and obtain:

Theorem 2.3. *The following conditions are equivalent:*

a) *For all $0 \leq k \leq l := \dim(supp\, \Phi_g \mathbb{F}) \cup (\overline{g}^{-1}(\{0\}) \cap Z_\infty)$, $0 < j_1 < \ldots < j_k \leq N$ there is a dense subset S_j of $\mathbb{C}^k$ such that for all $(s_{j_1}^*, \ldots, s_{j_k}^*) \in S_j$ the following holds: The number $\chi(\{z_{j_1} = s_{j_1}^*, \ldots, z_{j_k} = s_{j_k}^*, g = t\}, \mathbb{F})$ is independent of t, $|t| \ll 1$.*

b) *$\Phi_{\overline{g}} j_* \mathbb{F}$ is acyclic.*

Remark 2. Fix t, $|t| \ll 1$. For $0 < j_1 < \ldots < j_k \leq N$ with $0 \leq k \leq l$ put $\chi_t^{j_1, \ldots, j_k}(\mathbb{F}) := \chi(\{z_{j_1} = s_{j_1}^*, \ldots, z_{j_k} = s_{j_k}^*, g = t\}, \mathbb{F})$, where $(s_{j_1}^*, \ldots, s_{j_k}^*)$ is chosen general enough. Let $\chi_t^*(\mathbb{F})$ be the family of these numbers. Then we may replace a) by the following equivalent condition:

a) $* t \mapsto \chi_t^*(\mathbb{F})$ is constant for $|t| \ll 1$.

Here we argue as follows: Let $\delta > 0$ be sufficiently small and $j_1, \ldots, j_k$ as above. There is a value c such that $M := \{(s_{j_1}, \ldots, s_{j_k}, t) \mid \chi(\{z_{j_1} = s_{j_1}, \ldots, z_{j_k} = s_{j_k}, g = t\}, \mathbb{F}) = c\}$ is a constructible dense subset of $\mathbb{C}^k \times \{|t| < \delta\}$. Then the question is whether the intersection of M with $\mathbb{C}^k \times \{0\}$ is dense in $\mathbb{C}^k \times \{0\}$, too.

This condition a*) is similar to (but not the same as) the condition "polar numbers constant" of M. Tibăr, see [12].

Now look at the special case where $\mathbb{F} = \mathcal{L}$, $\mathcal{L}$ being a locally free R_Z−module of finite rank, e.g. $\mathcal{L} = R_Z$. Let $Sing\, g$ be the set of critical points of g in the stratified sense, with respect to the canonical Whitney stratification of Z. Then $supp\, \Phi_g \mathcal{L}$ is contained in $g^{-1}(\{0\} \cap Sing\, g$. So we get from Theorem 2.3:

Theorem 2.4. *Suppose that* $\dim Z = m$ *and* $rHd_R(Z) = m$ *(see Example 1.1). The following conditions are equivalent:*

a) *For all* $0 \leq k \leq l := \dim(g^{-1}(\{0\}) \cap Sing\, g) \cup (\overline{g}^{-1}(\{0\}) \cap Z_\infty), 0 < j_1 < \ldots < j_k \leq N$ *there is a dense subset* S_j *of* $\mathbb{C}^k$ *such that for all* $(s_{j_1}^*, \ldots, s_{j_k}^*) \in S_j$ *the following holds: The number* $\chi(\{z_{j_1} = s_{j_1}^*, \ldots, z_{j_k} = s_{j_k}^*, g = t\}, \mathcal{L})$ *is independent of* t, $|t| \ll 1$.
b) $\Phi_{\overline{g}} j_* \mathcal{L}$ *is acyclic.*

Note that $\overline{g^{-1}(\{0\})} = \overline{g}^{-1}(\{0\})$ if $Z = \mathbb{C}^m$. So we get:

Corollary 2.5. *Suppose that* $g : \mathbb{C}^2 \to \mathbb{C}$ *is a polynomial mapping such that* $g^{-1}(\{0\})$ *is generically reduced, i.e. reduced except for a nowhere dense subset. The following conditions are equivalent:*

a) *The number* $\chi(\{g = t\}, \mathcal{L})$ *is independent of* t, $|t| \ll 1$.
b) $\Phi_{\overline{g}} j_* \mathcal{L}$ *is acyclic.*

Remark 3. There are other cases which can be reduced to the situation considered in Theorem 2.4. In particular, we can include the case of rational functions:

a) Suppose that we look at a family of algebraic subvarieties of $\mathbb{C}^N$ depending on some parameter t instead of the fibres of some polynomial mapping: The varieties are of the form $\{z \in \mathbb{C}^N \mid f_1(z, t) = \ldots = f_k(z, t) = 0\}$, where $f_1, \ldots, f_k : \mathbb{C}^{N+1} \to \mathbb{C}$ are polynomial functions. Then these varieties can be identified with the fibres of the mapping $\{z \in \mathbb{C}^{N+1} \mid f_1(z, t) = \ldots = f_k(z, t) = 0\} \longrightarrow \mathbb{C}: (z, t) \mapsto t$.
b) Suppose that instead of Z we look at $\{z \in Z \mid h(z) \neq 0\}$, where h is some polynomial function: This space can be identified with $\{(z, t) \in Z \times \mathbb{C} \mid th(z) = 1\}$.
c) Suppose that we are looking at the fibres of a rational function f_1/f_2 on $\mathbb{C}^n$: These are of the form $\{z \mid f_1(z) = tf_2(z), f_2(z) \neq 0\}$. Using the approach of a) and b), we see that the fibres can be identified with the fibres of the mapping $\{(z, s, t) \in \mathbb{C}^{n+2} \mid f_1(z) = tf_2(z), sf_2(z) = 1\} \longrightarrow \mathbb{C}: (z, s, t) \mapsto t$.

This shows that it is important to look at polynomial functions on affine varieties, not just $\mathbb{C}^n$.

In fact the case where $\mathbb{C}^2$ is replaced by an affine surface has already been considered by Zaidenberg [14].

3. Topological triviality

3.1. Let g and $\overline{g}$ be as in section 1.1. We want to give a condition under which g defines a locally trivial topological fibre bundle. We assume that there is a principal ideal sheaf $\mathcal{I}$ in $\mathcal{O}_{\overline{Z}}$ whose zero set is Z_∞. If we choose a Whitney stratification of $\overline{Z}$ such that Z_∞ is a union of strata and if $\overline{g}$ is submersive in the stratified sense, i.e. the restriction of $\overline{g}$ to every stratum is submersive, we know that g defines a topological fibre bundle. We are, however, interested in a condition which does not necessitate to investigate Whitney conditions at infinity, i.e. along Z_∞.

Now we make the following

Assumption. Z is smooth and everywhere of dimension m, g is submersive.

In the following lemma the properness of $\overline{g}$ is not used. Let $p_0 \in Z_\infty \cap g^{-1}(\{0\})$. Let f be a representative of a generator of $\mathcal{I}_{p_0}$. Since the statement is local we may suppose that $\overline{Z}$ is an analytic subset of some open set U in $\mathbb{C}^l$, $p_0 = 0$ and that $\overline{g}$ and f are defined on U. We may assume that these are non-trivial linear functions, passing to the graph of the mapping $(\overline{g}, f)$.

Let $\mathcal{T}_{p_0}$ be the set of limits at p_0 of tangent spaces to $Z \cap \{f = const\}$ at points p near p_0.

Furthermore let R be a principal ideal domain and $\mathcal{L}$ a locally free R_Z-module of finite rank whose stalks are non-trivial.

Lemma 3.1. *Suppose that the complex* $\Phi_{\overline{g}} j_* \mathcal{L}$ *is acyclic. Then* $d\overline{g}_{p_0}|L \not\equiv 0$, $L \in \mathcal{T}_{p_0}$.

Proof. Since we may pass from R to its quotient field we may assume that R is a field.

Let us fix a Whitney stratification of $\overline{Z}$ such that Z_∞ is a union of strata. Note that automatically Thom's a_f-condition is satisfied because of [1] Th. 4.2.1.

Let k be the inclusion of $\overline{Z}$ in U. Remember that $j_* \mathcal{L}$ is perverse, so $k_* j_* \mathcal{L}$, too. By [8] Theorem 3.2 we know that $(d\overline{g})_0 \notin SS(k_* j_* \mathcal{L})$, where SS denotes the microsupport, see [7] V.1. Note that the microsupport is a closed subset of the cotangent bundle of U.

By [7] Prop. 8.6.4 we can conclude that the fact that $\Phi_{\overline{g}} j_* \mathcal{L}$ is acyclic is not disturbed if we replace $\overline{g}$ by a different function with the same differential at 0.

By adding a quadratic form $a_1 z_1^2 + \cdots + a_l z_l^2$, where $a_1, \ldots, a_l \in \mathbb{C}$ are general enough, we may achieve that we have an isolated critical point at 0 in the stratified sense. This is accomplished in the usual way: First pass to $\overline{Z} \times \mathbb{C}^l$. Then for each stratum S of $\overline{Z}$, $\{(z, a) \in (S \setminus \{0\}) \times \mathbb{C}^l \mid \overline{g}(z) + a_1 z_1^2 + \cdots +$

$a_l z_l^2 = 0\}$ is smooth, let $a = (a_1, \ldots, a_l)$ be a regular value for the restriction of $(z, a) \mapsto a$ to this subset.

Passing to new coordinates we may suppose that 0 is at most an isolated singular point from the very beginning.

Let $0 < \delta \ll \epsilon \ll 1$. Then Z_∞ and $\{\bar{g} = 0\}$ intersect transversally along the intersection with S_ϵ because we have at most an isolated singular point, and the intersection of $Z_\infty \cap \{\bar{g} = 0\}$ with S_ϵ is transversal, by the Curve Selection Lemma. We may pass here from Z_∞ to $\{f = s\}$ (by Thom's a_f-condition) and from $\{\bar{g} = 0\}$ to $\{\bar{g} = t\}$, where $|s|, |t| \leq \delta$. Finally, we may here replace $\bar{g}$ by a linear function v which is sufficiently near $\bar{g}$. $\qquad (*)$

Now fix $t_0, 0 < |t_0| < \delta$. Then $\{\bar{g} = t_0\}$ intersects the stratified space $D_\epsilon \cap \bar{Z}$ transversally. If v is sufficiently near to the linear function $\bar{g}$ we have that $\{v = t_0\}$ intersects $D_\epsilon \cap \bar{Z}$ still transversally and that $H^k(Z \cap D_\epsilon \cap \{|v| \leq \delta\}, Z \cap D_\epsilon \cap \{v = t_0\}; \mathcal{L}) = H^k(Z \cap D_\epsilon, Z \cap D_\epsilon \cap \{g = t_0\}; \mathcal{L}) = 0$ for all k. Because of $(*)$ we know that $\Phi_{v-t} j_* \mathcal{L}$ is acyclic near S_ϵ for $|t| \leq \delta$. So we can conclude that $H^k(Z \cap D_\epsilon \cap \{|v| \leq \delta\}, Z \cap D_\epsilon \cap \{v = t_0\}; \mathcal{L}) = \oplus (\Phi^k_{v-v(p)} j_* \mathcal{L})_p$, where $\|p\| < \epsilon, |v(p)| \leq \delta, v(p) \neq t_0$. Therefore $\Phi_{v-t} j_* \mathcal{L}$ must be acyclic for $|t| \leq \delta$ in $D_\epsilon \cap \bar{Z}$ (along $t = t_0$ this is clear). $\qquad (**)$

Let i and i_{p_0} denote the inclusion of $g^{-1}(0) \cap Z$ in Z and of p_0 in $g^{-1}(0) \cap Z$, respectively. By hypothesis, $\Phi_{\bar{g}} j_* \mathcal{L}$ is acyclic, so $i^!_{p_0} \Phi_{\bar{g}} j_* \mathcal{L}$, too. Now $i^!_{p_0} i^! j_* \mathcal{L}$ is acyclic, so $i^!_{p_0} \Psi_{\bar{g}} j_* \mathcal{L}$ is acyclic, too.

For $0 < |t| \leq \delta \ll \epsilon \ll 1$ we have $0 = h^k(i^!_{p_0} \Psi_{\bar{g}} j_* \mathcal{L}) = \mathbb{H}^k(D_\epsilon \cap Z \cap g^{-1}(t) \cap \{|f| \leq \delta\}, D_\epsilon \cap Z \cap g^{-1}(t) \cap \{|f| \leq \delta\} \cap (S_\epsilon \cup \{|f| = \delta\}); \mathcal{L})$.

So $0 = \mathbb{H}^k(D_\epsilon \cap Z \cap v^{-1}(t) \cap \{|f| \leq \delta\}, D_\epsilon \cap Z \cap v^{-1}(t) \cap \{|f| \leq \delta\} \cap (S_\epsilon \cup \{|f| = \delta\}); \mathcal{L})$ if v is near $d\bar{g}_{p_0}$ and $0 < |t| < \delta$. First, v should be near $d\bar{g}_{p_0}$ for given t but this restriction turns out to be unnecessary since the statement does not depend on t, because of $(**)$.

Now take $-|f|^2$ as Morse function on $D_\epsilon \cap Z \cap v^{-1}(t)$. The restriction to $S_\epsilon \cap Z \cap v^{-1}(t)$ has no critical point p with $|f(p)| \leq \delta$ because of $(*)$. If we have a critical point p in the interior with $|f(p)| \leq \delta$ and Milnor number μ_p we get a direct summand for the cohomology with $k = m - 1$ of the form $\mathcal{L}_p^{\mu_p} \neq 0$, which leads to a contradiction. So there are no such critical points.

Assume now that the assertion of the lemma is not true: then near p_0 there is a p and v close to $d\bar{g}_{p_0}$ such that p is a critical point of $f|\{v = v(p)\} \cap Z$. This is a contradiction. $\qquad \square$

As a first consequence we get that we have at least locally a fibre bundle:

Lemma 3.2. *Under the assumption of Lemma 3.1, $g | D_\epsilon \cap Z \cap g^{-1}(\{|t| < \alpha\})$ defines a C^∞ fibre bundle over $\{|t| < \alpha\}$ for $0 < \alpha \ll \epsilon \ll 1$.*

Proof. Let us choose a stratification and ϵ as in the preceding proof. Then we may find suitable vector fields whose flows lead to the desired trivialization. $\square$

Consequently we obtain:

Lemma 3.3. *Let $\mathcal{F}$ be a locally constant sheaf of R-modules on Z. Under the hypothesis of Lemma 3.1 $\Phi_{\overline{g}} j_* \mathcal{F}$ is acyclic.*

In order to handle the global case let us pass to some uniform version of Lemma 3.1:

Lemma 3.4. *Suppose that the complex $\Phi_{\overline{g}} j_* \mathcal{L}$ is acyclic, $M > 0$. Then $d\overline{g}_{p_0}|L \not\equiv 0$ whenever L is a limit at p_0 of tangent spaces to $Z \cap \{f(z)e^{<u,z>} = const\}$ at points p near p_0 where $||u|| < M$.*

Proof. We replace $\overline{Z}$ by $\overline{Z} \times \mathbb{C}^l$ with the induced stratification which is Whitney regular, too, f by $F : (z, u) \mapsto f(z)e^{<u,z>}$ and $\overline{g}$ by $(z, u) \mapsto \overline{g}(z)$. Then apply Lemma 3.1 to the new situation (the hypothesis is fulfilled).

Let us look at the tangent space to $Z \cap \{f(z)e^{<u,z>} = const\}$ at a point p. This corresponds to the tangent space to $(Z \times \mathbb{C}^l) \cap \{F = const\}$ at (p, u), intersected by the space $du = 0$. Finally we have Thom's a_F-condition which implies that the limits of tangent spaces to $(Z \times \mathbb{C}^l) \cap \{F = const\}$ at points $(0, u_0)$ are of the form $L \times \mathbb{C}^l$. Altogether this implies our assertion. $\square$

Theorem 3.5. *Assume that the complex $\Phi_{\overline{g}} j_* \mathcal{L}$ is acyclic. Then g defines a C^∞ fibre bundle over $\{|t| \ll 1\}$.*

Proof. Let p_0 and f be as above. By partition of unity we may construct a C^∞ function ϕ which is near p_0 of the form $\phi = |f|^2 \rho$ for some positive smooth function ρ. Then $(\partial \phi)_z = \overline{f(z)}\rho(z)df_z + |f(z)|^2 \partial \rho_z = \overline{f(z)}\rho(z)e^{-<u,z>}d(fe^{<u,z>})_z$ where u is determined by $\langle u, \ldots \rangle = \partial \rho_z / \rho(z)$. Therefore, by Lemma 3.4, the restriction of $\overline{g}$ to the fibres $\phi = const \neq 0$ near Z_∞ is submersive.

Now we may build a smooth vector field v on Z such that $d\phi(v) \equiv 0$, $dg(v) \equiv 1$ (resp. $\equiv i$). Integrating we obtain the desired smooth trivialization. $\square$

Corollary 3.6. *Let $\mathcal{F}$ be a locally constant sheaf of R-modules on Z. Under the assumption of Theorem 3.5 the sheaves $R^k g_* \mathcal{F}$ are constant with fibres isomorphic to $H^k(g^{-1}(t), \mathcal{F})$ over $\{|t| \ll 1\}$.*

Proof. This is a direct consequence of Theorem 3.5 but also of Lemma 1.3.

$\square$

Furthermore we can derive a result which has been proved in the case $Z = \mathbb{C}^m$ by Parusiński [9] who generalized an earlier result of Siersma and Tibăr [11]:

Corollary 3.7. *Assume that all complexes $\Phi_{\bar{g}} j_* \mathbb{C}_Z$ are acyclic outside some finite set and that the functions $t \mapsto \chi(g^{-1}(t))$ are constant. Then g defines a C^∞ fibre bundle over $\{|t| \ll 1\}$.*

Proof. This follows from Theorem 1.7c) and Theorem 3.5. □

3.2. From now on we drop the hypothesis that g is submersive but still make the following

Assumption. Z is smooth and everywhere of dimension m.

As before, let $\mathcal{L}$ be a locally free R_Z-module of finite rank with non-vanishing stalks.

Lemma 3.8. *The following conditions are equivalent:*

a) *g is submersive over $\{|t| \ll 1\}$,*
b) *$\Phi_g \mathcal{L}$ is acyclic.*

Proof. a) $\Rightarrow$ b) clear.

b) $\Rightarrow$ a): It is sufficient to show that $g^{-1}(\{0\})$ is a smooth hypersurface in Z. Take a Whitney regular stratification of Z such that $g^{-1}(\{0\})$ and $Sing\, g^{-1}(\{0\})$ are unions of strata. Assume $Sing\, g^{-1}(\{0\}) \neq \emptyset$: Let us take a point z in a stratum of $Sing\, g^{-1}(\{0\})$ of maximal dimension. Let Z' be a normal slice at z in Z. Then $(\Phi_g \mathcal{L})_z = (\Phi_{g|Z'} \mathcal{L}|Z')_z$. Now Z' is smooth, and z is at most an isolated singular point of $g|Z'$. The acyclicity of $(\Phi_{g|Z'} \mathcal{L}|Z')_z$ implies that the reduced Milnor number of $g|Z'$ at z vanishes, so $g|Z'$ is regular at z, hence g, too, which contradicts our assumption. □

Now we can show that Theorem 3.5 still holds without the assumption that g is submersive:

Theorem 3.9. *Assume that the complex $\Phi_{\bar{g}} j_* \mathcal{L}$ is acyclic. Then g defines a C^∞ fibre bundle over $\{|t| \ll 1\}$.*

Proof. Because of Lemma 3.8 g is submersive over $\{|t| \ll 1\}$; note that $(\Phi_{\bar{g}} j_* \mathcal{L})|g^{-1}(\{0\}) = \Phi_g \mathcal{L}$. Now apply Theorem 3.5. □

3.3. Now let us pass to the assumptions of section 2.2. As before let $\mathcal{L}$ be a locally free R_Z-module of finite rank with non-vanishing stalks. Furthermore let us suppose that Z is smooth (as before), pure-dimensional and that g does not annihilate any of the connected components of Z.

Theorem 3.10. *Suppose that* $\dim Z = m$. *Put* $l := \dim \operatorname{Sing} g^{-1}(\{0\}) \cup$ $(\overline{g}^{-1}(\{0\}) \cap Z_\infty) \leq m - 1$. *The following conditions are equivalent:*

a) *For all* $0 \leq k \leq l, 0 < j_1 < \ldots < j_k \leq N$ *there is a dense subset* S_j *of* $\mathbb{C}^k$ *such that for all* $(s_{j_1}^*, \ldots, s_{j_k}^*) \in S_j$ *the following holds: The number* $\chi(\{z_{j_1} = s_{j_1}^*, \ldots, z_{j_k} = s_{j_k}^*, z_{N+1} = t\}, \mathcal{L})$ *is independent of* t, $|t| \ll 1$.
b) $\Phi_{\overline{g}} j_* \mathcal{L}$ *is acyclic.*
c) *For all* $0 \leq k \leq l, 0 < j_1 < \ldots < j_k \leq N$ *there is a dense subset* S_j *of* $\mathbb{C}^k$ *such that for all* $(s_{j_1}^*, \ldots, s_{j_k}^*) \in S_j$ *the following holds: The map* $g | \{z_{j_1} = s_{j_1}^*, \ldots, z_{j_k} = s_{j_k}^*\}$ *defines a* C^∞ *fibre bundle over* $\{|t| \ll 1\}$.

Proof. a) $\Rightarrow$ b): see Theorem 2.1.
b) $\Rightarrow$ c): Apply Theorem 3.5.
c) $\Rightarrow$ a): clear.   □

Corollary 3.11. *Suppose that* $\dim Z = m$, *that* $\overline{g^{-1}(\{0\})} = \overline{g}^{-1}(\{0\})$ *(e.g.* $Z = \mathbb{C}^m$) *and that* $g^{-1}(\{0\})$ *is generically reduced. The following conditions are equivalent:*

a) *For all* $0 \leq k \leq m - 2, 0 < j_1 < \ldots < j_k \leq N$ *there is a dense subset* S_j *of* $\mathbb{C}^k$ *such that for all* $(s_{j_1}^*, \ldots, s_{j_k}^*) \in S_j$ *the following holds: The number* $\chi(\{z_{j_1} = s_{j_1}^*, \ldots, z_{j_k} = s_{j_k}^*, z_{N+1} = t\}, \mathcal{L})$ *is independent of* t, $|t| \ll 1$.
b $\Phi_{\overline{g}} j_* \mathcal{L}$ *is acyclic.*
c) *For all* $0 \leq k \leq m - 2, 0 < j_1 < \ldots < j_k \leq N$ *there is a dense subset* S_j *of* $\mathbb{C}^k$ *such that for all* $(s_{j_1}^*, \ldots, s_{j_k}^*) \in S_j$ *the following holds: The map* $g | \{z_{j_1} = s_{j_1}^*, \ldots, z_{j_k} = s_{j_k}^*\}$ *defines a* C^∞ *fibre bundle over* $\{|t| \ll 1\}$.

Proof. With the notations of Theorem 3.10 we have $l \leq m - 2$.   □

Remark 4. The condition c) in Theorem 3.10 resp. Corollary 3.11 does not depend on the choice of $\mathcal{L}$, so the same holds for a) and b) which is not obvious from the beginning!

Proof of the theorem of the introduction. This follows from Corollary 3.11.

                                                                        □

In the case $m = 2$ we obtain the following generalization of the theorem of Hà-Lê:

Corollary 3.12. *Suppose that* $\dim Z = 2$, *that* $\overline{g^{-1}(\{0\})} = \overline{g}^{-1}(\{0\})$ *(e.g.* $Z = \mathbb{C}^2$) *and that* $g^{-1}(\{0\})$ *is generically reduced. If* $\chi(\{g = t\})$ *is independent of* t *for* $|t| \ll 1$ *the mapping* g *defines a* C^∞ *fibre bundle over* $\{|t| \ll 1\}$.

Let us show that none of the hypotheses of Corollary 3.12 can be dropped.

First, the hypothesis that $g^{-1}(\{0\})$ is generically reduced cannot be omitted:

Example 3.13. Look at $\mathbb{C}^* \times \mathbb{C} \to \mathbb{C}: (z_1, z_2) \mapsto z_2^2$. The Euler characteristic of all fibres vanishes but the fibre over 0 is a multiple one and therefore nowhere reduced. The other assumptions of Corollary 3.12 are fulfilled: We take $Z = \{(z_1, z_2, z_3) \in \mathbb{C}^3 \mid z_1 z_2 = 1\}$, $g(z_1, z_2, z_3) = z_3^2$.

Second, we cannot renounce to the hypothesis that $\overline{g^{-1}(\{0\})} = \overline{g}^{-1}(\{0\})$ if we pass from $\mathbb{C}^2$ to a smooth surface:

Example 3.14. The mapping $\{z \in \mathbb{C}^2 \mid z_1 z_2 \neq 0\} \to \mathbb{C}: (z_1, z_2) \mapsto z_1 z_2$ has no critical points but does not define a topological fibre bundle since the fibre over 0 is empty, the others are isomorphic to $\mathbb{C}^*$. The Euler characteristic of all fibres is 0. According to section 2.2 we can rewrite our example as follows: $\{z \in \mathbb{C}^3 \mid z_1 z_2 z_3 = 1\} \longrightarrow \mathbb{C}: z \mapsto z_1 z_2$. The assumption $\overline{g^{-1}(\{0\})} = \overline{g}^{-1}(\{0\})$ is not fulfilled because $\overline{g^{-1}(\{0\})} = \emptyset$.

Third, the dimension assumption in Corollary 3.12 cannot be dropped:

Example 3.15. Let us modify the Broughton example $\mathbb{C}^2 \longrightarrow \mathbb{C}: (z_1, z_2) \longrightarrow z_1(z_1 z_2 - 1)$ by multiplying by a factor $\mathbb{C}^*$, i.e. $\mathbb{C}^2 \times \mathbb{C}^* \to \mathbb{C}: z \mapsto z_1(z_1 z_2 - 1)$. Since the Euler characteristic is multiplicative the Euler characteristic of every new fibre is 0. According to section 2.2 we may write our example as follows: $Z := \{z \in \mathbb{C}^4 \mid z_3 z_4 = 1\}$, $g : Z \longrightarrow \mathbb{C}: z \mapsto z_1^2 z_2 - z_1$. The Euler characteristic of the fibres of g is constant, as we have seen, but not the one of the fibres of $g|\{z \in Z \mid z_3 = z_3^*\}$ if $z_3^* \neq 0$, because these fibres correspond to the fibres of the original Broughton example. Of course 0 is an atypical value of g.

References

1. J. Briançon, P. Maisonobe and M. Merle. Localisation des systèmes différentiels, stratifications de Whitney et condition de Thom. *Invent. math.* **117** (1994) 531–550.
2. S. O. Broughton, On the topology of polynomial hypersurfaces. In: Singularities, Part 1 (Arcata, Calif. 1981), pp. 167–178. *Proc. Symp. Pure Math.* **40**, Amer. Math. Soc., Prov., RI 1983.
3. H. A. Hamm, On the cohomology of fibres of polynomial maps. In: *Trends in Singularities*, pp. 99–113. Trends Math., Birkhäuser, Basel 2002.
4. H. A. Hamm and Lê Dũng Tráng, Vanishing theorems for constructible sheaves I. *J. Reine Angew. Math.* **471** (1996) 115–138.
5. H. A. Hamm and Lê Dũng Tráng, Théorèmes d'annulation pour les faisceaux algébriquement constructibles. *C.R. Acad. Sci. Paris (I)* **327** (1998) 759–762.

6. H.V. Hà and Lê Dũng Tráng, Sur la topologie des polynômes complexes. *Acta Math. Vietnam.* **9** (1984) 21–32 .
7. M. Kashiwara and P. Schapira, *Sheaves on Manifolds.* Springer-Verlag: Berlin 1990.
8. D. Massey, Critical points of functions on singular spaces. *Topology and its Appl.* **103** (2000) 55–93.
9. A. Parusiński, A note on singularities at infinity of complex polynomials. In: Symplectic singularities and geometry of gauge fields, pp. 131-141. Banach Center Publ. **39** (1997).
10. J. Schürmann, Topology of singular spaces and constructible sheaves. *Monografie Mat. (N.S.)* **63**. Birkhäuser, Basel 2003.
11. D. Siersma and M. Tibăr, Singularities at infinity and their vanishing cycles. *Duke Math. J.* **80** (1995) 771–783.
12. M. Tibăr, Asymptotic Equisingularity and Topology of Complex Hypersurfaces. *Int. Math. Research Notices* **18** (1998) 979–990.
13. M. Tibăr, *Polynomials and Vanishing Cycles.* Cambridge Univ. Press: Cambridge 2007.
14. M. G. Zaidenberg, Isotrivial families of curves on affine surfaces, and the characterization of the affine plane (Russian). *Izv. Akad. Nauk SSSR Ser. Mat.* **51**, *no. 3, 534-567, 688 (1987) = Math. USSR - Izv.* **30** (1988) 503–532.

Helmut A. Hamm
Mathematisches Institut der WWU
Einsteinstr. 62
48149 Münster
Germany
hamm@uni-muenster.de

14

Answer to a question of Zariski

A. HEFEZ AND M. E. HERNANDES

Abstract

We give a negative answer to a question asked by Oscar Zariski in the book The Moduli Problem for Plane Branches, *about genericity, normal forms and analytic classification of plane branches.*

1. Introduction

The analytic classification program for plane branches, within an equisingularity class, started by Ebey [1] and Zariski [4], was finished by the authors of the present note in [3], whose results allow us also to answer, in the negative, a specific question about normal forms, genericity and analytic classification asked by Zariski in [4, *Remark 6.14*, Chapter VI]. The scope of this paper is widened by giving the complete analytic classification of the equisingularity class determined by the semigroup of values $\langle 7, 8 \rangle$, where our counterexample was picked. This may be useful in other contexts, since very few equisingularity classes are completely classified in the literature. Throughout this paper, we use the notation of [3].

2. Classification of the equisingularity class $\langle 7, 8 \rangle$

In this section we give the analytic classification of plane branches belonging to the equisingularity class given by the semigroup of values $\Gamma = \langle 7, 8 \rangle$.

2000 *Mathematics Subject Classification* 14H20 (primary), 14Q05, 14Q20, 32S10 (secondary). Both authors were partially supported by grants fromCNPq and a PROCAD/CAPES project.

Since the conductor of Γ is 42, we have that any plane branch in this equisingularity class is $\mathcal{A}$-equivalent to a branch in the family

$$\Sigma_\Gamma = \left\{ \left(t^7, t^8 + \sum_{8 < i < 42} a_i t^i \right); \ a_i \in \mathbb{C} \right\},$$

where $\mathcal{A}$-equivalence means equivalence under analytic changes of coordinates ρ in the source and σ in the target as germs of functions $(\mathbb{C}, 0) \to (\mathbb{C}^2, (0,0))$, given as follows (c.f. [3, (2.2)]):

$$\sigma(X, Y) = (r^7 X + p, r^8 Y + q)$$

and

$$t_1 = \rho(t) = rt \left(1 + \frac{p\left(t^7,\ t^8 + \sum_{8<i<42} a_i t^i\right)}{r^7 t^7} \right)^{\frac{1}{7}},$$

where $r \in \mathbb{C}^*$, p is in the ideal (X^2, Y) and q is in the ideal $(X, Y)^2$.

The application of [2, *Algorithm 4.10*] and [3, *Theorem 2.1*] can give the analytic classification of branches in the equisingularity class determined by any semigroup of values Γ.

In the case where $\Gamma = \langle 7, 8 \rangle$, the result is that any plane branch with this given semigroup of values is $\mathcal{A}$-equivalent to one normal form in **Table 1**, and two parametrizations, $\varphi = (t^7, t^8 + \sum_{i>8} a_i t^i)$ and $\varphi' = (t^7, t^8 + \sum_{i>8} a_i' t^i)$, belonging to the table, are equivalent if and only if they are homothetic, meaning that $a_i = r^{\lambda - i} a_i'$, for all i, where r is a $(\lambda - 8)$th root of unity and λ is the order of the first term in $\sum_{i>8} a_i t^i$.

3. The question of Zariski

Zariski dedicated [4, *Sections 4, 5 and 6*, Chapter VI] to the study of branches with semigroups of the form $\Gamma = \langle v_0, v_0 + 1 \rangle$, where the following result was proved:

Theorem 3.1 ([4, Theorem 6.12]). *Let $v_0 \geq 5$, and let*

$$\varphi = \left(t^{v_0}, t^{v_0+1} + a_{v_0+3} t^{v_0+3} + \cdots + a_{2v_0-1} t^{2v_0-1} + \sum_{i \in \cup_{s=2}^q \overline{L}_s} a_i t^i \right),$$

where

$$\overline{L}_s = \{s(v_0 + 1) + 1, \ldots, (s+1)v_0 - 1\} \text{ and } q = \left[\frac{v_0 - 3}{2}\right] \dagger$$

and such that $a_i = 0$, whenever i is one of the first $s + 1$ elements of $\overline{L}_s$ for all $2 \leq s \leq q$. Then two generic parametrizations of the above form (called canonical parametrizations) are $\mathcal{A}$-equivalent if and only if they are homothetic.

Zariski proved that the above theorem is true for $2 \leq v_0 \leq 6$ without the genericity condition on the parameters and asks the following question (c.f. [4, *Remark 6.14*]):

Is the above theorem true for $v_0 > 6$, without the genericity assumption on the coefficients of the parametrizations?

The answer is no! An example arises exactly in the first and simplest situation not analyzed by Zariski, namely for $\Gamma = \langle 7, 8 \rangle$, where Zariski's canonical form becomes:

$$(t^7, t^8 + at^{10} + bt^{11} + ct^{12} + dt^{13} + et^{20}).$$

The trick to build such an example consists in taking a branch which is simultaneously in one of our normal forms and in Zariski's canonical form. Then we transform it by $\mathcal{A}$-equivalence into a branch in canonical form, but with an extra term (hence not in normal form), so that the two branches are not homothetic.

To produce our example, we start with a branch:

$$\varphi = \left(t^7, t^8 + t^{10} + t^{11} + \frac{11}{4}t^{12} + a_{13}t^{13}\right),$$

where $a_{13} \neq \frac{21}{4}$. This corresponds to a branch in row 2 of table 14.1 and it is obviously in Zariski's canonical form.

Since any term of order greater than 20 is irrelevant for our analysis and may be eliminated by suitable changes of coordinates, it is enough to consider the changes of coordinates σ and ρ of section 2 with $p = b_1 Y + b_2 X^2 + b_3 XY + b_4 Y^2$ and $q = c_1 X^2 + c_2 XY + c_3 Y^2$. Computing the parameters b_i and c_j in order to keep Zariski's canonical form, we get

$$c_1 = b_1 = 0, \quad c_2 = \frac{8}{7}b_2, \quad c_3 = -\frac{12}{7}b_2, \quad b_3 = -\frac{3}{2}b_2, \quad b_4 = -\frac{1}{4}b_2.$$

$\dagger$ In view of [4, *Definition 6.10* and *Remark 6.11*], one should have $s < \frac{v_0 - 2}{2}$, which leads to $q = \left[\frac{v_0 - 3}{2}\right]$ and not $q = \left[\frac{v_0 - 4}{2}\right]$ as stated incorrectly in [4, *Proposition 6.12*] and not used as such in the sequel.

Table 14.1. *Normal forms for plane branches with* $\Gamma = \langle 7, 8 \rangle$

Condition	Normal Form
$a_{12} \neq \frac{13+9a_{11}^2}{8}$	$(t^7, t^8 + t^{10} + a_{11}t^{11} + a_{12}t^{12} + a_{13}t^{13} + a_{20}t^{20})$
$a_{13} \neq \frac{39}{10}a_{11} + \frac{27}{20}a_{11}^3$	$(t^7, t^8 + t^{10} + a_{11}t^{11} + \frac{13+9a_{11}^2}{8}t^{12} + a_{13}t^{13} + a_{19}t^{19})$
$a_{20} \neq A$	$(t^7, t^8 + t^{10} + a_{11}t^{11} + \frac{13+9a_{11}^2}{8}t^{12}$ $+ \left(\frac{39}{10}a_{11} + \frac{27}{20}a_{11}^3\right)t^{13} + a_{19}t^{19} + a_{20}t^{20})$
$a_{20} = A$	$(t^7, t^8 + t^{10} + a_{11}t^{11} + \frac{13+9a_{11}^2}{8}t^{12}$ $+ \left(\frac{39}{10}a_{11} + \frac{27}{20}a_{11}^3\right)t^{13} + a_{19}t^{19} + a_{20}t^{20} + a_{27}t^{27})$
	$(t^7, t^8 + t^{11} + a_{12}t^{12} + a_{13}t^{13} + a_{20}t^{20})$
	$(t^7, t^8 + t^{12} + a_{13}t^{13} + a_{18}t^{18})$
$a_{18} \neq -\frac{1}{2}$	$(t^7, t^8 + t^{13} + a_{18}t^{18} + a_{19}t^{19} + a_{26}t^{26})$
	$(t^7, t^8 + t^{13} - \frac{1}{2}t^{18} + a_{19}t^{19} + a_{26}t^{26})$
$a_{20} \neq \frac{121}{120}a_{19}^2$	$(t^7, t^8 + t^{18} + a_{19}t^{19} + a_{20}t^{20} + a_{27}t^{27})$
	$(t^7, t^8 + t^{18} + a_{19}t^{19} + \frac{121}{120}a_{19}^2t^{20} + a_{27}t^{27})$
	$(t^7, t^8 + t^{19} + a_{20}t^{20})$
	$(t^7, t^8 + t^{20} + a_{26}t^{26})$
	$(t^7, t^8 + t^{26} + a_{27}t^{27})$
	$(t^7, t^8 + t^{27})$
	$(t^7, t^8 + t^{34})$
	(t^7, t^8)

where

$$A = \frac{11}{4}a_{11}a_{19} - \frac{357}{512} - \frac{47399}{2560}a_{11}^2 - \frac{10097}{320}a_{11}^4 - \frac{17523}{1280}a_{11}^6 - \frac{2187}{1280}a_{11}^8.$$

In this way we get the following equivalent branch to φ:

$$\varphi_1 = \left(t_1^7, t_1^8 + t_1^{10} + t_1^{11} + \frac{11}{4}t_1^{12} + a_{13}t_1^{13} + 5b_2\left(\frac{3}{4} - \frac{1}{7}a_{13}\right)t_1^{20}\right).$$

Since $a_{13} \neq \frac{21}{4}$, for every choice of $b_2 \neq 0$ we get a branch φ_1 equivalent to φ and in Zariski's canonical form, but without being homothetically equivalent to it, giving a negative answer to Zariski's question.

References

1. S. Ebey, 'The classification of singular points of algebraic curves', *Trans. Amer. Math. Soc.* **118** (1965) 454–471.
2. A. Hefez and M. E. Hernandes, 'Standard bases for local rings of branches and their modules of differentials', *J. Symb. Comput.* **42** (2007) 178–191.
3. A. Hefez and M. E. Hernandes, 'The analytic classification of plane branches', *arXiv: 0707.4502.*
4. O. Zariski, 'The moduli problem for plane branches', *University Lecture Series* **39**, AMS (2006).

A. Hefez
Instituto de Matemática
Univ. Federal Fluminense
R. Mario Santos Braga, s/n
24020-140 Niterói, RJ
Brazil
hefez@mat.uff.br

M. E. Hernandes
Departamento de Matemática
Univ. Estadual de Maringá
Av. Colombo, 5790
87020-900 Maringá, PR
Brazil
mehernandes@uem.br

15

Projections of timelike surfaces in the de Sitter space

SHYUICHI IZUMIYA AND FARID TARI

Abstract

We study in this paper projections of embedded timelike hypersurfaces M in S_1^n along geodesics. We deal in more details with the case of surfaces in S_1^3, characterise geometrically the singularities of the projections and prove duality results analogous to those of Shcherbak for central projections of surfaces in $\mathbb{R}P^3$.

1. Introduction

We study in this paper the contact of timelike hypersurfaces in the de Sitter space S_1^n with geodesics. The contact is measured by the singularities of projections along geodesics to transverse sets. There are three types of geodesics in S_1^n, spacelike, timelike and lightlike ([13]). In the case of spacelike and timelike geodesics we project, respectively, to orthogonal hyperbolic and elliptic de Sitter hyperquadrics. For a lightlike geodesic, we project to a transverse space as the orthogonal space contains the geodesic. We give in section 3 the expressions for the families of projections along the three types of geodesics.

Given a point p on a timelike hypersurface $M \subset S_1^n$, there is a well defined unit normal vector $e(p) \in S_1^n$ to M at p; see [4] and section 2. If M is orientable, then $e(p)$ is globally defined. However, it is always locally defined and our investigation here is local in nature. We have the (de Sitter) Gauss map

$$\mathbb{E} : M \to S_1^n$$
$$p \mapsto e(p)$$

2000 *Mathematics Subject Classification* 53A35, 53B30, 58K05.
FT was partially supported by a Royal Society International Outgoing Short Visit grant.

with the property that its differential map (the Weingarten map) $-d\mathbb{E}_p$ is a self-adjoint operator on $T_p M$ ([4]). As M is timelike, the restriction of the pseudo-scalar product in the Minkowski space to $T_p M$ is also a pseudo scalar product. Therefore, $-d\mathbb{E}_p$ does not always have real eigenvalues. When these are real, we call the associated eigenvectors the *principal directions* of M at p. For timelike surfaces in S_1^3 there is a curve, labelled the *lightlike principal locus* in [7, 10] (LPL for short), that separates regions on M where there are two distinct principal directions and regions where there are none. On the LPL there is a unique double principal direction. One can also define the concept of an asymptotic direction on a surface M in S_1^n. We say that $v \in T_p M$ is an asymptotic direction at $p \in M$ if $\langle d\mathbb{E}_p(v), v \rangle = 0$, see section 2 for details.

We show in section 4 that the singularities of the projections of surfaces in S_1^3 along the three types of geodesics capture some aspects of the extrinsic geometry of the surface related to the Gauss map $\mathbb{E}$. Indeed, the singularity at $p \in M$ of a given projection is of type cusp or worse if and only if the tangent to the geodesic at p is an asymptotic direction (Theorems 4.2 and 4.3). We characterise geometrically in section 4 all the generic singularities of the projections along geodesics. For instance the LPL is picked up as the locus of points where the projections along the lightlike geodesics have singularities of type cusp. The projections also pick up special points on the LPL (Theorem 4.3), namely the singular points of the configuration of the lines of principal curvature.

The first author introduced duality concepts between hypersurfaces in the pseudo spheres in the Minkowski space [4, 5]; see section 6 for details. We use these concepts to prove in section 5 duality results between some surfaces associated to a timelike surface $M \subset S_1^3$ and special curves on the dual surface M^* of M. The results are analogous to those of Shcherbak in [16] for central projections of surfaces in $\mathbb{R}P^3$, and to those of Bruce-Romero Fuster in [2] for orthogonal projections of surfaces in the Euclidean space $\mathbb{R}^3$.

The work in this paper is part of a project on projections of submanifolds embedded in the pseudo-spheres in the Minkowski space $\mathbb{R}_1^n$ via singularity theory. We dealt in [8] with the contact of (hyper)surfaces with geodesics in the hyperbolic space (see also [11]) and in [9] with their contact with horocycles.

2. Preliminaries

We start by recalling some basic concepts in hyperbolic geometry (see for example [14] for details). The *Minkowski $(n+1)$-space* $(\mathbb{R}_1^{n+1}, \langle, \rangle)$ is the $(n+1)$-dimensional vector space $\mathbb{R}^{n+1}$ endowed by the *pseudo scalar* product $\langle x, y \rangle = -x_0 y_0 + \sum_{i=1}^{n} x_i y_i$,   for $x = (x_0, \ldots, x_n)$ and $y = (y_0, \ldots, y_n)$ in $\mathbb{R}_1^{n+1}$. We say that a vector x in $\mathbb{R}_1^{n+1} \setminus \{0\}$ is *spacelike, lightlike* or *timelike* if $\langle x, x \rangle > 0, = 0$ or < 0 respectively. The norm of a vector $x \in \mathbb{R}_1^{n+1}$ is defined by $\|x\| = \sqrt{|\langle x, x \rangle|}$. Given a vector $v \in \mathbb{R}_1^{n+1}$ and a real number c, a hyperplane with pseudo normal v is defined by

$$H P(v, c) = \{x \in \mathbb{R}_1^{n+1} \mid \langle x, v \rangle = c\}.$$

We say that $H P(v, c)$ is a *spacelike, timelike* or *lightlike hyperplane* if v is timelike, spacelike or lightlike respectively. We have the following three types of pseudo-spheres in $\mathbb{R}_1^{n+1}$:

$$\begin{aligned}
\textit{Hyperbolic n-space}: \quad & H^n(-1) = \{x \in \mathbb{R}_1^{n+1} \mid \langle x, x \rangle = -1\}, \\
\textit{de Sitter n-space}: \quad & S_1^n = \{x \in \mathbb{R}_1^{n+1} \mid \langle x, x \rangle = 1\}, \\
\textit{(open) lightcone}: \quad & LC^* = \{x \in \mathbb{R}_1^{n+1} \setminus \{0\} \mid \langle x, x \rangle = 0\}.
\end{aligned}$$

We also define the *lightcone $(n-1)$-sphere*

$$S_+^{n-1} = \{x = (x_0, \ldots, x_n) \mid \langle x, x \rangle = 0, \ x_0 = 1\}.$$

A hypersurface given by the intersection of S_1^n with a spacelike (resp. timelike) hyperplane is called an *elliptic hyperquadric* (resp. *hyperbolic hyperquadric*).

A smooth embedded hypersurface M in S_1^n is said to be timelike if its tangent space $T_p M$ at any point $p \in M$ is a timelike vector space. Some aspects of the extrinsic geometry of timelike hypersurfaces in S_1^n are studied in [4, 7, 10].

Let M be a timelike hypersurface embedded in S_1^n. Given a local chart $i : U \to M$, where U is an open subset of $\mathbb{R}^{n-1}$, we denote by $x : U \to S_1^n$ such embedding, identify $x(U)$ with U through the embedding x and write $M = x(U)$. Since $\langle x, x \rangle \equiv 1$, we have $\langle x_{u_i}, x \rangle \equiv 0$, for $i = 1, \ldots, n-1$, where $u = (u_1, \ldots, u_{n-1}) \in U$. We define the spacelike unit normal vector $e(u)$ to M at $x(u)$ by

$$e(u) = \frac{x(u) \wedge x_{u_1}(u) \wedge \ldots \wedge x_{u_{n-1}}(u)}{\|x(u) \wedge x_{u_1}(u) \wedge \ldots \wedge x_{u_{n-1}}(u)\|},$$

where $\wedge$ denotes the wedge product of n vectors in $\mathbb{R}_1^{n+1}$ (see for example [14]). The de Sitter Gauss map is defined in [4] by

$$\begin{aligned}
\mathbb{E} : M &\to \ S_1^n \\
p &\mapsto e(p)
\end{aligned}$$

At any $p \in M$ and $v \in T_p M$, one can show that $D_v \mathbb{E} \in T_p M$, where D_v denotes the covariant derivative with respect to the tangent vector v. The linear transformation $A_p = -d\mathbb{E}(p)$ is called the *de Sitter shape operator*. Because the surface M is timelike, the restriction of the pseudo scalar product in $\mathbb{R}_1^n$ to M is a pseudo scalar product. Therefore, the shape operator A_p does not always have real eigenvalues. When these are real, we call them the *principal curvatures* of M at p and the corresponding eigenvectors are called the *principal directions*.

We now review some concepts of the extrinsic geometry of embedded timelike surfaces M in S_1^3 (so $n = 3$ above). We denote by (u, v) the coordinates in $U \subset \mathbb{R}^2$. The first fundamental form of the surface M at a point p is the quadratic form $I_p : T_pM \rightarrow \mathbb{R}$ given by $I_p(v) = \langle v, v \rangle$. If $v = a\boldsymbol{x}_u + b\boldsymbol{x}_v \in T_pM$, then $I_p(v) = Ea^2 + 2Fab + Gb^2$, where

$$E = \langle \boldsymbol{x}_u, \boldsymbol{x}_u \rangle, \quad F = \langle \boldsymbol{x}_u, \boldsymbol{x}_v \rangle, \quad G = \langle \boldsymbol{x}_v, \boldsymbol{x}_v \rangle.$$

are the coefficients of the first fundamental form. Because M is timelike, we have $EG - F^2 < 0$, so at any point $p \in x(U) \subset M$ there are two lightlike directions in T_pM. These are the solutions of $I_p(v) = 0$.

The second fundamental form of the surface M at the point p is the quadratic form $II_p : T_pM \rightarrow \mathbb{R}$ given by $II_p(v) = \langle A_p(v), v \rangle$, with $A_p = -d\mathbb{E}(p)$. For $v = a\boldsymbol{x}_u + b\boldsymbol{x}_v \in T_pM$, we have $II_p(v) = la^2 + 2mab + nb^2$, where

$$
\begin{aligned}
l &= -\langle e_u, \boldsymbol{x}_u \rangle = \langle e, \boldsymbol{x}_{uu} \rangle \\
m &= -\langle e_u, \boldsymbol{x}_v \rangle = \langle e, \boldsymbol{x}_{uv} \rangle = \langle e, \boldsymbol{x}_{vu} \rangle = -\langle e_v, \boldsymbol{x}_u \rangle \\
n &= -\langle e_v, \boldsymbol{x}_v \rangle = \langle e, \boldsymbol{x}_{vv} \rangle
\end{aligned}
$$

The shape operator A determines pairs of foliations on M ([10]). A *line of principal curvature* is a curve on the surface whose tangent at all points is a principal direction. These form a pair of foliation in some region of M, given by the following binary differential equation

$$(Gm - Fn)dv^2 + (Gl - En)dvdu + (Fl - Em)du^2 = 0. \qquad (2.1)$$

The discriminant function of this equation is

$$\delta(u, v) = \left((Gl - En)^2 - 4(Gm - Fn)(Fl - Em)\right)(u, v). \qquad (2.2)$$

When $\delta(u, v) > 0$, there are two distinct principal directions at $p = x(u, v)$. These coincide at points where $\delta(u, v) = 0$. There are no principal directions at points where $\delta(u, v) < 0$. We labelled in [7, 10] the locus of points where $\delta(u, v) = 0$ the *Lightlike Principal Locus* (*LPL* for short).

Proposition 2.1 ([7, 10]). *(1) For a generic timelike surface $M \in S_1^3$, the LPL is a curve which is smooth except at isolated points where it has Morse singularities of type node. The singular points are where the shape operator is* *a multiple of the identity, and are labelled "timelike umbilic points". The LPL is also the set of points on M where the two principal directions coincide and become lightlike.*

(2) The LPL divides the surfaces into two regions. In one of them there are no principal directions and in the other there are two distinct principal directions at each point. In the latter case, the principal directions are orthogonal and one is spacelike while the other is timelike.

We also have the concept of asymptotic directions. A direction $v \in T_p M$ is called *asymptotic* if $\mathrm{II}_p(v) = \langle A_p(v), v \rangle = 0$ ([10]). An *asymptotic curve* is a curve on the surface whose tangent at all points is an asymptotic direction. The equation of the asymptotic curves is

$$ndv^2 + 2mdudv + ldu^2 = 0. \qquad (2.3)$$

The discriminant of equation (2.3) is the locus of points where $m^2 - nl$ vanishes. This is the set of points where the Gauss-Kronecker curvature $K = \det(A_p) = (m^2 - nl)/(F^2 - EG) = 0$ vanishes, and is labelled the *(de Sitter) parabolic set*. The parabolic set of a generic surface, when not empty, is a smooth curve. It meets the LPL at isolated points and the two curves are tangential at their points of intersection ([10]). In the region $K > 0$ there are two distinct asymptotic directions and there are no asymptotic directions in the region $K < 0$. On the parabolic set $K = 0$ there is a unique asymptotic direction. This direction is tangent to the parabolic set when it is lightlike and this occurs at the point of tangency of the parabolic set with the LPL. These points are the *folded singularities* of the asymptotic curves ([10]). On one side of such points, the unique asymptotic direction is spacelike and on the other side it is timelike ([10]). (On the LPL one of the asymptotic directions is lightlike and coincides with the unique principal direction there. The generic local topological configurations of the principal and asymptotic curves are studied in [10].)

Let $\gamma : I \to M \subset S_1^3$ be a regular curve on a timelike surface M. We can parametrise γ by arc-length and assume that $\gamma(s)$ is unit speed, that is, $\langle \gamma'(s), \gamma'(s) \rangle = \pm 1$. Let $t(s) = \gamma'(s)$ and $w(s) = \gamma(s) \wedge t(s) \wedge e(s)$, where $e(s) = e(\gamma(s))$. The acceleration vector $\gamma''(s)$ is written, in the frame $\{\gamma(s), t(s), e(s), w(s)\}$, in the form

$$\gamma''(s) = \mp \gamma(s) + \kappa_n(s)e(s) + \kappa_g(s)w(s)$$

where $\kappa_g(s)$ is the *geodesic curvature* of γ on M at $\gamma(s)$. When the curve γ is not parametrised by arc length, we re-parametrise by arc-length

$l(s) = \int_0^s ||\gamma'(t)|| dt$ and the formula for the curvature is

$$\kappa_g(t) = \frac{1}{l'(t)^3} \langle l'(t)\gamma''(t) - l''(t)\gamma'(t), w(t) \rangle.$$

A unit speed curve γ is geodesic on M if and only if $\kappa_g \equiv 0$. A point $\gamma(s)$ is called a *geodesic inflection* if $\kappa_g(s) = 0$.

Definition 2.2. The flecnodal curve of a timelike surface in S_1^3 is the locus of geodesic inflections of the of the asymptotic curves.

3. The family of projections along geodesics in S_1^n

We exhibit in this section the expressions for the family of projections along geodesics in S_1^n for $n \geq 3$ and deal in more details with the case $n = 3$ in the following section. We start with projections along timelike geodesics.

Let $HP(v, 0) \cap S_1^n$, $v \in H^n(-1)$, be (a flat) elliptic hyperquadric. Given a point $p \in S_1^n$, there is a unique timelike geodesic in S_1^n which intersects orthogonally the elliptic hyperquadric at some point $q(v, p)$. We call the point $q(v, p)$ the orthogonal projection of p in the direction v to the elliptic hyperquadric $HP(v, 0) \cap S_1^n$, and consider the fibre bundle

$$\pi_{12} : \Delta_1 = \{(v, q) \in H^n(-1) \times S_1^n \mid \langle v, q \rangle = 0\} \to S_1^n$$

where π_{12} is the canonical projection (see Appendix). By varying v, we obtain a family of orthogonal projections along timelike geodesics to elliptic hyperquadrics parametrised by vectors in $H^n(-1)$.

Theorem 3.1. *The family of orthogonal projections in S_1^n along timelike geodesics is given by*

$$P_T : H^n(-1) \times S_1^n \to \Delta_1$$
$$(v, p) \mapsto (v, q(v, p))$$

where $q(v, p)$ has the following expression

$$q(v, p) = \frac{1}{\sqrt{1 + \langle v, p \rangle^2}} (p + \langle v, p \rangle v).$$

Proof. Let $p \in S_1^n$ and $v \in H^n(-1)$. A timelike geodesic passing through p is parametrised by

$$c(t) = \cosh(t)p + \sinh(t)w, \tag{3.1}$$

for some $w \in H^n(-1)$ tangent to the geodesic at $c(0) = p$ and with $\langle w, p \rangle = 0$. At some t_0, we have $c(t_0) = q(v, p)$ and $c'(t_0) = \sinh(t_0)p + \cosh(t_0)w = v$. Thus, $\langle c'(t_0), p \rangle = \langle v, p \rangle$, which gives $\sinh(t_0) = \langle v, p \rangle$. Therefore, $\cosh(t_0) = \sqrt{1 + \langle v, p \rangle^2}$. From this we get

$$w = \frac{1}{\sqrt{1 + \langle v, p \rangle^2}} \left(v - \langle v, p \rangle p \right).$$

Substituting in (3.1) for $t = t_0$ yields $q(v, p) = (p + \langle v, p \rangle v) / \sqrt{1 + \langle v, p \rangle^2}$.
$\square$

We consider next projections along spacelike geodesics. Let $HP(v, 0) \cap S_1^n$ be a hyperbolic hyperquadric, so $v \in S_1^n$. Given a point $p \in S_1^n - \{\pm v\}$, there is a unique spacelike geodesic in S_1^n which intersects orthogonally $HP(v, 0) \cap S_1^n$ at two points $q^\pm(v, p)$. The points $p = \pm v$ are excluded as all the spacelike geodesics orthogonal to $HP(v, 0) \cap S_1^n$ pass through these two points. Therefore, their projection is not well defined. We call the points $q^\pm(v, p)$ the orthogonal projection of p in the direction v to the hyperbolic hyperquadric $HP(v, 0) \cap S_1^n$. We consider the fibre bundle

$$\pi_{52} : \Delta_5 = \{(v, q) \in S_1^n \times S_1^n \mid \langle v, q \rangle = 0\} \to S_1^n$$

with π_{52} the canonical projection to the second component (see Appendix). By varying v, we obtain a family of orthogonal projections along spacelike geodesics to hyperbolic hyperquadrics parametrised by vectors in S_1^n.

Theorem 3.2. *The family of orthogonal projections in S_1^n along spacelike geodesics is given by*

$$P_S : S_1^n \times S_1^n - \{(v, \pm v), v \in S_1^n\} \to \Delta_5$$
$$(v, p) \quad \mapsto (v, q^\pm(v, p))$$

where $q^\pm(v, p)$ has the following expression

$$q^\pm(v, p) = \pm \frac{1}{\sqrt{1 - \langle v, p \rangle^2}} \left(p - \langle v, p \rangle v \right).$$

Proof. Let $(v, p) \in S_1^n \times S_1^n$. A spacelike geodesic passing through p is parametrised by

$$c(t) = \cos(t)p + \sin(t)w, \tag{3.2}$$

for some $w \in S_1^n$ tangent to the geodesic at $c(0) = p$ and with $\langle w, p \rangle = 0$. At some t_0, we have $c(t_0) = q(v, p)$ and $c'(t_0) = -\sin(t_0)p + \cos(t_0)w = v$.

So $\langle c'(t_0), p \rangle = \langle v, p \rangle$, which gives $\sin(t_0) = -\langle v, p \rangle$. Therefore, $\cos(t_0) = \pm\sqrt{1 - \langle v, p \rangle^2}$. We have $\langle v, p \rangle^2 = 1$ if and only if $p = \pm v$ and this is excluded. Hence,

$$w = \pm \frac{1}{\sqrt{1 - \langle v, p \rangle^2}} (v - \langle v, p \rangle \, p).$$

By substituting in (3.2) for $t = t_0$ we get

$$q^{\pm}(v, p) = \pm (p - \langle v, p \rangle \, v) / \sqrt{1 - \langle v, p \rangle^2}.$$

The hyperbolic hyperquadric $HP(v, 0) \cap S_1^n$ has two connected components, $q^+(v, p)$ lies on one component and $q^-(v, p)$ on the other. □

We consider now projections along lightlike geodesics, which are lines in S_1^n parallel to lightlike vectors. An orthogonal space to a lightlike geodesic contains the geodesic, so we cannot define projections along lightlike geodesics to orthogonal spaces (which are cylinders). We shall fix instead a space transverse to all lightlike geodesics in S_1^n and project to this space. We denote by $\{e_0, \ldots, e_n\}$ the canonical basis of $\mathbb{R}_1^{n+1}$. Any lightlike geodesic intersects transversally the elliptic de Sitter quadric $S^{n-1} = HP(e_0, 0) \cap S_1^n$, so we take S^{n-1} as the space to project to.

We fix a point in S^{n-1}, say $e_1 = (0, 1, 0, \ldots, 0)$. A lightlike line through e_1 is parametrised by $e_1 + tv$, $t \in \mathbb{R}$, where $v \in S_+^{n-1} \subset LC^*$. This line lies is S_1^n if and only if $\langle v, e_1 \rangle = 0$. Thus, the lightlike geodesics in S_1^n that pass through e_1 can be parametrised by

$$S_+^{n-2} = \left\{ (1, 0, v_2, \ldots, v_n) \in \mathbb{R}^{n+1} : v_2^2 + \ldots + v_n^2 = 1 \right\} \subset S_+^{n-1} \subset LC^*.$$

Any lightlike geodesic in S_1^n can be obtained by rotating a lightlike geodesic through e_1. The rotation is in the form $A = \mathrm{Id}_{e_0} \times B$, where B is a rotation in $S^{n-1} = HP(e_0, 0) \cap S_1^n$. Given $v \in S_+^{n-2}$, the geodesics $A(e_1) + tA(v)$, $t \in \mathbb{R}$, obtained by varying $B \in SO(n-1)$ foliate S_1^n. We can now define the projection in S_1^n along lightlike geodesics as follows.

Given a point $p \in S_1^n$ and $v \in S_+^{n-2}$, there exists a unique $A = \mathrm{Id}_{e_0} \times B$, with $B \in SO(n-1)$, such that $A(p)$ belongs to the lightlike geodesic $e_1 + tv$. Then $A(p) = e_1 + \langle e_0, p \rangle \, v$. We define the lightlike projection of p to S^{n-1} along the direction v as the point

$$q(v, p) = A^{-1}(e_1) = A^{-1}(A(p) - \langle e_0, p \rangle \, v) = p - \langle e_0, p \rangle \, A^{-1}(v).$$

Definition 3.3. The family of projections along lightlike geodesics to the de Sitter elliptic hyperquadric S^{n-1} is defined by

$$P_L : S_+^{n-2} \times S_1^n \to S^{n-1}$$
$$(v, p) \mapsto q(v, p)$$

where $q(v, p) = p - \langle e_0, p \rangle A^{-1}(v)$, and $A = \mathrm{Id}_{e_0} \times B$, with $B \in \mathrm{SO}(n-1)$, is the unique rotation taking p to a point on the lightlike geodesic $e_1 + tv$, $t \in \mathbb{R}$.

Given an embedded submanifold M in S_1^n, the family of projections of M along geodesics refer to the restriction of the families P_L, P_S and P_T to M. We still denote this restriction by P_L, P_S and P_T respectively. We have the following result where the term generic is defined in terms of transversality to submanifolds of multi-jet spaces (see for example [3]).

Theorem 3.4. *For a residual set of embeddings* $x : M \to S_1^n$, *the families* P_L, P_S *and* P_T *are generic families of mappings.*

Proof. The theorem follows from Montaldi's result in [12] and the fact that $P_L|_{S_0^{n-2} \times M}$, $P_S|_{S_1^n \times M - \{(v,v) \in S_1^n \times M\}}$ and $P_T|_{H^n(-1) \times M}$ are stable maps. □

4. Projections of timelike surfaces in S_1^3

A projection along a geodesic is singular at $p \in M$ if and only if the geodesic is tangent to M at p. Therefore, for spacelike surfaces (whose tangent spaces at all points are spacelike) the projections along timelike and lightlike geodesics are always local diffeomorphisms. The study of projections of spacelike surfaces along spacelike geodesics is similar to that of projections of surfaces in $H^3(-1)$ [9]. So we deal here with embedded timelike surfaces M in S_1^3. The projection of M at $p_0 \in M$ along a given geodesic can be represented locally by a map-germ from the plane to the plane. These map-germs are extensively studied. We refer to [15] for the list of the $\mathcal{A}$-orbits with $\mathcal{A}_e$-codimension ≤ 6, where $\mathcal{A}$ denotes the Mather group of smooth changes of coordinates in the source and target. In Table 15.1, we reproduce from [15] the list of local singularities of $\mathcal{A}_e$-codimension ≤ 3.

We study the local singularities of the projections along the three types of geodesics and characterise them geometrically.

Table 15.1. $\mathcal{A}_e$-codimension ≤ 3 *local singularities of map-germs*
$\mathbb{R}^2, 0 \to \mathbb{R}^2, 0$ *([15]).*

Name	Normal form	$\mathcal{A}_e$-codimension
Immersion	(x, y)	0
Fold	(x, y^2)	0
Cusp	$(x, xy + y^3)$	0
4_k ($k = 2$ lips/beaks;	$(x, y^3 \pm x^k y), k = 2, 3, 4$	$k - 1$
$k = 3$ goose)		
5 (swallowtail)	$(x, xy + y^4)$	1
6 (butterfly)	$(x, xy + y^5 \pm y^7)$	2
7	$(x, xy + y^5)$	3
11_{2k+1} ($k = 2$ gulls)	$(x, xy^2 + y^4 + y^{2k+1}), k = 2, 3$	k
12	$(x, xy^2 + y^5 + y^6)$	3
16	$(x, x^2 y + y^4 \pm y^5)$	3

4.1. Projections along timelike and spacelike geodesics

It follows from Theorem 3.4 that, for a generic embedding of a timelike surface in S_1^3, only singularities of $\mathcal{A}_e$-codimension ≤ 3 can occur in the members of the family of orthogonal projections of the surface along spacelike and timelike geodesics (3 being the dimension of the parameter spaces S_1^3 and $H^3(-1)$ respectively). We denote by P_S^v (resp. P_T^v) the map $M \to H P(v, 0) \cap S_1^3$ given by $P_S^v(p) = \pi \circ P_S(v, p)$ (resp. $P_T^v(p) = \pi \circ P_T(v, p)$), where π is the projection to the second component. The following result follows from Theorem 3.4.

Proposition 4.1. *For a residual set of embeddings* $x : M \to S_1^3$, *the projections* P_S^v *(resp.* P_T^v*) in the family* P_S *(resp.* P_T*) have local singularities* $\mathcal{A}$*-equivalent to one in Table 1. Moreover, these singularities are versally unfolded by the family* P_S *(resp.* P_T*).*

We seek to derive geometric information on M from the local singularities of the projections. We deal with the members of the family P_T and make an observation about those of P_S.

Given $v \in H^3(-1)$ and $p \in S_1^3$, we denote by v^* the parallel transport of v to p along a geodesic orthogonal to $H P(v, 0) \cap S_1^3$. From the proof of Theorem 3.1 we have $v^* = (v - \langle v, p \rangle p) / \sqrt{1 + \langle v, p \rangle^2}$. We observe that the map $H^3(-1) \to T_p S_1^3 \cap H^3(-1)$ given by $v \mapsto v^* / \|v^*\|$ is a submersion, and the pre-image of a vector w is the curve

$$C_w(t) = \cosh(t)w + \sinh(t)p, \, t \in \mathbb{R}.$$

(There is a pencil of hyperplanes defining the same elliptic quadric in S_1^3.) We have the following result where the names of the singularities of $\tilde{P}_T^v$ are those in Table 15.1.

Theorem 4.2. *Let M be an embedded timelike surface in S_1^3 and $v \in H^3(-1)$.*

(1) *The projection P_T^v is singular at a point $p \in M$ if and only if $v^* \in T_p M$.*

(2) *The singularity of P_T^v at p is of type cusp or worse if and only if v^* is a timelike asymptotic direction at p.*

(3) *The singularity of P_T^v at p is of Type 5 (i.e., swallowtail) if and only if v^* is a timelike asymptotic direction and p is on the flecnodal curve. The singularity is of Type 6 if and only if v^* is tangent to the flecnodal curve at p. At these tangency points, there are generically up to 8 directions on the curve $C_{v^*} \subset H^3(-1)$ where the singularity becomes of Type 7.*

(4) *The singularities of P_T^v at p is of type 4_k, $k = 2, 3, 4$, if and only if p is a parabolic point but not a folded singular point of the asymptotic curves and v^* is the unique timelike asymptotic direction there. There are up to 12 directions on the curve C_{v^*} where the singularity becomes of type 4_3. There are isolated points on the parabolic set where the singularity of the projection along these special directions becomes of Type 4_4.*

(5) *At folded singularity of the asymptotic curves there are up to 12 directions on the curve C_{v^*} where the singularity is of Type 16. Away from these directions the singularity is generically of Type 11_5, and for up to 38 directions on C_{v^*} it becomes of Type 11_7. The singularities of Type 12 do not occur in general.*

Proof. As our study is local in nature, we can make some assumptions about the position of the of surface patch and the choice of the geodesic. We shall assume that the surface patch is at some point p_0 and that this point is taken by a geodesic C_1 to $e_1 = (0, 1, 0, 0) \in HP(e_0, 0) \cap S_1^3$. We also suppose that $e_0 = (1, 0, 0, 0)$ is tangent to C_1 at e_1. The geodesic C_1 can then be parametrised by

$$c_1(t) = \cosh(t)e_1 + \sinh(t)e_0, \, t \in \mathbb{R}.$$

We take, without loss of generality, the point on the surface to be $p_0 = (1, \sqrt{2}, 0, 0)$. The tangent to C_1 at p_0 is parallel to $w = (\sqrt{2}, 1, 0, 0)$. Then the vectors $v \in H^3(-1)$ satisfying $w = v^*/||v^*||$ are in the form $v = (v_0, v_1, 0, 0)$ with $-v_0^2 + v_1^2 = -1$.

The surface patch at p_0 can be taken in Monge form

$$\phi(x, y) = (1 + x, \sqrt{1 + (1 + x)^2 - f^2(x, y) - y^2}, y, f(x, y))$$

where f is a smooth function in some neighbourhood U of the origin in $\mathbb{R}^2$, $(x, y) \in U$ and $f(0, 0) = 0$. (There is nothing special about the above setting, the results are local in nature and are valid for any $v \in H^3(-1)$ and at any point $p_0 \in M$.)

The projection to the elliptic quadric $HP(v, 0) \cap S_1^3 = HP(e_0, 0) \cap S_1^3$ in the direction v (with v as above) is then given by

$$P_T^v(x, y) = q(v, \phi(x, y)) = \frac{1}{\sqrt{1 + \langle v, \phi(x, y)\rangle^2}} (\phi(x, y) + \langle v, \phi(x, y)\rangle v)$$

(see Theorem 3.1). As we are interested in the $\mathcal{A}$-singularities of the projection, we can simplify the expression of P_T^v by projecting further to the tangent space of the elliptic quadric $HP(e_0, 0) \cap S_1^3$ at e_1. This tangent space is generated by $(0, 0, 1, 0)$ and $(0, 0, 0, 1)$ and we take the projection to this space to be the restriction of the canonical projection $\pi : \mathbb{R}^4 \to \mathbb{R}^2$, with $\pi(x_0, x_1, x_2, x_3) = (x_2, x_3)$. Therefore, the modified projection $\tilde{P}_T^{e_0} = \pi \circ P_T^{e_0}$ is a map-germ from the plane to the plane given by

$$\tilde{P}_T^v(x, y) = \frac{1}{\lambda(x, y, v)} (y, f(x, y)),$$

with $\lambda(x, y, v) = (1 + (-v_0(1 + x) + v_1\sqrt{1 + (1 + x)^2 - f^2(x, y) - y^2})^2)^{1/2}$. The map-germ $\tilde{P}_T^v$ is singular at the origin if and only if $f_x(0, 0) = 0$, if and only if $v^* = w \in T_{p_0}M$.

We can make successive changes of coordinates in the sources and target and write the appropriate k-jet of $\tilde{P}_T^v$ in the form $(y, g(x, y))$. We can then obtain the conditions on the coefficients of the Taylor expansion of f for $\tilde{P}_T^v$ to have a given singularity at the origin. The calculations are carried out using Maple. For example, the 2-jet of $\tilde{P}_T^v$ is $\mathcal{A}$-equivalent to $(y, a_{20}x^2 + a_{21}xy)$. We have a fold singularity if and only if $a_{20} \neq 0$. The condition $a_{20} = 0$ means that $\phi_x(0, 0) = \sqrt{2}w$ is an asymptotic direction at p_0. The remaining calculations are done similarly but are too lengthy to reproduce here. □

For projections along a spacelike geodesic, we choose the geodesic $\mathcal{C}_2$ given by $c_2(t) = \cos(t)e_3 + \sin(t)e_1$ and project to the hyperbolic quadric $HP(e_1, 0) \cap S_1^3$. We take the point $p_0 = (0, \sqrt{2}/2, 0, \sqrt{2}/2)$ on the surface (and on $\mathcal{C}_2$) and project the surface patch around p_0 along geodesics parallel to

C_2. We take the surface in Monge form

$$\phi(x, y) = \left(f(x, y), \sqrt{1 + f^2(x, y) - \left(\frac{\sqrt{2}}{2} + x\right)^2 - y^2}, \ y, \ \frac{\sqrt{2}}{2} + x \right)$$

with $f(0, 0) = f_x(0, 0) = 0$. The modified projection is then a map-germ from the plane to the plane and is given by

$$\tilde{P}_S^{e_1}(x, y) = \frac{1}{\sqrt{\left(\frac{\sqrt{2}}{2} + x\right)^2 + y^2 - f^2(x, y)}} (f(x, y), y).$$

We can obtain the conditions for $\tilde{P}_S^{e_1}$ to have a given singularity at the origin from the coefficients of the Taylor expansion of f and interpret these geometrically. The results are similar to those in Theorem 4.2. One needs to take $v \in S_1^3$ and replace timelike asymptotic direction by spacelike asymptotic direction in the statement of Theorem 4.2. The numbers of directions in the statements also need changing. In statement (3) we have up to 4 directions on C_{v^*} giving singularities of Type 7; in statement (4), there are up to 2 directions on C_{v^*} giving singularities of Type 4_3; in statement (5), there up to 16 directions on C_{v^*} where the singularity becomes of Type 11_7. There are generically no singularities of type 12 or 16.

4.2. Projections along lightlike geodesics

We shall give an explicit expression for the projection P_L (Definition 3.3) in S_1^3. Consider the sphere $S^2 = \{(0, v_1, v_2, v_3) : v_1^2 + v_2^2 + v_3^2 = 1\}$ (we shall drop the first coordinate of points in S^2). Let

$$T_\theta = \begin{pmatrix} \cos\theta & -\sin\theta & 0 \\ \sin\theta & \cos\theta & 0 \\ 0 & 0 & 1 \end{pmatrix}, \quad T_\phi = \begin{pmatrix} \cos\phi & 0 & -\sin\phi \\ 0 & 1 & 0 \\ \sin\phi & 0 & \cos\phi \end{pmatrix}$$

and their composite

$$T_{(\theta,\phi)} = T_\theta \circ T_\phi = \begin{pmatrix} \cos\theta\cos\phi & -\sin\theta & -\cos\theta\sin\phi \\ \sin\theta\cos\phi & \cos\theta & -\sin\theta\sin\phi \\ \sin\phi & 0 & \cos\phi \end{pmatrix}.$$

Any point on $S^2 - (0, 0, \pm 1)$ is the image of the point $e_1 = (1, 0, 0)$ by a rotation $T(\theta, \phi)$, for some $(\theta, \phi) \in [0, 2\pi] \times] - \frac{\pi}{2}, \frac{\pi}{2}[$. (One can consider other rotations to cover the points $(0, 0, \pm 1)$.)

We consider in S_1^3 the rotation $A = \mathrm{Id}_{e_0} \times T_{(\theta,\phi)}$. Let $v \in S_+^1 = \{(1, 0, v_2, v_3) : v_2^2 + v_3^2 = 1\} \subset LC^*$. A point $p = (p_0, p_1, p_2, p_3)$ is projected to $q(v, p) \in S^2$ along the lightlike geodesic determined by v (see Definition 3.3). The point $q(v, p)$ is the image of e_1 by a rotation A for some (θ, ϕ). The point p is on the line $q(v, p) + t A(v)$, and we have $p = q(v, p) + p_0 A(v)$, that is,

$$
\begin{pmatrix} p_0 \\ p_1 \\ p_2 \\ p_3 \end{pmatrix} = \begin{pmatrix} 0 \\ \cos\theta \cos\phi \\ \sin\theta \cos\phi \\ \sin\phi \end{pmatrix} + p_0 \begin{pmatrix} 1 \\ -v_2 \sin\theta - v_3 \cos\theta \sin\phi \\ v_2 \cos\theta - v_3 \sin\theta \sin\phi \\ v_3 \cos\phi \end{pmatrix}.
$$

We suppose that $q(v, p) \neq (0, 0, \pm 1)$, this implies that $p_1^2 + p_2^2 \neq 0$. We can then solve the above system for θ and ϕ and get

$$
\cos\phi = \frac{p_0 p_3 v_3 + \sqrt{1 - p_3^2 + p_0^2 v_3^2}}{1 + p_0^2 v_3^2} \qquad \cos\theta = \frac{p_0 p_2 v_2 + p_1 \sqrt{1 - p_3^2 + p_0^2 v_3^2}}{p_1^2 + p_2^2}
$$

$$
\sin\phi = \frac{p_3 - p_0 v_3 \sqrt{1 - p_3^2 + p_0^2 v_3^2}}{1 + p_0^2 v_3^2} \qquad \sin\theta = \frac{-p_0 p_1 v_2 + p_2 \sqrt{1 - p_3^2 + p_0^2 v_3^2}}{p_1^2 + p_2^2}
$$

To analyse the singularities of P_L^v, we take $v = (1, 0, 0, 1)$, the point $p_0 = e_1 + v$ so that $q(p_0, v) = e_1$. Then, a local parametrisation of the surface can be taken in the form

$$
\phi(x, y) = (1 + x, g(x, y), y, f(x, y))
$$

with (x, y) is in some neighbourhood of the origin in $\mathbb{R}^2$, $f(0, 0) = 1$, $f_x(0, 0) = 1$ and $g(x, y) = \sqrt{1 + (1 + x)^2 - y^2 - f^2(x, y)}$. We consider the modified projection $\tilde{P}_L^v$ by projecting further to the tangent space of $T_{e_1} S^2 = \{(0, 0, x_2, x_3) : x_2, x_3 \in \mathbb{R}\}$, which we identify with $\mathbb{R}^2$. Then, the resulting map-germ from the plane to the plane is given by the last two components of $q(v, p)$, that is $(\sin\theta \cos\phi, \sin\phi)$, with $\sin\theta$, $\cos\phi$, $\sin\phi$ as above. That is,

$$
\tilde{P}_L^v = \left(\frac{y g(x, y)((1 + x) f(x, y) + g(x, y))}{(1 + (1 + x)^2) g^2(x, y)}, \frac{f(x, y) - (1 + x) g(x, y)}{1 + (1 + x)^2} \right).
$$

We take the 4-jet of f in the form $j^4 f(x, y) = 1 + x + a_{11} y + a_{20} x^2 + a_{21} x y + a_{22} y^2 + \sum_{i=0}^3 a_{3i} x^{3-i} y^i + \sum_{i=0}^4 a_{4i} x^{4-i} y^i$.

The surface patch parametrised by ϕ is timelike if and only if $a_{11} \neq 0$. A short calculation shows that the 4-jet of the projection is $\mathcal{A}$-equivalent to the map-germ given by

$$
\left(y, a_{20} x^2 + a_{21} x y + a_{30} x^3 + a_{31} x^2 y - \left(\frac{1}{2} a_{11}^2 + a_1 a_{21} - a_{32} \right) x y^2 + f_4(x, y) \right),
$$

with

$$f_4(x, y) = \left(a_{40} + \tfrac{1}{2}a_{20}^2\right)x^4 + a_{41}x^3y - \left(\tfrac{1}{2}a_{11}^2 a_{20} + a_{11}(a_{31} + a_{21}) + a_{20}a_{22}\right.$$
$$\left. + \tfrac{1}{2}a_{21}^2 - a_{42}\right)x^2y^2 + \left(\tfrac{1}{2}a_{43} - a_{21}a_{22} - a_{11}(a_{22} + a_{32})\right)xy^3.$$

We have a fold singularity if and only if $a_{20} \neq 0$; a cusp singularity if and only if $a_{20} = 0$ and $a_{21}a_{30} \neq 0$; a lips/beaks singularity if and only if $a_{20} = a_{21} = 0$ and $a_{30}(3(a_1^2 + 2a_1a_{21} - 2a_{32})a_{30} + 2a_{31}^2) \neq 0$; a swallowtail singularity if and only if $a_{20} = a_{30} = 0$, and $a_{40} \neq 0$.

To interpret these conditions geometrically we consider the LPL, given by $\delta(x, y) = 0$ in expression (2.2) in section 2. We calculate the coefficients of the first and second fundamental forms and find that the point p_0 is on the LPL if and only if

$$a_{20}\left(a_{20} - 2a_{11}a_{21} + 4a_{20}a_{11}^4 + 4a_{11}^2 a_{20} - 4a_{11}^3 a_{21} + 4a_{11}^2 a_{22}\right) = 0.$$

The asymptotic directions at p_0 are given by

$$a_{22}dy^2 + a_{21}dxdy + a_{20}dx^2 = 0.$$

In particular, the singularity of P_L^v is worse than fold ($a_{20} = 0$) if and only if $p_0 \in LPL$ and $v = \phi_x(0, 0)$ is a lightlike asymptotic direction. (Then v is also the double principal direction at p_0, see section 2.) The other asymptotic direction is not lightlike unless the LPL is singular.

When $a_{20} = 0$, we assume that $2a_{11}a_{22} - a_{21} - 2a_{11}^2 a_{21} \neq 0$, otherwise the point p_0 is a timelike umbilic point so the LPL is singular there. At such points both asymptotic directions are lightlike and the projection P_L^v along these directions has a cusp singularity.

Suppose that $a_{20} = 0$. Then the 1-jet of the equation of the LPL is given by

$$\left(2a_{11}a_{22} - 2a_{11}^2 a_{21} - a_{21}\right)\left(-3a_{11}a_{30}x - \left(-a_{21}^2 + a_{11}a_{31}\right)y\right).$$

As the surface is timelike, $a_{11} \neq 0$. Therefore, the singularity of the projection is of type swallowtail at p_0 or worse (i.e., $a_{30} = 0$) if and only if the lightlike direction v is tangent to the LPL. (This occur at precisely the folded singularities of the configuration of the principal curves [10].)

The point p_0 is on the parabolic set if and only if $a_{21}^2 - 4a_{20}a_{22} = 0$. Therefore, we have a lips/beaks singularity ($a_{20} = a_{21} = 0$) at the point of tangency of the LPL with the parabolic set. We have thus the following result.

Theorem 4.3. *The projection P_L^v can have generically the local codimension ≤ 1 singularities in Table 1. The singularity of P_L^v at $p_0 \in M$ is of type*

(1) *fold if and only if v is not a lightlike asymptotic direction at p_0;*
(2) *cusp if and only if v is a lightlike asymptotic direction at p_0 and is transverse to the LPL;*
(3) *swallowtail if and only if v is lightlike asymptotic direction at p_0 and is tangent to the LPL;*
(4) *lips/beaks if and only if v is a lightlike asymptotic direction at p_0 and p_0 is the point of tangency of the LPL with the parabolic curve.*

5. Duality

We prove in this section duality result similar to those in [8, 9], and to those in [16] for central projections of surfaces in $\mathbb{R}P^3$ and in [2] for orthogonal projections of surfaces in $\mathbb{R}^3$.

Let M be an embedded timelike surface in S_1^3. We shall use the duality concepts in [4, 5, 6], see §6 for details. We denote by A_2^{par} the ruled surface in S_1^3 swept out by the geodesics in S_1^3 passing through a parabolic point of M and with tangent direction there its unique asymptotic direction. We assume that the unique asymptotic direction is not lightlike, so the point is not on the LPL.

Let $p(t), t \in I = (-a, a), a > 0$, be a local parametrisation of the parabolic set of M and $u(t)$ be the unique unit asymptotic direction at $p(t)$. Recall that if $p_0 = p(t_0)$ is on the LPL, $u(t)$ is spacelike on one side of p_0 and timelike on the other side ([10] and section 2). The surface A_2^{par} has two connected components given by

$$A_2^{S-par} = \{\cos(s)p(t) + \sin(s)u(t), (s, t) \in I \times J, \langle u(t), u(t) \rangle = 1\}$$
$$A_2^{T-par} = \{\cosh(s)p(t) + \sinh(s)u(t), (s, t) \in I \times J, \langle u(t), u(t) \rangle = -1\}$$

with $J = (-b, b)$, for some $b > 0$. We also denote by $A_1||A_1$ the ruled surface swept out by the spacelike or timelike geodesics in S_1^3 that are tangent to M at two points where the normals to M at such points are parallel (i.e., the projection P_S or P_T has a multi-local singularity of type double tangent fold). The surface $A_1||A_1$ has also two component determined by the type of bi-tangent geodesics. We have the following result, whose proof is similar to that of Theorem 3.8 in [9].

Theorem 5.1. *Let M^* be the Δ_5-dual of a timelike surface M embedded in S_1^3.*

(1) *The Δ_5-dual of the surface A_2^{par} is the cuspidaledge of M^*.*
(2) *The Δ_5-dual of the surface $A_1||A_1$ is the self-intersection line of M^*.*

Proof. (1) We deal with the A_2^{S-par} component (the calculations are similar for the component A_2^{T-par} and are omitted) and parametrise it as above by $y(s, t) = \cos(s)p(t) + \sin(s)u(t)$. The normal to the surface A_2^{S-par} is along

$$y \wedge y_s \wedge y_t = \cos^3(s)p(t) \wedge u(t) \wedge p'(t) + \sin^3(s)p(t) \wedge u(t) \wedge u'(t).$$

At a generic point p on the parabolic set, the asymptotic direction is transverse to the parabolic set, so $p(t) \wedge u(t) \wedge p'(t)$ is along $e(p(t))$. One can prove, following the same arguments in the proof of Lemma 3.11 in [8], that $p(t) \wedge u(t) \wedge u'(t)$ is also along $e(p(t))$. Therefore, $y \wedge y_s \wedge y_t$ is along $e(p(t))$, and it follows from this that the normal to the ruled surface A_2^{S-par} is constant along the rulings and is given by the normal vector $e(p(t))$ to M at $p(t)$. This means that A_2^{S-par} is a de Sitter developable surface (i.e., $K \equiv 0$ on A_2^{S-par}). Therefore, the Δ_5-dual of A_2^{S-par} is $\{e(p), \ p$ a parabolic point$\}$. This is precisely the singular set (i.e., the cuspidaledge) of M^*, the Δ_5-dual surface of M.

(2) Suppose a multi-local singularity (double tangent fold) occurs at two points p_1 and p_2 on M. The surface $A_1||A_1$ is then a ruled surface generated by spacelike geodesics along a curve C_1 on M through p_1, or a curve C_2 on M through p_2. The normals to the surface at points on C_1 and C_2 that are on the same ruling of $A_1||A_1$ are parallel. Let $q(t)$ be a local parametrisation of the curve C_1 and $u(t)$ be the unit tangent direction to the ruling in $A_1||A_1$ through $q(t)$. A parametrisation of $A_1||A_1$ is given by

$$w(s, t) = \cos(s)q(t) + \sin(s)u(t).$$

The normal to this surface is along $\cos^3(s)V_1(t) + \sin^3(s)V_2(t)$ with $V_1(t) = q(t) \wedge u(t) \wedge q'(t)$ and $V_2(t) = q(t) \wedge u(t) \wedge u'(t)$. These normals are parallel at two points on any ruling, one point being on the curve C_1 and the other on C_2. Therefore, $V_1(t)$ and $V_2(t)$ are parallel, so the normal to the surface $A_1||A_1$ is constant along the rulings of this surface. As these are along the normal to the surface at $q(t)$, it follows that the Δ_5-dual of $A_1||A_1$ is $\{e(p), \ p \in C_1\} = \{e(p), \ p \in C_2\}$. This is precisely the self-intersection line of M^*, the Δ_5-dual surface of M. $\qquad\square$

We consider now some components of the bifurcation sets of the families of projections P_S and P_T.

Theorem 5.2. *Let M^* be the Δ_5-dual of a timelike surface M embedded in S_1^3. Then,*

(1) *The local stratum $Bif(P_S, lips/beaks)$ of the bifurcation set of P_S, which consists of vectors $v \in S_1^3$ for which the projection P_S^v has a lips/beaks*

singularity, is a ruled surface. The Δ_5-dual of $Bif(P_S, lips/beaks)$ is the cuspidaledge of M^*.

(2) The multi-local stratum $Bif(P_S, DTF)$ of the bifurcation set of P_S, which consists of vectors $v \in S_1^3$ for which the projection P_S^v has a multi-local singularity of type double tangent fold, is a ruled surface. The Δ_5-dual of this ruled surface is the self-intersection line of M^*.

(3) The local stratum $Bif(P_T, lips/beaks)$ of the bifurcation set of P_T, which consists of vectors $v \in H^3(-1)$ for which the projection P_T^v has a lips/beaks singularity, is a ruled surface. The Δ_1-dual of $Bif(P_T, lips/beaks)$ is the cuspidaledge of M^*.

(4) The multi-local stratum $Bif(P_T, DTF)$ of the bifurcation set of P_T, which consists of vectors $v \in H^3(-1)$ for which the projection P_T^v has a multi-local singularity of type double tangent fold, is a ruled surface. The Δ_5-dual of this ruled surface is the self-intersection line of M^*.

Proof. We prove (1) as the proof of (2) is similar. It follows from Theorem 4.2(5) that the lips/beaks stratum $Bif(P_S, lips/beaks)$ of the family P_S is given by the set of $v \in S_1^3$ such that v^* is an asymptotic direction at a parabolic point p, where v^* denotes the parallel transport of v to p. Thus, $v^* = u(t)$ when $v \in Bif(P_S, lips/beaks)$, where $u(t)$ is the unique asymptotic direction at $p(t)$.

We have then

$$u(t) = v^* = \frac{1}{\sqrt{1 - \langle v, p(t) \rangle^2}} (v - \langle v, p(t) \rangle \, p(t))$$

and hence

$$v = \sqrt{1 - \langle v, p(t) \rangle^2} u(t) + \langle v, p(t) \rangle \, p(t).$$

If we set $\sin(s) = \langle v, p(t) \rangle$ we get

$$Bif(P_P, lips/beaks) = \{\cos(s)u(t) + \sin(s)p(t), t \in I, s \in \mathbb{R}\},$$

which shows that $Bif(P_P, lips/beaks)$ is a ruled surface. For the duality result, following Remark 2, we need to find the unit normal vector to $Bif(P_S, lips/beaks)$. Following the same argument in the proof of Theorem 5.1, we find that the normal vector is constant along the rulings of the surface $Bif(P_S, lips/beaks)$ and is along $e(t)$, and the result follows. □

Remark 1. It is shown in [16] that other strata of the bifurcation set of the family of central projections of a surface in $\mathbb{R}P^3$ are also self-dual. For instance, the strata A_3, A_1^3 and $A_1 \times A_2$ are all self-dual. These results do not hold in our context. If we define the A_3 set as the surface formed by geodesics through points on the flecnodal curve and with tangent at these points along the associated asymptotic direction, then this surface is not in general a ruled surface. This means that the dual of the A_3-set is not the flecnodal curve on the Δ_5-dual surface of M. The situation is similar for the other strata.

6. Appendix

We require some properties of contact manifolds and Legendrian submanifolds for the duality results in this paper (for more details see for example [1]). Let N be a $(2n + 1)$-dimensional smooth manifold and K be a field of tangent hyperplanes on N. Such a field is locally defined by a 1-form α. The tangent hyperplane field K is said to be *non-degenerate* if $\alpha \wedge (d\alpha)^n \neq 0$ at any point on N. The pair (N, K) is a *contact manifold* if K is a non-degenerate hyperplane field. In this case K is called a *contact structure* and α a *contact form*.

A submanifold $i : L \subset N$ of a contact manifold (N, K) is said to be *Legendrian* if dim $L = n$ and $di_x(T_x L) \subset K_{i(x)}$ at any $x \in L$. A smooth fibre bundle $\pi : E \to M$ is called a *Legendrian fibration* if its total space E is furnished with a contact structure and the fibres of π are Legendrian submanifolds. Let $\pi : E \to M$ be a Legendrian fibration. For a Legendrian submanifold $i : L \subset E$, $\pi \circ i : L \to M$ is called a *Legendrian map*. The image of the Legendrian map $\pi \circ i$ is called a *wavefront set* of i and is denoted by $W(i)$.

The duality concepts we use in this paper is one of those introduced in [4, 5, 6], where five Legendrian double fibrations are considered on subsets of the product of two of the pseudo spheres $H^n(-1)$, S_1^n and LC^*. We recall here only those that are needed in this paper:

(1) (a) $H^n(-1) \times S_1^n \supset \Delta_1 = \{(v, w) \mid \langle v, w \rangle = 0\}$,
   (b) $\pi_{11} : \Delta_1 \to H^n(-1)$, $\pi_{12} : \Delta_1 \to S_1^n$,
   (c) $\theta_{11} = \langle dv, w \rangle | \Delta_1$, $\theta_{12} = \langle v, dw \rangle | \Delta_1$.
(5) (a) $S_1^n \times S_1^n \supset \Delta_5 = \{(v, w) \mid \langle v, w \rangle = 0\}$,
   (b) $\pi_{51} : \Delta_5 \to S_1^n$, $\pi_{52} : \Delta_5 \to S_1^n$,
   (c) $\theta_{51} = \langle dv, w \rangle | \Delta_5$, $\theta_{52} = \langle v, dw \rangle | \Delta_5$.

Here, $\pi_{i1}(v, w) = v$ and $\pi_{i2}(v, w) = w$ for $i = 1, 5$, $\langle dv, w \rangle = -w_0 dv_0 + \sum_{i=1}^n w_i dv_i$ and $\langle v, dw \rangle = -v_0 dw_0 + \sum_{i=1}^n v_i dw_i$. The 1-forms θ_{i1} and θ_{i2}, $i = 1, 5$, define the same tangent hyperplane field over Δ_i which is denoted by K_i.

Theorem 6.1. ([4, 5, 6]) *The pairs* (Δ_i, K_i), $i = 1, 5$, *are contact manifolds and* π_{i1} *and* π_{i2} *are Legendrian fibrations.*

Remark.

(1) Given a Legendrian submanifold $i : L \to \Delta_i$, $i = 1, 5$, Theorem 6.1 states that $\pi_{i1}(i(L))$ is dual to $\pi_{i2}(i(L))$ and vice-versa. We shall call this duality Δ_i-duality.

(2) If $\pi_{11}(i(L))$ is smooth at a point $\pi_{11}(i(u))$, then $\pi_{12}(i(u))$ is the normal vector to the hypersurface $\pi_{11}(i(L)) \subset H_+^n(-1)$ at $\pi_{11}(i(u))$. Conversely, if $\pi_{12}(i(L))$ is smooth at a point $\pi_{12}(i(u))$, then $\pi_{11}(i(u))$ is the normal vector to the hypersurface $\pi_{12}(i(L)) \subset S_1^n$. The same properties hold for the Δ_5-duality.

References

1. V. I. Arnold, S. M. Gusein-Zade and A. N. Varchenko, *Singularities of Differentiable Maps vol. I*. Birkhäuser, 1986.
2. J. W. Bruce and M. C. Romero-Fuster, Duality and orthogonal projections of curves and surfaces in Euclidean 3-space. *Quart. J. Math. Oxford* **42** (1991), 433–441.
3. M. Golubitsky and V. Guillemin, *Stable mappings and their singularities*. Graduate Texts in Mathematics, **14**. Springer-Verlag, New York-Heidelberg, 1973.
4. S. Izumiya, Timelike hypersurfaces in de Sitter space and Legendrian singularities. *J. Math. Sciences.* **144** (2007) 3789–3803.
5. S. Izumiya, Legendrian dualities and spacelike hypersurfaces in the lightcone. *Mosc. Math. J.* **9** (2009) 325–357.
6. S. Izumiya and M. Takahashi, Spacelike parallels and evolutes in Minkowski pseudo-spheres. *J. Geometry and Physics* **57** (2007) 1569–1600.
7. S. Izumiya, M. Takahashi and F. Tari, Folding maps on spacelike and timelike surfaces and duality. To appear in *Osaka J. Math.*
8. S. Izumiya and F. Tari, Projections of surfaces in the hyperbolic space to hyper-horospheres and hyperplanes. *Rev. Mat. Iberoam.* **24** (2008) 895–920.
9. S. Izumiya and F. Tari, Projections of surfaces in the hyperbolic space along horo-cycles. *Proc. Roy. Soc. Edinburgh Sect. A* **140** (2010) 399–418.
10. S. Izumiya and F. Tari, Self-adjoint operators on surfaces with singular metrics. To appear in *J. Dyn. Control Syst.* Preprint, 2008.
11. B. Karlığa and M. Savas, Orthogonal projections to a k-plane in hyperbolic, de Sitter and spherical n-spaces. Preprint, 2007.
12. J. A. Montaldi, On generic composites of maps. *Bull. London Math. Soc.* **23** (1991) 81–85.
13. B. O'Neill, *Semi-Riemannian geometry. With applications to relativity*. Pure and Applied Mathematics, 103. Academic Press, Inc., New York, 1983.
14. J. G. Ratcliffe, *Foundations of hyperbolic manifolds*. Graduate Texts in Mathematics, **149**. Springer-Verlag, New York, 1994.

15. J. H. Rieger, Families of maps from the plane to the plane. *J. London Math. Soc.* **36** (1987) 351–369.
16. O. P. Shcherbak, Projectively dual space curves and Legendre singularities. *Sel. Math. Sov* **5** (1986) 391–421.

S. Izumiya
Department of Mathematics
Hokkaido University
Sapporo 060-0810
Japan
izumiya@math.sci.hokudai.ac.jp

F. Tari
Department of Mathematical Sciences
Durham University
Science Laboratories, South Road
Durham DH1 3LE, UK
farid.tari@durham.ac.uk

16

Spacelike submanifolds of codimension at most two in de Sitter space

M. KASEDOU

Abstract

The aim of this paper is to provide a description of the main geometrical properties of spacelike submanifolds of codimension at
most two in de Sitter space, that have been studied by the author
with full details in other papers, as an application of the theory
of Legendrian singularities. We analyze the geometrical meaning
of the singularities of lightcone Gauss images, lightcone Gauss
maps and lightlike hypersurfaces of generic spacelike surfaces in
de Sitter 3-space and de Sitter 4-space.

1. Introduction

In this paper we consider de Sitter space, which is a Lorentzian space form
with positive curvature defined by a pseudo n-sphere in Minkowski space. The
spacelike curves in de Sitter 3-space are investigated in [5] and the lightlike
surface of the spacelike curves are constructed from the Frenet-Serret type
formula. In [9] the differential geometry of the timelike surfaces in de Sitter
space are discussed, and the singularities of de Sitter Gauss images of timelike
surfaces in de Sitter 3-space are classified. The principal, asymptotic and characteristic curves associated to the de Sitter Gauss maps are investigated in [10],
and the contact of timelike surfaces with geodesic loci are investigated in [11].
In [12] we investigated the lightcone Gauss image of spacelike hypersurface
in de Sitter space, which is the analogous tool in [6]. The singularities of the
Gauss images correspond to the parabolic sets of spacelike hypersurfaces. In

2000 *Mathematics Subject Classification* 53A35 (primary), 53B30, 58C25 (secondary).
This work was supported by the JSPS International Training Program(ITP).

the case of spacelike submanifold of codimension two, the normal direction of spacelike submanifold cannot be chosen uniquely. However, we can determine the lightcone normal frames. In [13] we investigated the singularities of Gauss maps and lightlike hypersurfaces by using analogous tools to those applied in [7, 8] to the study of spacelike submanifolds in Minkowski space. The singularities of lightlike hypersurfaces have a relation with the principal curvatures of spacelike submanifolds.

This paper is organized as follows. In §2 we introduce the notion of the lightcone Gauss images of spacelike hypersurfaces. We define the Gauss-Kronecker curvature with respect to the lightcone Gauss image. In §3 we introduce a family of functions that is called the lightcone height function. The singular set of the lightcone Gauss image is the lightcone parabolic set of the spacelike hypersurface and this can be interpreted as the discriminant set of the family of height functions. We give a proof of Proposition 3.3. In §4 we discuss the contact between hypersurfaces and de Sitter hyperhorospheres. We apply the theory of Legendrian singularities to study the lightcone Gauss images of generic hypersurfaces. In §5 we give the classification of the singularities of lightcone Gauss images of generic spacelike surfaces in de Sitter 3-space. In this case we have two singular types of lightcone Gauss images, which are the cuspidal edge and the swallowtail. In §6 we introduce the notion of the lightcone Gauss maps, the normalized lightcone Gauss-Kronecker curvatures and principal curvatures of spacelike submanifold of codimension two. In §7 we introduce the notion of the lightlike hypersurfaces and a family of functions that is called the Lorentzian distance squared functions. In §8 we apply the Legendrian singularity theory for the study of lightcone Gauss images of generic spacelike submanifolds. In §9 we classify the singular types of lightcone Gauss maps and lightlike surfaces.

2. Lightcone Gauss images of spacelike hypersurfaces

In this section we discuss the differential geometry of spacelike hypersurfaces in de Sitter space and introduce the notion of the lightcone Gauss image by using analogous tools to those of [6]. Let $n > 0$ be an integer and $\mathbb{R}^{n+1} = \{\mathbf{x} = (x_0, \ldots, x_n) \mid x_i \in \mathbb{R}\}$ be an $(n+1)$-dimensional vector space. For any vector $\mathbf{x}$ and $\mathbf{y}$, the *pseudo scalar product* is defined by $\langle \mathbf{x}, \mathbf{y} \rangle = -x_0 y_0 + \sum_{i=1}^{n} x_i y_i$. We call $(\mathbb{R}^{n+1}, \langle, \rangle)$ a *Minkowski $(n+1)$-space* and write $\mathbb{R}_1^{n+1}$ instead of $(\mathbb{R}^{n+1}, \langle, \rangle)$. We say that a vector $\mathbf{x} \in \mathbb{R}_1^{n+1} \setminus \{\mathbf{0}\}$ is *spacelike*, *lightlike* or *timelike* if $\langle \mathbf{x}, \mathbf{x} \rangle > 0$, $\langle \mathbf{x}, \mathbf{x} \rangle = 0$ or $\langle \mathbf{x}, \mathbf{x} \rangle < 0$ respectively. The norm of the vector $\mathbf{x}$ is defined by $||x|| = \sqrt{|\langle \mathbf{x}, \mathbf{x} \rangle|}$. For a vector

$\mathbf{v} \in \mathbb{R}_1^{n+1} \setminus \{\mathbf{0}\}$ and a real number c, a *hyperplane* with pseudo normal $\mathbf{v}$ is defined by $HP(\mathbf{v}, c) = \{\mathbf{x} \in \mathbb{R}_1^{n+1} \mid \langle \mathbf{x}, \mathbf{v} \rangle = c\}$. We call $HP(\mathbf{v}, c)$ a *spacelike hyperplane*, *timelike hyperplane* or *lightlike hyperplane* if $\mathbf{v}$ is timelike, spacelike or lightlike respectively. We define *hyperbolic n-space* and *de Sitter n-space* by

$$H_+^n(-1) = \left\{\mathbf{x} \in \mathbb{R}_1^{n+1} \mid \langle \mathbf{x}, \mathbf{x} \rangle = -1, x_0 \geq 1\right\},$$
$$S_1^n = \left\{\mathbf{x} \in \mathbb{R}_1^{n+1} \mid \langle \mathbf{x}, \mathbf{x} \rangle = 1\right\}.$$

For any $\mathbf{x}_1, \mathbf{x}_2, \ldots, \mathbf{x}_n \in \mathbb{R}_1^{n+1}$, we can define a vector $\mathbf{x}_1 \wedge \mathbf{x}_2 \wedge \ldots \wedge \mathbf{x}_n$ with the property $\langle \mathbf{x}, \mathbf{x}_1 \wedge \ldots \wedge \mathbf{x}_n \rangle = \det(\mathbf{x}, \mathbf{x}_1, \ldots, \mathbf{x}_n)$, so that $\mathbf{x}_1 \wedge \ldots \wedge \mathbf{x}_n$ is pseudo-orthogonal to any $\mathbf{x}_i$ (for $i = 1, \ldots, n$). Let $\lambda \in \mathbb{R}_1^{n+1}$, we define a set $LC_\lambda = \{\mathbf{x} \in \mathbb{R}_1^{n+1} \mid \langle \mathbf{x} - \lambda, \mathbf{x} - \lambda \rangle = 0\}$, which is called a *closed lightcone* with vertex λ. We denote $LC_\pm^* = \{\mathbf{x} = (x_0, \ldots, x_n) \in LC_0 \mid x_0 > 0 \ (x_0 < 0)\}$ and call it the *future* (resp. *past*) *lightcone* at the origin.

We now study the extrinsic differential geometry of spacelike hypersurfaces in S_1^n. Let $\mathbf{X} : U \longrightarrow S_1^n$ be an embedding, where $U \subset \mathbb{R}^{n-1}$ is an open subset. We say $\mathbf{X}$ is a *spacelike hypersurface* in S_1^n if every non zero vector generated by $\{\mathbf{X}_{u_i}(u)\}_{i=1}^{n-1}$ is always spacelike, where $u = (u_1, \ldots, u_{n-1})$ is an element of U and $\mathbf{X}_{u_i}$ is a partial derivative of $\mathbf{X}$ with respect to u_i. We denote $M = \mathbf{X}(U)$ and identify M with U through the embedding $\mathbf{X}$. Since $\langle \mathbf{X}, \mathbf{X} \rangle \equiv 1$, we have $\langle \mathbf{X}_{u_i}, \mathbf{X} \rangle \equiv 0$ (for $i = 1, \ldots, n - 1$). It follows that a hyperplane spanned by $\{\mathbf{X}, \mathbf{X}_{u_1}, \ldots, \mathbf{X}_{u_{n-1}}\}$ is spacelike. We define a vector

$$\mathbf{e}(u) = \frac{\mathbf{X}(u) \wedge \mathbf{X}_{u_1}(u) \wedge \ldots \wedge \mathbf{X}_{u_{n-1}}(u)}{||\mathbf{X}(u) \wedge \mathbf{X}_{u_1}(u) \wedge \ldots \wedge \mathbf{X}_{u_{n-1}}(u)||}.$$

Then we have $\langle \mathbf{e}, \mathbf{X}_{u_i} \rangle \equiv \langle \mathbf{e}, \mathbf{X} \rangle \equiv 0$ and $\langle \mathbf{e}, \mathbf{e} \rangle \equiv -1$ for $i = 1, \ldots, n - 1$. Therefore we have $\mathbf{X}(u) \pm \mathbf{e}(u) \in LC_\pm^*$. We define a map $\mathbb{L}^\pm : U \longrightarrow LC_\pm^*$ by

$$\mathbb{L}^\pm(u) = \mathbf{X}(u) \pm \mathbf{e}(u),$$

which is called the *lightcone Gauss image* of $\mathbf{X}$.

We now consider a hypersurface defined by $HP(\mathbf{v}, c) \cap S_1^n$. We say that $HP(\mathbf{v}, c) \cap S_1^n$ is an *elliptic hyperquadric* or *hyperbolic hyperquadric* if $HP(\mathbf{v}, c)$ is spacelike or timelike. We say that $HP(\mathbf{v}, 1) \cap S_1^n$ is a *de Sitter hyperhorosphere* if $HP(\mathbf{v}, 1)$ is lightlike. We have the following proposition analogous to ([6] Proposition 2.2).

Proposition 2.1 ([12]). *Let* $\mathbf{X} : U \longrightarrow S_1^n$ *be a spacelike hypersurface in* S_1^n. *The lightcone Gauss image* $\mathbb{L}^\pm$ *is constant if and only if the spacelike hypersurface* $M = \mathbf{X}(U)$ *is a part of a de Sitter hyperhorosphere.*

We now define the lightcone Gauss-Kronecker curvature and the lightcone mean curvature of the hypersurface $M = \mathbf{X}(U)$. Under the identification of U and M, we can identify the derivative $d\mathbf{X}(u)$ with the identity mapping id_{T_pM} on the tangent space T_pM for any $p = \mathbf{X}(u)$. This means that $d\mathbb{L}^{\pm}(u) = \mathrm{id}_{T_pM} \pm de(u)$. For any $p = \mathbf{X}(u_0) \in M$ and $\mathbf{v} \in T_pM$, we can show $D_{\mathbf{v}}\mathbf{e}, D_{\mathbf{v}}\mathbb{L}^{\pm} \in T_pM$, where $D_{\mathbf{v}}$ denotes the covariant derivative with respect to the tangent vector $\mathbf{v}$. So that $de(u)$ is a linear transformation on the tangent space T_pM. We call a linear transformation $S_p^{\pm} = -d\mathbb{L}^{\pm}(u) : T_pM \longrightarrow T_pM$ the *lightcone shape operator* of $M = \mathbf{X}(U)$ of at $p = \mathbf{X}(u)$, and denote the eigenvalue of $S_p^{\pm}$ by $\bar{\kappa}_p^{\pm}$.

The *lightcone Gauss-Kronecker curvature* of $M = \mathbf{X}(U)$ at $p = \mathbf{X}(u)$ is defined to be $K_{\ell}^{\pm}(u) = \det S_p^{\pm}$. We say that a point $u \in U$ or $p = \mathbf{X}(u)$ is an *umbilic point* if $S_p^{\pm} = \bar{\kappa}_p^{\pm}\mathrm{id}_{T_pM}$. The spacelike hypersurface M is called *totally umbilic* if all points on M are umbilic. Let $p = \mathbf{X}(u) \in M$ be an umbilic point, we say that a point p is a *lightcone flat point* (briefly an $L^{\pm}$-*flat point*) if $\bar{\kappa}_p^{\pm} = 0$. The following proposition is analogous to ([6], Proposition 2.3).

Proposition 2.2 ([12]). *Suppose that $M = \mathbf{X}(U)$ is totally umbilic. Then $\bar{\kappa}_p^{\pm}$ are constant $\bar{\kappa}^{\pm}$. Under this condition, we have the following classification.*

(i) *If $0 \leq |\bar{\kappa}^{\pm} + 1| < 1$, then M is a part of a hyperbolic hyperquadric.*
(ii) *If $1 < |\bar{\kappa}^{\pm} + 1|$, then M is a part of an elliptic hyperquadric.*
(iii) *If $\bar{\kappa}^{\pm} = 0$, then M is a part of a de Sitter hyperhorosphere.*

Since $\mathbf{X}_{u_i}$ (for $i = 1, \ldots, n-1$) are spacelike vectors, we have the Riemannian metric (the *first fundamental form*) $ds^2 = \sum_{i,j=1}^{n-1} g_{ij}du_i du_j$ on $M = \mathbf{X}(U)$, where $g_{ij}(u) = \langle \mathbf{X}_{u_i}(u), \mathbf{X}_{u_j}(u) \rangle$ for any $u \in U$. We also define a *positive* (or *negative*) *lightcone second fundamental form* by $\bar{h}_{ij}^{\pm}(u) = \langle -\mathbb{L}_{u_i}^{\pm}(u), \mathbf{X}_{u_j}(u) \rangle$ for any $u \in U$. We have the following *Weingarten formula* which is analogous to ([6], Proposition 2.4).

$$\mathbb{L}_{u_i}^{\pm} = -\sum_{j=1}^{n-1} (\bar{h}^{\pm})_i^j \mathbf{X}_{u_j},$$

where $((\bar{h}^{\pm})_i^j) = (\bar{h}_{ik}^{\pm})(g^{kj})$ and $(g^{kj}) = (g_{kj})^{-1}$. Therefore we have an explicit expression for the lightcone Gauss-Kronecker curvature $K_{\ell}^{\pm} = \det(\bar{h}_{ij}^{\pm})/\det(g_{\alpha\beta})$ in terms of the Riemannian metric and the lightcone second fundamental invariant. We say that $p = \mathbf{X}(u)$ is a *positive* (or *negative*) *lightcone parabolic point* (briefly an $L^{\pm}$-*parabolic point*) of $\mathbf{X}$ if $K_{\ell}^{\pm}(u) = 0$.

3. Lightcone height functions

In this section we introduce families of functions on a spacelike hypersurface in de Sitter space, which are useful for the study of singularities of lightcone Gauss maps. Let $\mathbf{X} : U \longrightarrow S_1^n$ be a spacelike hypersurface in S_1^n. We define a family of functions $H : U \times LC^* \longrightarrow \mathbb{R}$ by

$$H(u, \mathbf{v}) = \langle \mathbf{X}(u), \mathbf{v} \rangle - 1.$$

We call H a *lightcone height function* on $\mathbf{X} : U \longrightarrow S_1^n$. We denote the Hessian matrix of the lightcone height function $h_{\mathbf{v}_0}^{\pm}(u) = H(u, \mathbf{v}_0)$ at u_0 by $\mathrm{Hess}\,(h_{\mathbf{v}_0}^{\pm})(u_0)$. We have the following lemma analogous to ([6], Propositions 3.1, 3.2).

Proposition 3.1 ([12]). *Let* $\mathbf{X} : U \longrightarrow S_1^n$ *be a hypersurface in* S_1^n, *then* $H(u, \mathbf{v}) = 0$ *and* $\partial H(u, \mathbf{v})/\partial u_i = 0$ $(i = 1, \ldots, n - 1)$ *if and only if* $\mathbf{v} = \mathbb{L}^{\pm}(u)$. *Under this condition, we have:*

(i) $p = \mathbf{X}(u_0)$ *is an* $L^{\pm}$-*parabolic point if and only if* $\det \mathrm{Hess}\,(h_{\mathbf{v}_0}^{\pm})(u_0) = 0$.
(ii) $p = \mathbf{X}(u_0)$ *is an* $L^{\pm}$-*flat point if and only if* $\mathrm{rank}\,\mathrm{Hess}\,(h_{\mathbf{v}_0}^{\pm})(u_0) = 0$.

We now interpret the lightcone Gauss image of a spacelike hypersurface in S_1^n as a wave front set in the theory of Legendrian singularities. Let $\pi^{\pm} : PT(LC_{\pm}^*) \longrightarrow LC_{\pm}^*$ be the projective cotangent bundles with canonical contact structures. Consider the tangent bundles $\tau^{\pm} : TPT^*(LC_{\pm}^*) \longrightarrow PT^*(LC_{\pm}^*)$ and the differential maps $d\pi^{\pm} : TPT(LC_{\pm}^*) \longrightarrow T(LC_{\pm}^*)$ of $\pi^{\pm}$. For any $X \in TPT^*(LC_{\pm}^*)$, there exists an element $\alpha \in T^*(LC_{\pm}^*)$ such that $\tau^{\pm}(X) = [\alpha]$. For an element $V \in T_x(LC_{\pm}^*)$, the property $\alpha(V) = 0$ does not depend on the choice of representative of the class $[\alpha]$. Thus, we can define the canonical contact structure on $PT^*(LC_{\pm}^*)$ by $K = \{X \in TPT^*(LC_{\pm}^*) \mid \tau^{\pm}(X)(d\pi^{\pm}(X)) = 0\}$.

On the other hand, we consider a point $\mathbf{v} = (v_0, v_1, \ldots, v_n) \in LC_{\pm}^*$, then we have the relation $v_0 = \pm\sqrt{v_1^2 + \cdots + v_n^2}$. So we adopt the coordinate system $(v_1, \ldots, v_n)$ of the manifold $LC_{\pm}^*$. Then we have the trivialization $PT^*(LC_{\pm}^*) \equiv LC_{\pm}^* \times P\mathbb{R}^{n-1}$, and call $((v_0, \ldots, v_n), [\xi_1 : \ldots : \xi_n])$ homogeneous coordinates of $PT^*(LC_{\pm}^*)$, where $[\xi_1 : \ldots : \xi_n]$ are the homogeneous coordinates of the dual projective space $P\mathbb{R}^{n-1}$. It is easy to show that $X_\bullet \in K_\bullet^{\pm}$ if and only if $\sum_{i=1}^n \mu_i \xi_i = 0$, where $\bullet = (v, [\xi])$ and $d\pi_\bullet^{\pm}(X_\bullet) = \sum_{i=1}^n \mu_i \partial/\partial v_i \in T_\bullet LC_{\pm}^*$. An immersion $i : L \longrightarrow PT^*(LC_{\pm}^*)$ is said to be a *Legendrian immersion* if $\dim L = n - 1$ and $di_q(T_q L) \subset K_{i(q)}$ for any $q \in L$. The map $\pi \circ i$ is also called the *Legendrian map* and the image $W(i) = \mathrm{im}(\pi \circ i)$, the *wave* *front* of i. Moreover, i (or the image of i) is called the *Legendrian lift* of $W(i)$.

Let $F : (\mathbb{R}^{n-1} \times \mathbb{R}^k, (u_0, \mathbf{v}_0)) \longrightarrow (\mathbb{R}, 0)$ be a function germ. We say that F is a *Morse family* of hypersurfaces if the map germ $\Delta^* F : (\mathbb{R}^{n-1} \times \mathbb{R}^k, (u_0, \mathbf{v}_0)) \longrightarrow (\mathbb{R}^n, \mathbf{0})$ defined by $\Delta^* F = (F, \partial F / \partial u_1, \ldots, \partial F / \partial u_{n-1})$ is non singular. In this case, we have a smooth $(k-1)$-dimensional smooth submanifold, $\Sigma_*(F) = \{(u, \mathbf{v}) \in (\mathbb{R}^{n-1} \times \mathbb{R}^k, (u_0, \mathbf{v}_0)) \mid \Delta^* F(u, \mathbf{v}) = 0\}$, and the Legendrian immersion germ $\mathcal{L}_F : (\Sigma_*(F), (u_0, \mathbf{v}_0)) \longrightarrow PT^* \mathbb{R}^k$ defined by $\mathcal{L}_F(u, \mathbf{v}) = (v, [\partial F / \partial v_1(u, \mathbf{v}) : \ldots : \partial F / \partial v_k(u, \mathbf{v})])$. Then we have the following fundamental theorem of Arnold and Zakalyukin [1, 19].

Proposition 3.2. *All Legendrian submanifold germs in $PT^* \mathbb{R}^k$ are constructed by the above method.*

We call F a generating family of $\mathcal{L}_F(\Sigma_*(F))$. Therefore the wave front is $W(\mathcal{L}_F) = \{\mathbf{v} \in \mathbb{R}^k \mid \exists u \in \mathbb{R}^{n-1} \text{ such that } F(u, \mathbf{v}) = \partial F / \partial u_1(u, \mathbf{v}) = \cdots = \partial F / \partial u_{n-1}(u, \mathbf{v}) = 0\}$. We call it the *discriminant set* of F. We now give a proof of the following proposition.

Proposition 3.3. *The lightcone height function $H : U \times LC^* \longrightarrow \mathbb{R}$ is a Morse family.*

Proof. We denote $\mathbf{X}(u) = (x_0(u), \ldots, x_n(u))$ and $\mathbf{X}_{u_i}(u) = (x_{0,u_i}(u) \ldots, x_{n,u_i}(u))$. For any $\mathbf{v} = (v_0, \ldots, v_n) \in LC_{\pm}^*$, we have $v_0 \neq 0$. Without loss of generality, we assume that $v_0 = \sqrt{v_1^2 + \cdots + v_n^2} > 0$, so that we have $H(u, \mathbf{v}) = -1 - x_0(u)\sqrt{v_1^2 + \cdots + v_n^2} + \sum_{k=1}^n x_k(u)v_k$. We have to prove that the mapping $\Delta^* H : U \times LC_{\pm}^* \longrightarrow \mathbb{R}^n$ is non-singular on $(\Delta^* H)^{-1}(0)$. Therefore it is sufficient to show that the Jacobian matrix of $\Delta^* H$

$$J\Delta^* H(u, v) = \begin{pmatrix} \left(-x_0 \dfrac{v_j}{v_0} + x_j\right)_{j=1,\ldots,n} \\ \hline \left(-x_{0,u_i} \dfrac{v_j}{v_0} + x_{j,u_i}\right)_{\substack{j=1,\ldots,n \\ i=1,\ldots,n-1}} \end{pmatrix}$$

is regular on $\Sigma_*(H)$. We denote vectors $\bar{a}, \bar{b}_i$ $(i = 1, \ldots, n)$ by $\bar{a} = {}^t(x_0, x_{0,u_1}, \ldots, x_{0,u_{n-1}})$ and $\bar{b}_j = {}^t(x_j, x_{j,u_1}, \ldots, x_{j,u_{n-1}})$ for $j = 1, \ldots, n$. Then the determinant of $J\Delta^*H$ is

$$\det J\Delta^*H = \det\left(-\bar{a}\frac{v_1}{v_0} + \bar{b}_1, \ldots, -\bar{a}\frac{v_n}{v_0} + \bar{b}_n\right)$$

$$= \det(\bar{b}_1, \ldots, \bar{b}_n) - \frac{v_1}{v_0}\det(\bar{a}, \bar{b}_2, \ldots, \bar{b}_n) - \cdots$$

$$-\frac{v_n}{v_0}\det(\bar{b}_1, \ldots, \bar{b}_{n-1}, \bar{a})$$

$$= \left\langle\left(\frac{v_0}{v_0}, \ldots, \frac{v_n}{v_0}\right), \mathbf{X}(u) \wedge \mathbf{X}_{u_1}(u) \wedge \ldots \wedge \mathbf{X}_{u_{n-1}}(u)\right\rangle$$

$$= \frac{1}{v_0}\langle\mathbb{L}^\pm(u), e(u)\rangle.$$

Since $v = \mathbb{L}^\pm(u)$, so that we have $\det P(u, v) = \mp 1/v_0 \neq 0$. This completes the proof. $\qquad\square$

The Legendrian immersion germs $\mathcal{L}^\pm : (\Sigma_*^\pm(H), (u_0, \mathbb{L}^\pm(u_0))) \longrightarrow PT^*(LC_\pm^*)$ are obtained by $\mathcal{L}^\pm(u, \mathbf{v}) = (\mathbf{v}, [\partial H/\partial v_1(u, \mathbf{v}) : \ldots : \partial H/\partial v_n(u, \mathbf{v})])$ where $\Sigma_*^\pm(H)$ are singular sets of H

$$\Sigma_*^\pm(H) = \{(u, \mathbf{v}) \in U \times LC_\pm^* \mid \mathbf{v} = \mathbb{L}^\pm(u)\}.$$

Therefore we have the Legendrian immersion germs $\mathcal{L}^\pm$ whose wave front sets are the lightcone Gauss image germs $\mathbb{L}^\pm$.

4. Generic properties and de Sitter hyperhorospheres

In this section we review the theory of contact due to Montaldi [15] to study the contact between spacelike hypersurfaces and de Sitter hyperhorospheres. Let X_i and Y_i (for $i = 1, 2$) be submanifolds of $\mathbb{R}^n$ with $\dim X_1 = \dim X_2$ and $\dim Y_1 = \dim Y_2$. We say that the contact of X_1 and Y_1 at y_1 is the same type as the contact of X_2 and Y_2 at y_2 if there is a diffeomorphism germ $\Phi : (\mathbb{R}^n, y_1) \longrightarrow (\mathbb{R}^n, y_2)$ such that $\Phi(X_1) = X_2$ and $\Phi(Y_1) = Y_2$. In this case we write $K(X_1, Y_1; y_1) = K(X_2, Y_2; y_2)$. Function germs $g_1, g_2 : (\mathbb{R}^n, a_i) \longrightarrow (\mathbb{R}, 0)$ $(i = 1, 2)$ are $\mathcal{K}$-equivalent if there are a diffeomorphism germ $\Phi : (\mathbb{R}^n, a_1) \longrightarrow (\mathbb{R}^n, a_2)$ and a function germ $\lambda : (\mathbb{R}^n, a_1) \longrightarrow \mathbb{R}$ with $\lambda(a_1) \neq 0$ such that $f_1 = \lambda \cdot (g_2 \circ \Phi)$. In [15] Montaldi has shown the following theorem.

Theorem 4.1 (Montaldi [15]). *Let X_i and Y_i (for $i = 1, 2$) be submanifolds of $\mathbb{R}^n$ with $\dim X_1 = \dim X_2$ and $\dim Y_1 = \dim Y_2$. Let $g_i : (X_1, x_1) \longrightarrow (\mathbb{R}^n, y_i)$ be immersion germs and $f_i : (\mathbb{R}^n, y_i) \longrightarrow (\mathbb{R}^p, \mathbf{0})$ be submersion germs with*

$(Y_i, y_i) = (f_i^{-1}(\mathbf{0}), y_i)$. *Then* $K(X_1, Y_1; y_1) = K(X_2, Y_2; y_2)$ *if and only if* $f_1 \circ g_1$ *and* $f_2 \circ g_2$ *are* $\mathcal{K}$-*equivalent.*

We apply the above theorem to our case. Let $\mathbf{v}_0 \in LC^*$, we define $\mathfrak{h}_{\mathbf{v}_0} : S_1^n \longrightarrow \mathbb{R}$ by $\mathfrak{h}_{\mathbf{v}_0}(w) = \langle w, \mathbf{v}_0 \rangle - 1$. Then we have a de Sitter hyperhorosphere $\mathfrak{h}_{\mathbf{v}_0}^{-1}(0) = HP(\mathbf{v}_0, +1) \cap S_1^n$, and write $HS(\mathbf{v}_0, +1) = HP(\mathbf{v}_0, +1) \cap S_1^n$. For any $u_0 \in U$, we consider the lightlike vector $\mathbf{v}_0^{\pm} = \mathbb{L}^{\pm}(u_0)$. Then we have $(\mathfrak{h}_{\mathbf{v}_0^{\pm}} \circ \mathbf{X})(u_0) = H(u_0, \mathbb{L}^{\pm}(u_0)) = 0$ and $(\partial \mathfrak{h}_{\mathbf{v}_0^{\pm}} \circ \mathbf{X})(u_0)/\partial u_i = \langle \mathbf{X}_{u_i}(u_0), \mathbb{L}^{\pm}(u_0) \rangle = 0$ for $i = 1, \ldots, n-1$. This means that the de Sitter hyperhorosphere is tangent to the spacelike hypersurface M at $p_0 = \mathbf{X}(u_0)$. In this case, we call $HS(\mathbf{v}_0^{\pm}, +1)$ the *tangent de Sitter hyperhorosphere* of M at p_0 (or u_0), which we write $HS^{\pm}(\mathbf{X}, u_0)$. Let $\mathbf{v}_1$ and $\mathbf{v}_2$ be lightlike vectors. We say that de Sitter hyperhorospheres $HS(\mathbf{v}_1, +1)$ and $HS(\mathbf{v}_2, +1)$ are *parallel* if $\mathbf{v}_1$ and $\mathbf{v}_2$ are linearly independent. Then we have the following proposition which is analogous to ([6] Lemma 6.2).

Proposition 4.2 ([12]). *Let* $\mathbf{X} : U \longrightarrow S_1^n$ *be a hypersurface. Consider two points* $u_1, u_2 \in U$. *Then* $\mathbb{L}^{\pm}(u_1) = \mathbb{L}^{\pm}(u_2)$ *if and only if* $HS^{\pm}(\mathbf{X}, u_1) = HS^{\pm}(\mathbf{X}, u_2)$.

We now review some notions of Legendrian singularity theory. We say that Legendrian immersion germs $i_j : (U_j, u_j) \longrightarrow (PT^*\mathbb{R}^n, p_j)$ $(j = 1, 2)$ are *Legendrian equivalent* if there exists a contact diffeomorphism germ $H : (PT^*\mathbb{R}^n, p_1) \longrightarrow (PT^*\mathbb{R}^n, p_2)$ such that H preserves fibers of π and $H(U_1) = U_2$. A Legendrian immersion germ at a point is said to be *Legendrian stable* if for every map with the given germ there are a neighborhood in the space of Legendrian immersions (in the Whitney C^∞-topology) and a neighborhood of the original point such that each Legendrian map belonging to the first neighborhood has in the second neighborhood a point at which its germ is Legendrian equivalent to the original germ.

Proposition 4.3 (Zakalyukin [20]). *Let* i_1, i_2 *be Legendrian immersion germs such that regular sets of* $\pi \circ i_1$ *and* $\pi \circ i_2$ *are respectively dense. Then* i_1, i_2 *are Legendrian equivalent if and only if corresponding wave front sets* $W(i_1)$ *and* $W(i_2)$ *are diffeomorphic as set germs.*

Let $F_i : (\mathbb{R}^n \times \mathbb{R}^k, (a_i, b_i)) \longrightarrow (\mathbb{R}, 0)$ $(i = 1, 2)$ be k-parameter unfoldings of function germs f_i. We say F_1 and F_2 are $\mathcal{P}$-$\mathcal{K}$-equivalent if there exists a diffeomorphism germ $\Phi : (\mathbb{R}^n \times \mathbb{R}^k, (a_1, b_1)) \longrightarrow (\mathbb{R}^n \times \mathbb{R}^k, (a_2, b_2))$ of the form $\Phi(u, x) = (\phi_1(u, x), \phi_2(x))$ for $(u, x) \in \mathbb{R}^n \times \mathbb{R}^k$ and a function germ $\lambda : (\mathbb{R}^n \times \mathbb{R}^k, (a_1, b_1)) \longrightarrow \mathbb{R}$ such that $\lambda(a_1, b_1) \neq 0$ and $F_1(u, x) = \lambda(u, x) \cdot (F_2 \circ \Phi)(u, x)$.

Theorem 4.4 (Arnold, Zakalyukin [1, 19]). *Let $F, G : (\mathbb{R}^k \times \mathbb{R}^n, \mathbf{0}) \longrightarrow$ $(\mathbb{R}, \mathbf{0})$ be Morse families and denote their Legendrian immersion by $\mathcal{L}_F, \mathcal{L}_G$. Then*

  (i) *$\mathcal{L}_F$ and $\mathcal{L}_G$ are Legendrian equivalent if and only if F and G are $\mathcal{P}$-$\mathcal{K}$-equivalent.*

  (ii) *$\mathcal{L}_F$ is Legendrian stable if and only if F is $\mathcal{K}$-versal deformation.*

Let $\mathbb{L}_i^\pm : (U, u_i) \longrightarrow (LC_\pm^*, \mathbf{v_i}^\pm)$ (for $i = 1, 2$) be de Sitter Gauss image germs of hypersurface germs $\mathbf{X}_i : (U, u_i) \longrightarrow (S_1^n, p_i)$. We say $\mathbb{L}_1^\pm$ and $\mathbb{L}_2^\pm$ are $\mathcal{A}$-equivalent if and only if there exist diffeomorphism germs $\phi : (U, u_1) \longrightarrow$ (U, u_2) and $\Phi : (LC_\pm^*, \mathbf{v_1}^\pm) \longrightarrow (LC_\pm^*, \mathbf{v_2}^\pm)$ such that $\Phi \circ \mathbb{L}_1^\pm = \mathbb{L}_2^\pm \circ \phi$.

We denote $h_{i,\mathbf{v_i}^\pm} : (U, u_i) \longrightarrow (\mathbb{R}, 0)$ by $h_{i,\mathbf{v_i}^\pm}(u) = H_i(u, \mathbf{v_i}^\pm)$. Then we have $h_{i,\mathbf{v_i}^\pm}(u) = (\mathfrak{h}_{\mathbf{v_i}^\pm}) \circ \mathbf{X}_i(u)$. By Theorem 4.1, $K(\mathbf{X}_1(U), HS^\pm(\mathbf{X}_1, u_1); p_1) = K(\mathbf{X}_2(U), HS^\pm(\mathbf{X}_2, u_2); p_2)$ if and only if $h_{1,\mathbf{v_1}^\pm}$ and $h_{2,\mathbf{v_2}^\pm}$ are $\mathcal{K}$-equivalent. We denote $Q^\pm(\mathbf{X}, u_0)$ the local ring of the function germ $h_{\mathbf{v_0}^\pm} : (U, u_0) \longrightarrow \mathbb{R}$ by $Q^\pm(\mathbf{X}, u_0) = C_{u_0}^\infty(U)/\langle h_{\mathbf{v_0}} \rangle_{C_{u_0}^\infty}$, where $\mathbf{v_0}^\pm = \mathbb{L}^\pm(u_0)$ and $C_{u_0}^\infty(U)$ is the local ring of function germs at u_0 with the unique maximal ideal $\mathfrak{M}$. By the above arguments, we have the following theorem which is analogous to ([6] Theorem 6.3).

Theorem 4.5 ([12]). *Let $\mathbf{X}_i : (U, u_i) \longrightarrow (S_1^n, p_i)$ (for $i = 1, 2$) be spacelike hypersurface germs such that the corresponding Legendrian immersion germs are Legendrian stable. Then the following conditions are equivalent:*

  (i) *Lightcone Gauss image germs $\mathbb{L}_1^\pm$ and $\mathbb{L}_2^\pm$ are $\mathcal{A}$-equivalent.*

  (ii) *Legendrian immersion germs $\mathcal{L}_1$ and $\mathcal{L}_2$ are Legendrian equivalent.*

  (iii) *Lightcone height function germs H_1 and H_2 are $\mathcal{P}$-$\mathcal{K}$-equivalent.*

  (iv) *$h_{1,\mathbf{v_1}^\pm}$ and $h_{2,\mathbf{v_2}^\pm}$ are $\mathcal{K}$-equivalent.*

  (v) *$K(\mathbf{X}_1(U), HS^\pm(\mathbf{X}_1, u_1); p_1) = K(\mathbf{X}_2(U), HS^\pm(\mathbf{X}_2, u_2); p_2)$*

  (v) *$Q^\pm(\mathbf{X}_1, u_1)$ and $Q^\pm(\mathbf{X}_2, u_2)$ are isomorphic as $\mathbb{R}$-algebras.*

We consider the space of spacelike embeddings Sp-Emb(U, S_1^n) with the Whitney C^∞-topology. The assumption of Theorem 4.5 is a generic property if $n \leq 6$ (see [6] §7).

Theorem 4.6. *If $n \leq 6$, there exists an open dense subset $\mathcal{O} \subset$ Sp-Emb(U, S_1^n) such that for any $\mathbf{X} \in \mathcal{O}$, corresponding Legendrian immersion germ $\mathcal{L}$ is Legendrian stable.*

In general we have the following proposition.

Proposition 4.7 ([12]). *Let* $X_i : (U, u_i) \longrightarrow (S_1^n, p_i)$ *(for* $i = 1, 2$*) be hypersurface germs such that their* $L^\pm$*-parabolic sets have no interior points as subspaces of* U*. If the lightcone Gauss image germs* $\mathbb{L}_1^\pm$ *and* $\mathbb{L}_2^\pm$ *are* $\mathcal{A}$*-equivalent, then*

$$K(X_1(U), HS^\pm(X_1, u_1); p_1) = K(X_2(U), HS^\pm(X_2, u_2); p_2).$$

In this case, $(X_i^{-1}(HS(\mathbb{L}_i^\pm(u_i), +1)), u_i)$ *(* $i = 1, 2$*) are diffeomorphic as set germs.*

For a hypersurface germ X, we call $(X^{-1}(HS(\mathbb{L}^\pm(u_0), +1)), u_0)$ the *tangent de Sitter hyperhorospherical indicatrix germ* of X. By Proposition 4.7, the diffeomorphism type of the tangent de Sitter horospherical indicatrix germ is an invariant under the $\mathcal{A}$-equivalence among lightcone Gauss image germs.

5. Spacelike surfaces in de Sitter 3-space

In this section we consider $n = 3$. In this case, we call $X : U \longrightarrow S_1^3$ a *spacelike surface* and $HS(v_0, +1)$ a *de Sitter horosphere*. We say that a map germ $L : (\mathbb{R}^2, a) \longrightarrow (\mathbb{R}^3, b)$ is the *cuspidal edge* if it is $\mathcal{A}$-equivalent to the germ (u_1, u_2^2, u_2^3) and that L is the swallowtail if it is $\mathcal{A}$-equivalent to the germ $(3u_1^4 + u_1^2 u_2, 4u_1^3 + 2u_1 u_2, u_2)$. By Theorem 4.6 and the classification of function germs [1], we have the following theorem.

Theorem 5.1 ([12]). *There exists an open dense subset* $\mathcal{O} \subset Sp\text{-}Emb(U, S_1^3)$ *such that for any* $X \in \mathcal{O}$*, the* $L^\pm$*-parabolic set* $(K_\ell^\pm)^{-1}(0)$ *is a regular curve. Let* $X \in \mathcal{O}$*,* $v_0^\pm = \mathbb{L}^\pm(u_0)$ *and* $h_{v_0^\pm} : (U, u_0) \longrightarrow \mathbb{R}$ *be the lightcone height function germ at* u_0*. Then we have the following.*

(i) *The point* u_0 *is an* $L^\pm$*-parabolic point of* X *if and only if* $H\text{-}corank^\pm(X, u_0) = 1$ *(that is,* u_0 *is not an* $L^\pm$*-flat point). In this case,* $h_{v_0^\pm}$ *has the* $\mathcal{A}_k$*-type singularity for* $k = 2, 3$*.*

(ii) *If* $h_{v_0^\pm}$ *has the* $\mathcal{A}_2$*-type singularity, then* $\mathbb{L}^\pm$ *has the cuspidal edge at* u_0*. The tangent de Sitter horospherical indicatrix germ is diffeomorphic to the curve given by* $\{(u_1, u_2) \mid u_1^2 - u_2^3 = 0\}$*. We call the set germ an ordinary cusp.*

(iii) *If* $h_{v_0^\pm}$ *has the* $\mathcal{A}_3$*-type singularity, then* $\mathbb{L}^\pm$ *has the swallowtail at* u_0*. The tangent de Sitter horospherical indicatrix germ is a point or diffeomorphic to* $\{(u_1, u_2) \mid u_1^2 - u_2^4 = 0\}$*. We call the set germ a tacnode. In this case,* *for any open neighborhood U of the point u_0, there exists two non $L^{\pm}$-parabolic points $u_1, u_2 \in U'$ such that the tangent de Sitter horospheres $HS^{\pm}(\mathbf{X}_i, u_i)$ $(i = 1, 2)$ are equal.*

6. Spacelike submanifolds of codimension 2

In this section we discuss the differential geometry of spacelike submanifolds of codimension two in de Sitter space which is analogous to [8]. Let $\mathbf{X} : U \longrightarrow S_1^n$ be an embedding from an open set $U \subset \mathbb{R}^{n-2}$. We say that $\mathbf{X}$ is *spacelike* if every non zero vector generated by $\{\mathbf{X}_{u_i}(u)\}_{i=1}^{n-2}$ is always spacelike, where $u \in U$. We identify $M = \mathbf{X}(U)$ with U through the embedding $\mathbf{X}$ and call M a *spacelike submanifold of codimension two* in S_1^n. Since $\langle \mathbf{X}, \mathbf{X} \rangle \equiv 1$, we have $\langle \mathbf{X}_{u_i}, \mathbf{X} \rangle \equiv 0$ (for $i = 1, \ldots, n-1$). In this case, for any $p = \mathbf{X}(u)$, the pseudo-normal space $N_p M$ is a timelike plane. We can choose a *future directed unit normal section* $\mathbf{n}^T(u) \in N_p M$ satisfying $\langle \mathbf{n}^T(u), \mathbf{X}(u) \rangle = 0$. Therefore we can construct a spacelike unit normal section $\mathbf{n}^S(u) \in N_p M$ by

$$\mathbf{n}^S(u) = \frac{\mathbf{n}^T(u) \wedge \mathbf{X}_{u_1}(u) \wedge \ldots \wedge \mathbf{X}_{u_{n-2}}(u)}{||\mathbf{n}^T(u) \wedge \mathbf{X}_{u_1}(u) \wedge \ldots \wedge \mathbf{X}_{u_{n-2}}(u)||}.$$

We remark that $\mathbf{n}^T(u) \pm \mathbf{n}^S(u)$ are lightlike. We call $(\mathbf{n}^T(u), \mathbf{n}^S(u))$ a *future directed normal frame along* $M = \mathbf{X}(U)$. The system $\{\mathbf{X}(u), \mathbf{n}^T(u), \mathbf{n}^S(u), \mathbf{X}_{u_1}(u), \ldots, \mathbf{X}_{u_{n-2}}(u)\}$ is a basis of $T_p \mathbb{R}_1^{n+1}$. We have the following lemma which is analogous to Lemma 3.1 in [8].

Lemma 6.1 ([13]). *Given two future directed unit timelike normal sections $\mathbf{n}^T(u), \bar{\mathbf{n}}^T(u) \in N_p M$, the corresponding lightlike normal sections $\mathbf{n}^T(u) \pm \mathbf{n}^S(u), \bar{\mathbf{n}}^T(u) \pm \bar{\mathbf{n}}^S(u)$ are parallel.*

Under the identification of M and U through $\mathbf{X}$, we have the linear mapping $d_p(\mathbf{n}^T \pm \mathbf{n}^S)$ provided by the derivative of the *lightlike normal sections*. We consider orthonormal projection $\pi^t : T_p \mathbb{R}_1^{n+1} \longrightarrow T_p M$ and call the linear transformation

$$S_p^{\pm}(\mathbf{n}^T, \mathbf{n}^S) = -\pi^t \circ d_p(\mathbf{n}^T \pm \mathbf{n}^S)$$

an $(\mathbf{n}^T, \mathbf{n}^S)$-*shape operator of* $M = \mathbf{X}(U)$ at $p = \mathbf{X}(u)$. The eigenvalues of $S_p^{\pm}(\mathbf{n}^T, \mathbf{n}^S)$ denoted by $\{\kappa_i^{\pm}(\mathbf{n}^T, \mathbf{n}^S)(p)\}_{i=1}^{n-2}$ are called the *lightcone principal curvatures* with respect to $(\mathbf{n}^T, \mathbf{n}^S)$ at p. Then the *lightcone Gauss-Kronecker curvature* with respect to $(\mathbf{n}^T, \mathbf{n}^S)$ at p is defined by $K_{\ell}^{\pm}(\mathbf{n}^T, \mathbf{n}^S)(p) = \det S_p^{\pm}(\mathbf{n}^T, \mathbf{n}^S)$.

Since $\mathbf{X}_{u_i}(u)$ $(i = 1, \ldots, n-2)$ are spacelike vectors, we have a *Riemannian metric* (or the *first fundamental form*) on M defined by $ds^2 = \sum_{i,j=1}^{n-2} g_{ij} du_i du_j$, where $g_{ij}(u) = \langle \mathbf{X}_{u_i}, \mathbf{X}_{u_j} \rangle$ for any $u \in U$. We also have a *lightcone second fundamental form* (or the *lightcone second fundamental invariant*) with respect to the normal vector field $(\mathbf{n}^T, \mathbf{n}^S)$ defined by $h_{ij}^{\pm}(u) = -\langle (\mathbf{n}^T \pm \mathbf{n}^S)_{u_i}, \mathbf{X}_{u_j} \rangle$ for any $u \in U$. We have the following *lightcone Weingarten formula* which is analogous to Proposition 3.2 in [8].

$$\pi^t \circ (\mathbf{n}^T \pm \mathbf{n}^S)_{u_i} = -\sum_{j=1}^{n-2} h_i^{j\pm}(\mathbf{n}^T, \mathbf{n}^S)\mathbf{X}_{u_j},$$

where $(h_i^{j\pm}(\mathbf{n}^T, \mathbf{n}^S))_{ij} = \left(h_{ik}^{\pm}(\mathbf{n}^T, \mathbf{n}^S)\right)_{ik}(g_{kj})^{-1}$.

For a lightlike vector $v = (v_0, v_1, \ldots, v_n)$ we define $\widetilde{v} = (1, v_1/v_0, \ldots, v_n/v_0)$. By Lemma 6.1, the normalized lightcone normal directions $\mathbf{n}^T(u) \pm \mathbf{n}^S(u)$ are independent on the choice of the lightcone normal directions $\mathbf{n}^T(u)$, $\mathbf{n}^S(u)$. Therefore we define the *lightcone Gauss map* $\widetilde{\mathbb{L}}^{\pm} : U \longrightarrow S_+^{n-1}$ of M as

$$\widetilde{\mathbb{L}}^{\pm}(u) = \widetilde{\mathbf{n}^T(u) \pm \mathbf{n}^S(u)}.$$

The lightcone Gauss map is analogous in this case to the one defined in [8] for codimension two submanifolds in Minkowski space. This induces a linear mapping $d\widetilde{\mathbb{L}}^{\pm} : T_p M \longrightarrow T_p \mathbb{R}_1^{n+1}$ under the identification of U and M, where $p = \mathbf{X}(u)$. We have the following normalized lightcone Weingarten formula:

$$\pi^t \circ \widetilde{\mathbb{L}}_{u_i}^{\pm} = \frac{1}{\ell_0^{\pm}}\left(\pi^t \circ \mathbb{L}_{u_i}^{\pm}\right) = -\sum_{j=1}^{n-2} \frac{1}{\ell_0^{\pm}} h_i^{\pm j}(\mathbf{n}^T, \mathbf{n}^S)\mathbf{X}_{u_j},$$

where $\mathbb{L}^{\pm}(u) = (\ell_0^{\pm}(u), \ldots, \ell_n^{\pm}(u))$. We call a linear transformation $S_p^{\pm} = -\pi^t \circ d\widetilde{\mathbb{L}}_p^{\pm} : T_p M \longrightarrow T_p M$ the *normalized lightcone shape operator* of M at p. The eigenvalues $\{\widetilde{\kappa}_i^{\pm}(p)\}_{i=1}^{n-2}$ of $\widetilde{S}_p^{\pm}$ are called *normalized lightcone principal curvatures*. By the above arguments, we have $\widetilde{\kappa}_i^{\pm}(p) = (1/\ell_0^{\pm}(u))\kappa_i^{\pm}(\mathbf{n}^T, \mathbf{n}^S)(p)$. The *normalized lightcone Gauss-Kronecker curvature of M* is defined to be $\widetilde{K}_\ell^{\pm}(u) = \det \widetilde{S}_p^{\pm}$, and we have the following relation between the normalized lightcone Gauss-Kronecker curvature and the lightcone Gauss-Kronecker curvature

$$\widetilde{K}_\ell^{\pm}(u) = K_\ell^{\pm}(\mathbf{n}^T, \mathbf{n}^S)(u)/\left(\ell_0^{\pm}(u)\right)^{n-2}.$$

We say that a point $u \in U$ or $p = \mathbf{X}(u)$ is a *lightlike umbilic point* if $\widetilde{S}_p^{\pm} = \widetilde{\kappa}_p^{\pm}(p)\mathrm{id}_{T_p M}$. Spacelike submanifold M is called *totally lightlike umbilic* if all points on M are lightlike umbilic. We also say that p is a *lightlike parabolic point* (briefly $\widetilde{L}^{\pm}$-*parabolic*) if $\widetilde{K}_{\ell}^{\pm}(u) = 0$. Moreover, p is called a *lightlike flat point* if $\widetilde{S}_{p}^{\pm} = O_{T_pM}$. The spacelike submanifold M in S_1^n is called *totally lightlike flat* if every point in M is lightlike flat.

7. Lightlike hypersurfaces and contact with lightcones

In this section we define the Lorentzian distance squared function in order to study the singularities of lightlike hypersurfaces. We use the theory of contacts between submanifolds due to Montaldi [15]. We define a hypersurface $LH_M^{\pm} : U \times \mathbb{R} \longrightarrow S_1^n$ by

$$LH_M^{\pm}(u, \mu) = \mathbf{X}(u) + \mu\widetilde{\mathbb{L}}^{\pm}(u).$$

We call $LH_M^{\pm}$ a *lightlike hypersurface along M*. This notion, introduced by Izumiya and Fusho [5] for the case of spacelike curves for surfaces in 3-dimensional de Sitter space, is analogous to the one studied in [7] for surfaces in Minkowski four space. We introduce the notion of Lorentzian distance squared functions on spacelike submanifold of codimension two, which is useful for the study of singularities of lightlike hypersurfaces. We define a family of functions $G : U \times S_1^n \longrightarrow \mathbb{R}$ on a spacelike submanifold M by

$$G(u, \lambda) = \langle \mathbf{X}(u) - \lambda, \mathbf{X}(u) - \lambda \rangle,$$

where $p = \mathbf{X}(u)$. We call G a *Lorentzian distance squared function* on the spacelike submanifold M. For any fixed $\lambda_0 \in S_1^n$, we write $g_{\lambda_0}(u) = G(u, \lambda_0)$ and have following proposition.

Proposition 7.1 ([13]). *Let $G : U \times S_1^n \longrightarrow \mathbb{R}$ be the Lorentzian distance squared function on M. Suppose that $p_0 = \mathbf{X}(u_0) \neq \lambda_0$ and have the following:*

(i) *$g_{\lambda_0}(u_0) = \partial g_{\lambda_0}(u_0)/\partial u_i = 0$ $(i = 1, \ldots, n - 2)$ if and only if $\lambda_0 = LH_M^{\pm}(u_0, \mu_0)$ for some $\mu_0 \in \mathbb{R} \setminus \{0\}$.*

(ii) *$g_{\lambda_0}(u_0) = \partial g_{\lambda_0}(u_0)/\partial u_i = 0$ $(i = 1, \ldots, n - 2)$ and $\det \text{Hess}(g_{\lambda_0})(u_0) = 0$ if and only if $\lambda_0 = LH_M^{\pm}(u_0, \mu_0)$ for some $\mu_0 \in \mathbb{R} \setminus \{0\}$ and $-1/\mu_0$ is one of the non-zero normalized lightcone principal curvatures $\widetilde{\kappa}_i^{\pm}(p_0)$.*

(iii) *G is a Morse family of hypersurfaces around $(u_0, \lambda_0) \in \Delta^* G^{-1}(\mathbf{0})$.*

Therefore Legendrian immersion germs $\mathcal{L}_G^{\pm} : (\Sigma_*(G), (u_0, \lambda_0)) \longrightarrow PT^*(S_1^n)$ are defined by

$$\mathcal{L}_G^{\pm}(u, \lambda) = (\lambda, [\partial G/\partial\lambda^0(u, \lambda) : \ldots : \widehat{\partial G/\partial\lambda^k}(u, \lambda) : \ldots : \partial G/\partial\lambda^n(u, \lambda)])$$

where $\lambda = (\lambda^0, \ldots, \lambda^n)$ is a local coordinate around $\lambda_0 \in S_1^n$. and $\Sigma_*(G) = (\Delta^* G)^{-1}(0) = \{(u, \lambda) \in U \times S_1^n \mid \lambda = LH_M^{\pm}(u, \mu), \ \mu \in \mathbb{R}\}$. The Lorentzian distance squared function G is a generating family of the Legendrian immersion $\mathcal{L}_G^{\pm}$ whose wave front set is the image of $LH_M^{\pm}$.

We now apply the theory of contacts between spacelike submanifolds. We call $LC_{\lambda_0} \cap S_1^n$ a *de Sitter lightcone*. The following proposition is generalization of Proposition 4.1 in [7].

Proposition 7.2 ([13]). *Let $\lambda_0 \in S_1^n$ and M be a spacelike submanifold of codimension two without umbilic points satisfying $\widetilde{K}_\ell \neq 0$. Then $M \subset LC_{\lambda_0} \cap S_1^n$ if and only if λ_0 is an isolated singular value of the lightlike hypersurface $LH_M^{\pm}$ and $LH_M^{\pm}(U \times \mathbb{R}) \subset LC_{\lambda_0} \cap S_1^n$.*

We consider the contact of spacelike submanifolds of codimension two with lightcones due to Montaldi's result [15]. The function $\mathcal{G} : S_1^n \times S_1^n \longrightarrow \mathbb{R}$ defined by $\mathcal{G}(x, \lambda) = \langle x - \lambda, x - \lambda \rangle$. For a given $\lambda_o \in S_1^n$, we denote $\mathfrak{g}_{\lambda_0}(x) = \mathcal{G}(x, \lambda_0)$, then we have $\mathfrak{g}_{\lambda_0}^{-1}(0) = LC_{\lambda_0} \cap S_1^n$. For any $u_0 \in U$, we take the point $\lambda_0^{\pm} = \mathbf{X}(u_0) + \mu_0 \widetilde{\mathbb{L}}^{\pm}(u_0)$, the lightcone $LC_{\lambda_0} \cap S_1^n$ is tangent to M at $p_0 = \mathbf{X}(u_0)$. In this case, we call each $LC_{\lambda_0} \cap S_1^n$ a *tangent lightcone* of M at p_0. From the previous arguments in §4, we have following theorem.

Theorem 7.3 ([13]). *Let $\mathbf{X}_i : (U, u_i) \longrightarrow (S_1^n, p_i)$ (for $i = 1, 2$) be spacelike submanifold germs such that the corresponding Legendrian immersion germs are Legendrian stable. Then the following conditions are equivalent:*

  (i) *Lightlike hypersurface germs $LH_{M,1}^{\pm}$ and $LH_{M,2}^{\pm}$ are $\mathcal{A}$-equivalent.*
 (ii) *Legendrian immersion germs $\mathcal{L}_{G,1}^{\pm}$ and $\mathcal{L}_{G,2}^{\pm}$ are Legendrian equivalent.*
(iii) *Lorentzian distance squared function germs G_1 and G_2 are $\mathcal{P}$ $\mathcal{K}$-equivalent.*
 (iv) *$\mathfrak{g}_{1,\lambda_1}^{\pm}$ and $\mathfrak{g}_{2,\lambda_2}^{\pm}$ are $\mathcal{K}$-equivalent.*
  (v) *$K(\mathbf{X_1}(U), LC_{\lambda_1} \cap S_1^n; p_1) = K(\mathbf{X_2}(U), LC_{\lambda_2} \cap S_1^n; p_2)$*

From the arguments in [6], if $n \leq 6$, there exists an open subset $\mathcal{O} \subset$ Sp-Emb(U, S_1^n) such that for any $x \in \mathcal{O}$, the corresponding Legendrian immersion germ $\mathcal{L}_G$ is Legendrian stable.

8. Lightcone Gauss maps and lightcone height functions

In this section, we define the lightcone height function whose wave front set is the image of the lightcone Gauss map, and describe contacts of submanifolds with lightlike cylinders by applying Montaldi's theory. We define a *lightcone height function* $H : U \times S_+^{n-1} \longrightarrow \mathbb{R}$ by $H(u, v) = \langle X(u), v \rangle$. For $v_0 \in S_+^{n-1}$, we write $h_{v_0}(u) = H(u, v_0)$ and have following proposition.

Proposition 8.1 ([13]). *Let H be the lightcone height function of a spacelike submanifold* **X**, *then we have the following:*

(i) $H(u_0, v_0) = H_{u_i}(u_0, v_0) = 0$ $(i = 1, \ldots, n-2)$ *if and only if* $v_0 = \widetilde{\mathbb{L}}^{\pm}(u_0)$.

(ii) $H(u_0, v_0) = H_{u_i}(u_0, v_0) = 0$ $(i = 1, \ldots, n-2)$ *and* $\det \text{Hess}(h_{v_0})$ $(u_0) = 0$ *if and only if* $v_0 = \widetilde{\mathbb{L}}^{\pm}(u_0)$ *and* $\widetilde{K}_\ell^{\pm}(u_0) = 0$.

(iii) *H is a Morse family of hypersurfaces around* $(u, v) \in \Delta^* H^{-1}(0)$.

By the above proposition, the discriminant set of the lightcone height function is given by $D_H = \{v \in S_+^{n-1} \mid v = \widetilde{\mathbb{L}}^{\pm}(u), \ u \in U\}$, which is the image of the lightcone Gauss map of M. The singular set of the lightcone Gauss map is the normalized lightcone parabolic set of M.

The Legendrian immersion germs $\mathcal{L}_H^{\pm} : (\Sigma_*(H), (u_0, v_0)) \longrightarrow PT^*(S_+^{n-1})$ are defined by $\mathcal{L}_H^{\pm}(u, v) = (v, [\partial H/\partial v_1 : \ldots : \widehat{\partial H/\partial v_k} : \ldots : \partial H/\partial v_n])$ where $v = (1, v_1, \ldots, v_n) \in S_+^{n+1}$ and $\Sigma_*(H) = \{(u, v) \in U \mid v = \widetilde{\mathbb{L}}^{\pm}(u)\}$. The lightcone height function H is the generating family of the Legendrian immersion germs $\mathcal{L}_H^{\pm}$ whose wave front set are the lightcone Gauss maps $\widetilde{\mathbb{L}}^{\pm}$.

We apply the theory of Montaldi again to describe contacts of submanifolds with lightlike cylinders. For any $v \in S_+^{n-1}$, we define a *lightlike cylinder* along v by $HP(v, 0) \cap S_1^n$. It is an $(n-1)$-dimensional submanifold in S_1^n which is isomorphic to $S^{n-2} \times \mathbb{R}$, where S^{n-2} is an $(n-2)$-sphere. We observe that its tangent space at each point has lightlike directions.

Proposition 8.2 ([13]). *Let* $\widetilde{\mathbb{L}}^{\pm}$ *be a lightcone Gauss map of* **X**. *Then* $\widetilde{\mathbb{L}}^{\pm}$ *is a constant map if and only if M is a part of lightlike cylinder* $HP(v, 0) \cap S_1^n$ *for some* $v \in S_+^{n-1}$.

We consider the function $\mathcal{H} : S_1^n \times S_+^{n-1} \longrightarrow \mathbb{R}$ defined by $\mathcal{H}(x, v) = \langle x, v \rangle$. Given $v_0 \in S_+^{n-1}$, we denote $\mathfrak{h}_{v_0}(x) = \mathcal{H}(x, v_0)$, so that we have $\mathfrak{h}_{v_0}^{-1}(0) = HP(v_0, 0) \cap S_1^n$. For any $u_0 \in U$ and $v_0^{\pm} = \widetilde{\mathbb{L}}^{\pm}(u_0)$, the lightcone cylinder $HP(v_0^{\pm}, 0) \cap S_1^n$ is tangent to M at $p_0 = \mathbf{X}(u_0)$. In this case, we call $HP(v_0^{\pm}, 0) \cap S_1^n$ a *tangent lightlike cylinder* of M at p_0. From the previous arguments in §4, we have following theorem.

Theorem 8.3 ([13]). $\mathbf{X}_i : (U, u_i) \longrightarrow (S_1^n, p_i)$ $(i = 1, 2)$ *be spacelike submanifold germs and* $v_i = \widetilde{\mathbb{L}}_i^{\pm}(u_i)$. *If the corresponding Legendrian immersion germs are Legendrian stable. Then the following conditions are equivalent:*

(i)   Lightcone Gauss map germs $\widetilde{\mathbb{L}}_1^{\pm}$ and $\widetilde{\mathbb{L}}_2^{\pm}$ are $\mathcal{A}$-equivalent.

(ii) Legendrian immersion germs $\mathcal{L}_{H,1}^{\pm}$ and $\mathcal{L}_{H,2}^{\pm}$ are Legendrian equivalent.

(iii) Lightcone height function germs H_1 and H_2 are $\mathcal{P}$-$\mathcal{K}$-equivalent.

(iv) $h_{1,v_1}^{\pm}$ and $h_{2,v_2}^{\pm}$ are $\mathcal{K}$-equivalent.

(v) $K(\mathbf{X}_1(U), HP(v_1, 0) \cap S_1^n; p_1) = K(\mathbf{X}_2(U), HP(v_2, 0) \cap S_1^n; p_2)$

We call $(\mathbf{X}_i^{-1}(HP(v_i^{\pm}, 0) \cap S_1^n), u_i)$ a tangent lightlike cylindrical indicatrix germ of M_i at p_i.

9. Classification in de Sitter 4-space

In this section we consider the case of $n = 4$ and classify singularities of lightlike hypersurfaces and lightcone Gauss maps. We now define $\mathcal{K}$-invariants of spacelike surfaces in de Sitter 4-space. For open subset $U \subset \mathbb{R}^2$ and spacelike submanifold $X : U \longrightarrow S_1^4$, we define the $\mathcal{K}$-codimension (or Tyurina number) of the function germs $h_{v_0^{\pm}}$, $g_{\lambda_0^{\pm}}$ and corank of $h_{v_0^{\pm}}$, $g_{\lambda_0^{\pm}}$ by

$$\text{H-ord}^{\pm}(\mathbf{X}, u_0) = \dim C_{u_0}^{\infty}/\langle h_{v_0^{\pm}}(u_0), \partial h_{v_0^{\pm}}(u_0)/\partial u_i \rangle_{C_{u_0}^{\infty}},$$

$$\text{H-corank}^{\pm}(\mathbf{X}, u_0) = 2 - \text{rank Hess}(h_{v_0^{\pm}}(u_0)),$$

$$\text{G-ord}^{\pm}(\mathbf{X}, u_0) = \dim C_{u_0}^{\infty}/\langle g_{\lambda_0^{\pm}}(u_0), \partial g_{\lambda_0^{\pm}}(u_0)/\partial u_i \rangle_{C_{u_0}^{\infty}},$$

$$\text{G-corank}^{\pm}(\mathbf{X}, u_0) = 2 - \text{rank Hess}(g_{\lambda_0^{\pm}}(u_0)),$$

where $v_0^{\pm} = \widetilde{\mathbb{L}}^{\pm}(u_0)$ and $\lambda_0^{\pm} = \mathbf{X}(u_0) + t_0\widetilde{\mathbb{L}}^{\pm}(u_0)$.

Theorem 9.1 ([13]). *Let $Sp\text{-}Emb(U, S_1^n)$ be the set of spacelike submanifolds. We have open dense subset $\mathcal{O} \subset Sp\text{-}Emb(U, S_1^n)$ such that for $\mathbf{X} \in \mathcal{O}$ and $\lambda_0^{\pm} = LH_M^{\pm}(u_0, t_0)$, $\lambda_0^{\pm}$ is a singular value of $LH_M^{\pm}$ if and only if $G\text{-corank}^{\pm}(\mathbf{X}, u_0) = 1$ or 2.*

(i)   *If $G\text{-corank}^{\pm}(\mathbf{X}, u_0) = 1$ then there are distinct normalized lightcone principal curvatures $\widetilde{\kappa}_1^{\pm}$, $\widetilde{\kappa}_2^{\pm}$ such that $\widetilde{\kappa}_1^{\pm} \neq 0$ and $t_0 = -1/\widetilde{\kappa}_1^{\pm}$. In this case we have $G\text{-}ord^{\pm}(\mathbf{X}, u_0) = k$ $(k = 2, 3$ or $4)$ and the lightlike hypersurface $LH_M^{\pm}$ has $\mathcal{A}_k$ type singularity:*

$$(\mathcal{A}_2)f(u_1, u_2, u_3) = \left(3u_1^2, 2u_1^3, u_1, u_2\right)$$

$$(\mathcal{A}_3)f(u_1, u_2, u_3) = \left(4u_1^3 + 2u_1u_2, 3u_1^4 + u_2u_1^2, u_2, u_3\right)$$

$$(\mathcal{A}_4)f(u_1, u_2, u_3) = \left(5u_1^4 + 3u_2u_1^2 + 2u_1u_3, 4u_1^5 + 2u_2u_1^3 + u_3u_1^2, u_2, u_3\right).$$

(ii) *If $G\text{-corank}^{\pm}(\mathbf{X}, u_0) = 2$ then u_0 is a non-flat umbilic point and $t_0 = -1/\widetilde{\kappa}_1^{\pm}$. In this case we have $G\text{-}ord^{\pm}(\mathbf{X}, u_0) = 4$ and $LH_M^{\pm}$ has $\mathcal{D}_4^+$ or $\mathcal{D}_4^-$*

$$\mathcal{A}_2 \qquad \mathcal{A}_3 \qquad \mathcal{A}_4 \qquad \mathcal{D}_4^+ \qquad \mathcal{D}_4$$

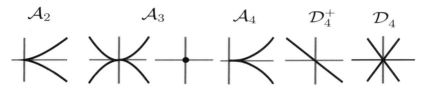

Figure 16.1. The list of tangent lightcone indicatrix germs

type singularity:

$$(\mathcal{D}_4^+)f(u_1, u_2, u_3) = \left(2(u_1^3 + u_2^3) + u_1 u_2 u_3, 3u_1^2 + u_2 u_3, 3u_2^2 + u_1 u_3, u_3\right)$$

$$(\mathcal{D}_4^-)f(u_1, u_2, u_3) = \left(2(u_1^3 - u_1 u_2^2) + (u_1^2 + u_2^2)u_3, u_2^2 - 3u_1^2\right.$$
$$\left. -2u_1 u_3, u_1 u_2 - u_2 u_3, u_3\right).$$

The corresponding tangent lightcone indicatrix germ is listed in Figure 16.1

Theorem 9.2 ([13]). *There exists an open dense subset $\mathcal{O}' \subset Sp\text{-}Emb(U, S_1^n)$ such that for any $\mathbf{X} \in \mathcal{O}'$, $u_0 \in U$ is an $\widetilde{L}^\pm$-parabolic point if and only if $H\text{-}corank^\pm(\mathbf{X}, u_0) = 1$. That is, M has non flat point and $\widetilde{K}_\ell^{-1}(0)$ is a regular curve.*

(i) *If $H\text{-}ord^\pm(\mathbf{X}, u_0) = 2$, then $\widetilde{\mathbb{L}}^\pm$ has the cuspidal edge point at u_0, and the tangent lightlike cylindrical indicatrix germ is an ordinary cusp.*

(ii) *If $H\text{-}ord^\pm(\mathbf{X}, u_0) = 3$, then $\widetilde{\mathbb{L}}^\pm$ has the swallowtail point at u_0, and the tangent lightlike cylindrical indicatrix germ is a tacnode or a point.*

References

1. V. I. Arnold, S. M. Gusein-Zade and A. N. Varchenko, Singularities of Differential Maps, Volume I, Birkhäuser, Basel, 1986.
2. T. Banchoff, T. Gaffney and C. McCrory, Cusps of Gauss mappings, Research Notes in Mathematics 55, Pitman, London, 1982.
3. D. Bleeker and L. Wilson, Stability of Gauss maps, *Illinois J. Math.* **22** (1978) 279–289.
4. J. W. Bruce, The dual of generic hypersurfaces, *Math. Scand.* **49** (1981) 36–60.
5. T. Fusho and S. Izumiya, Lightlike surfaces of spacelike curves in de Sitter 3-space, *J. Geom.* **88** (2008) 19–29.
6. S. Izumiya, D. Pei and T. Sano, Singularities of hyperbolic Gauss maps, *Proc. London Math Soc.* **86** (2003) 485–512.
7. S. Izumiya, M. Kossowski, D. Pei and M. C. Romero Fuster, Singularities of Lightlike hypersurfaces in Minkowski four-space, *Tohoku Math J.* **58** (2006) 71–88.

8. S. Izumiya and M. C. Romero Fuster, The lightlike flat geometry on spacelike submanifolds of codimension two in Minkowski space, *Sel. Math. NS.* **13** (2007) 23–55.
9. S. Izumiya, Timelike hypersurfaces in de Sitter space and Legendrian singularities, *J. Math. Sciences.* **144** (2007), 3789–3803.
10. S. Izumiya and F. Tari, Pairs of foliations on timelike surfaces in the de Sitter space S_1^3. Preprint, available from http://eprints3.math.sci.hokudai.ac.jp/view/type/preprint.html
11. S. Izumiya and F. Tari, Projections of timelike surfaces in the de Sitter space. In this Proceedings.
12. M. Kasedou, Singularities of lightcone Gauss images of spacelike hypersurfaces in de Sitter space, *J. Geom* **94**(1) (2009) 107–121.
13. M. Kasedou, Spacelike submanifolds of codimension two in de Sitter space, *J. of Geom. and Physics* **60**(1) (2010) 31–42.
14. E. E. Landis, Tangential singularities, *Funct. Anal. Appl.* **15** (1981) 103–114.
15. J. A. Montaldi, On contact between submanifolds, *Michigan Math. J.* **33** (1986) 195–199.
16. O. A. Platonova, Singularities in the problem of the quickest way round an obstruct, *Funct. Anal. Appl.* **15** (1981) 147–148.
17. M. C. Romero-Fuster, Sphere stratifications and the Gauss map, *Proc. Roy. Soc. Edinburgh Sect. A* **95** (1983) 115–136.
18. G. Wassermann, Stability of Caustics, *Math. Ann.* **216** (1975) 43–50.
19. V. M. Zakalyukin, Lagrangian and Legendrian singularities, *Funct. Anal. Appl.* **10** (1976) 26–36.
20. V. M. Zakalyukin, Reconstructions of fronts and caustics depending one parameter and versality of mappings, *J. Soviet. Math.* **27** (1984) 2713–2735.

Masaki Kasedou
Department of Mathematics
Hokkaido University
Sapporo 060-0810
Japan
kasedou@math.sci.hokudai.ac.jp

17

The geometry of Hopf and saddle-node bifurcations for waves of Hodgkin-Huxley type

ISABEL S. LABOURIAU AND PAULO R.F. PINTO

Abstract

We study a class of ordinary differential equations extending the Hodgkin-Huxley equations for the nerve impulse under a travelling wave condition. We obtain a geometrical description of the subset in parameter space were the equilibria lose stability. This may happen in two ways: first, the linearisation around the equilibrium may have a pair of purely imaginary eigenvalues with multiplicity j. This happens for parameters in the set N_j, the possible sites for Hopf bifurcation, where a branch of periodic solutions is created. Second, a real eigenvalue may change sign at the site for a possible saddle-node bifurcation. This happens at parameter values in the sets K_l where there is a zero eigenvalue of multiplicity l. We show that the sets N_j and K_l are singular ruled submanifolds of the parameter space $\mathbf{R}^{M+3}$ with rulings contained in codimension 1 affine subspaces. The subset of regular points in N_j has codimension $2j - 1$ in $\mathbf{R}^{M+3}$ and its rulings have codimension $2j + 1$. The subset of regular points in K_l has codimension l in $\mathbf{R}^{M+3}$ with rulings of codimension $l + 1$. We illustrate the result with a numerical plot of the surface N_2 for the Hodgkin-Huxley equations simplified to have $M = 2$.

2000 *Mathematics Subject Classification* 92C20 (primary), 37G10, 37G15, 58K45 (secondary).
Both authors had financial support from Fundação para a Ciência e a Tecnologia (FCT), Portugal, through the programs POCTI and POSI of Quadro Comunitário de Apoio III (2000–2006) with European Union funding (FEDER) and national funding.

1. Introduction – equations of Hodgkin-Huxley type

The Hodgkin-Huxley model [2] describes the variation of the difference $x \in \mathbf{R}$ of the electrical potential across a nerve cell membrane, as a function of the distance $s \in \mathbf{R}$ along an axon and of the time $t \in \mathbf{R}$, for an electrical stimulus of intensity $I \in \mathbf{R}$. Changes in the voltage x are due to the active transport of ions across the membrane through N *ionic channels* whose dynamics is controlled by M independent *gates* that open with probabilities y_i, $i = 1, \ldots, M$. Equations of Hodgkin-Huxley type generalise the original Hodgkin-Huxley equations that have $N = 2$ channels controlled by $M = 3$ gates. The general model, first studied with Ruas [4], is a reaction-diffusion equation:

$$
\text{(HH)} \quad
\begin{cases}
C_m \dfrac{\partial x}{\partial t} = a \dfrac{\partial^2 x}{\partial s^2} - I - c_0(x - V_0) - \displaystyle\sum_{j=1}^{N} c_j u_j(y)(x - V_j) \\[2ex]
\dfrac{\partial y_i}{\partial t} = (\gamma_i(x) - y_i)\, \tau_i(x) , \qquad i = 1, \ldots, M
\end{cases}
$$

where $y = (y_1, \ldots, y_M)$, the constant $C_m \geq 0$ is the membrane capacity and $a > 0$ is half the axon radius divided by the electrical resistance of the axoplasm. The voltage x is the only dependent variable that may be observed directly in experiments.

In many experimental settings the observed response to stimulation is a voltage pulse that propagates along the axon with constant speed. Experimental observations range from quiescent behaviour (equilibria) to different types of periodic response to stimulation. Our aim is to describe parts of parameter space where solutions corresponding to these different types of behaviour may be found, by locating the parameter values where these solutions bifurcate from equilibria.

In this article we impose a travelling wave condition on (HH) and study the boundary of stability of its equilibria under very general assumptions, suitable for applications to different types of excitable tissue. The geometrical structure is similar to that of the case $a = 0$ called clamped equations (section 1.2 below) discussed in [3, 4].

Hodgkin and Huxley originally imposed a travelling wave condition on (HH) before integrating it numerically and used a shooting method to find the propagation speed [2]. They also integrated the clamped equations in order to compare it to experimental data. Later, the clamped equations have been mostly used: they describe what was initially the most common experimental setting and they avoid having to handle the propagation speed as an additional parameter. The present work allows a comparison of the two approaches.

We describe the geometry of the set of parameters where there is a change in the stability of equilibria of (HH) under a travelling wave condition. This happens when there is a change of sign on the real part of an eigenvalue of the linearisation around the equilibrium. If the eigenvalue is real, then generically the number of equilibrium points changes at a fold point. For complex eigenvalues generically there is a Hopf bifurcation, where a periodic solution is created. Parameter values where the linearisation has multiple complex eigenvalues could lead to periodic solutions having two or more frequencies, that may correspond to some of the complicated periodic behaviour observed in experiments. Thus, the analysis of constant solutions yields some information on the dynamics that is relevant for applications, since it helps to characterise specific models and parameter ranges where periodic solutions of different types exist.

1.1. Overview

In the remainder of this section we state explicitly our assumptions on (HH), rewrite it under a travelling wave condition (HHW) and describe its equilibria. The parameter dependence of the linearisation around equilibria and of the coefficients of its characteristic polynomial are described in section 2 where we introduce new parameters to simplify the expressions. In section 3 we describe the exchange of stability set, for a general monic polynomial of fixed degree, in terms of its coefficients, extending the description of [3]. The main result appears in section 4: a description of the geometry of the exchange of stability set for (HHW). In section 5 we illustrate the result with an example of an equation with one channel and two gates, obtained as a simplification of the original Hodgkin-Huxley model.

We use the techniques for handling multiparameter nonlinear equations first developed in [4] and similar to [1].

1.2. General properties

(1) The expressions $\gamma_i(x)$ and $\tau_i(x)$ are fitted to experimental data, with $\tau_i(x) \geq 0$ and $\gamma_i(x) \in [0, 1]$. It follows that if $y_i(t_0, s) \in [0, 1]$ then $y_i(t, s) \in [0, 1]$ for $t > t_0$,

(2) In the original Hodgkin-Huxley model the terms $c_j u_j(y)(x - V_j)$ are called the *ionic channels*, where the functions $u_j(y)$ are monomials and $c_j > 0$, V_j are constant. In some cases the form $c_j(u_j(y) - f_j(x))(x - V_j)$ is used instead, where the $f_j(x)$ are fitted to experimental data. The term $c_0(x - V_0)$ is called the *leakage channel*.

1.3. Ordinary differential equations

There are two standard ways of obtaining an ordinary differential equation (ODE) from (HH). The first consists in taking $a = 0$ and reduces (HH) to ODEs in $\mathbf{R} \times [0, 1]^M$, called the *clamped* equations of Hodgkin-Huxley type. The bifurcation of its equilibria was studied in [4] and the boundary of stability in [3].

Another way of obtaining an ODE from (HH) is to consider a solution $x(t, s)$ that is a wave propagating with constant speed. In this case, we may write $x(t, s) = X(\hat{t})$, $y_i(t, s) = Y_i(\hat{t})$ with $\hat{t} = \theta t + \sigma s$ and the wave propagates forward if $\theta\sigma < 0$. For $\dot{\xi} = d\xi/d\hat{t}$ we have

$$\frac{\partial^2 x}{\partial s^2} = \sigma^2 \ddot{X} \qquad \frac{\partial x}{\partial t} = \theta \dot{X} \qquad \frac{\partial y_i}{\partial t} = \theta \dot{Y}_i$$

and the first equation in (HH) takes the form:

$$-C_m \theta \dot{X} = a\sigma^2 \ddot{X} - I - c_0(X - V_0) - \sum_{j=1}^{N} c_j u_j(Y)(X - V_j).$$

Rewriting in lower case we get:

$$\text{(HHW)} \quad \begin{cases} \dot{x} = -z \\ \dot{y}_i = (\gamma_i(x) - y_i) \, \tau_i(x)/\theta \,, \qquad i = 1, \ldots, M \\ a\sigma^2 \dot{z} = C_m \theta z - I - c_0(x - V_0) - \sum_{j=1}^{N} c_j u_j(y)(x - V_j) \end{cases}$$

which defines an ODE in $\mathbf{R}^{M+2}$ when $\sigma \neq 0$. For $\sigma = 0$ (HHW) is equivalent to the clamped equations, after a time rescaling. To simplify the notation, from now on we write $\tau_i(x)$ for $\tau_i(x)/\theta$, bearing in mind that θ may be negative.

1.4. Equilibria

We are interested in studying (HHW) for different values of the stimulus intensity I, treating I as a special bifurcation parameter. Equilibria of (HHW) satisfy $\dot{x} = -z = 0$ and $y_i = \gamma_i$. Thus $a\sigma^2 \dot{z} = 0$ if and only if

$$I = \eta(x, c, V) = -c_0(x - V_0) - \sum_{j=1}^{N} c_j u_j(\gamma(x))(x - V_j) \qquad (1.1)$$

where $\gamma(x) = (\gamma_1(x), \ldots, \gamma_M(x))$, $c = (c_0, \ldots, c_N)$, $V = (V_0, \ldots, V_N)$.

Equilibria may thus be parametrised by x and we may use the value of x at equilibrium as a new bifurcation parameter μ. The intensity I may be computed from the expression $\eta(\mu, c, V) = I$ depending on the $2N + 2$ parameters c and V.

1.5. Bifurcation

A geometrical description of the subset in parameter space where the number of multiple solutions of (1.1) changes locally is given in [4] in the context of the clamped equations. This description may be applied to equilibria of (HHW) without any changes. It follows that for an equation with N channels and for generic ion dynamics (i.e. for generic functions $\psi_j(\mu) = u_j(\gamma(\mu))$), equilibria of (HHW) have multiplicity at most $2N + 2$ and that there are always equilibria of multiplicity $2N + 1$. Moreover, if for a fixed value of parameters $\eta(\mu, c, V) - I$ has a zero of order $k \leq 2N + 2$ at $\mu = \mu_0$ then, for any $l \leq k$, there are parameter values arbitrarily close to the initial one where $\eta - I$ has a zero of order l at a point μ near μ_0. In particular, it follows that there are nearby parameter values where the equation (1.1) has k simple zeros near μ_0.

2. Linearisation

We are concerned with the way an equilibrium may lose stability when we vary the parameters in (HHW). This happens at parameter values where the linearisation around the equilibrium has an eigenvalue whose real part changes sign. We are interested in the geometry of these parameter sets. Whether the equilibrium is initially stable depends on the functions $\gamma_i(x)$, $\tau_i(x)$ and $u_j(y)$ used in a specific model, but the geometry of the bifurcation set may be described for generic functions.

For $\sigma \neq 0$, the linearisation of (HHW) around the equilibrium $(\mu, \gamma(\mu))$ is

$$
L = \begin{pmatrix}
0 & 0 & 0 & \cdots & -1 \\
\gamma_1' \tau_1 & -\tau_1 & 0 & \cdots & 0 \\
\vdots & \vdots & \ddots & \vdots & \vdots \\
\gamma_M' \tau_M & 0 & \cdots & -\tau_M & 0 \\
R_0 & R_1 & \cdots & R_M & -R_{M+1}
\end{pmatrix},
$$

where

$$
R_0 = \frac{1}{a\sigma^2}\left(-c_0 - \sum_{j=1}^{N} c_j u_j(\gamma(\mu))\right) \qquad R_{M+1} = \frac{C_m \theta}{a\sigma^2}
$$

and

$$
R_i = \frac{1}{a\sigma^2}\left(-\sum_{j=1}^{N} c_j \frac{\partial u_j}{\partial y_i}(\gamma(\mu))(\mu - V_j)\right) \qquad i = 1, \ldots, M.
$$

We consider new parameters $R = (R_0, R_1, \ldots, R_M, R_{M+1})$ given by the expressions above. The characteristic polynomial of L is given by

$$P_L(X) = \det(L - XI) = (-1)^M X (R_{M+1} + X) \prod_{i=1}^{M} (\tau_i(\mu) + X) + r(X)$$

where $r(X) = \det \begin{pmatrix} \gamma_1' \tau_1 & -\tau_1 - X & \ldots & 0 \\ \vdots & \vdots & \ddots & \vdots \\ \gamma_M' \tau_M & 0 & \ldots & -\tau_M - X \\ R_0 & R_1 & \ldots & R_M \end{pmatrix}$.

Writing

$$P_L(X) = (-1)^M \left(X^{M+2} + \sum_{i=0}^{M+1} \alpha_i X^i \right)$$

and defining the symmetric functions $s_0(\tau) = 1$ and

$$s_n(\tau) = \sum_{1 \leq i_1 < \cdots < i_n \leq M} \tau_{i_1} \cdots \tau_{i_n}, \qquad n = 1, \ldots, M,$$

the coefficients α_0 and α_1 are given by

$$\alpha_0 = R_0 s_M(\tau) + \sum_{i=1}^{M} R_i \gamma_i' \tau_i \frac{\partial s_M(\tau)}{\partial \tau_i},$$

$$\alpha_1 = R_0 s_{M-1}(\tau) + \sum_{i=1}^{M} R_i \gamma_i' \tau_i \frac{\partial s_{M-1}(\tau)}{\partial \tau_i} + R_{M+1} s_M(\tau),$$

for $1 < j < M$

$$\alpha_j = R_0 s_{M-j}(\tau) + \sum_{i=1}^{M} R_i \gamma_i' \tau_i \frac{\partial s_{M-j}(\tau)}{\partial \tau_i} + R_{M+1} s_{M-j+1}(\tau) + s_{M-j+2}(\tau),$$

and

$$\alpha_M = R_0 s_0(\tau) + R_{M+1} s_1(\tau) + s_2(\tau),$$

$$\alpha_{M+1} = R_{M+1} s_0(\tau) + s_1(\tau).$$

The expression $\alpha = (\alpha_0, \alpha_1, \ldots, \alpha_{M+1})$ can be written in the form $\alpha^T = D + E \cdot R^T$, where $D^T = (0, 0, s_M, s_{M-1}, \ldots, s_1)$, and

$$
E = \begin{pmatrix}
s_M(\tau) & \gamma'_1 \tau_1 \dfrac{\partial s_M(\tau)}{\partial \tau_1} & \cdots & \gamma'_M \tau_M \dfrac{\partial s_M(\tau)}{\partial \tau_M} & 0 \\[2ex]
s_{M-1}(\tau) & \gamma'_1 \tau_1 \dfrac{\partial s_{M-1}(\tau)}{\partial \tau_1} & \cdots & \gamma'_M \tau_M \dfrac{\partial s_{M-1}(\tau)}{\partial \tau_M} & s_M(\tau) \\[2ex]
\vdots & \vdots & & \vdots & \vdots \\[2ex]
s_1(\tau) & \gamma'_1 \tau_1 \dfrac{\partial s_1(\tau)}{\partial \tau_1} & \cdots & \gamma'_M \tau_M \dfrac{\partial s_1(\tau)}{\partial \tau_M} & s_2(\tau) \\[2ex]
s_0(\tau) & 0 & \cdots & 0 & s_1(\tau) \\[2ex]
0 & 0 & \cdots & 0 & s_0(\tau)
\end{pmatrix}.
$$

Lemma 2.1. *The determinant of E satisfies*

$$
\det(E) = \pm \left(\prod_{i=1}^{M} \gamma'_i \tau_i \right) \prod_{1 \le i < j \le M} (\tau_i - \tau_j).
$$

Proof. Developing the determinant along the last row of E and again along the last row of the minor we obtain

$$
\det(E) = \pm s_0(\tau)^2 \left(\prod_{j=1}^{M} \gamma'_j \tau_j \right) \det(E')
$$

where

$$
E' = \begin{pmatrix}
\dfrac{\partial s_M(\tau)}{\partial \tau_1} & \cdots & \dfrac{\partial s_M(\tau)}{\partial \tau_M} \\[2ex]
\dfrac{\partial s_{M-1}(\tau)}{\partial \tau_1} & \cdots & \dfrac{\partial s_{M-1}(\tau)}{\partial \tau_M} \\[2ex]
\vdots & & \vdots \\[2ex]
\dfrac{\partial s_1(\tau)}{\partial \tau_1} & \cdots & \dfrac{\partial s_1(\tau)}{\partial \tau_M}
\end{pmatrix}.
$$

Note that $\det E'$ is a homogeneous polynomial of degree $\frac{M(M-1)}{2}$ on the variables τ_j and that it is divisible by all the terms $(\tau_i - \tau_j), i \ne j$. Therefore, by Lemma 1 of [3], $\det(E') = \pm \prod_{1 \le i < j \le M} (\tau_i - \tau_j)$, completing the proof. $\square$

3. The exchange of stability set

The loss of stability of an equilibrium point happens where the linearisation around the equilibrium has an eigenvalue whose real part changes sign. For complex eigenvalues this may happen at parameter values where the linearisation has a pair of purely imaginary eigenvalues. These points are the possible sites for Hopf bifurcations, where a branch of periodic solutions is created. The other possibility is that a real eigenvalue changes sign at the site for a possible saddle-node bifurcation.

We address the general problem of describing the set of coefficients where a real polynomial $P(X)$ has purely imaginary roots in a form suitable for application to our problem. When $P(X)$ is the characteristic polynomial of a Jacobian matrix these roots indicate a change in stability. Consider the polynomial $P_\alpha(X)$ of degree $n \geq 2$ with real coefficients

$$P_\alpha(X) = \alpha_0 + \alpha_1 X + \cdots + \alpha_n X^n.$$

We want to study the *exchange of stability set* Σ_0 of parameters $\alpha \in \mathbf{R}^{n+1}$ such that $P_\alpha(i\sqrt{B}) = 0$ for some $B \geq 0$. There is a natural decomposition $\Sigma_0 = A_1 \cup S_1$ where the subset

$$A_1 = \left\{ \alpha \in \mathbf{R}^{n+1} : \exists B > 0 : P_\alpha(i\sqrt{B}) = 0 \right\},$$

corresponds to possible Hopf bifurcations and S_1 is the subspace $\alpha_0 = 0$, that corresponds to possible saddle-node bifurcations. Denoting the integer part of x by $[x]$, we define the *singular set A_j*, $1 \leq j \leq [n/2]$ as the set of parameters α where $P_\alpha(X)$ has roots $X = i\sqrt{B}$, $B > 0$ of multiplicity at least j. Using the symbol | for polynomial divisibility we have

$$A_j = \left\{ \alpha \in \mathbf{R}^{n+1} : \exists B > 0 : (X^2 + B)^j | P_\alpha(X) \right\}.$$

A *ruled submanifold of $\mathbf{R}^l$ with codimension j rulings* is a submanifold of $\mathbf{R}^l$ that is also a parametrised family of affine subspaces of codimension j. A *cone* $C \subset \mathbf{R}^l$ is a set invariant under scalar multiplication. The cone C is *regular* if $C - \{0\}$ is a smooth manifold, otherwise C is a *singular cone*. If every point of the cone lies in a vector subspace of $\mathbf{R}^l$ of codimension j, contained in the cone, then we say that the cone has *rulings of codimension j*.

The next result extends Proposition 2 in [3].

Proposition 3.1. *The exchange of stability set Σ_0 for a polynomial of degree $n \geq 2$ is the union of the hyperplane $\{\alpha_0 = 0\} = S_1 \subset \mathbf{R}^{n+1}$ and the singular cone A_1 of codimension 1 in $\mathbf{R}^{n+1}$ that has rulings of codimension 2. Each A_j, $1 \leq j \leq [n/2]$ is a singular cone of codimension $2j - 1$ with rulings of codimension $2j$ and, for $j < [n/2]$ singularities lying in A_{j+1}. The cone $A_{[n/2]}$ is regular. The closure $\overline{A_j}$ of A_j meets the hyperplane S_1 at the subspace $S_{2j} = \{\alpha_0 = \cdots = \alpha_{2j-1} = 0\}$ and is given by $\overline{A_j} = S_{2j} \cup A_j$.*

Proof. The projection $\pi : \mathbf{R}^+ \times \mathbf{R}^{n+1} \longrightarrow \mathbf{R}^{n+1}$, given by $\pi(B, \alpha) = \alpha$, maps each set

$$\Lambda_j = \left\{ (B, \alpha) \in \mathbf{R}^+ \times \mathbf{R}^{n+1} : (X^2 + B)^j \,|\, P_\alpha(X) \right\}$$

into $\pi(\Lambda_j) = A_j$. These sets may be rewritten in a more convenient way if we divide $\mathbf{R}^{n+1}$ into $n_E = \left[\frac{n}{2} \right]$ even coordinates and $n_O = \left[\frac{n-1}{2} \right]$ odd coordinates using the maps $q_E : \mathbf{R}^{n+1} \longrightarrow \mathbf{R}^{n_E+1}$ and $q_O : \mathbf{R}^{n+1} \longrightarrow \mathbf{R}^{n_O+1}$ given by

$$
\begin{aligned}
\alpha_E &= q_E(\alpha) &= (\alpha_0, -\alpha_2, \alpha_4, \ldots, (-1)^{n_E}\alpha_{2n_E}) \\
\alpha_O &= q_O(\alpha) &= (\alpha_1, -\alpha_3, \alpha_5, \ldots, (-1)^{n_O}\alpha_{2n_O+1}).
\end{aligned}
$$

Then $P_\alpha(i\sqrt{B}) = P_{\alpha_E}(B) + i\sqrt{B}\,P_{\alpha_O}(B)$ and $P_\alpha(i\sqrt{B}) = 0$ with $B > 0$ if and only if $P_{\alpha_E}(B) = P_{\alpha_O}(B) = 0$: in order to decide the divisibility of $P_\alpha(X)$ by powers of $X^2 - B$ it is sufficient to study the common positive real roots B of $P_{\alpha_E}(X)$ and $P_{\alpha_O}(X)$.

We claim that

$$\Lambda_1 = \left\{ (B, \alpha) \in \mathbf{R}^+ \times \mathbf{R}^{n+1} : r_1^{n_E}(B) \cdot \alpha_E = 0 = r_1^{n_O}(B) \cdot \alpha_O \right\}$$

and that

$$\Lambda_j = \left\{ (B, \alpha) \in \Lambda_{j-1} : r_j^{n_E}(B) \cdot \alpha_E = 0 = r_j^{n_O}(B) \cdot \alpha_O \right\},$$

for $1 < j \leq [n/2]$, where

$$r_1^m(B) = \left(1, B, B^2, \ldots, B^m \right) \qquad \text{and} \qquad r_{j+1}^m(B) = \frac{d r_j^m(B)}{dB}.$$

Using $P_{\alpha_E}(B) = r_1^{n_E}(B) \cdot \alpha_E$ and $P_{\alpha_O}(B) = r_1^{n_O}(B) \cdot \alpha_O$ we obtain the expression for Λ_1. For Λ_j, $j > 1$, note that $P_\gamma(B)$ is divisible by $(X - B)^j$ if and only if $P_\gamma(B) \in \Lambda_{j-1}$ and $\dfrac{d^j P_\gamma}{dB^j} = 0$, for $\gamma = \alpha_E, \alpha_O$.

Let $f_j : \mathbf{R}^+ \times \mathbf{R}^{n+1} \longrightarrow \mathbf{R}^{2j}$, $1 \leq j \leq n/2$ be the maps

$$f_1(B, \alpha) = \left(r_1^{n_E}(B) \cdot \alpha_E, r_1^{n_O}(B) \cdot \alpha_O \right)$$

$$f_j(B, \alpha) = \left(f_{j-1}(B, \alpha), r_j^{n_E}(B) \cdot \alpha_E, r_j^{n_O}(B) \cdot \alpha_O \right)$$

Since $\dfrac{\partial f_j}{\partial \alpha}$ has maximum rank $2j$, using the implicit function theorem it follows that the Λ_j are smooth codimension $2j$ submanifolds of $\mathbf{R}^{n+2}$.

From the expressions for Λ_j it follows that, for fixed B, the set $f_j^{-1}(0)$ is an affine subspace of $\mathbf{R}^+ \times \mathbf{R}^{n+1}$ of codimension $2j + 1$, having the form $\{B\} \times V^\perp$, where V is a $2j$-dimensional subspace of $\mathbf{R}^{n+1}$. Thus the sets Λ_j are smooth ruled manifolds, with rulings of codimension $2j + 1$ contained in hyperplanes B=constant.

The projection π is singular at the points in Λ_j where $\dfrac{\partial f_j}{\partial B} = 0$. Since

$$f_{j+1}(B, \alpha) = \left(f_j(B, \alpha), r_{j+1}^{n_E}(B) \cdot \alpha_E, r_{j+1}^{n_O}(B) \cdot \alpha_O \right)$$

$$= \left(f_1(B, \alpha), \frac{\partial f_j}{\partial B}(B, \alpha) \right)$$

it follows that singular points of $\pi|_{\Lambda_j}$ lie on Λ_{j+1} for $1 \leq j < \left[\frac{n}{2}\right]$ and that π is regular on $\Lambda_{[n/2]}$.

Each one of the rulings of Λ_j is projected into a vector subspace of codimension $2j$ orthogonal to the subspaces V discussed above. $\square$

For the study of the characteristic polynomial of a matrix we are interested in polynomials $P_\alpha(X)$ with $\alpha_n = (-1)^n$. Let $\hat\Sigma \subset \mathbf{R}^n$ and H_j, $1 \leq j \leq [n/2]$, be given by

$$\hat\Sigma = \{\alpha \in \mathbf{R}^n : \exists B \geq 0 : P_{(\alpha, -1^n)}(i\sqrt{B}) = 0\}$$

$$H_j = \{\alpha \in \mathbf{R}^n : \exists B > 0 : (X^2 + B)^j | P_{(\alpha, -1^n)}(X)\}.$$

Then, $\hat\Sigma$ and H_j are the transverse intersection of Σ_0 and A_j, respectively, with the hyperplane $\alpha_n = (-1)^n$.

Corollary 3.2. *The sets $H_j \subset \mathbf{R}^n$, $1 \leq j < [n/2]$ are singular ruled submanifolds of codimension $2j - 1$ with rulings of codimension $2j$. All the singularities of H_j lie in H_{j+1} and $\overline{H_j} = H_j \cup S_{2j}$. The set $H_{[n/2]}$ is regular: a curve for n even and a ruled surface for n odd. In particular $\hat\Sigma = S_1 \cup H_1$ is a ruled manifold of codimension 1, with rulings of codimension 2. Its singularities lie in $H_2 \cup S_2$.*

4. The exchange of stability set in (HHW)

The description of the set of parameters μ, R where the linearisation of (HHW) has either pure imaginary or zero eigenvalues can now be completed. For $j = 1, \ldots, [M/2] + 1$, let N_j be given by

$$N_j = \left\{ (\mu, R) \in \mathbf{R} \times \mathbf{R}^{M+2} : \exists B > 0 : (X^2 + B)^j | P_L(X) \right\}$$

and let

$$K_j = \left\{ (\mu, R) \in \mathbf{R} \times \mathbf{R}^{M+2} : X^j | P_L(X) \right\}$$

then

$$\tilde{\Sigma} = \left\{ (\mu, R) \in \mathbf{R} \times \mathbf{R}^{M+2} : \exists B \geq 0 : P_L(i\sqrt{B}) = 0 \right\} = N_1 \cup K_1.$$

Theorem 4.1. *For a residual set of functions $\tau_i(\mu) > 0$ and for all functions $\gamma_i(\mu)$ such that $\gamma_i'(\mu) \neq 0$, where $i = 1, \ldots, M$, the sets N_j, where $j = 1, \ldots, [M/2] + 1$, and K_l, where $l = 1, \ldots, M + 1$, are singular ruled submanifolds of $\mathbf{R} \times \mathbf{R}^{M+2}$ with rulings contained in the codimension 1 affine subspaces where μ is constant. Singular points of N_j occur both for $(\mu, R) \in N_{j+1}$ and for isolated values of $\mu = \mu_\star$, where $\tau_i(\mu_\star) = \tau_k(\mu_\star)$ for some $i \neq k$. Similarly, singular points of K_l occur both for $(\mu, R) \in K_{l+1}$ and for isolated values of $\mu = \mu_\star$. Moreover, $\overline{N_j} = N_j \cup K_{2j}$.*

The subset of regular points in N_j has codimension $2j - 1$ in $\mathbf{R} \times \mathbf{R}^{M+2}$ and its rulings have codimension $2j + 1$.

The subset of regular points in K_l has codimension l in $\mathbf{R} \times \mathbf{R}^{M+2}$ with rulings of codimension $l + 1$. The sets K_1 and K_2 are singular cones and K_{M+2} is the union of a set of isolated lines and a curve whose singular points lie on those lines.

Proof. From Lemma 2.1 it follows that the set $T = \left\{ \tau \in \mathbf{R}^M : \det(E) = 0 \right\}$ is the union of M hyperplanes in $\mathbf{R}^M$. By a direct application of Thom's transversality theorem, the set of functions $\tau_i(\mu)$ whose 0-jet is transverse to T is residual in $C^\infty(\mathbf{R}, \mathbf{R})$. Since T has codimension one, transversality to T means that $\det(E) = 0$ only at isolated values $\mu_\star$ of μ. Therefore, for generic functions $\tau_i(\mu) > 0$ and for all functions $\gamma_i(\mu)$, $i = 1, \ldots, M$, such that $\gamma_i'(\mu) \neq 0$, the map $R \mapsto E \cdot R^T$ is invertible, except at isolated values $\mu_\star$ of μ such that $\tau_i(\mu_\star) = \tau_k(\mu_\star)$, $i \neq k$. Moreover, the subset of points in T where more than two τ_i have the same value has codimension one in T, and therefore transversality to T means the set of functions may be refined to ensure that $\tau(\mu)$ does not meet this subset. Thus, E has corank at most 1 for all values of μ. From now on, we suppose the $\tau_i(\mu)$ are in this residual set.

The parameters (μ, R) are mapped into the coefficients of the characteristic polynomial of L by $F(\mu, R) = E(\mu) \cdot R^T + D(\mu)$ with $N_j = F^{-1}(H_j)$. For most values of μ the affine map $F_\mu : \mathbf{R}^{M+2} \longrightarrow \mathbf{R}^{M+2}$, $F_\mu(R) = F(\mu, R)$ is invertible, the rulings of N_j are the preimage by F_μ of the rulings of H_j and the result follows from Corollary 3.2.

In order to complete the description of the N_j it remains to see that at the isolated values $\mu_\star$ of μ where $\det E(\mu_\star) = 0$ the preimage of the rulings still has the correct dimension. At these values of μ the image of $F_{\mu_\star}$ is an affine hyperplane in $\mathbf{R}^{M+2}$. We claim that the residual set of functions $\tau(\mu)$ may be refined in such a way that the image of $F_{\mu_\star}$ meets all the rulings of H_j in general

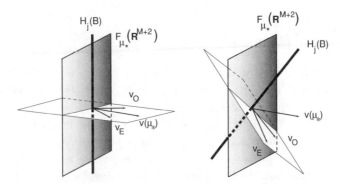

Figure 17.1. The situation on the left, where the rulings of A_j are contained in the image of F_{μ_*}, may be avoided by a small perturbation of the $\tau_i(\mu)$ (right) that tilts it away from the rulings.

position, and their inverse image by F are rulings of N_j of codimension $2j$, as for other values of μ.

To see this we look again at the proof of Proposition 3.1 to obtain $2j$ vectors $v_l^E(B)$ and $v_l^O(B) \in \mathbf{R}^{M+3}$ that generate the subspace orthogonal to the rulings of $A_j = \pi(\Lambda_j)$ given by $v_1^E(B) = (1, 0, -B, 0, B^2, 0, \ldots)$, $v_1^O(B) = (0, 1, 0, -B, 0, B^2, 0, \ldots)$ with $v_{l+1}^E(B)$ and $v_{l+1}^O(B)$ their l^{th} derivatives. The cone swept by these subspaces as B varies has codimension at least one in $\mathbf{R}^{M+3}$, for $j \leq [M/2] + 1$.

The claim will be proved if we show that the image of F_{μ_*} meets all the rulings of H_j in general position. Let $v(\mu_*) \neq 0$ be a vector in the kernel of $(E(\mu_*))^T$, so that the image of F_{μ_*} is orthogonal to $v(\mu_*)$ (Figure 17.1).

Since $H_j = A_j \cap \{\alpha_{M+2} = (-1)^{M+2}\}$, it is sufficient to show that $v(\mu_*)^\perp$ is transverse to the rulings of A_j inside $\{\alpha_{M+2} = (-1)^{M+2}\}$, or equivalently, that $(v(\mu_*), 0)$ does not meet the cone swept by $v_l^E(B)$ and $v_l^O(B)$ as B varies. The residual set of functions $\tau(\mu)$ may be refined to satisfy this condition. Therefore we may assume that the rulings of A_j are transverse to the image of F_{μ_*} and that the codimension of the intersection of the image of F_{μ_*} with the rulings of A_j is one plus the codimension of the rulings of A_j.

Intersecting the rulings of H_j with $v(\mu_*)^\perp$ increases their codimension by one, by transversality to $\{\alpha_{M+2} = (-1)^{M+2}\}$. Therefore the codimension of the preimage of $H_j \cap v(\mu_*)^\perp$ by F_{μ_*} is still $2j$ as we had claimed.

Finally, we note that $K_l = F^{-1}(S_l)$. Refining the residual set again, the proofs for K_l are entirely analogous (and simpler) to those for H_j. The only point that remains is to check that both K_1 and K_2 are cones, which follows from the fact that the first two coordinates of $D(\mu)$ are zero, so each $F_\mu^{-1}(S_l)$ is a vector subspace for $l = 1, 2$.

The remaining assertions follow directly from Corollary 3.2.   □

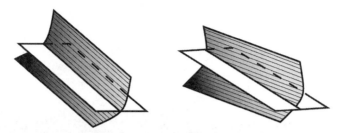

Figure 17.2. Left: rulings of H_j are not in general position with respect to the image of F_{μ_*} (white plane); on the right they are, after a small perturbation of the $\tau_i(\mu)$.

Note that if $\det E(\mu) \neq 0$ then for $j = [M/2] + 1$ and for M even N_j is a regular surface. For M odd $N_{[M/2]+1}$ is a regular ruled 3-manifold, with one-dimensional rulings.

5. Example: single channel Hodgkin-Huxley equations

For the original Hodgkin-Huxley equations there are two channels, corresponding to Na and K ions. The first equation in (HH) is:

$$C_m \frac{\partial x}{\partial t} = a \frac{\partial^2 x}{\partial s^2} - c_{Na} m^3 h(x - V_{Na}) - c_K n^4 (x - V_K) - c_0 (x - V_0) - I.$$

As a simple example, we discuss the case when only the Na channel is present, so $N = 1$, $M = 2$ and we assume $c_K = 0$. In our notation, $y_1 = m$, $y_2 = h$ and $u_1(y_1, y_2) = y_1^3 y_2 = u_{Na}(m, h) = m^3 h$. In the Hodgkin and Huxley 1952 article [2] the functions γ_j and τ_j are fitted to experimental data and are given in the form

$$\gamma_j(x) = \frac{\alpha_j(x)}{\alpha_j(x) + \beta_j(x)} \qquad \tau_j(x) = \alpha_j(x) + \beta_j(x)$$

for $j = m, h$, where

$$\alpha_m(x) = \varphi((v + 25)/10) \quad \beta_m(x) = 4e^{v/18}$$
$$\alpha_h(x) = 0.07 e^{v/20} \qquad \beta_h(x) = \left(1 + e^{(v+30)/10}\right)^{-1}$$

$$\text{with} \qquad \varphi(x) = \begin{cases} x/(e^x - 1) & \text{if } x \neq 0 \\ 1 & \text{if } x = 0 \end{cases}$$

Notice that $\alpha_j(x) + \beta_j(x) \neq 0$ for all x and j.

The functions $\gamma_j(x)$ of Hodgkin and Huxley [2] were chosen to be almost constant outside an interval. They are monotonic, asymptotically 1 or 0 at $\pm\infty$.

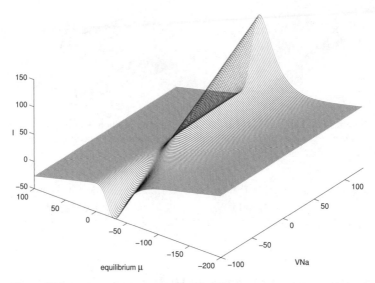

Figure 17.3. Graph of $I = \eta(x, V_{Na})$ for single-channel HHW, other parameters
with original Hodgkin-Huxley values: $c_{Na} = 120$, $c_0 = 0.3$ and $V_0 = 10.6$.

Their derivatives tend to zero very fast as x tends to $\pm\infty$ and the same is true
a fortiori of derivatives of $u_1(\gamma(x))$. We restrict our study to values of x where
results may be significant both from a physiological and from a numerical point
of view.

Figure 17.3 shows equilibria for the original Hodgkin Huxley with parameter
values $c_{Na} = 120$, $c_0 = 0.3$ and $V_0 = 10.6$. Recall that equilibria of (HHW)
are given by $I = \eta(x, V_N)$ where η was defined in (1.1). Note the two fold
points for each V_{Na}.

For $M = 2$ the parameters R lie in $\mathbf{R}^{M+2} = \mathbf{R}^4$. We have

$$H_2 = \{R = (B^2, 0, 2B, 0) \; : \; B > 0\}$$

and K_4 is the origin of $\mathbf{R}^4$, corresponding to $B = 0$ in H_2. The genericity conditions in Theorem 4.1 hold here, since by Lemma 2.1 $\det E(\mu) =
\gamma_1'\gamma_2'\tau_1\tau_2(\tau_2 - \tau_1) \neq 0$, see Figure 17.4. Numerical plots of the surface $N_2 \subset \mathbf{R}^5$
and its boundary K_4 are shown in Figure 17.5 as a B-parametrised family of
curves projected into the three-dimensional subspace (R_0, R_1, R_3) of $\mathbf{R}^4$. Each
curve is parametrised by μ as $F_\mu^{-1}(B^2, 0, 2B, 0)$ for a fixed value of B. with
$B > 0$ for N_2 and $B = 0$ for K_4. A numerical plot of the projection of K_4 into the
three-dimensional subspace (μ, R_1, R_2) is shown in Figure 17.6, where it may
be seen that for almost all values of μ the variables R_1 and R_2 have a constant

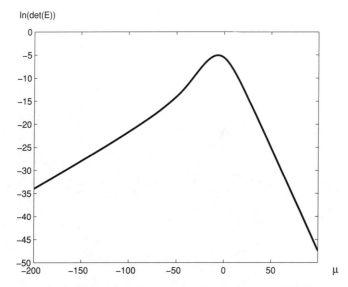

Figure 17.4. Graph of $\ln(\det E(\mu))$ for single-channel HHW.

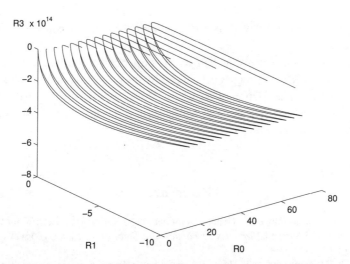

Figure 17.5. Projection of N_2 and K_4 into the space (R_0, R_1, R_3) for single-channel HHW. Leftmost curve is K_4 corresponding to $B = 0$. Each one of the other curves corresponds to a fixed value of $B > 0$. All curves parametrised by μ, the equilibrium value of x.

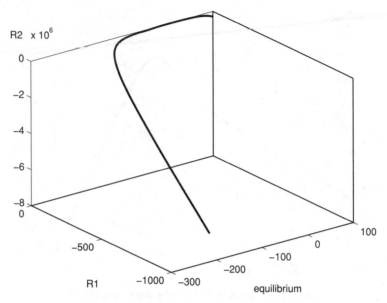

Figure 17.6. Projection of the curve $K_4 = \{(\mu, R) : F_\mu(R) = 0\}$ into the space (μ, R_1, R_2) for single-channel HHW.

relation and moreover that R_2 is very small. For the clamped equation both N_2 and K_4 are empty.

Since N_1 has codimension 1 in $\mathbf{R}^5$, its projection into R-space covers an open region in $\mathbf{R}^4$. This region is bounded by the projection of $N_2 \cup K_4 \cup \overline{K_2}$, plus points in the projection of N_1 where $\partial F(\mu, R)/\partial \mu$ is orthogonal to both v_1^E and v_1^O. For each R within this region there is a value of the equilibrium μ with a Hopf bifurcation.

References

1. D. R. J. Chillingworth, Generic multiparameter bifurcation from a manifold, *Dynamical Systems (formerly Dynamics and Stability of Systems)* **15(2)** (2000) 101–137.

2. A. L. Hodgkin and A. F. Huxley, A quantitative description of membrane current and its application to conduction and excitation in nerve, *Journal of Physiology* **117** (1952) 500–544.

3. I. S. Labouriau and C. M. S. G. Rito, Stability of equilibria in equations of Hodgkin-Huxley type, *Real and Complex Singularities* (eds T. Gaffney and M.A.S. Ruas), Contemporary Mathematics, **354** (American Mathematical Society, Providence, RI, USA, 2004) 137–143.

4. I. S. Labouriau and M. A. S. Ruas, Singularities of equations of Hodgkin-Huxley type, *Dynamical Systems (formerly Dynamics and Stability of Systems)* **11(2)** (1996) 91–108.

I. S. Labouriau and P. R. F. Pinto
Centro de Matemática
Universidade do Porto
Rua do Campo Alegre, 687
4169-007 Porto
Portugal
islabour@fc.up.pt
paulo.pinto@fc.up.pt

18

Global classifications and graphs

J. MARTÍNEZ-ALFARO, C. MENDES DE JESUS
AND M. C. ROMERO-FUSTER

Abstract

We review some of the problems where the graphs have been applied to the study of the global classification of stable maps.

AMS Classification: 57R45, 57M15, 57R65
Key words: Stable maps; foliations; graphs; global classification

1. Introduction

Let M and N be manifolds, M compact and $f : M \to N$. Assume that f defines an extra structure on any neighborhood of a point $p \in M$. The point p is said to be regular with respect to the extra structure if there exists a neighborhood U_p of p such that for any $q \in U$ there exists also a neighborhood U_q so that the extra structure is equivalent on the two neighborhoods. The study of the structure on U_p leads to a local problem. The study of the decomposition of M in maximal subsets with an homogeneous structure leads to a global problem. One way to achieve the global problem is to associate a graph to this decomposition.

This method has been applied to the study of the global classification of flows and maps. The approach in each case goes as follows: Once the local and multi-local behaviour of the critical set has been described, the relevant global topological information is codified in a graph, possibly with labels in either the vertices, the edges, or both. The typical questions then are:

1. Determine all the (labelled) abstract graphs that can be associated to some of the objects under study (Realization Problem).

2000 *Mathematics Subject Classification* 00000.

2. Is the graph a complete invariant for the given problem? In the negative case: Which are all the topological classes associated to a given graph? Determine further invariants in order to distinguish among them.

From a historical point of view, the first structures to be classified by graphs were the stable ones. M. Peixoto [40] was the first to relate graphs and flows. As a consequence he obtained a global classification for Morse-Smale flows in the plane. Other well known applications of graphs are the Lyapunov graphs in the study of non singular Morse-Smale flows on 3-manifolds, introduced by J. Franks [17] and the graphs introduced by A. T. Fomenko [14] in the analysis of Integrable Hamiltonian flows. In the case of stable maps, the graphs were first introduced by A. S. Kronrod in order to describe the global behaviour of smooth functions defined on the 2-sphere [27]. More recently, V. I. Arnold has pushed forward the study of the graphs associated to Morse functions on closed surfaces [4].

We review here some of the problems where this procedure has been used paying a special attention to the description of the problem for stable maps between surfaces that have been recently studied by the second and third authors. Following Arnold's ideas on counting the topological classes of Morse functions with a fixed number of saddle points described in §3.1, we push forward in §3.2 the first steps in the study of the determination of the number of topological classes of stable maps from surfaces to the plane with prefixed Vassiliev type invariants. We also describe, as an illustrative example of a different viewpoint, how graphs can be used in the global study of foliations in the plane.

2. Surfaces and graphs

2.1. Graphs of Morse functions

Given a Morse function $f : M \to \mathbb{R}$, we associate to it the space whose points are connected components of the level spaces of f. If M is a closed (compact without boundary) manifold, this space is a finite 1-dimensional complex known as the graph of f. We denote it by $\mathcal{G}_f$.

The vertices of $\mathcal{G}_f$ are the critical points of f. These vertices may either be extremal points (maxima and minima), or triple points (saddle points). We consider the following order relation is imposed on the set of vertices of the graph:

$$v_1 < v_2 \Leftrightarrow f(v_1) < f(v_2).$$

It is not difficult to see that a necessary condition for a graph to be attached to a Morse function is that the neighbours of a vertex of valence 3 are neither all below nor all above it. Graphs satisfying this condition are called *regularly ordered graphs*. Clearly, these graphs are topological invariants of Morse functions. For Morse functions on a surface of genus g, the graph has g independent cycles. In the case of the 2-sphere, the graphs are trees and the following results hold (see [4]):

1. The regularly ordered trees are topological invariants for Morse functions on S^2.
2. All regular orderings (on any tree) with n vertices can be realized as the graphs of Morse functions with exactly n critical points.

An interesting problem discussed by V. I. Arnol'd in [4] (see also [5]) is the following: *Count the number $\varphi(T)$ of regularly ordered trees with T triple points (= number of diffeomorphism classes of Morse functions on S^2 having $2T + 2$ critical points, including T saddle points).*

The first values for $\varphi(T)$ were calculated by Arnol'd in [4]:

$$\varphi(1) = 2; \quad \varphi(2) = 19; \quad \varphi(3) = 428; \quad \varphi(4) = 17746.$$

Moreover, he proved:

Theorem 2.1. *There exist positive constants a and b such that for any T we have the inequalities*

$$aT^T < \varphi(T) < bT^{2T}.$$

L. I. Nicolaescu has studied with detail the asymptotic behaviour of the function φ in [37], proving Arnold's conjecture that the upper bound of $\varphi(T)$ is close to the asymptotic value.

An important particular case is that of Morse polynomials in two variables (on S^2). A Morse polynomial of degree m has at most $(m - 1)^2$ critical points. For $m = 4$ we get that the number T of saddle points is at most 4. According to Arnol'd ([5]), among the 17746 topologically different classes of Morse functions on S^2 with at most 4 saddles, there are (at least) 1000 with a polynomial representative of degree 4. It is then natural to pose the following question: Which is the growth rate with respect to the growth of T in this case? Some answers to this question can be found in [5]. One can also consider the corresponding problem on the torus. In this case, the graph has one cycle. Several interesting results concerning the growth rate of $\varphi(T)$ with respect to the growth of T for periodic Morse functions and trigonometric polynomials are also discussed by Arnol'd in [5] and [6].

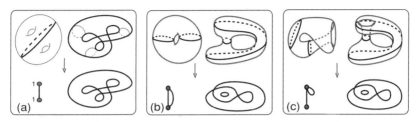

Figure 18.1. Graph and stable map to the plane from the: a) bi-torus, b) torus and c) Klein bottle.

Remark 1. Given a vector field, for instant a Hamiltonian field H, whose first integral (i.e. Hamiltonian function) is a Morse function, the above graph provides an invariant for H. In many cases this situation can be generalized, explaining why similar tools may work as well in flows than in maps. See, for instance, [12].

2.2. Graphs of stable maps between surfaces

Let M and N be orientable surfaces, where M is closed (compact without boundary). By Whitney's theorem ([18]), the singular set Σf of any stable smooth map $f : M \to N$ consists of curves of double points, possibly containing isolated cusp points. The branch set or apparent contour of f (i.e. the image of the singular set of f) consists of a number of immersed curves in the surface N (possibly with cusps) whose self-intersections are all transverse and disjoint from the cusps (if any). The complement of Σf is a disjoint union of open orientable surfaces, separated by closed curves lying in Σf. The topological type of $M - \Sigma f$ is a topological invariant ($\mathcal{A}$-invariant). We can encode the information provided by this invariant in a weighted graph $\mathcal{G}_f$. This is defined as follows:

- Each path-component of $M - \Sigma f$ determines a vertex and each curve of Σf an edge.
- A vertex v and an edge a are incident if and only if the curve represented by a lies in the boundary of the region represented by v.
- We attach to each vertex v a weight given by the genus $g(v)$ of the region of $M - \Sigma f$ represented by v.

It is not difficult to see that this graph is an $\mathcal{A}$-invariant. Figure 18.1 illustrates an example of stable map from the bi-torus to the plane with a unique singular curve showing both its graph and its apparent contour.

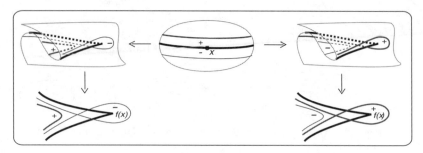

Figure 18.2. Example of negative and positive cusps.

The singular set of a stable map from an orientable surface M to S^2 separates M into two collections of submanifolds M^+ and M^- made of regions inducing opposite orientations on their images by f. So we can conclude that the graph of any stable map from an orientable surface to S^2 is bipartite. Moreover, we observe that in the orientable case the cusp points may be *positive and negative* as illustrated in Figure 18.2

J. R. Quine proved in [42] that given orientable surfaces M and N and a stable map $f : M \to N$, we have

$$\chi(M) - 2\chi(M^-) + (C^+ - C^-) = deg(f)\chi(N),$$

where $deg(f)$ is the degree of f, and C^+ and C^- are respectively the numbers of positive and negative cusp points of f. For a stable map $f : M \to S^2$ from an orientable surface to S^2, Quine's formula can be re-written as

$$\chi(M^+) - \chi(M^-) + (C^+ - C^-) = 2deg(f).$$

Moreover, we have that $\chi(M^\pm) = 2(V^\pm - g^\pm - m)$. Therefore we get the following expression for the degree of f:

$$deg(f) = (V^+ - V^-) - (g^+ - g^-) - (C^+ - C^-),$$

where V^+ and V^- respectively represent the numbers of (vertices corresponding to) positive and negative regions, and g^+ and g^- the respective genus of M^+ and M^-.

Some of the basic questions arising in the study of the graphs are the following:

(1) *Determine necessary and sufficient conditions for a graph to be associated to stable maps from closed orientable surfaces to S^2 (or to $\mathbb{R}^2$).*

(2) *Which of these graphs can be attached to fold maps (i.e. stable maps without cusps)?*

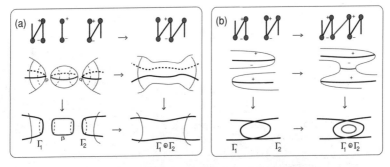

Figure 18.3. Surgeries: (a) horizontal, (b) vertical.

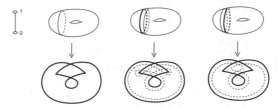

Figure 18.4. Stable map with two cusps on the torus.

(3) *Which of these graphs can be attached to stable maps (or fold maps) with a given degree d?*

We have the following answer to the first question:

Realization of weighted graphs: *Any weighted bipartite graph $\mathcal{G}$ may be realized by a stable map of any degree. The domain M of this map must satisfy:*

$$\chi(M) = 2(\chi(\mathcal{G}) - p),$$

where $\chi(M)$ denotes the Euler characteristic of M and p is the sum of all the weights of $\mathcal{G}$.

The basic tools in the proof of this result are:

- The use of codimension one transitions that affect the graphs of the stable maps, such as lips and beaks.
- Performing horizontal and vertical surgeries, whose effect on the graphs is shown in Figure 18.3 (see [22]).
- Realization of a basic graph with a unique edge an weights zero and one on the two vertices by a stable map with two cusps from the torus to the plane as illustrated by Figure 18.4.

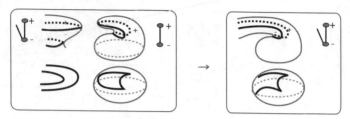

Figure 18.5. Altering the degree and preserving the graph.

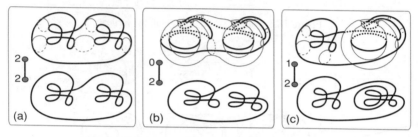

Figure 18.6. Examples of fold maps and their graphs.

The strategy of the proof consists in

- Step 1: Prove that any tree with zero weights can be realized by a stable map from S^2 to the plane (see [21]).
- Step 2: Given any graph with arbitrary weights in its vertices we can choose a maximal subtree. By putting zero weights in all its vertices we can realize it by means of a stable map from S^2 to the plane. Then add edges and weights by means of vertical and horizontal surgeries respectively, in order to recover the original graph (see [22]).
- Step 3: In order to obtain a stable map of any degree with target S^2 we can perform horizontal surgeries with a basic stable graph of degree one from S^2 to S^2 (See figure 18.5). This process alters the degree preserving the graph.

An important class of stable maps are the *fold maps* [1, 2, 43, 45]. Their apparent contours are regular closed curves. Such curves may be seen as the images of the boundary of immersed regions in $\mathbb{R}^2$ (or S^2). We illustrate in Figure 18.6 some fold maps from an orientable surface of genus 4 to the 2-sphere. Example a) shows a map of degree 0, whereas examples b) and c) respectively correspond to maps of degree 2 and 1.

The following answer to the second and third questions above is based in a constructive process through which a fold map can be obtained by conveniently assembling surfaces with prescribed immersions on their boundaries.

The corresponding result for fold maps from S^2 to the plane has been proven in [21] and for planar fold maps (i. e. with graphs of zero weights) in [22]. The proof of the general case will be given in a forthcoming paper by D. Hacon, C. Mendes de Jesus and M. C. Romero Fuster [23].

Realization of weighted graphs by fold maps: *Any bipartite weighted graph may be realized by a fold map from a surface into the sphere. The degree of this map is given by*

$$(V^+ - V^-) - (g^+ - g^-).$$

If we define a graph to be *balanced* provided $(V^+ - V^-) - (g^+ - g^-) = 0$, then we can state:

Corollary 2.2. *A bipartite weighted graph may be realized by a fold map from a surface to the plane if and only if it is balanced.*

We remark that a general result due to Y. Eliashberg (Theorem B, [13]) implies that given any closed non necessarily connected curve C separating a closed orientable surface M into two surfaces M^+ and M^- with common boundary C, there exists a fold map whose singular set is C if and only if $\chi(M^+) = \chi(M^-)$. Now, there is a $1 - 1$ relation between topological classes of curves in a surface M and weighted graphs satisfying the relation $\chi(M) = 2(\chi(\mathcal{G}) - p)$ (see [20]), moreover, it can be seen that the condition $\chi(M^+) = \chi(M^-)$ is equivalent to asking that the corresponding graph is bipartite and balanced. It thus follows that the above Corollary could also be obtained as a consequence of Eliashberg's result. Nevertheless, we emphasize that whereas Eliashberg's techniques guarantee the existence of such a map, those presented here furnish a method to construct it.

Once studied the realization problem of the graphs in this context, we can analyze their role in the global classification of stable maps from surfaces to S^2 or $I\!R^2$. Given an orientable surface of genus g, suppose that $f : M \rightarrow S^2$ is a stable map of degree d. According to the above results we have that the graph $\mathcal{G}_f$ of f has the following topological constraint:

$$g = 1 + E - V + P,$$

where E and V are respectively the numbers of edges (i.e. the number of singular curves of f) and vertices of $\mathcal{G}_f$ and P its total weight. In the case that f is a fold map, we must also have:

$$d = (V^+ - V^-) - (g^+ - g^-).$$

Following Arnold's line on the statistics of functions on surfaces, some natural questions that one can propose in the case of stable maps between

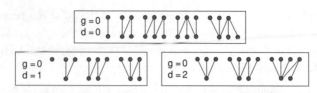

Figure 18.7. Graphs of fold maps from S^2 to S^2 with $E \leq 5$ and $d = 0, 1, 2$.

surfaces are the following: *Consider the numerical global invariants of stable maps on closed orientable surfaces provided by: E = number of singular curves; C = number of cusp points and D = number of double points in the apparent contour. How many topological classes of stable maps of degree d from an orientable surface M of genus g to S^2, with fixed values of E, C and D can be found? How many classes of fold maps ($C = 0$)?*

The strategy to study this problem consists in

1. Determine the number of bipartite weighted graphs with E edges satisfying the above constraints (for g and d).
2. For each one of the bipartite weighted graphs satisfying the g-constraint, determine all the possible apparent contours with prefixed values C and D that can be associated to it. Where, for $C = 0$ we must also impose the d-constraint.
3. Investigate all the possible topological classes of stable maps that may be associated to each couple (Graph, Apparent contour).

Denote by $\varphi_g(E)$ the number of bipartite weighted graphs with E edges satisfying the g-constraint, and by and $\phi_{g,d}(E)$ the number of bipartite weighted graphs satisfying the g- and d-constraints (i.e. corresponding to fold maps). Then we have

$$\varphi_0(1) = 1, \varphi_0(2) = 1, \varphi_0(3) = 2, \varphi_0(4) = 3, \varphi_0(5) = 6, \varphi_0(6) = 10.$$

Figure 18.7 displays all the graphs of fold maps with $E \leq 5$ and $d = 0, 1, 2$ from the 2-sphere to the 2-sphere. From this we get,

$$\varphi_{0,0}(1) = 1, \varphi_{0,0}(2) = 0, \varphi_{0,0}(3) = 1, \varphi_{0,0}(4) = 0, \varphi_{0,0}(5) = 2,$$

$$\varphi_{0,1}(0) = 0, \varphi_{0,1}(1) = 0, \varphi_{0,1}(2) = 1, \varphi_{0,1}(3) = 0, \varphi_{0,1}(4) = 2,$$

$$\varphi_{0,2}(1) = 0, \varphi_{0,2}(2) = 0, \varphi_{0,2}(3) = 1, \varphi_{0,2}(4) = 0, \varphi_{0,2}(5) = 2.$$

On the other hand, Figures 18.8 and 18.9 show all the possible graphs of fold maps respectively from the torus and the bi-torus to the 2-sphere. From these, we can get the values of $\varphi_{g,d}(E)$, $g = 1, 2$, for $E \leq 5$ and $d = 0, 1, 2$.

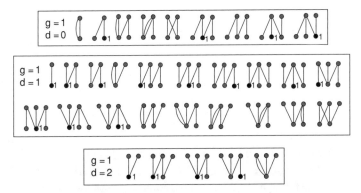

Figure 18.8. Graphs of fold maps from the torus to S^2 with $E \leq 5$ and $d = 0, 1, 2$.

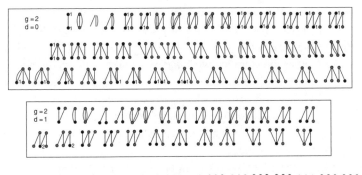

Figure 18.9. Graphs of fold maps from the bi-torus to S^2 with $E \leq 5$ and $d = 0, 1, 2$.

Now, in order to investigate all the possible isotopy classes of apparent contours $\mathcal{C}$, with fixed values C and D, that can be associated to a given graph with E edges we must first distinguish among the possible collections of E curves that may be the apparent contour of some stable map. The following result proven in ([22]) is a first step in this direction.

Proposition 2.3. *Any branch curve of a fold map of the sphere to the plane has odd winding number (or, equivalently, an even number of double points).*

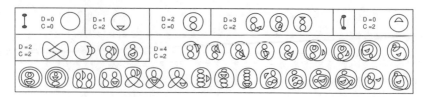

Figure 18.10. Apparent contours of stable maps from the sphere to the plane.

We must now determine all the isotopy types of closed curves that can appear in a branch set. This problem was initially put by Haefliger [24] who reduced it to the following: *Given an immersion f of the boundary of a compact surface M to the plane, under which conditions does f extend to an immersion over M?*

S. Blank solved the problem for the immersions of the disc by introducing an algebraic algorithm which is nowadays known as Blank's word ([9]). This was extended by S. Troyer [46] to the disc with holes and by K. Bailey [7] to the disc with handles. Subsequently, S. Troyer and G. Francis [16] used Blank's technique in order to determine when a closed curve with a finite number of cusps and transverse self-intersections may be the singular set of some stable maps from a closed surface to the plane, providing a classification of the stable maps that can be associated to a given curve. A further generalization of these works for non necessarily compact surfaces, with clearer and simpler methods, was obtained by V. C. Zanetic ([52]).

We observe that Blank's words can be geometrically interpreted in terms of decompositions of oriented (non necessarily connected) curves as sums of embedded circles with the same orientations in the sense of [33]. By applying this to the possible combinations of the isotopy classes of closed plane curves determined by Arnold in [3], we can determine all the possible isotopy classes of apparent contours of fold maps from the sphere to the plane and by applying convenient lips transitions we can introduce couples of cusps in these maps. We include in Figure 18.10 those corresponding to stable maps with $C = 0$, $E \leq 2$ and $D \leq 4$ from S^2 to $\mathbb{R}^2$.

Finally, a further problem resides in the fact that the pair (graph, apparent contour) is not enough to determine the class of a stable map, as shown by Milnor's example (illustrated in Figure 18.11). This example shows two different immersions of the 2-disc in the plane that coincide in a neighborhood of the boundary. If we define a mapping $f : S^2 \to \mathbb{R}^2$ by putting it equal to one of these immersions on the lower hemisphere and to the other on the upper hemisphere, we obtain a fold map from S^2 to $\mathbb{R}^2$. On the other hand, by choosing the same immersion on both hemispheres we get a new fold map from S^2 to

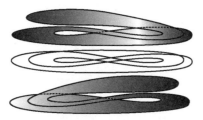

Figure 18.11. Two non equivalent immersions of a disc in the plane with prefixed boundary.

$I\!R^2$ (which is equivalent to the orthogonal projection of the unit 2-sphere in $I\!R^3$ on the equatorial plane). These two maps although share both, their graph and their apparent contour, are not topologically equivalent ([13]).

We thus need to add some extra information which is encoded in the Blank's word of the apparent contour:

Given a pair (graph, apparent contour) together with a bijection between the edges of the graph and the curves in the apparent contour, the number v of stable maps sharing this pair can be obtained in terms of all the possible groupings of the Blank's words attached to the ("ordered") apparent contour.

It is not difficult to see that for $E = 1$ and $D \le 4$, the Blank's word of the possible apparent contours (as shown in Figure 18.7) just admits one grouping. Therefore, if we denote by $\Phi_g(E, D)$ the number of topological classes of fold maps from a surface of genus g to the plane with exactly E singular curves and D double fold points, we get:

$$\Phi_0(1, 0) = 1; \quad \Phi_0(1, 2) = 1; \quad \Phi_0(1, 4) = 6.$$

We finally observe that the number of groupings for each pair $(\mathcal{G}_f, \mathcal{C}_f)$ associated to a stable map f is a stable isotopy invariant of global type. So we can ask: *Is there a relation between the number of groupings and the Ohmoto-Aicardi invariants for stable maps between orientable surfaces ([39])?*

It is also possible to attach graphs to stable maps from a non-orientable surface M to the plane (or the 2-sphere). Given a finite collection $\mathcal{C}$ of closed curves in M, for a curve $c \in \mathcal{C}$, we may have that either a) a neighborhood of c is an cylinder, or b) a neighborhood of c is a Möbius Band, as shown in Figure 18.12. Then we associate a weighted graph to $(\mathcal{C}, M)$ in the following way: Each path-component of $M \setminus \mathcal{C}$ determines a *vertex* and each curve of $\mathcal{C}$ an *edge*. A vertex u and an edge a are incident if and only if the curve represented by a lies in the boundary of the region represented by u. We attach to each vertex u_i a weight $(w_i^o, 0)$ (resp. $(0, w_i^n)$) provided the region, $M_i \subset M \setminus \mathcal{C}$, represented by u_i, is orientable with genus w_i^o (resp. M_i is non-orientable with

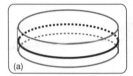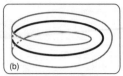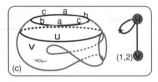

Figure 18.12. Neighborhood of a curve in the surface.

genus w_i^n). The edge $a \in \mathcal{G}$ receives a asterisk, $*$, if the neighborhood of curve represented by a is a Möbius Band.

Figure 18.12(c) gives an example of graph associated to a collection of closed curves on the Klein bottle. The definition of graph associated to a stable map follows analogously to the orientable case. A realization theorem for this kind of graphs will be given in a forthcoming paper by the second and third authors of this paper.

2.3. Particular case I: Stable Gauss maps on surfaces

We can view the Gauss map $\Gamma : M \to S^2$ of an immersed surface $f : M \to \mathbb{R}^3$ as the lagrangian map whose generating family is the *Height functions family*,

$$\lambda(f) : M \times S^2 \longrightarrow \mathbb{R}$$

$$(x, v) \longmapsto \langle f(x), v \rangle = f_v(x).$$

It is well known that the *stability of Γ corresponds to the structural stability of* $\lambda(f)$ [48].

Moreover, in the (2-dimensional) case, the stable Gauss maps are stable as maps between surfaces [8]. The singular set $\Sigma\Gamma$ of the Gauss map is the parabolic curve. For a generic closed surface in $\mathbb{R}^3$, this is a closed curve with finitely many components. We define the graph of Γ as the dual graph of $\Sigma\Gamma$ (as in the general case of a stable map).

Figure 18.13 illustrates a generic embedding of the torus in $\mathbb{R}^3$ in which the parabolic curve has two connected components. The corresponding graph has two vertices of weight zero with a loop connecting them. By means of convenient beaks transitions we can obtain the two lower embeddings with a connected parabolic curve. Their corresponding Gauss maps have one-edge graphs with weight 1 in one of their vertices that may correspond either to the hyperbolic or to the elliptic one. The codimension one transitions of Gauss maps have been described by J. W. Bruce and F. Tari [10]. The only ones of these that affect the graphs are, as in the general case of stable maps, lips and beaks. By means of these transitions we can introduce certain surgeries between Gauss

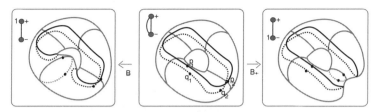

Figure 18.13. Examples of parabolic curves on the warped torus and the corresponding graphs.

maps that allow us to conveniently manipulate their corresponding graphs. As a consequence we prove the following result, whose proof will appear in a forthcoming paper by C. Mendes de Jesus, S. Moraes and M. C. Romero Fuster [32].

Theorem 2.4. *Any weighted bipartite graph $\mathcal{G}$ can be realized as the graph of a stable Gauss map of a closed orientable surface M, with*

$$\chi(M) = 2 - 2\big(\beta_1(\mathcal{G}) + \omega(\mathcal{G})\big),$$

where $\beta_1(\mathcal{G})$ is the number of independent cycles of $\mathcal{G}$ and $\omega(\mathcal{G})$ is the total weight of $\mathcal{G}$.

2.4. Particular case II: Linking numbers of spacial curves and graphs

Given closed curves, $\alpha, \beta : S^1 \longrightarrow \mathbb{R}^3$, their normalized secant map (or zodiacal map) is given by:

$$S_{\alpha,\beta} : S^1 \times S^1 \longrightarrow S^2$$
$$(s, t) \longmapsto \frac{(\alpha(s) - \beta(t))}{||\alpha(s) - \beta(t)||}$$

The "generic" singularities of this map have been studied by J.W. Bruce who proved that *for almost any pair of curves (α, β), the map $S_{\alpha,\beta}$ is locally stable.*

This result has a global version. The singular set of $S_{\alpha,\beta}$ is the bitangency curve consisting of pairs $(s, t) \in S^1 \times S^1$ such that there is a plane in $\mathbb{R}^3$ tangent to α at $\alpha(s)$ and to β at $\beta(t)$. Define a function $\mathcal{B}_{\alpha,\beta} : S^1 \times S^1 \to \mathbb{R}$ by

$$\mathcal{B}(s, t) = det(\beta(t) - \alpha(s), \beta'(t), \alpha'(s)),$$

then the bitangency curve is the zero locus of $\mathcal{B}_{\alpha,\beta}$, and for generic (α, β) it is a closed regular curve (see [38] for the study of the particular case $\alpha = \beta$).

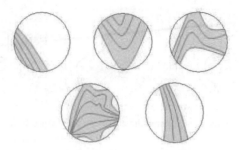

Figure 18.14. Canonical regions.

Recall that the linking number $\mathcal{L}(\alpha, \beta)$ of the pair (α, β) coincides with the degree of the map $S_{\alpha,\beta}$. Therefore, we obtain the following expression for the linking number of the curves α and β in terms of information relative to the graph of their secant map.

$$\mathcal{L}(\alpha, \beta) = (V^+ - V^-) - (g^+ - g^-) - (C^+ - C^-).$$

2.5. Invariants of pairs of transverse foliations of the disc

The characterization of one dimensional foliations on the plane has been developed by W. Kaplan ([25], [26]), L. Markus ([31]) and X. Wang ([50]) among others. Since any foliation without singularities on the plane is orientable, also the results of [36] can be applied here. In this section we consider that the foliation is defined on the disc D^2.

We first recall some basic results. This description will give an idea of the connections between the classification of maps and flows. In order to simplify the characterization of transverse foliations we shall reformulate some of the well known results.

Each leaf of a foliation divides the plane in two regions and two such leaves determine a unique region. The set of regions limited by two leaves constitute a base for a topology of D^2. In the space of leaves we consider the quotient topology. Let S_0 be the set of leaves that are not closed with respect to this topology. Any leaf that belongs to the closure of S_0 is a separatrix. Let S be the set of separatrices. A leaf of $S - S_0$ is a limit separatrix. We will always assume that S, as a subspace of the space of leaves, is a Hausdorff space. The associated invariant will be a distinguished tree ([49]).

Any connected component of D^2 minus the set of separatrices is a canonical region, O. In the Figure 18.14 we have schematized the five types of canonical regions.

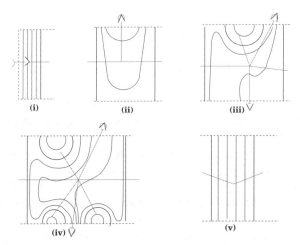

Figure 18.15. Canonical regions and graphs.

For each canonical region O we associate a vertex, $v(O)$. For two adjacent canonical regions we have an edge incident to the corresponding vertex. In this way we define a tree associated to the canonical regions.

To complete the description, we associate to $v(O)$ another edge for each open side of O and for each of this new edges we add another vertex on the opposite side (point at infinity). If O is of type i), we give an orientation to this edge so that the final point is $v(O)$, otherwise we consider the inverse orientation on the edge. Finally, we associate a vertex to each limit separatrix.

We can consider a representation of this tree on $\overline{D^2}$; $v(O)$ is any interior point of O. The edges between canonical regions cut the common separatrix at a point which is equidistant to the intersection points of the separatrix with the boundary of $\overline{D^2}$.

In many situations we do not have a unique foliation, but the union of n distinct foliations. For instance, we obtain 2-foliations in the case of the implicit differential equations corresponding to the curvature or asymptotic lines on a smooth surface immersed in $\mathbb{R}^3$. For these, just a local analysis of the generic situations is known ([19]). Usually, these foliation have some intrinsic relation and providing the topological invariants of each one of them separately is not enough in order to describe the complete situation. This is the case, for instance, of the foliations defined by a positive quadratic form. Consider the general case of $n = 2$ where both foliations are assumed to be *transverse*. As in the case of a unique foliation we define a canonical region as any component of the disc minus the set of separatrices of each foliation. But here, the set of canonical regions is described by a graph. Now, if there are no regions iii) and iv), we define the reduced 2-foliation as the foliation obtained from the initial one by collapsing all the regions of type v) (Fig. 18.2) to a separatrix, we have that in the reduced foliation all canonical regions have non empty intersection with $\partial \overline{D^2}$. Then we have that a transverse pair of reduced foliations on the disk is characterized by a distinguished topological $I\!R$-tree.

The case of a finite number of separatrices is more simple. We can consider that the foliations are defined over the open unit disc. Any leaf cuts the boundary of the disc in two points. These two points can be considered as a S^0 knot over the disc. Then we obtain the following result whose proof will be given in a forthcoming paper by R. López and J. Martínez Alfaro:

Theorem 2.5. *An invariant of a pair of transverse 2-foliations on the unit disc with a finite number of separatrices is the link defined by the separatrices of each foliation and a representative of each canonical region of each foliation.*

We observe at this point that by taking a first integral of each foliation we obtain a map from the disc to the plane. Again, the classification of this map is relevant for the system of foliations. An interesting problem would consist in studying the case in which this is a stable map. Now, similar techniques to those applied to stable maps on compact surfaces without boundary would lead to a new graph with valuable information on the system of foliations on the disc (or on any other compact surface). This is an open problem.

3. 3-Manifolds and graphs

3.1. Stable maps from closed 3-manifolds to $I\!R^3$

The local singularities of a stable map f from a closed (compact without boundary) orientable 3-manifold M to $I\!R^3$ are also well known: fold points along surfaces, cusp points along curves and isolated swallowtail points. From a multi-local viewpoint, we may also have curves of double folds, isolated triple points and isolated crossings of a folds surface with a cusps curve in the apparent contour. In order to obtain global information on these maps we can also associate a graph to each one of them ([34]). The singular set of a stable map f from an orientable compact connected 3-manifold without boundary M to $I\!R^3$ is made of a finite collection of regular surfaces embedded in M. These surfaces separate M into different 3-manifolds with boundary which are embedded by f into $I\!R^3$. Denote by $b_i = rkH_i(M)$ the i-th Betti number of M. It is clear that $b_0 = b_3 = 1$. Moreover, it is also a well known fact that $\chi(M) = 0$, where χ stands for the Euler characteristic. Since $\chi(M) = \sum_{i=0}^{3} b_i$, it follows that $b_1 = b_2$. Let $\bigcup_{i=1}^{n} S_i \subset M$ be a collection of disjoint embedded oriented surfaces in a compact 3-manifold. We define the *weighted graph* $\mathcal{G}$ associated to this collection of surfaces in M as follows: to each surface S_i we associate an edge and to each component M_j of $M - \bigcup_{i=1}^{n} S_i$ a vertex. We then have that an edge is incident to a vertex if and only if the surface corresponding to this edge lies in the boundary of the 3-manifold represented by the vertex. The weights are attached as follows: given a vertex v_j (corresponding to the region M_j) we define its weight as $c_j := b_2(M_j) - s_j + 1$, where s_j stands for the number of connected components of the boundary of M_j. Intuitively, c_j can be seen as the number of generators of $H_2(M)$ which are not determined by the boundary of M_j. For instance, if $M = S^3$ it can be seen, by using basic homology techniques, that all the vertices must have zero weight. To each edge we associate a weight given by the genus, g_i, of the surface S_i that it represents. We will denote by μ the number of edges (which is the number of surfaces S_i) of the graph and by V the number of its vertices. The well known decomposition of S^3 as two solid tori conveniently glued by their boundary leads to a graph made of a unique edge with weight 1 and two vertices with zero weights. A first result in the study of the graphs that can be attached to stable maps on 3-manifolds, proven in [34] by using standard topological techniques such as tubular neighbourhoods and surgeries, is:

Theorem 3.1. *Any graph with weights in its vertices and edges can be realized as the graph of some collection of closed embedded surfaces in some 3-manifold.*

We can now use the well known result on existence fold maps due to Eliashberg (Theorem B, [13]), that for the case of an orientable 3-manifold M that can be embedded in $\mathbb{R}^4$ (i.e. M is stably parallelizable) asserts that *any closed orientable (non necessarily connected) embedded surface S splitting M into two (possibly disconnected) manifolds M_+ and M_- with common boundary S can be the singular set of a fold map on M.* Then by taking into account that the 3-manifolds constructed in the above theorem are all stably parallelizable we obtain:

Corollary 3.2. *Any graph with weights in its vertices and edges can be realized as the graph of a fold map from some 3-manifold to $\mathbb{R}^3$.*

Now, an important question, related to the global classification of stable maps, consists in determining all the possible graphs attached to stable maps on a given 3-manifold. In this sense, a first step is the following:

Theorem 3.3 ([34]). *The graph of a stable map from S^3 to $I\!R^3$ is a tree whose vertices have zero weight. Moreover, any tree with $c_j = 0, \forall j = 1, \ldots, V$ is realizable as the graph of a stable map $f : S^3 \longrightarrow I\!R^3$.*

We observe that a tree with zero weights in the vertices may also be the graph of a stable map from a non simply connected 3-manifold to $I\!R^3$. A construction of this together with a more detailed study of the (topological) constraints on graphs that can be attached to stable maps from a given 3-manifold to $I\!R^3$ will be described in a forthcoming paper by R. Oset and M.C. Romero Fuster.

3.2. Comments on stable maps from closed 3-manifolds to $I\!R^2$

The stable maps from a closed 3-manifold M to the plane, in particular those known as special generic maps, have been treated by several authors ([11], [44], [28], [41], [35]). The Vassiliev invariants of these maps have been determined by M. Yamamoto [51]). The singularities of these maps are fold points (along curves) and isolated cusps. Their singular sets are disjoint unions of embedded circles in M. In this case, there is not a natural way of attaching a graph to such maps. Nevertheless, there is a well known tool that codifies the topological information related to the behavior of the singular set in the 3-manifold:

A stable map $f : M \to I\!R^2$ induces a *Stein factorization* $f = q_f \cdot f_0$, where q_f maps M to the quotient space W_f formed by identifying points belonging to the same connected component of the fiber of f. The space W_f is a two-dimensional complex whose local behaviour is described in [28]. We can view this complex as a generalization of the graph of a Morse function. In the case of a special generic map (i.e. those whose only singularities are definite folds) the complex W_f is a smooth compact surface whose boundary is diffeomorphic to the singular set of the map. Moreover, the 3-manifold M can be recovered from W_f as a circle bundle with the fibers collapsed to a point along the boundary of W_f. It is also shown in [28] that given any compact surface with boundary W, it is possible to find a 3-manifold M and a special generic map $f : M \to I\!R^2$ such that $W_f = W$. This procedure can be also applied to stable maps from m-manifolds to the plane, for $m > 3$.

Acknowledgements: Work partially supported by DGCYT and FEDER grant no. MTM2009-08933.

References

1. Y. Ando, Existence theorems of fold maps. *Japan J. Math.* **30** (2004) 29–73.
2. Y. Ando, Fold-maps and the space of base point preserving maps of spheres. *J. Math. Kyoto Univ.* **41** (2002) 691–735.

3. V. I. Arnol'd, Plane curves, their invariants, perestroikas and classifications, *Advances in Soviet Math.* **21** AMS, Providence RI, (1994) 33–91.
4. V. I. Arnol'd, Smooth functions statistics. *Funct. Anal. Other Math.* **1(2)** (2006) 125à-133.
5. V. I. Arnol'd, Topological classification of Morse functions and generalisations of Hilbert's 16-th problem. *Math. Phys. Anal. Geom.* **10(3)** (2007) 227–236.
6. V. I. Arnol'd, Statistics and classification of topologies of periodic functions and trigonometric polynomials. *Proc. Steklov Inst. Math.* 2006, Dynamical Systems: Modeling, Optimization, and Control, suppl. 1, S13–S23.
7. K. D. Bailey, Extending closed plane curves to immersions of a disc with n handles. *Trans. AMS* **206** (1975), 1–24.
8. Th. Banchoff, T. Gaffney and C. McCrory, *Cusps of Gauss mappings.* Research Notes in Mathematics, **55**. Pitman (Advanced Publishing Program), Boston, Mass.-London, 1982.
9. S. J. Blank, *Extending immersions of the circle.* PhD Thesis, Brandeis University, Waltham, Mass., 1967.
10. J. W. Bruce and F. Tari, Families of surfaces: height functions and projections to planes. *Math. Scand.* **82(2)** (1998) 165–185.
11. O. Burlet and G. de Rham, Sur certaines applications génériques d'une variété close à 3 dimensions dans le plan. *Enseignement Math. (2)* **20** (1974) 275–292.
12. J. Casasayas, J. Martinez Alfaro and A. Nunes, Knots and Links in Integrable Hamiltonian Systems. *Knot Theory and its Ramifications* **7(2)** (1998), 123–153.
13. Y. Eliashberg, On singularities of folding type. *Math. USSR-Izv.* **4** (1970) 1119–1134.
14. A. T. Fomenko, *Integrability and Nonintegrability in Geometry and Mechanics*, Kluwer, 1988.
15. G. K. Francis, Assembling compact Riemann surfaces with given boundary curves and branch points on the sphere. *Illinois J. Math.* **20(2)** (1976), 198–217.
16. G. K. Francis and S.F. Troyer, Excellent maps with given folds and cusps. *Houston J. Math.* **3(2)** (1977) 165–194.
17. J. Franks, Nonsingular Smale flows on S^3. *Topology* **24** (1985) 265–282.
18. M. Golubitsky and V. Guillemin, *Stable Mappings and Their Singularities*, Springer Verlag, Berlin, 1976.
19. C. Gutierrez, J. Sotomayor, Structurally stable configurations of lines of curvature, *Asterisque* **98-99** (1982) 195–215.
20. D. Hacon, C. Mendes de Jesus and M. C. Romero Fuster, Topological invariants of stable maps from a surface to the plane from a global viewpoint. *Proceedings of the 6th Workshop on Real and Complex Singularities*. Lecture Notes in Pure and Applied Mathematics, **232**, Marcel and Dekker, (2003) 227–235.
21. D. Hacon, C. Mendes de Jesus and M. C. Romero Fuster, Fold maps from the sphere to the plane. *Experimental Maths* **15** (2006) 491–497.
22. D. Hacon, C. Mendes de Jesus and M. C. Romero Fuster, Stable maps from surfaces to the plane with prescribed branching data *Topology and Its Appl.* **154** (2007) 166–175.
23. D. Hacon, C. Mendes de Jesus and M. C. Romero Fuster, Graphs of stable maps from closed orientable surfaces to the 2-sphere. Preprint available at http://www.singularidadesvalencia.com/s.asp

24. A. Haefliger, Quelques remarques sur les applications differentables d'une surface dans le plan. *Ann. Inst. Fourier* **10** (1960) 47–60.
25. W. Kaplan, Regular curve families filling the plane, I. *Duke Math. J.* **7** (1940) 154–185.
26. W. Kaplan, Regular curve families filling the plane, II. *Duke Math. J.* **8** (1941) 11–46.
27. A.S. Kronrod, On functions of two variables. *Uspehi Matem Nauk (NS)* **5** (1950) 24–134.
28. L. Kushner, H. Levine and P. Porto, Mapping three-manifolds into the plane. I. *Bol. Soc. Mat. Mexicana (2)* **29** (1984) 11–33.
29. H. Levine, Stable maps: an introduction with low dimensional examples. *Bol. Soc. Bras. Mat.* **7** (1976) 145–184.
30. I. P. Malta, N. C. Saldanha and C. Tomei, *Geometria e Análise Numérica de Func cões do Plano no Plano.* 19 Colóquio Brasileiro de Matemática, IMPA, 1993.
31. L. Markus, Global structure of ordinary differential equations in the plane. *Trans. Amer. Math. Soc.* **76** (1954) 137–148.
32. C. Mendes de Jesus, S. M. Moraes and M. C. Romero Fuster, Stable Gauss maps from a global viewpoint. Preprint available at http://www. singularidadesvalen-cia.com/s.asp
33. C. Mendes de Jesus and M. C. Romero Fuster, Bridges, channels and Arnold's invariants for generic plane curves. *Topology Appl.* **125(3)** (2002) 505–524.
34. C. Mendes de Jesus, R. Oset and M.C. Romero Fuster Global Topological Invariants of Stable Maps from 3-Manifolds to $I\!R^3$. To appear in Proc. Steklov Institute of Mathematics **267** (2009).
35. W. Motta, P. Porto and O. Saeki, Stable maps of 3-manifolds into the plane and their quotient spaces. *Proc. London Math. Soc.* **71** (1995) 158à–174.
36. D. Neumann and T. O'Brien, Global structure of Continuous flows on 2-Manifolds. *J. of Differential Equations* **22** (1976) 89–110.
37. L. I. Nicolaescu, Morse functions statistics. *Funct. Anal. Other Math.* **1(1)** (2006) 97à-103.
38. J. J. Nuño and M. C. Romero Fuster, Global bitangency properties of generic closed space curves. *Math. Proc. Cambridge Philos. Soc.* **112(3)** (1992) 519–526.
39. T. Ohmoto and F. Aicardi, First order local invariants of apparent contours. *Topology* **45(1)** (2006) 27–45.
40. M. M. Peixoto, On the classification of flows on 2-manifolds. *Proc. Symp. Dynamical Systems* Academic Press (1973) 389–419.
41. P. Porto and Y. Furuya, On special generic maps from a closed manifold into the plane. *Topology Appl.* **35(1)** (1990) 41–52.
42. J. R. Quine, A global theorem for singularities of maps between oriented 2-manifolds. *Trans. AMS* **236** (1978) 307–314.
43. O. Saeki, Fold maps on 4-manifolds. *Comment. Math. Helv.* **78** (2003) 627–647.
44. O. Saeki, Topology of special generic maps of manifolds into Euclidean spaces. *Topology Appl.* **49** (1993) 265à–293.
45. K. Sakuma, On the topology of simple fold maps. *Tokyo J. Math.* **17** (1994) 21–31.
46. S. F. Troyer, Extending on boundary inmersion to the disc with n holes. PhD Thesis, Northeastern University, Boston, 1973.

47. V. A. Vassiliev, *Complements of discriminants of smooth maps: topology and applications*, AMS, Providence, RI, 1992.

48. C. T. C. Wall, Geometric properties of generic differentiable manifolds. *Geometry and topology (Proc. III Latin Amer. School of Math., Inst. Mat. Pura Aplicada CNPq, Rio de Janeiro, 1976)*, pp. 707–774. *Lecture Notes in Math.* **597**, Springer, Berlin, 1977.

49. X. Wang, On the C^*-Algebras of foliations in the plane. *Lecture Notes in Mathematics* **1257** Springer, Berlin, 1987.

50. X. Wang, The C^*-algebras of Morse-Smale flows on two manifolds. *Ergod. Th. Dynam. Sys.* **10**(1990) 565–597.

51. M. Yamamoto, First order semi-local invariants of stable maps of 3-manifolds into the plane. *Proc. London Math. Soc* **92** (2006) 471–504.

52. V. C. Zanetic, A extensão em Codimensão Dois de Imersões e as Funções Diferenciáveis com Imagem do Conjunto Singular Especificada. Tese, IME-USP, São Paulo, 1984.

J. Martinez Alfaro
Dep. de Matemática Aplicada
Facultat de Matemàtiques
Universitat de València
46100 Burjassot (València)
Spain
email: martinja@uv.es

C. Mendes de Jesus
Dep. de Matemática
Univ. Federal de Viçosa
36570-000 Viçosa-MG
Brazil
email: cmendes@ufv.br

M. C. Romero Fuster
Dep. Geometria i Topologia
Facultat de Matemàtiques
Universitat de València
46100 Burjassot (València)
Spain
carmen.romero@uv.es

19

Real analytic Milnor fibrations and a strong Łojasiewicz inequality

DAVID B. MASSEY

Abstract

We give a strong version of a classic inequality of Łojasiewicz; one which collapses to the usual inequality in the complex analytic case. We show that this inequality for real analytic functions allows us to construct a real Milnor fibration inside a ball.

1. Introduction

Suppose that $\mathcal{U}$ is an open neighborhood of the origin in $\mathbb{C}^N$, and that $f_{\mathbb{c}} : (\mathcal{U}, \mathbf{0}) \to (\mathbb{C}, 0)$ is a complex analytic function.

In the now-classic book [15], Milnor shows that one has what is now called the Milnor fibration of $f_{\mathbb{c}}$ (at $\mathbf{0}$). The Milnor fibration is **the** fundamental device in the study of the topology of the hypersurface X defined by the vanishing of $f_{\mathbb{c}}$.

In fact, there are two Milnor fibrations associated with $f_{\mathbb{c}}$: one defined on small spheres, and one defined inside small open balls. Both of these are referred to as **the** Milnor fibration because the two fibrations are diffeomorphic. We wish to be precise.

For $\epsilon > 0$, let S_ϵ (resp., B_ϵ, $\overset{\circ}{B}_\epsilon$) denote the sphere (resp., closed ball, open ball) of radius ϵ, centered at the origin. In the special case of balls in $\mathbb{R}^2 \cong \mathbb{C}$, we write $\mathbb{D}_\epsilon$ in place of B_ϵ and $\overset{\circ}{\mathbb{D}}_\epsilon$ in place of $\overset{\circ}{B}_\epsilon$. Finally, we write $\mathbb{D}_\epsilon^*$ in place of $\mathbb{D}_\epsilon - \{\mathbf{0}\}$, and $\overset{\circ}{\mathbb{D}}_\epsilon^*$ in place of $\overset{\circ}{\mathbb{D}}_\epsilon - \{\mathbf{0}\}$.

One version of the Milnor fibration (at the origin) is given by: there exists $\epsilon_0 > 0$ such that, for all ϵ such that $0 < \epsilon \leq \epsilon_0$, the map $f_{\mathbb{c}}/|f_{\mathbb{c}}| : S_\epsilon - S_\epsilon \cap$

2000 *Mathematics Subject Classification* 14P15, 14B05, 58K05.

$X \to S^1 \subseteq \mathbb{C}$ is a smooth, locally trivially fibration, whose diffeomorphism-type is independent of the choice of ϵ (see [15]).

The second, diffeomorphic version of the Milnor fibration is given by: there exists $\epsilon_0 > 0$ such that, for all ϵ such that $0 < \epsilon \leq \epsilon_0$, there exists a $\delta_0 > 0$ such that, for all δ such that $0 < \delta \leq \delta_0$, the map $f_{\mathbb{c}} : {}^{\circ}B_{\epsilon} \cap f_{\mathbb{c}}^{-1}(\partial \mathbb{D}_\delta) \to \partial \mathbb{D}_\delta \cong S^1 \subseteq \mathbb{C}$ is a smooth, locally trivial fibration, whose diffeomorphism-type is independent of the choice of ϵ and (sufficiently small) δ (see Theorem 5.11 of [15] and [10]). The primary advantage to this second characterization of the Milnor fibration is that it compactifies nicely to yield a locally trivial fibration $f_{\mathbb{c}} : B_{\epsilon} \cap f_{\mathbb{c}}^{-1}(\partial \mathbb{D}_\delta) \to \partial \mathbb{D}_\delta$, which, up to homotopy, is equivalent to either of the two previously-defined Milnor fibrations.

We shall not summarize the important properties of the Milnor fibration here, but refer the reader to [15], [3], [17], and the Introduction to [13]. We wish to emphasize that our discussion of the Milnor fibration above assumes that $f_{\mathbb{c}}$ is a complex analytic function.

Of course, a complex analytic function yields a pair of real analytic functions, coming from the real and imaginary parts of the complex function, and one can ask the more general question: **when does a pair of real analytic functions possess one or both types of Milnor fibrations?**

This topic of real analytic Milnor fibrations is complicated and interesting, and gives rise to many questions. The reader may wish to consult the very nice survey article [2] in this volume.

In Chapter 11 of [15], Milnor discusses, fairly briefly, some results in the case of the real analytic function $f = (g, h)$. He considers the very special case where f has an isolated critical point at the origin, and shows that, while the restriction of f still yields a fibration over a small circle inside the ball, $f/|f|$ does not necessarily yield the projection map of a fibration from $S_\epsilon - S_\epsilon \cap X$ to S^1; see p. 99 of [15]. Can one relax the condition that f has an isolated critical point and still obtain a locally trivial fibration $f : B_{\epsilon} \cap f^{-1}(\partial \mathbb{D}_\delta) \to \partial \mathbb{D}_\delta$? Are there reasonable conditions that guarantee that $f/|f| : S_\epsilon - S_\epsilon \cap X \to S^1$ is a locally trivial fibration which is diffeomorphic to the fibration inside the ball? Such questions have been investigated by a number of researchers; see [15], [9], [16], [20], [19], and [4].

An obvious approach to answering the question about the existence of real analytic Milnor fibrations is simply to try to isolate what properties of a complex analytic function are used in proofs that Milnor fibrations exist. We write "proofs" and not "the proof" here because we will not follow Milnor's proof, but rather follow Lê's proof in [10] of the existence of Milnor fibrations inside the ball for complex analytic functions.

Lê's proof consists almost solely of using the existence of a Thom (or a_f, or *good*) stratification. In [7], Hamm and Lê, following a suggestion of Pham, used the complex analytic Łojasiewicz inequality (see Corollary 1.2 below) and a "trick" to show that Thom stratifications exist.

Our goal in this paper is very modest: we will give the "correct" generalization of the complex analytic Łojasiewicz inequality, and then show that if a pair (or k-tuple) of real analytic functions satisfies this new *strong Łojasiewicz inequality*, then the Milnor fibration inside a ball exists.

In the remainder of the introduction, we will summarize our primary definition and result.

Let $\mathcal{U}$ now denote an open subset of $\mathbb{R}^n$, and p denote a point in $\mathcal{U}$. Let g and h be real analytic functions from $\mathcal{U}$ to $\mathbb{R}$, and let $f := (g, h) : \mathcal{U} \to \mathbb{R}^2$. Recall the classic inequality of Łojasiewicz [11] (see also [1]).

Theorem 1.1. *There exists an open neighborhood $\mathcal{W}$ of p in $\mathcal{U}$, and $c, \theta \in \mathbb{R}$ such that $c > 0$, $0 < \theta < 1$, and, for all $x \in \mathcal{W}$,*

$$|g(x) - g(p)|^\theta \le c|\nabla g(x)|,$$

where ∇g is the gradient vector.

Remark 1. The phrasing above is classical, and convenient in some arguments. However, one can also fix the value of the constant c above to be any $c > 0$, e.g., $c = 1$. In other words, one may remove the reference to c in the statement Theorem 1.1 and simply use the inequality

$$|g(x) - g(p)|^\theta \le |\nabla g(x)|.$$

The argument is easy, and simply requires one to pick a larger θ (still less than 1). As we shall not use this "improved" statement, we leave the proof as an exercise.

Theorem 1.1 implies a well-known complex analytic version of itself. One can easily obtain this complex version by replacing g by the square of the norm of the complex analytic function. However, we shall prove two not so well-known more general corollaries, Corollary 3.1 and Corollary 3.2, which yield the complex analytic statement.

If n is even, say $n = 2m$, then we may consider the complexified version of f by defining f_c by

$$f_c(x_1 + iy_1, \ldots, x_m + iy_m) := g(x_1, y_1, \ldots, x_m, y_m) + ih(x_1, y_1, \ldots, x_m, y_m).$$

From Corollary 3.2, we immediately obtain:

Corollary 1.2. *Suppose that f_{c} is complex analytic. Then, there exist an open neighborhood $\mathcal{W}$ of p in $\mathcal{U} \subseteq \mathbb{C}^n$, and $c, \theta \in \mathbb{R}$ such that $c > 0$, $0 < \theta < 1$, and, for all $z \in \mathcal{W}$,*

$$|f_{\mathrm{c}}(z) - f_{\mathrm{c}}(p)|^\theta \le c |\nabla f_{\mathrm{c}}(z)|,$$

where ∇f_{c} denotes the complex gradient.

Our generalization of this complex analytic Łojasiewicz inequality is:

Definition 1.3. We say that $f = (g, h)$ satisfies the **strong Łojasiewicz inequality at** p or that f **is Ł-analytic at** p if and only if there exist an open neighborhood $\mathcal{W}$ of p in $\mathcal{U}$, and $c, \theta \in \mathbb{R}$ such that $c > 0$, $0 < \theta < 1$, and, for all $x \in \mathcal{W}$,

$$|f(x) - f(p)|^\theta \le c \cdot \min_{|(a,b)|=1} \left| a\nabla g(x) + b\nabla h(x) \right|.$$

Main Result. *Suppose that $f(\mathbf{0}) = 0$, that f is not locally constant near the origin, and that f is Ł-analytic at $\mathbf{0}$.*

Then, for all $0 < \delta \ll \epsilon \ll 1$, $f : B_\epsilon \cap f^{-1}(\partial\mathbb{D}_\delta) \to \partial\mathbb{D}_\delta$ is a proper, stratified submersion, and so $f : B_\epsilon \cap f^{-1}(\partial\mathbb{D}_\delta) \to \partial\mathbb{D}_\delta$ and $f : \overset{\circ}{B}_\epsilon \cap f^{-1}(\partial\mathbb{D}_\delta) \to \partial\mathbb{D}_\delta$ are locally trivial fibrations.

Moreover, the diffeomorphism-type of $f : \overset{\circ}{B}_\epsilon \cap f^{-1}(\partial\mathbb{D}_\delta) \to \partial\mathbb{D}_\delta$ is independent of the appropriately small choices of ϵ and δ.

We have presented the main definition and result above in the case of real analytic functions into $\mathbb{R}^2$; this was for simplicity of the discussion. In fact, in Definition 3.3 and Corollary 5.6, we give our main definition and result when the dimension of the codomain is arbitrary.

We thank T. Gaffney for pointing out the existence of nap-maps (see Definition 5.2) in all dimensions.

2. Singular values of matrices

In this section, we wish to recall some well-known linear algebra, and establish/recall some inequalities that we will need in later sections.

Let $\vec{v}_1, \ldots \vec{v}_k$ be vectors in $\mathbb{R}^n$. Let A denote the $n \times k$ matrix whose i-th column is $\vec{v}_i$. Let M denote the $k \times k$ matrix $A^t A$. Consider the function v from the unit sphere centered at the origin in $\mathbb{R}^k$ into the non-negative real numbers given by $v(t_1, \ldots, t_k) := |t_1 \vec{v}_1 + \cdots + t_k \vec{v}_k|$. The critical values of v are the *singular values of A*, and they are the square roots of the necessarily non-negative eigenvalues of M. It is traditional to index the singular values in a decreasing manner, i.e., we let the singular values of A (which need not be distinct) be denoted by $\sigma_1, \ldots, \sigma_k$, where $\sigma_1 \geq \cdots \geq \sigma_k$. We denote the eigenvalues of M by $\lambda_i := \sigma_i^2$. The singular value σ_1 is the norm of A. The minimum singular value σ_k will be of particular interest throughout this paper. Note that

(†)
$$\sigma_k = \min_{|(t_1,\ldots,t_k)|=1} |t_1\vec{v}_1 + \cdots + t_k\vec{v}_k|.$$

Also note that the trace of M, tr$M$, is equal to $|\vec{v}_1|^2 + \cdots + |\vec{v}_k|^2$, which is equal to $\lambda_1 + \cdots + \lambda_k = \sigma_1^2 + \cdots + \sigma_k^2$. Hence, tr$M = 0$ if and only if $A = 0$ if and only if $\sigma_1 = 0$. It will be important for us later that the singular values of A vary continuously with the entries of A. Also, as the non-zero eigenvalues of $A^t A$ and $A A^t$ are the same, the non-zero singular values of A are equal to the non-zero singular values of A^t; in particular, $\sigma_1(A) = \sigma_1(A^t)$, i.e., the norm of a matrix is equal to the norm of its transpose.

The following proposition contains the fundamental results on singular values that we shall need; these results are known, but we include brief proofs for the convenience of the reader.

Proposition 2.1.

(i) $\sigma_k = 0$ *if and only if* $\vec{v}_1, \ldots, \vec{v}_k$ *are linearly dependent;*
(ii) *if* $|\vec{v}_1|^2 + \cdots + |\vec{v}_k|^2 \neq 0$, *then*

$$\frac{1}{\sqrt{k}} \leq \frac{\sigma_1}{\sqrt{|\vec{v}_1|^2 + \cdots + |\vec{v}_k|^2}} \leq 1;$$

and
(iii) $k(\det M)^{1/k} \leq |\vec{v}_1|^2 + \cdots + |\vec{v}_k|^2$, *with equality holding if and only if* $\vec{v}_1, \ldots, \vec{v}_k$ *all have the same length and are pairwise orthogonal.*

Proof. Item 1 is immediate from (†).

Assume that $|\vec{v}_1|^2 + \cdots + |\vec{v}_k|^2 \neq 0$. As $|\vec{v}_1|^2 + \cdots + |\vec{v}_k|^2 = \sigma_1^2 + \cdots + \sigma_k^2$, the right-hand inequality in Item 2 is immediate. For all i, let

$$r_i := \frac{\sigma_i}{\sqrt{|\vec{v}_1|^2 + \cdots + |\vec{v}_k|^2}}.$$

Then, $r_1^2 + \cdots + r_k^2 = 1$ and $r_1^2 \geq \cdots \geq r_k^2$. It follows that $r_1^2 \geq 1/k$; this proves the left-hand inequality in Item 2.

The inequality in Item 3 is nothing more than the fact that the geometric mean of the eigenvalues of M is less than or equal to the arithmetic mean, where we have again used that the trace of M is $|\vec{v}_1|^2 + \cdots + |\vec{v}_k|^2$. In addition, these two means are equal if and only if all of the eigenvalues of M are the same. This occurs if and only if all of singular values of A are the same, which would mean that the function $v(t_1, \ldots, t_k) := |t_1 \vec{v}_1 + \cdots + t_k \vec{v}_k|$ is constant on the unit sphere. The remainder of Item 3 follows easily. $\qquad \square$

We need some results for dealing with compositions. Let $\sigma_i(C)$ denote the i-th singular value of a matrix C, indexed in descending order. Let $\lambda_i(C) := \sigma_i^2(C)$, i.e., $\lambda_i(C)$ is the i-th eigenvalue of $C^t C$, where the eigenvalues are indexed in descending order.

Let B be an $m \times n$ matrix.

Proposition 2.2.

$$\sigma_k(BA) \geq \sigma_n(B)\sigma_k(A).$$

Hence,

$$\left[\det(A^t B^t B A)\right]^{1/k} \geq \sigma_k^2(BA) \geq \sigma_n^2(B)\sigma_k^2(A).$$

Proof. Let $\vec{u}$ be a unit vector in $\mathbb{R}^k$, written as a $k \times 1$ matrix. If $A\vec{u} = 0$, then $\sigma_k(A) = 0$ and the first inequality is immediate. So, assume that $A\vec{u} \neq 0$. Then,

$$|BA\vec{u}| = \left|B\left(\frac{A\vec{u}}{|A\vec{u}|}\right)\right| \cdot |A\vec{u}| \cdot \geq \sigma_n(B)\sigma_k(A).$$

This proves the first inequality.

As $\left[\det(A^t B^t B A)\right]^{1/k}$ is the geometric mean of the $\sigma_i^2(BA)$, the second set of inequalities is immediate. $\qquad \square$

Lemma 2.3. *Suppose that P and Q are $r \times r$ real matrices, and that P is diagonalizable (over the reals) and has no negative eigenvalues. Then,*

$$\lambda_r(P)\text{tr}(Q) \leq \text{tr}(PQ) \leq \lambda_1(P)\text{tr}(Q).$$

Proof. Suppose that $P = S^{-1}DS$, where D is diagonal. Then, $\text{tr}(PQ) = \text{tr}(S^{-1}DSQ) = \text{tr}(DSQS^{-1})$. Let $B = SQS^{-1}$. Hence, $\text{tr}(B) = \text{tr}(Q)$, and

$$\text{tr}(PQ) = \text{tr}(DB) = \sum_{i=1}^{r}(DB)_{i,i} = \sum_{i=1}^{r}\sum_{j=1}^{r} D_{i,j} B_{j,i} = \sum_{i=1}^{r} D_{i,i} B_{i,i}.$$

As $\lambda_r(P) \leq D_{i,i} \leq \lambda_1(P)$, the conclusion follows. $\qquad \square$

Proposition 2.4.

$$n\sigma_n^2(A')\sigma_n^2(B) \leq \text{tr}(A^t B^t B A) \leq n\sigma_1^2(A)\sigma_1^2(B).$$

Proof. As $\text{tr}(A^t B^t BA) = \text{tr}(AA^t B^t B)$, we may apply Lemma 2.3 with $P = AA^t$ and $Q = B^t B$, and conclude that

$$\sigma_n^2(A^t)[n\sigma_n^2(B)] \leq \sigma_n^2(A^t)\text{tr}(B^t B) \leq \text{tr}(A^t B^t BA) \leq \sigma_1^2(A^t)\text{tr}(B^t B)$$

$$\leq \sigma_1^2(A^t)[n\sigma_1^2(B)].$$

Now, use that $\sigma_1(A) = \sigma_1(A^t)$. $\qquad\square$

Corollary 2.5. *If $BA \neq 0$, then $\text{tr}(A^t B^t BA)$, $\sigma_1^2(B)$, and $\sigma_1^2(A)$ are not zero and*

$$\frac{k\left[\det(A^t B^t BA)\right]^{1/k}}{\text{tr}(A^t B^t BA)} \geq \frac{k\sigma_n^2(B)\sigma_k^2(A)}{n\sigma_1^2(B)\sigma_1^2(A)}.$$

Proof. Combine Proposition 2.4 with Proposition 2.2. $\qquad\square$

3. Ł-maps and Ł-weights

As in the introduction, let $\mathcal{U}$ be an open subset of $\mathbb{R}^n$. Let $f = (g, h)$ be a C^1 map from $\mathcal{U}$ to $\mathbb{R}^2$ and let $G := (g_1, \ldots, g_k)$ be a C^1 map from $\mathcal{U}$ to $\mathbb{R}^k$. By the *critical locus of G, ΣG*, we mean the set of points where G is not a submersion (this is reasonable since our main hypothesis in most results will imply that $n \geq k$ or that G is locally constant). A number of our results will apply only with the stronger assumption that f and G are real analytic, but we shall state that hypothesis explicitly as needed.

Throughout this section, we let $A = A(x)$ denote the $n \times k$ matrix which has the gradient vector of $g_i(x)$ as its i-th column (i.e., A is the transpose of the derivative matrix $[d_x G]$ of $(g_1, \ldots, g_k)$), and let $M := A^t A$. We will be applying the results of Section 2 to A and M. Let $\sigma_1(x), \ldots, \sigma_k(x)$ denote the singular values of $A(x)$, indexed in decreasing order.

We have the following corollary to Theorem 1.1.

Corollary 3.1. *Suppose that G is real analytic. Then, there exists an open neighborhood $\mathcal{W}$ of p in $\mathcal{U}$, and $c, \theta \in \mathbb{R}$ such that $c > 0, 0 < \theta < 1$, and, for all $x \in \mathcal{W}$,*

$$|G(x) - G(p)|^\theta \leq c\sqrt{|\nabla g_1(x)|^2 + \cdots + |\nabla g_k(x)|^2}.$$

Proof. For notational convenience, we shall prove the result for $k = 2$, using $f = (g, h)$, and we shall assume that $f(p) = 0$; the proof of the general case proceeds in exactly the same manner.

We will prove that there exist $\mathcal{W}$, c, and θ as in the statement such that the inequality holds at all $x \in \mathcal{W}$ such that $f(x) \neq 0$. This clearly suffices to prove the corollary. We will also assume that $|\nabla g(p)|^2 + |\nabla h(p)|^2 = 0$ for, otherwise, the result is trivial.

Apply Theorem 1.1 to the function from $\mathcal{U} \times \mathcal{U}$ to $\mathbb{R}$ given by $g^2(x) + h^2(w)$. We conclude that there exists an open neighborhood $\mathcal{W}$ of p in $\mathcal{U}$, and $c, \theta \in \mathbb{R}$ such that $c > 0$, $0 < \theta < 1$, and, for all $(x, w) \in \mathcal{W} \times \mathcal{W}$,

$$|g^2(x) + h^2(w)|^\theta \leq c|2g(x)(\nabla g(x), 0) + 2h(w)(0, \nabla h(w)|.$$

Restricting to the diagonal, we obtain that, for all $x \in \mathcal{W}$,

$$|g^2(x) + h^2(x)|^\theta \leq 2c\sqrt{g^2(x)|\nabla g(x)|^2 + h^2(x)|\nabla h(x)|^2}.$$

Applying the Cauchy-Schwarz inequality, we have, for all $x \in \mathcal{W}$,

$$|g^2(x) + h^2(x)|^\theta \leq 2c\sqrt{|(g^2(x), h^2(x))| \cdot |(|\nabla g(x)|^2, |\nabla h(x)|^2)|}$$
$$= 2c\left(g^4(x) + h^4(x)\right)^{1/4}\left(|\nabla g(x)|^4 + |\nabla h(x)|^4\right)^{1/4}.$$

For $a, b, x > 0$, $(a^x + b^x)^{1/x}$ is a decreasing function of x. Therefore, we conclude that, for all $x \in \mathcal{W}$,

$$|(g(x), h(x))|^{2\theta} = |g^2(x) + h^2(x)|^\theta$$
$$\leq 2c\left(g^2(x) + h^2(x)\right)^{1/2}\left(|\nabla g(x)|^2 + |\nabla h(x)|^2\right)^{1/2},$$

and so, for all $x \in \mathcal{W}$ such that $f(x) \neq 0$,

$$|f(x)|^{2\theta-1} \leq 2c\sqrt{|\nabla g(x)|^2 + |\nabla h(x)|^2}.$$

This proves the result, except for the bounds on the exponent. As $-1 < 2\theta - 1 < 1$, we have only to eliminate the possibility that $-1 < 2\theta - 1 \leq 0$. However, this is immediate, as we are assuming that $f(p) = 0$ and $|\nabla g(p)|^2 + |\nabla h(p)|^2 = 0$. $\qquad\square$

The following corollary follows at once.

Corollary 3.2. *Suppose that G is real analytic and, for all $x \in \mathcal{U}$, $\nabla g_1(x), \ldots, \nabla g_k(x)$ have the same magnitude.*

Then, there exists an open neighborhood $\mathcal{W}$ of p in $\mathcal{U}$, and $c, \theta \in \mathbb{R}$ such that $c > 0$, $0 < \theta < 1$, and, for all $x \in \mathcal{W}$,

$$|G(x) - G(p)|^\theta \leq c|\nabla g_1(x)|.$$

We now give the fundamental definition of this paper. Our intention is to isolate the properties of a complex analytic function that are used in proving the existence of the Milnor fibration inside a ball.

Definition 3.3. We say that G satisfies the **strong Łojasiewicz inequality at** p or is an **Ł-map at** p if and only if there exists an open neighborhood $\mathcal{W}$ of p in $\mathcal{U}$, and $c, \theta \in \mathbb{R}$ such that $c > 0, 0 < \theta < 1$, and, for all $x \in \mathcal{W}$,

$$|G(x) - G(p)|^{\theta} \leq c \cdot \min_{|(a_1,\ldots,a_k)|=1} \left| a_1 \nabla g_1(x) + \cdots + a_k \nabla g_k(x) \right| = c\sigma_k(x).$$

If G satisfies the strong Łojasiewicz inequality at p and is real analytic in a neighborhood of p, then we say that G **is Ł-analytic at** p.

We say that G is **Ł-analytic** if and only if G is Ł-analytic at each point $p \in \mathcal{U}$.

Remark 2. Note that if $\nabla g_1(x), \ldots, \nabla g_k(x)$ are always pairwise orthogonal and have the same length, then the inequality in Definition 3.3 collapses to

$$|G(x) - G(p)|^{\theta} \leq c|\nabla g_1(x)|,$$

which, as we saw in Corollary 3.2, is automatically satisfied if G is real analytic.

In the case $k = 2$, and $f = (g, h)$, one easily calculates the eigenvalues of the matrix

$$\begin{bmatrix} |\nabla g|^2 & \nabla g \cdot \nabla h \\ \nabla g \cdot \nabla h & |\nabla h|^2 \end{bmatrix}$$

and finds that $\sigma_2(x)$ is

$$\sqrt{\frac{|\nabla g(x)|^2 + |\nabla h(x)|^2 - \sqrt{(|\nabla g(x)|^2 + |\nabla h(x)|^2)^2 - 4\left[|\nabla g(x)|^2|\nabla h(x)|^2 - (\nabla g(x) \cdot \nabla h(x))^2\right]}}{2}}.$$

Definition 3.4. When G is such that $\nabla g_1(x), \ldots, \nabla g_k(x)$ are always pairwise orthogonal and have the same length, we say that G is a **simple Ł-map**.

Thus, G is a simple Ł-map if and only if M is a scalar multiple of the identity, i.e., for all $x \in \mathcal{U}$, $M(x) = |\nabla g_1(x)|^2 I_k$.

Remark 3. Note that the pair of functions given by the real and imaginary parts of a holomorphic or anti-holomorphic function yields a simple Ł-analytic map.

Definition 3.5. At each point $x \in \mathcal{U}$ where $d_x G \neq 0$ (i.e., where $|\nabla g_1(x)|^2 + \cdots + |\nabla g_k(x)|^2 \neq 0$), define

$$\rho_G(x) := \frac{k (\det M)^{1/k}}{\operatorname{tr} M} = \frac{k (\det M)^{1/k}}{|\nabla g_1(x)|^2 + \cdots + |\nabla g_k(x)|^2}.$$

If $d_x G$ is not identically zero, we define the **Ł-weight of** G, $\rho_G^{\inf}$, to be the infimum of $\rho_G(x)$ over all $x \in \mathcal{U}$ such that $|\nabla g_1(x)|^2 + \cdots + |\nabla g_k(x)|^2 \neq 0$.

If $d_x G$ is identically zero, so that G is locally constant, we set ρ_G^{inf} equal to 1.

We say that G **has positive Ł-weight at** x if and only if there exists an open neighborhood $\mathcal{W}$ of x such that $G_{|_{\mathcal{W}}}$ has positive Ł-weight.

Remark 4. If G is a submersion at x, then it is clear that G has positive Ł-weight at x. However, the converse is not true; for instance, a simple Ł-map has positive Ł-weight (see below). Thus, in a sense, having positive Ł-weight is a generalization of being a submersion.

Note that, if $\rho_G^{\text{inf}} > 0$, we must have either $n \geq k$ or that G is locally constant; furthermore, it must necessarily be true that $x \in \Sigma G$ implies that $\nabla g_1(x) = \cdots = \nabla g_k(x) = \mathbf{0}$.

For a pair of functions $f = (g, h)$, one easily calculates that

$$\rho_f(x) = \frac{2|\nabla g(x)||\nabla h(x)| \sin \eta(x)}{|\nabla g(x)|^2 + |\nabla h(x)|^2},$$

where $\eta(x)$ equals the angle between $\nabla g(x)$ and $\nabla h(x)$.

One immediately concludes from Proposition 2.1:

Proposition 3.6. *Suppose that* $d_x G \neq 0$. *Then,*

(i) $0 \leq \rho_G(x) \leq 1$;

(ii) $\rho_G(x) = 0$ *if and only if* $\nabla g_1(x), \ldots, \nabla g_k(x)$ *are linearly dependent;*

(iii) $\rho_G(x) = 1$ *if and only if* $\nabla g_1(x), \ldots, \nabla g_k(x)$ *all have the same length and are pairwise orthogonal.*

Thus, $0 \leq \rho_G^{\text{inf}} \leq 1$, *and* $\rho_G^{\text{inf}} = 1$ *if and only if* G *is a simple Ł-map.*

In the case where $k = 2$, there is a very precise relation between $\rho_f(x)$ and the strong Ł-inequality.

Proposition 3.7. *There exist positive constants* α *and* β *such that, at all points* $x \in \mathcal{U}$ *such that* $d_x f \neq 0$,

$$\alpha \sigma_2(x) \leq \rho_f(x)\sqrt{|\nabla g(x)|^2 + |\nabla h(x)|^2} \leq \beta \sigma_2(x).$$

Hence, f *satisfies the strong Łojasiewicz inequality at* p *if and only if there exists an open neighborhood* $\mathcal{W}$ *of* p *in* $\mathcal{U}$, *and* $c, \theta \in \mathbb{R}$ *such that* $c > 0$, $0 < \theta < 1$, *and, for all* $x \in \mathcal{W}$ *such that* $d_x f \neq 0$,

$$|f(x) - f(p)|^\theta \leq c\rho_f(x)\sqrt{|\nabla g(x)|^2 + |\nabla h(x)|^2},$$

and, for all $x \in \mathcal{W}$ *such that* $d_x f = 0$, $f(x) = f(p)$.

Proof. Assume that $d_x f \neq 0$. Then, the first set of inequalities is trivially true if $\sigma_2(x) = 0$; so, assume $\sigma_2(x) \neq 0$. We need to show that we can find $\alpha, \beta > 0$ such that

$$\alpha \leq \frac{\rho_f(x)\sqrt{|\nabla g(x)|^2 + |\nabla h(x)|^2}}{\sigma_2(x)} \leq \beta.$$

This follows immediately from Item 2 of Proposition 2.1 since

$$\frac{\rho_f(x)\sqrt{|\nabla g(x)|^2 + |\nabla h(x)|^2}}{\sigma_2(x)} = \frac{2\sqrt{\sigma_1^2(x)\sigma_2^2(x)}}{\sigma_2(x)\sqrt{|\nabla g(x)|^2 + |\nabla h(x)|^2}}$$

$$= \frac{2\sigma_1(x)}{\sqrt{|\nabla g(x)|^2 + |\nabla h(x)|^2}}.$$

$\square$

Proposition 3.8. *The following are equivalent:*

(i) $\rho_G^{\inf} > 0$;

(ii) *there exists $b > 0$ such that, for all $x \in \mathcal{U}$, $b \cdot \sqrt{|\nabla g_1(x)|^2 + \cdots + |\nabla g_k(x)|^2} \leq \sigma_k(x)$;*

(iii) *there exists $b > 0$ such that, for all $x \in \mathcal{U}$, $b \cdot \sigma_1(x) \leq \sigma_k(x)$.*

Proof. Throughout this proof, when we take an infimum, we mean the infimum over all $x \in U$ such that $d_x G \neq 0$; this implies that $\sigma_1(x) \neq 0$.

Note that Items 2 and 3 simply say that

$$0 < \inf \frac{\sigma_k(x)}{\sqrt{|\nabla g_1(x)|^2 + \cdots + |\nabla g_k(x)|^2}}$$

and

$$0 < \inf \frac{\sigma_k(x)}{\sigma_1(x)}.$$

Now,

$$\rho_G(x) = \frac{k\,(\det M)^{1/k}}{|\nabla g_1(x)|^2 + \cdots + |\nabla g_k(x)|^2} = \frac{k\left(\sigma_1^2(x)\cdots\sigma_k^2(x)\right)^{1/k}}{|\nabla g_1(x)|^2 + \cdots + |\nabla g_k(x)|^2}$$

$$= k\left[\frac{\sigma_1(x)}{\sqrt{|\nabla g_1(x)|^2 + \cdots + |\nabla g_k(x)|^2}} \cdots \frac{\sigma_k(x)}{\sqrt{|\nabla g_1(x)|^2 + \cdots + |\nabla g_k(x)|^2}}\right]^{2/k},$$

and the factors are non-negative, at most 1, and are in decreasing order. This immediately yields the equivalence of Items 1 and 2.

The equivalence of Items 2 and 3 follows at once from

$$\frac{\sigma_k(x)}{\sqrt{|\nabla g_1(x)|^2 + \cdots + |\nabla g_k(x)|^2}} = \frac{\sigma_k(x)}{\sqrt{\sigma_1^2(x) + \cdots + \sigma_k^2(x)}}$$

$$= \frac{\sigma_k(x)/\sigma_1(x)}{\sqrt{1 + (\sigma_2(x)/\sigma_1(x))^2 + \cdots + (\sigma_k(x)/\sigma_1(x))^2}}.$$

$\square$

Our primary interest in Ł-weights is due to the following:

Theorem 3.9. *Suppose that G is real analytic and $\rho_G^{\inf} > 0$. Then, G is Ł-analytic.*

Proof. By Corollary 3.1, if x is near p and $d_x G = 0$, then $G(x) = G(p)$. Thus, it suffices to verify that the strong Łojasiewicz inequality holds near any point at which $d_x G \neq 0$. The desired inequality follows at once from Item 2 of Proposition 3.8, combined with Corollary 3.1. $\square$

Let $H := (h_1, \ldots, h_n)$ be a C^1 map from an open subset $\mathcal{W}$ of $\mathbb{R}^m$ to $\mathcal{U}$. For $x \in \mathcal{W}$, we let $A = A(H(x))$ be the matrix $[d_{H(x)} G]^t$, and set $B = B(x) := [d_x H]^t$. Let $C = C(x)$ denote the matrix $[d_x(G \circ H)]^t = BA$.

Theorem 3.10. *Suppose that $q \in \mathcal{W}$, that H has positive Ł-weight at q, and that G has positive Ł-weight at $H(q)$. Then, $G \circ H$ has positive Ł-weight at q.*

Proof. At a point $x \in \mathcal{W}$ where $C \neq 0$, Corollary 2.5 tells us that

$$\rho_{G \circ H}(x) \geq \frac{k\sigma_n^2(B(x))\sigma_k^2(A(H(x)))}{n\sigma_1^2(B(x))\sigma_1^2(A(H(x)))}.$$

By Item 3 of Proposition 3.8, there exists an open neighborhood $\mathcal{W}'$ of q on which the infimum b_H of $\sigma_n(B)/\sigma_1(B)$ is positive, and an open neighborhood $\mathcal{U}'$ of $H(q)$ on which the infimum b_G of $\sigma_k(A)/\sigma_1(A)$ is positive. Therefore, the infimum of $\rho_{G \circ H}(x)$ over all $x \in \mathcal{W}' \cap H^{-1}(\mathcal{U}')$ such that $C(x) \neq 0$ is at least $k b_H^2 b_G^2/n$. $\square$

Theorem 3.10 gives us an easy way of producing new non-simple Ł-analytic maps: take a map into $\mathbb{R}^2$ consisting of the real and imaginary parts of a holomorphic or anti-holomorphic function, and then compose with a real analytic change of coordinates on an open set in $\mathbb{R}^2$. Actually, we can give a much more precise result when H is a simple Ł-map.

Proposition 3.11. *Suppose that H is a simple Ł-map. Then, for all $x \in \mathcal{W}$ at which $d_x(G \circ H) \neq 0$, $\rho_{G \circ H}(x) = \rho_G(H(x))$.*

Proof.   Suppose that $d_x(G \circ H) \neq 0$. As H is a simple Ł-map, $B^t B = \lambda I_n$, where $\lambda = |\nabla h_1(x)|^2$. Thus,

$$\rho_{G \circ H}(x) = \frac{k \left[\det(A^t B^t B A) \right]^{1/k}}{\mathrm{tr}(A^t B^t B A)} = \frac{k \left[\lambda^k \det(A^t A) \right]^{1/k}}{\lambda \, \mathrm{tr}(A^t A)} = \frac{k \left[\det(A^t A) \right]^{1/k}}{\mathrm{tr}(A^t A)}$$

$$= \rho_G(H(x)).$$

$\square$

We give two quick examples of how to produce interesting Ł-analytic maps.

Example 3.12.   As we mentioned above, if g and h are the real and imaginary parts of a holomorphic or anti-holomorphic function, then $f = (g, h)$ is a simple Ł-analytic function.

One can also mix holomorphic and anti-holomorphic functions. Let $z = x + iy$ and $w = u + iv$, and consider the real and imaginary parts of $\bar{z}w^2$, i.e., let $g = x(u^2 - v^2) + 2yuv$ and $h = 2xuv - y(u^2 - v^2)$. It is trivial to verify that (g, h) is simple Ł-analytic.

More generally, if (g_1, h_1) and (g_2, h_2) are simple Ł-analytic, then the real and imaginary parts of

$$(g_1(\mathbf{z}) + i h_1(\mathbf{z}))(g_2(\mathbf{w}) + i h_2(\mathbf{w}))$$

will yield a new simple Ł-analytic function.

Example 3.13.   Suppose that $f = (g, h)$ is a simple Ł-analytic map.

Consider the linear map L of $\mathbb{R}^2$ given by $(u, v) \mapsto (au + bv, cu + dv)$. Then, at every point, the matrix of the derivative of L is $\begin{bmatrix} a & b \\ c & d \end{bmatrix}$, and one calculates that ρ_L is constantly $2|ad - bc|/(a^2 + b^2 + c^2 + d^2)$. Thus, if $ad - bc \neq 0$, then $\rho_L > 0$ and, unless $ab + cd = 0$ and $a^2 + c^2 = b^2 + d^2$, we have that $\rho_L < 1$.

Therefore, if L is an isomorphism which is not an orthogonal transformation composed with scalar multiplication, then Proposition 3.11 tells us that $(ag + bh, cg + dh)$ has positive Ł-weight less than one and, hence, is an Ł-analytic map which is not simple and, therefore, cannot arise from a holomorphic or anti-holomorphic complex function.

Relationship to Jacquemard's Conditions

For the remainder of this section, we will restrict ourselves to considering a real analytic map $f = (g, h)$ into $\mathbb{R}^2$. In [9], Jacquemard investigates such f satisfying three conditions. We will refer to these conditions as $J0$, $J1$ and $J2$;

actually, our condition $J2$ will be the weaker condition given by Ruas, Seade, and Verjovsky in [18].

Definition 3.14. The **Jacquemard conditions** are:

($J0$) the origin is an isolated critical point of f;

($J1$) there exist an open neighborhood $\mathcal{W}$ of 0 in $\mathcal{U}$ and a real number $\tau > 0$ such that, if $x \in \mathcal{W}$ is such that $\nabla g(x) \neq \mathbf{0}$ and $\nabla h(x) \neq \mathbf{0}$, then

$$\left| \frac{\nabla g(x)}{|\nabla g(x)|} \cdot \frac{\nabla h(x)}{|\nabla h(x)|} \right| \leq 1 - \tau;$$

($J2$) the real integral closures of the Jacobian ideals of g and h, inside the ring of real analytic germs at the origin, are equal.

The reader should note that we may discuss the Jacquemard conditions holding independently; in particular, we will assume in some settings that $J1$ and $J2$ hold, without assuming $J0$.

The importance of the Jacquemard conditions stems from the following theorem, which is essentially proved in [9], and improved using J2 in [18].

Theorem 3.15. *If Jacquemard's conditions hold, then f satisfies the strong Milnor condition, i.e., there exists $\epsilon_0 > 0$ such that, for all ϵ such that $0 < \epsilon \leq \epsilon_0$, the map $f/|f|$ from S_ϵ (the sphere of radius ϵ, centered at the origin) to $\mathbb{R}^2$ is a smooth, locally trivial fibration.*

We wish to see that, if we assume $J1$, greatly weaken $J2$, and omit $J0$, then f has positive Ł-weight at 0 and, hence, is Ł-analytic at 0. It will then follow from Corollary 5.6 that the Milnor fibration inside a ball, centered at 0, exists. First, we need a lemma.

Lemma 3.16. *Condition J2 implies that there exists an open neighborhood $\mathcal{W}$ of the origin in $\mathcal{U}$ in which ∇g and ∇h are comparable in magnitude, i.e., such that there exist $A, B > 0$ such that, for all $x \in \mathcal{W}$,*

$$A|\nabla g(x)| \leq |\nabla h(x)| \leq B|\nabla g(x)|.$$

Proof. By Proposition 4.2 of [6], J2 is equivalent to: there exists a neighborhood $\mathcal{W}$ of the origin in $\mathcal{U}$ and $C_1, C_2 > 0$ such that, at all points in $\mathcal{W}$, for all i such that $1 \leq i \leq n$, $|\partial g/\partial x_i| \leq C_1 \max_j |\partial h/\partial x_j|$ and $|\partial h/\partial x_i| \leq C_2 \max_j |\partial g/\partial x_j|$. One quickly concludes that, at all points of $\mathcal{W}$, $|\nabla g| \leq \sqrt{n} C_1 |\nabla h|$ and $|\nabla h| \leq \sqrt{n} C_2 |\nabla g|$. The lemma follows. $\qquad \square$

Proposition 3.17. *Suppose there exists an open neighborhood $\mathcal{W}$ of the origin in $\mathcal{U}$ in which ∇g and ∇h are comparable in magnitude. Then, Condition J1 is satisfied if and only if f has positive Ł-weight at 0.*

Proof. First, note that, when ∇g and ∇h are comparable in magnitude, then one of them equals zero at a point x if and only if both of them equal zero at x. Suppose that $A, B > 0$ are such that, for all $x \in W$,

$$A|\nabla g(x)| \leq |\nabla h(x)| \leq B|\nabla g(x)|.$$

Let $\eta(x)$ equal the angle between $\nabla g(x)$ and $\nabla h(x)$. Then,

$$\left| \frac{\nabla g(x)}{|\nabla g(x)|} \cdot \frac{\nabla h(x)}{|\nabla h(x)|} \right| = |\cos \eta(x)|,$$

and it is trivial to conclude that Condition J1 holds if and only if there exists an open neighborhood of the origin $W' \subseteq W$ such that $0 < \inf_{x \in W'} (\sin \eta(x))$.

Recall from Remark 4 that, at a point x where $|\nabla g(x)|^2 + |\nabla h(x)|^2 \neq 0$,

$$\rho_f(x) = \frac{2|\nabla g(x)||\nabla h(x)| \sin \eta(x)}{|\nabla g(x)|^2 + |\nabla h(x)|^2}.$$

Now, $A|\nabla g(x)| \leq |\nabla h(x)| \leq B|\nabla g(x)|$ implies that

$$\frac{2A}{1 + B^2} \leq \frac{2|\nabla g(x)||\nabla h(x)|}{|\nabla g(x)|^2 + |\nabla h(x)|^2} \leq \frac{2B}{1 + A^2}.$$

The conclusion follows. $\square$

Remark 5. In light of Lemma 3.16 and Proposition 3.17, we see that the Jacquemard conditions are far stronger than what one needs to conclude that f has positive Ł-weight at the origin.

4. Milnor's conditions (a) and (b)

In this section, we will discuss general conditions which allow us to conclude that Milnor fibrations exist.

We continue with our notation from the previous section, except that we now assume only that G is C^∞. For notational convenience, we restrict our attention to the case where the point p is the origin and where $G(\mathbf{0}) = \mathbf{0}$. We also assume that G is not locally constant near $\mathbf{0}$. Let $X := G^{-1}(\mathbf{0}) = V(G)$.

Let $\mathfrak{A} = \Sigma G$, so that $\mathfrak{A}$ is the closed set of points in $\mathcal{U}$ at which the gradients $\nabla g_1, \nabla g_2, \ldots \nabla g_k$ are linearly dependent. Let r denote the function given by the square of the distance from the origin, and let $\mathfrak{B}$ denote the closed set of points in $\mathcal{U}$ at which the gradients $\nabla r, \nabla g_1, \nabla g_2, \ldots \nabla g_k$ are linearly dependent. Of course, $\mathfrak{A} \subseteq \mathfrak{B}$.

We wish to give names to the submersive conditions necessary to apply Ehresmann's Theorem [5] (in the case of manifolds with boundary).

Definition 4.1. We say that the map G satisfies **Milnor's condition (a) at 0** (or that **0 is an isolated critical value of G near 0**) if and only if $0 \notin \overline{\mathfrak{A} - X}$, i.e., if $\Sigma G \subseteq V(G)$ near **0**.

We say that the map G satisfies **Milnor's condition (b) at 0** if and only if **0** is an isolated point of (or, is not in) $X \cap \overline{\mathfrak{B} - X}$.

If G satisfies Milnor's condition (a) (respectively, (b)), then we say that $\epsilon > 0$ is a **Milnor (a) (respectively, (b)) radius for G at 0** provided that $B_\epsilon \cap (\overline{\mathfrak{A} - X}) = \emptyset$ (respectively, $B_\epsilon \cap X \cap (\overline{\mathfrak{B} - X}) \subseteq \{0\}$). We say simply that $\epsilon > 0$ **is a Milnor radius for G at 0** if and only if ϵ is both a Milnor (a) and Milnor (b) radius for G at **0**.

Remark 6. Using our notation from the introduction, if $f_{\mathbf{c}}$ is complex analytic, then $f = (g, h)$ satisfies Milnor's conditions (a) and (b); in this case, Milnor's condition (a) is well-known and follows easily from a curve selection argument or from Corollary 1.2, and Milnor's condition (b) follows from the existence of good (or a_f) stratifications of $V(f)$ (see [7] and [10], and below).

Lemma 4.2. *Suppose that the map G satisfies Milnor's condition (b) at* **0** *and that $\epsilon_0 > 0$ is a Milnor (b) radius for G at* **0**. *Then, for all ϵ such that $0 < \epsilon < \epsilon_0$, there exists $\delta_\epsilon > 0$ such that the map*

$$H : \left(\mathring{B}_{\epsilon_0} - B_\epsilon \right) \cap G^{-1}\left(\mathring{B}_{\delta_\epsilon} - \{\mathbf{0}\} \right) \to \left(\mathring{B}_{\delta_\epsilon} - \{\mathbf{0}\} \right) \times \left(\epsilon^2, \epsilon_0^2 \right)$$

given by $H(x) = (G(x), |x|^2)$ is a proper submersion.

In particular, for all ϵ' such that $0 < \epsilon' < \epsilon_0$, there exists $\delta_{\epsilon'} > 0$ such that

$$G : \partial B_{\epsilon'} \cap G^{-1}\left(\mathring{B}_{\delta_{\epsilon'}} - \{\mathbf{0}\} \right) \to \mathring{B}_{\delta_{\epsilon'}} - \{\mathbf{0}\}$$

is a proper submersion.

Proof. That H is proper is easy. Let $\pi : \left(\mathring{B}_{\delta_\epsilon} - \{\mathbf{0}\} \right) \times (\epsilon^2, \epsilon_0^2) \to (\epsilon^2, \epsilon_0^2)$ denote the projection. Suppose that $C \subseteq \left(\mathring{B}_{\delta_\epsilon} - \{\mathbf{0}\} \right) \times (\epsilon^2, \epsilon_0^2)$ is compact. Then, $\pi(C)$ is compact, and $H^{-1}(C)$ is a closed subset of the compact set $\{x \in \mathring{B}_{\epsilon_0} \mid |x|^2 \in \pi(C)\}$. Thus, $H^{-1}(C)$ is compact.

Now, a critical point of H is precisely a point in $\mathfrak{B} \cap \left(\mathring{B}_{\epsilon_0} - B_\epsilon \right) \cap G^{-1}\left((\mathring{B}_{\delta_\epsilon} - \{\mathbf{0}\}) \times (\epsilon^2, \epsilon_0^2) \right)$. Suppose that, for all $\delta_\epsilon > 0$, such a point exists. Then, we would have a sequence $x_i \in (\mathfrak{B} - X) \cap \left(\mathring{B}_{\epsilon_0} - B_\epsilon \right)$ such that $G(x_i) \to 0$. As the x_i are in the compact set $B_{\epsilon_0} - \mathring{B}_\epsilon$, by taking a subsequence if necessary, we may assume that $x_i \to x \in B_{\epsilon_0} - \mathring{B}_\epsilon$. As $G(x_i) \to 0$, $x \in X$. Therefore, $x \in (B_{\epsilon_0} - \mathring{B}_\epsilon) \cap X \cap \overline{(\mathfrak{B} - X)}$; a contradiction, since ϵ_0 is a Milnor (b) radius. Hence, H is a submersion.

The last statement follows at once from this, since one need only pick an ϵ such that $0 < \epsilon < \epsilon'$ and apply that H is a submersion. $\square$

Theorem 4.3. *Suppose that G satisfies Milnor's conditions (a) and (b) at $\mathbf{0}$, and let ϵ_0 be a Milnor radius for G at $\mathbf{0}$.*

Then, for all ϵ such that $0 < \epsilon < \epsilon_0$, there exists $\delta_\epsilon > 0$ such that the map

$$G : B_\epsilon \cap G^{-1}\big(\mathring{B}_{\delta_\epsilon} - \{\mathbf{0}\}\big) \to \mathring{B}_{\delta_\epsilon} - \{\mathbf{0}\}$$

is a proper, stratified submersion and, hence, a locally trivial fibration, in which the local trivializations preserve the strata.

In addition, for all such $(\epsilon, \delta_\epsilon)$ pairs, for all δ such that $0 < \delta < \delta_\epsilon$, the map

$$G : B_\epsilon \cap G^{-1}\big(\partial B_\delta\big) \to \partial B_\delta$$

is a proper, stratified submersion and, hence, a locally trivial fibration, whose diffeomorphism-type is independent of the choice of such ϵ and δ.

It follows that, for all such $(\epsilon, \delta_\epsilon)$ pairs, for all δ such that $0 < \delta < \delta_\epsilon$, the map

$$G : \mathring{B}_\epsilon \cap G^{-1}\big(\partial B_\delta\big) \to \partial B_\delta$$

is a locally trivial fibration, whose diffeomorphism-type is independent of the choice of such ϵ and δ.

Proof. The map $G : B_\epsilon \cap G^{-1}\big(\mathring{B}_{\delta_\epsilon} - \{\mathbf{0}\}\big) \to \mathring{B}_{\delta_\epsilon} - \{\mathbf{0}\}$ is clearly proper. Milnor's condition (a) tells us that

$$G : \mathring{B}_\epsilon \cap G^{-1}\big(\mathring{B}_{\delta_\epsilon} - \{\mathbf{0}\}\big) \to \mathring{B}_{\delta_\epsilon} - \{\mathbf{0}\},$$

is a submersion regardless of the choice of $\delta_\epsilon > 0$.

The last line of Lemma 4.2 tells us that we may pick $\delta_\epsilon > 0$ such that the map

$$G : \partial B_\epsilon \cap G^{-1}\big(\mathring{B}_{\delta_\epsilon} - \{\mathbf{0}\}\big) \to \mathring{B}_{\delta_\epsilon} - \{\mathbf{0}\}$$

is a submersion.

Therefore,

$$G : B_\epsilon \cap G^{-1}\big(\mathring{B}_{\delta_\epsilon} - \{\mathbf{0}\}\big) \to \mathring{B}_{\delta_\epsilon} - \{\mathbf{0}\}$$

is a proper, stratified submersion, and so, by Ehresmann's Theorem (with boundary) [5] or Thom's first isotopy lemma [14], this map is a locally trivial fibration, in which the local trivializations preserve the strata.

It follows at once that for a fixed such ϵ, for all δ such that $0 < \delta < \delta_\epsilon$, the map

$$G : B_\epsilon \cap G^{-1}\big(\partial B_\delta\big) \to \partial B_\delta$$

is a proper, stratified submersion and, hence, a locally trivial fibration, whose diffeomorphism-type is independent of the choice of such δ.

It remains for us to show that, if we pick $0 < \epsilon < \epsilon' < \epsilon_0$, then there exists $\delta > 0$ such that $\delta < \min\{\delta_\epsilon, \delta_{\epsilon'}\}$, and such that the fibrations $G : B_\epsilon \cap G^{-1}(\partial B_\delta) \to \partial B_\delta$ and $G : B_{\epsilon'} \cap G^{-1}(\partial B_\delta) \to \partial B_\delta$ have the same diffeomorphism-type.

Let $\hat{\epsilon}$ be such that $0 < \hat{\epsilon} < \epsilon$, let $\delta_{\hat{\epsilon}} > 0$ be as in the first part of Lemma 4.2, and let $\delta > 0$ be less than $\min\{\delta_\epsilon, \delta_{\epsilon'}, \delta_{\hat{\epsilon}}\}$. Then, as the interval $(\hat{\epsilon}^2, \epsilon_0^2)$ is contractible, Lemma 4.2 tells us immediately that $G : B_\epsilon \cap G^{-1}(\partial B_\delta) \to \partial B_\delta$ and $G : B_{\epsilon'} \cap G^{-1}(\partial B_\delta) \to \partial B_\delta$ have the same diffeomorphism-type. $\quad\square$

Remark 7. One might hope that, if n and k are even, then Milnor's conditions (a) and (b) are satisfied by maps G which come from complex analytic maps. This is **not** the case.

Even in the nice case of a complex analytic isolated complete intersection singularity, the set of critical values would not locally consist solely of the origin, but would instead be a hypersurface in an open neighborhood of the origin in $\mathbb{C}^{k/2}$; see [12]. Thus, Milnor's condition (a) does not hold.

This means that the types of Milnor fibrations that we obtain when we assume Milnor's conditions (a) and (b) are extremely special.

Definition 4.4. If G satisfies Milnor's conditions (a) and (b) at $\mathbf{0}$, we refer to the two fibrations over spheres (or their diffeomorphism classes) in Theorem 4.3, $G : B_\epsilon \cap G^{-1}(\partial B_\delta) \to \partial B_\delta$ and $G : \overset{\circ}{B}_\epsilon \cap G^{-1}(\partial B_\delta) \to \partial B_\delta$, as the **compact Milnor fibration of G at 0 inside a ball** and the **Milnor fibration of G at 0 (inside a ball)**, respectively.

Corollary 4.5. *Suppose that G satisfies Milnor's conditions (a) and (b) at $\mathbf{0}$, and that $k > 1$. Then, G maps an open neighborhood of the origin onto an open neighborhood of the origin.*

Proof. Recall that $G(\mathbf{0}) = \mathbf{0}$ and that G is not locally constant by assumption. As $k > 1$, $\overset{\circ}{B}_{\delta_\epsilon} - \{\mathbf{0}\}$ is connected, and Theorem 4.3 implies that $G : \overset{\circ}{B}_\epsilon \cap G^{-1}(\overset{\circ}{B}_{\delta_\epsilon}) \to \overset{\circ}{B}_{\delta_\epsilon}$ is surjective. $\quad\square$

Definition 4.6. Suppose that G satisfies Milnor's conditions (a) and (b) at $\mathbf{0}$, and that ϵ is a Milnor radius for G at $\mathbf{0}$. If $\delta > 0$, then (ϵ, δ) is a **Milnor pair for G at 0** if and only if there exists a $\hat{\delta} > \delta$ such that $G : B_\epsilon \cap G^{-1}(\overset{\circ}{B}_{\hat{\delta}}^*) \to \overset{\circ}{B}_{\hat{\delta}}^*$ is a stratified submersion (which is, of course, smooth and proper).

Remark 8. Whenever we write that (ϵ, δ) is a Milnor pair for G at $\mathbf{0}$, we are assuming that G satisfies Milnor's conditions (a) and (b) at $\mathbf{0}$.

5. The main theorem

Now that we have finished our general discussion of Ł-analytic functions and Milnor's conditions, we wish to investigate how they are related to each other.

The following proposition is trivial to conclude.

Proposition 5.1. *If G is a C^1 Ł-map at p, then, near p, $\Sigma G \subseteq G^{-1}(G(p))$, i.e., G satisfies Milnor's condition (a) at p.*

Our goal is to prove that if G is Ł-analytic at p, then G also satisfies Milnor's condition (b) at p, for then Theorem 4.3 will tell us that the Milnor fibrations inside a ball exist.

For this purpose, we need a special type of auxiliary function.

Definition 5.2. Let $\mathcal{W}$ be an open neighborhood of the origin in $\mathbb{R}^k$. Let P be an infinite subset of $\mathbb{N}$ (the **admissible powers**). A function $M : P \times \mathcal{W} \to \mathbb{R}^k$, where $M(p, y)$ is written as $M_p(y)$, is a **normed analytic power map** (a **nap-map**) if and only if, for all $p \in P$,

   (i) M_p is real analytic;
  (ii)   for all $y \in \mathcal{W}$, $|M_p(y)| = |y|^p$;
 (iii)   the image of M_p contains an open neighborhood of the origin; and
  (iv)   there exists a constant $K_p > 0$ such that, if $M_p = (m_1, \ldots, m_k)$, then, for all $y \in \mathcal{W}$,

$$\max_{|(t_1,\ldots,t_k)|=1} \left| t_1 \nabla m_1(y) + \cdots + t_k \nabla m_k(y) \right| \leq K_p |y|^{p-1}.$$

Remark 9. It is easy to show that Items 1 and 2 in the above definition (or even replacing real analytic by C^1) imply that

$$\min_{|(t_1,\ldots,t_k)|=1} \left| t_1 \nabla m_1(y) + \cdots + t_k \nabla m_k(y) \right| \leq p|y|^{p-1}$$

$$\leq \max_{|(t_1,\ldots,t_k)|=1} \left| t_1 \nabla m_1(y) + \cdots + t_k \nabla m_k(y) \right|.$$

We will not use these inequalities.

Originally, we used multiplication in the reals, complex numbers, quaternions, and octonions to produce nap-maps from $\mathbb{R}^k$ to $\mathbb{R}^k$ when $k = 1, 2, 4,$ or 8. T. Gaffney produced the easier nap-maps below **for all** k.

Proposition 5.3. *For all k, there exists a nap-map from $\mathbb{R}^k$ to $\mathbb{R}^k$. In particular, if P is the set of odd natural numbers, then, for all k, the function $M : P \times \mathbb{R}^k \to \mathbb{R}^k$ given by $M(p, y) = M_p(y) = |y|^{p-1}y$ is a nap-map.*

Proof. Let P and M_p be as in the statement of the proposition. Since $p \in P$ is odd, Items 1, 2, and 3 in the definition of a nap-map are trivially satisfied. We need to prove the inequality in Item 4 of Definition 5.2.

Let $M_p = (m_1, \ldots, m_k)$. One calculates that

$$\frac{\partial m_i}{\partial y_j} = |y|^{p-1}\delta_{i,j} + (p-1)|y|^{p-3}y_i y_j,$$

where $\delta_{i,j}$ is the Kronecker delta function. Thus, considering y as a column-vector and denoting its transpose by y^t, the transpose of the derivative matrix of M_p at y is

$$L := |y|^{p-1}I + (p-1)|y|^{p-3}y^t y.$$

Now, suppose that T is a column vector of unit length in $\mathbb{R}^k$. Our goal is to show that $|LT| \leq K_p|y|^{p-1}$ for some constant K_p. We find

$$|LT| = \left| |y|^{p-1}T + (p-1)|y|^{p-3}y^t yT \right|$$
$$\leq \left| |y|^{p-1}T \right| + \left| (p-1)|y|^{p-3}y^t yT \right| \leq |y|^{p-1} + (p-1)|y|^{p-1},$$

where we use repeatedly in the last inequality that for matrices A and B, $|AB| \leq |A||B|$. Thus, $\displaystyle\max_{|(t_1,\ldots,t_k)|=1} \left| t_1 \nabla m_1(y) + \cdots + t_k \nabla m_k(y) \right| = |L| \leq p|y|^{p-1}.$ $\square$

We can now prove our main lemma.

Lemma 5.4. *Suppose that G is Ł-analytic at $\mathbf{0}$, and $G(\mathbf{0}) = \mathbf{0}$. Then, exists an open neighborhood $\mathcal{W}$ of $\mathbf{0}$ in $\mathcal{U}$ and a Whitney stratification $\mathfrak{S}$ of $\mathcal{W} \cap X$ such that, for all $S \in \mathfrak{S}$, the pair $(\mathcal{W} - X, S)$ satisfies Thom's a_G condition, i.e., if $p_j \to p \in S \in \mathfrak{S}$, where $p_j \in \mathcal{W} - X$, and $T_{p_j}G^{-1}G(p_j)$ converges to some linear subspace $\mathcal{T}$ in the Grassmanian of $(n-k)$-dimensional linear subspaces of $\mathbb{R}^n$, then $T_p S \subseteq \mathcal{T}$.*

Proof. This proof follows that of Theorem 1.2.1 of [7].

Let $\mathcal{W}$, c, and θ be as in Definition 3.3: $\mathcal{W}$ is an open neighborhood of $\mathbf{0}$ in $\mathcal{U}$, and $c, \theta \in \mathbb{R}$ are such that $c > 0, 0 < \theta < 1$, and, for all $x \in \mathcal{W}$,

$$|G(x))|^\theta \leq c \cdot \min_{|(a_1,\ldots,a_k)|=1} \left| a_1 \nabla g_1(x) + \cdots + a_k \nabla g_k(x) \right| = c\sigma_k(x).$$

Let M be a nap-map on a neighborhood $\mathcal{V}$ of the origin in $\mathbb{R}^k$ (which exist by Proposition 5.3), and let $P \subseteq \mathbb{N}$ denote the set of admissible powers. Let $\pi \in P$ be such that $\pi > 1/(1-\theta)$, so that the image of M_π contains a neighborhood of the origin. Let $M_\pi = (m_1, \ldots, m_k)$.

Consider the real analytic map $\widehat{G} : \mathcal{W} \times \mathcal{V} \to \mathbb{R}^k$ given by

$$\widehat{G}(x, y) := (g_1(x) + m_1(y), \ldots, g_k(x) + m_k(y)).$$

Let $\widehat{\mathfrak{S}}$ be a Whitney stratification of $V(\widehat{G})$ such that $V(G) \times \{0\}$ is a union of strata. Let $\mathfrak{S} := \{S \mid S \times \{0\} \in \widehat{\mathfrak{S}}\}$. We claim that $\mathcal{W}$ and $\mathfrak{S}$ satisfy the conclusion of the lemma.

It will be convenient to deal with the conormal formulation of the a_G condition. Suppose that we have a sequence of points $p_j \to p \in S \in \mathfrak{S}$, where $p_j \in \mathcal{W} - X$, and a sequence ${}^j\mathbf{a} := ({}^ja_1, \ldots {}^ja_k) \in \mathbb{R}^k$ such that ${}^ja_1 d_{p_j} g_1 + \cdots + {}^ja_k d_{p_j} g_k$ converges to a cotangent vector η (in the fiber of $T^*\mathcal{W}$ over p). We wish to show that $\eta(T_p S) \equiv 0$.

If $\eta \equiv 0$, there is nothing to show; so we assume that $\eta \not\equiv 0$.

We may assume that, for all j, ${}^j\mathbf{a} \neq \mathbf{0}$. Also, as $p_j \to p \in V(G)$, we may assume that, for all j, $-G(p_j)$ is in the image of M_π. Let ${}^j\mathbf{u} := ({}^ju_1, \ldots, {}^ju_k) \in M_\pi^{-1}(-G(p_j))$, so that $q_j := (p_j, {}^j\mathbf{u}) \in V(\widehat{G})$. Since $G(p) = 0$ and $\widehat{G}(q_j) = 0$, $M_\pi({}^j\mathbf{u}) \to \mathbf{0}$; by Item 2 in the definition of a nap-map, it follows that ${}^j\mathbf{u} \to \mathbf{0}$, and so $q_j \to (p, \mathbf{0}) \in S \times \{0\} \in \widehat{\mathfrak{S}}$.

By taking a subsequence, if necessary, and using that $\eta \not\equiv 0$, we may assume that the projective class

$$\left[{}^ja_1 d_{q_j}(g_1 + m_1) + \cdots + {}^ja_k d_{q_j}(g_k + m_k) \right]$$

converges to some $[\omega]$, where $\omega = b_1 dx_1 + \ldots b_n dx_n + s_1 dy_1 + \cdots + s_k dy_k \neq 0$. By Whitney's condition (a), we have that $\omega(T_p S \times \{0\}) \equiv 0$.

Let us reformulate part of our discussion above using vectors, instead of covectors. In terms of vectors, we are assuming that ${}^ja_1 \nabla g_1(p_j) + \cdots + {}^ja_k \nabla g_k(p_j) \to \mathbf{w} \neq \mathbf{0}$, and that

$$\left[{}^ja_1\left(\nabla g_1(p_j), \nabla m_1({}^j\mathbf{u}) \right) + \cdots + {}^ja_k\left(\nabla g_k(p_j), \nabla m_k({}^j\mathbf{u}) \right) \right]$$
$$\to [(b_1, \ldots, b_n, s_1, \ldots, s_k)] \neq \mathbf{0}.$$

If we could show that the projective classes $[\mathbf{w} \times \{0\}]$ and $[(b_1, \ldots, b_n, s_1, \ldots, s_k)]$ are equal, then we would be finished.

To show that $[\mathbf{w} \times \{0\}]$ and $[(b_1, \ldots, b_n, s_1, \ldots, s_k)]$ are equal, it clearly suffices to show that

$$\frac{\left| {}^ja_1 \nabla m_1({}^j\mathbf{u}) + \cdots + {}^ja_k \nabla m_k({}^j\mathbf{u}) \right|}{\left| {}^ja_1 \nabla g_1(p_j) + \cdots + {}^ja_k \nabla g_k(p_j) \right|} \to 0.$$

Dividing the numerator and denominator by $|{}^j\mathbf{a}|$, we see that we may assume that $|{}^j\mathbf{a}| = 1$.

Using that G is an Ł-map at $\mathbf{0}$ and Item 4 in the definition of a nap-map, we obtain that

$$\frac{\left| {}^ja_1 \nabla m_1({}^j\mathbf{u}) + \cdots + {}^ja_k \nabla m_k({}^j\mathbf{u}) \right|}{\left| {}^ja_1 \nabla g_1(p_j) + \cdots + {}^ja_k \nabla g_k(p_j) \right|} \leq \frac{K_p |{}^j\mathbf{u}|^{\pi-1}}{|G(p_j)|^\theta}.$$

Now, $|^j\mathbf{u}|^\pi = |M_\pi(^j\mathbf{u})| = |G(p_j)|$. Thus, $|^j\mathbf{u}|^{\pi-1} = |G(p_j)|^{(\pi-1)/\pi}$, and we would like to show that

$$\frac{|G(p_j)|^{(\pi-1)/\pi}}{|G(p_j)|^\theta} \to 0.$$

However, this follows at once, since $G(p_j) \to \mathbf{0}$ and, by our choice of π, $(\pi - 1)/\pi > \theta$. $\qquad\square$

Remark 10. In Lemma 5.4, we assumed that G was real analytic and used that a normed, analytic power map M_p exists on $\mathbb{R}^k$. We assumed analyticity for both maps so that $V(\widehat{G})$ would have a Whitney stratification. However, it is enough to assume that G and M_p are *subanalytic*. See [8].

The main theorem follows easily. Note that, when $k = 1$, Theorem 1.1 implies that, if G is real analytic, then G is Ł-analytic.

Theorem 5.5. *Suppose that G is Ł-analytic at p. Then, Milnor's conditions (a) and (b) hold at p.*

Proof. We assume, without loss of generality, that $p = \mathbf{0}$ and $G(\mathbf{0}) = \mathbf{0}$.

Milnor's condition (a) is immediate, as we stated in Proposition 5.1. Milnor's condition (b) follows from Lemma 5.4 by using precisely the argument of Lê in [10].

Let $\epsilon_0 > 0$ be a Milnor (a) radius for G at $\mathbf{0}$ such that, for all ϵ' such that $0 < \epsilon' \leq \epsilon_0$, the sphere $S_{\epsilon'}$ transversely intersects all of the strata of the Whitney a_G stratification whose existence is guaranteed by Lemma 5.4. We claim that ϵ_0 is a Milnor (b) radius.

Suppose not. Then, there would be a sequence of points $p_i \in B_{\epsilon_0} - V(G)$ such that $G(p_i) \to 0$ and $T_{p_i}(G^{-1}G(p_i)) \subseteq T_{p_i}S_{\epsilon_i}$, where ϵ_i denotes the distance of p_i from the origin (we use Milnor's condition (a) here to know that $G^{-1}G(p_i)$ is smooth). By taking a subsequence, we may assume that the p_i approach a point $p \in S_\epsilon \cap X$ and that $T_{p_i}(G^{-1}G(p_i))$ approaches a limit $\mathcal{T}$. If we let M denote the stratum of our a_G stratification which contains p, then we find that $T_pM \subseteq \mathcal{T}$ and $\mathcal{T} \subseteq T_pS_\epsilon$. As S_ϵ transversely intersects M, we have a contradiction, which proves the desired result. $\qquad\square$

Corollary 5.6. *Suppose that G is Ł-analytic at p. Then, the Milnor fibrations for $G - G(p)$ inside a ball, centered at p, exist.*

Proof. As we saw in Section 4, this follows immediately from Milnor's conditions (a) and (b). $\qquad\square$

We include the following just to emphasize that Ł-analytic have many properties that one associates with complex analytic functions.

Corollary 5.7. *Suppose that G is an Ł-analytic function which is nowhere locally constant, and that $k \geq 2$. Then, G is an open map.*

Proof.   This follows at once from Theorem 5.5 and Corollary 4.5.   □

6. Comments and questions

In many places throughout this paper, we assumed that G was real analytic. However, as we mentioned in Remark 10, we do not actually need analyticity to conclude Theorem 5.5, subanalyticity is enough. We can also ask:

Question 1. If we assume that G satisfies the strong Łojasiewicz inequality, do we, in fact, need to assume that G is subanalytic in order to obtain the existence of the Milnor fibration inside a ball?

We have many other basic questions.

Question 2. Are there real analytic maps which are Ł-maps but do not have positive Ł-weight?

We suspect that the answer to the above question is "yes".
Given Theorem 3.10, it is natural to ask:

Question 3. Is the composition of two Ł-maps an Ł-map? What if the maps are Ł-analytic?

Considering Example 3.12, we are led to ask:

Question 4. If one takes two simple Ł-maps, in disjoint variables, into $\mathbb{R}^4$ or $\mathbb{R}^8$ and multiplies them using quaternionic or octonionic multiplication (analogous to the complex situation in Example 3.12), does one obtain a new simple Ł-map?

We feel certain that the answer to the above question is "yes", but we have not actually verified this.

It is not difficult to produce examples of real analytic $f = (g, h)$ which possess Milnor fibrations inside balls, and yet are **not** Ł-analytic. It is our hope that the Ł-analytic condition is strong enough to allow one to conclude that the Milnor fibration inside an open ball is equivalent to the Milnor fibration **on the sphere** (the fibration analogous to $f_c/|f_c| : S_\epsilon - S_\epsilon \cap X \to S^1 \subseteq \mathbb{C}$, which exists when f_c is complex analytic).

As we discussed in the introduction, in the real analytic setting, $f/|f|$ does not necessarily yield the projection map of a fibration from $S_\epsilon - S_\epsilon \cap X$ to S^1; when this map is, in fact, a locally trivial fibration (for all sufficiently small ϵ), one says that f satisfies the **strong Milnor condition** at the origin (see VII.2 of [20]).

It is not terribly difficult to try to mimic Milnor's proof that the Milnor fibration inside the ball and the one on the sphere are diffeomorphic; one wants to integrate an appropriate vector field. In the case where $f = (g, h)$ is real analytic and satisfies Milnor's conditions (a) and (b), one finds that one needs a further condition:

Definition 6.1. Let $\omega := -h\nabla g + g\nabla h$, and assume that $f(0) = 0$. We say that f **satisfies Milnor's condition (c) at 0** if and only if there exists an open neighborhood $\mathcal{W}$ of 0 in $\mathcal{U}$ such that, at all points $\mathbf{x}$ of $\mathcal{W} - X$, if $\mathbf{x}$ is a linear combination of ∇g and ∇h, then

$$|\omega(\mathbf{x})|^2 \big(\mathbf{x} \cdot \nabla |f|^2(\mathbf{x})\big) > \big(\nabla |f|^2(\mathbf{x}) \cdot \omega(\mathbf{x})\big)\big(\mathbf{x} \cdot \omega(\mathbf{x})\big).$$

It is not too difficult to show that, if f is a simple Ł-analytic function, there exists an open neighborhood $\mathcal{W}$ of 0 in $\mathcal{U}$ such that, at all points $\mathbf{x}$ of $\mathcal{W} - X$, $\mathbf{x} \cdot \nabla |f|^2(\mathbf{x}) > 0$. It is also trivial that a simple Ł-analytic function has $\nabla |f|^2(\mathbf{x}) \cdot \omega(\mathbf{x}) = 0$. Milnor's condition (a) implies that $\mathcal{W}$ can be chosen so that $\omega(\mathbf{x}) \neq 0$ for $\mathbf{x} \in \mathcal{W} - X$. Thus, we conclude that simple Ł-analytic functions satisfy the strong Milnor condition.

However, the real question is:

Question 5. Do general Ł-analytic maps into $\mathbb{R}^2$ satisfy Milnor's condition (c), and so satisfy the strong Milnor condition? What about real analytic maps with positive Ł-weight? In light of Proposition 3.17, what about real analytic maps with positive Ł-weight where ∇g and ∇h have comparable in magnitude?

Finally, having seen in this paper that Ł-analytic pairs of functions share some important properties with complex analytic functions, one is led to ask a very general question:

Question 6. Do Ł-analytic maps form an interesting class of functions to study for reasons having nothing to do with Milnor fibrations?

References

1. Bierstone, E. and Milman, P. Semianalytic and subanalytic sets. *Publ. math. de I.H.É.S.*, **67** (1988) 5–42.

2. Cisneros-Molina, J. L. and Araújo dos Santos, R. N. About the Existence of Milnor Fibrations. In this volume, 2010.
3. Dimca, A. *Singularities and Topology of Hypersurfaces*. Universitext. Springer-Verlag 1992.
4. dos Santos, R. Topological triviality of families of real isolated singularities and their Milnor fibrations. *Math. Scand.*, **96(1)** (2005) 96–106.
5. Ehresmann, C. Les connexions infinitésimales dans un espace fibré différentiable. *Colloque de Topologie, Bruxelles* (1950) 29–55.
6. Gaffney, T. Integral closure of modules and Whitney equisingularity. *Invent. Math.*, **107** (1992) 301–322.
7. Hamm, H. and Lê D. T. Un théorème de Zariski du type de Lefschetz. *Ann. Sci. Éc. Norm. Sup.*, **6** (series 4) (1973) 317–366.
8. Hardt, R. Topological properties of subanalytic sets. *Trans. AMS* **211** (1975) 57–70.
9. Jacquemard, A. Fibrations de Milnor pour des applications réelles. *Boll. Un. Mat. Ital.B. Serie VII*, **3** (1989) 591–600.
10. Lê, D. T. Some remarks on Relative Monodromy. In P. Holm, editor, *Real and Complex Singularities, Oslo 1976*, pp. 397–404. Nordic Summer School/NAVF, 1977.
11. Łojasiewicz, S. *Ensemble semi-analytiques*. IHES Lecture notes, 1965.
12. Looijenga, E. *Isolated Singular Points on Complete Intersections*. Cambridge Univ. Press, 1984.
13. Massey, D. *Lê Cycles and Hypersurface Singularities* **1615** *Lecture Notes in Math*. Springer-Verlag, 1995.
14. Mather, J. Notes on Topological Stability. Notes from Harvard Univ., 1970.
15. Milnor, J. *Singular Points of Complex Hypersurfaces*, **77** *Annals of Math. Studies*. Princeton Univ. Press, 1968.
16. Pichon, A. and Seade, J. Real singularities and open-book decompositions of the 3-sphere. *Annales de la faculté des sciences de Toulouse, Sér. 6*, **12(2)** (2003) 245–265.
17. Randell, R. On the Topology of Non-isolated Singularities. *Proc. Georgia Top. Conf., Athens, Ga.* (1979) 445–473.
18. Ruas, M. A. S., Seade, J., and Verjovsky, A. *On real singularities with a Milnor fibration*, pp. 191–213. Birkhäuser, 2002.
19. Ruas, M. and dos Santos, R. Real Milnor fibrations and (c)-regularity. *Manuscripta Math.* **17(2)** (2005) 207–218.
20. Seade, J. *On the topology of isolated singularities in analytic spaces*, **241** *Progress in Mathematics*. Birkhäuser, 2005.

David Massey
Department of Mathematics
Northeastern University
Boston, MA 02115
USA
masseydb@gmail.com

20

An estimate of the degree of $\mathcal{L}$-determinacy by the degree of $\mathcal{A}$-determinacy for curve germs

T. NISHIMURA

Abstract

We give an estimate of the degree of $\mathcal{L}$-determinacy by the degree of $\mathcal{A}$-determinacy for both of C^∞ curve germs and holomorphic curve germs simultaneously. In the complex case, our result is an effective version of Gaffney's rigidity theorem on finite determinacy of curve germs.

1. Introduction

Let $\mathbb{K}$ be $\mathbb{R}$ or $\mathbb{C}$. Two map-germs $f, g : (\mathbb{K}^n, 0) \to (\mathbb{K}^p, 0)$ are said to be $\mathcal{A}$-*equivalent* if there exist germs of C^∞ diffeomorphisms or biholomorphic map-germs $h : (\mathbb{K}^n, 0) \to (\mathbb{K}^n, 0)$ and $H : (\mathbb{K}^p, 0) \to (\mathbb{K}^p, 0)$ such that $H \circ f = g \circ h$, and $\mathcal{L}$-*equivalent* if there exists a germ of C^∞ diffeomorphism or biholomorphic map-germ $H : (\mathbb{K}^p, 0) \to (\mathbb{K}^p, 0)$ such that $H \circ f = g$. For a C^∞ or holomorphic map-germ $f : (\mathbb{K}^n, 0) \to (\mathbb{K}^p, 0)$, the *r-jet of f at the origin* (denoted by $j^r f(0)$) is the Maclaurin series of f up to order r, and the least order of non-zero terms of the Maclaurin series of f is called the *multiplicity of f*. For $\mathcal{G} = \mathcal{A}$ or $\mathcal{L}$, a C^∞ or holomorphic map-germ $f : (\mathbb{K}^n, 0) \to (\mathbb{K}^p, 0)$ is said to be *r-$\mathcal{G}$-determined* if f is $\mathcal{G}$-equivalent to any C^∞ or holomorphic map-germ g with $j^r f(0) = j^r g(0)$ (the least such r is called the *degree of $\mathcal{G}$-determinacy of f* and is denoted by $\deg(f, \mathcal{G})$); and *finitely $\mathcal{G}$-determined* if f is r-$\mathcal{G}$-determined for some integer r.

In his celebrated thesis [5], Gaffney showed a rigidity theorem on finite determinacy which states that any finitely $\mathcal{A}$-determined holomorphic map-germ $f : (\mathbb{C}^n, 0) \to (\mathbb{C}^p, 0)$ is finitely $\mathcal{L}$-determined if $p \geq 2n$; and hence the

2000 *Mathematics Subject Classification* 58K40.

problem to obtain effective versions of it, which has been posed by Wall (p. 512 of [7]), is an important problem. In this short note, as a partial answer to this problem, we give an estimate of the degree of $\mathcal{L}$-determinacy by the degree of $\mathcal{A}$-determinacy for both of C^∞ curve germs and holomorphic curve germs simultaneously.

Theorem 1.1. *Let $f : (\mathbb{K}, 0) \to (\mathbb{K}^p, 0)$ $(p \geq 2)$ be an r-$\mathcal{A}$-determined map-germ. Then, we have the following.*

(i) *The map-germ f is r-$\mathcal{L}$-determined if the multiplicity of f is less than 3.*

(ii) *The map-germ f is $(2r^2 - 6r + 3)$-$\mathcal{L}$-determined if the multiplicity of f is more than or equal to 3.*

Note that our estimate depends only on r. This might be a somewhat surprising fact since in general $T\mathcal{L}(f)$ is a proper subset of $T\mathcal{A}(f)$ for a finitely $\mathcal{A}$-determined curve germ f although $TC(f)$ is always equal to $TK(f)$ for a finitely $\mathcal{K}$-determined curve germ f (for the definitions of $T\mathcal{L}(f)$, $T\mathcal{A}(f)$, $TC(f)$ and $TK(f)$, see [7]).

On the other hand, note also that our estimate in the assertion 2 of Theorem 1.1 is not the best possible in general since for instance we have that $\deg(f, \mathcal{L}) = \deg(f, \mathcal{A}) = 2i - 1$ for the curve germ $f(x) = (x^i, x^{i+1}, \ldots, x^{2i-1}, 0, \ldots, 0)$ $(3 \leq i \leq p)$. Even if we consider only finitely $\mathcal{A}$-determined curve germs f such that $\deg(f, \mathcal{L}) \neq \deg(f, \mathcal{A})$ our estimate is still not the best in general. For instance, the curve germs $g(x) = (x^{p+1}, x^{p+2}, \ldots, x^{2p})$ and $\widetilde{g}(x) = (x^i, x^{i+1})$ $(i \geq 4)$ are examples under this restriction. For the curve germ g, we have that $\deg(g, \mathcal{L}) = 2p + 1$, $\deg(g, \mathcal{A}) = 2p$ and $2(2p)^2 - 6(2p) + 3 > 2p + 1$ since $p \geq 2$. For the curve germ $\widetilde{g}$, we have that $\deg(\widetilde{g}, \mathcal{L}) = i(i - 1) - 1$ and $\deg(\widetilde{g}, \mathcal{A}) = i(i - 2) - 1$ since $i \geq 4$. We see that $2\deg(\widetilde{g}, \mathcal{A})^2 - 6\deg(\widetilde{g}, \mathcal{A}) + 3 > \deg(\widetilde{g}, \mathcal{L})$.

2. Proof of Theorem 1.1

In the following, we prove Theorem 1.1 in the case that $\mathbb{K} = \mathbb{R}$. By replacing the Malgrange preparation theorem with the Weierstrass preparation theorem, we have the proof of Theorem 1.1 in the case that $\mathbb{K} = \mathbb{C}$ (for the Malgrange preparation theorem and the Weierstrass preparation theorem, see [3]).

Before giving the proof, we prepare several notations and one elementary lemma. Let $\mathcal{E}_q$ be the $\mathbb{R}$-algebra of C^∞ function-germs $(\mathbb{R}^q, 0) \to \mathbb{R}$ with usual operations, m_q^k be the ideal of $\mathcal{E}_q$ such that $\varphi \in m_q^k$ if and only if $j^{k-1}\varphi(0) = 0$. For a C^∞ map-germ $g : (\mathbb{R}^n, 0) \to (\mathbb{R}^q, 0)$ and an ideal $\mathcal{I}$ of $\mathcal{E}_n$, we set $g^*m_q^k = \{\varphi \circ g \mid \varphi \in m_q^k\}$ and let $g^*m_q^k\mathcal{I}$ be the ideal of $\mathcal{E}_n$ generated by products of elements of $g^*m_q^k$ and $\mathcal{I}$.

Lemma 2.1. *Let* $g : (\mathbb{R}^n, 0) \to (\mathbb{R}^q, 0)$ *be a* C^∞ *map-germ and* $h : (\mathbb{R}^n, 0) \to (\mathbb{R}^n, 0)$, $H : (\mathbb{R}^q, 0) \to (\mathbb{R}^q, 0)$ *be germs of* C^∞ *diffeomorphisms. Suppose that the inclusion* $m_n^k \subset g^* m_q^\ell$ *holds. Then, the inclusion* $m_n^k \subset \left(H \circ g \circ h^{-1}\right)^* m_q^\ell$ *holds.*

[*Proof of Lemma 2.1*]. Let ψ be an element of m_n^k. Then, $\psi \circ h$ is also an element of m_n^k. Thus, by the assumption, there exists an element $\varphi \in m_q^\ell$ such that $\psi \circ h = \varphi \circ g$. Since H is a germ of C^∞ diffeomorphism, we have that $\varphi \circ H^{-1} \in m_q^\ell$. Therefore, we have that $\psi = \left(\varphi \circ H^{-1}\right) \circ \left(H \circ g \circ h^{-1}\right) \in \left(H \circ g \circ h^{-1}\right)^* m_q^\ell$. $\qquad\square$

Now, we start to prove Theorem 1.1. For the given $f : (\mathbb{R}, 0) \to (\mathbb{R}^p, 0)$, by composing appropriate germs of C^∞ diffeomorphisms $\widetilde{h} : (\mathbb{R}, 0) \to (\mathbb{R}, 0)$ and $\widetilde{H}^{-1} : (\mathbb{R}^p, 0) \to (\mathbb{R}^p, 0)$ to f, the map-germ f can be transformed to $\widetilde{f} = \widetilde{H}^{-1} \circ f \circ \widetilde{h}$ which has the following form:

$$\widetilde{f}(x) = \left(x^\delta, f_2(x), \dots, f_p(x)\right),$$

where δ is the multiplicity of f and $f_j(x) = o(x^\delta)$ for $j = 2, \dots, p$.

[*Proof of the assertion 1 of Theorem 1.1*]. In the case that $\delta = 1$, f is a germ of immersion and thus 1-$\mathcal{G}$-determined for $\mathcal{G} = \mathcal{A}$ and $\mathcal{L}$. Thus we assume that $\delta = 2$ in the following. Since $p \geq 2$, there must exist an integer $i \geq 2$ such that $2i - 1 \leq r$ and $\widetilde{f}$ is $(2i - 1)$-$\mathcal{A}$-determined but not $(2i - 2)$-$\mathcal{A}$-determined. We see easily that

$$m_1^{2i} \subset \widetilde{f}^* m_p^2 + \widetilde{f}^* m_p m_1^{2i}.$$

Thus, by the Malgrange preparation theorem we have that $m_1^{2i} \subset \widetilde{f}^* m_p^2$. Therefore, by Lemma 2.1 we have that

$$m_1^{2i} \subset f^* m_p^2.$$

Let g be a C^∞ map-germ such that $j^{2i-1} f(0) = j^{2i-1} g(0)$. Then, as in [4], since any component of $g - f$ belongs to m_1^{2i}, from the above inclusion we see that there exists a germ of C^∞ diffeomorphism $H : (\mathbb{R}^p, 0) \to (\mathbb{R}^p, 0)$ such that $g = f + (g - f) = H \circ f$. Therefore, in the case that $\delta = 2$ we see that any r-$\mathcal{A}$-determined curve germ $f : (\mathbb{R}, 0) \to (\mathbb{R}^p, 0)$ $(p \geq 2)$ is r-$\mathcal{L}$-determined. $\qquad\square$

[*Proof of the assertion 2 of Theorem 1.1*]. Let k be the integer such that $r \leq k\delta \leq r + \delta - 1$ and let $G : (\mathbb{R}, 0) \to (\mathbb{R}^p, 0)$ be the map-germ given by $G(x) = \left(x^\delta, x^{k\delta+1}, 0, \dots, 0\right)$.

Lemma 2.2. *The inclusion* $m_1^{(\delta-1)k\delta} \subset G^*m_p^2$ *holds.*

[*Proof of Lemma 2.2*]. Since δ and $k\delta + 1$ are relatively prime, by Sylvester's duality on the numerical semigroup generated by δ and $k\delta + 1$ ([6], see also [1], comment on 1999-8 of [2]) we see that for any integer $N \geq (\delta - 1)k\delta$ there exist non-negative integers ℓ_1, ℓ_2 such that $N = \ell_1\delta + \ell_2(k\delta + 1)$. Since $\delta \geq 3$ we have that $k\delta + 1 < (\delta - 1)k\delta$, which implies that $\ell_1 + \ell_2 \geq 2$ for the above ℓ_1, ℓ_2. Therefore, we have that

$$m_1^{(\delta-1)k\delta} \subset G^*m_p^2 + G^*m_p m_1^{(\delta-1)k\delta}.$$

Thus, by using the Malgrange preparation theorem we have that $m_1^{(\delta-1)k\delta} \subset G^*m_p^2$. $\square$

Let $F : (\mathbb{R}, 0) \to (\mathbb{R}^p, 0)$ be the map-germ given by

$$F(x) = \left(x^\delta, f_2(x) + x^{k\delta+1}, f_3(x), \ldots, f_p(x)\right).$$

Lemma 2.3. *For any germ of C^∞ diffeomorphism $h : (\mathbb{R}, 0) \to (\mathbb{R}, 0)$, the inclusion* $m_1^{(\delta-1)k\delta} \subset \left(F \circ h, \tilde{f}\right)^* m_{2p}^2$ *holds.*

[*Proof of Lemma 2.3*]. Since h is a germ of C^∞ diffeomorphism, $(h, id._{\mathbb{R}})^* : (F, F - G)^* m_{2p}^2 \to (F \circ h, F - G)^* m_{2p}^2$ is a linear isomorphism of vector spaces, where $id._{\mathbb{R}}(x) = x$. Consider the restriction of $(h, id._{\mathbb{R}})^*$ to the finite dimensional vector subspace $V = \{\psi \in (F, F - G)^* m_{2p}^2 \mid \psi \in m_1^{(\delta-1)k\delta}$ is a polynomial with degree less than $(\delta - 1)k\delta + \delta\}$. Since h is a germ of C^∞ diffeomorphism, the image of the restriction is $V_h = \{\psi \in (F \circ h, F - G)^* m_{2p}^2 \mid \psi \in m_1^{(\delta-1)k\delta}, j^{(\delta-1)k\delta+\delta-1}\psi(0) \neq 0\}$. Since $G^*m_p^2 \subset (F, F - G)^*m_{2p}^2$, by Lemma 2.2 we see that $\dim_{\mathbb{R}} V = \delta$. Therefore, we see that $\dim_{\mathbb{R}} V_h = \delta$ and thus we have the following inclusion:

$$m_1^{(\delta-1)k\delta} \subset (F \circ h, F - G)^* m_{2p}^2 + m_1^{(\delta-1)k\delta+\delta}.$$

Since $(F \circ h, F - G)^* m_{2p}\mathcal{E}_1 = m_1^\delta$ and $(F \circ h, F - G)^*m_{2p}^2 \subset (F \circ h, \tilde{f})^*m_{2p}^2$, by using the Malgrange preparation theorem again we have the desired inclusion. $\square$

Lemma 2.4. *The inclusion* $m_1^{(\delta-1)k\delta} \subset f^*m_p^2$ *holds.*

[*Proof of Lemma 2.4*]. Since $\tilde{f}$ is r-$\mathcal{A}$-determined and $j^r\tilde{f}(0) = j^r F(0)$ there exist germs of C^∞ diffeomorphisms $h : (\mathbb{R}, 0) \to (\mathbb{R}, 0)$ and $H : (\mathbb{R}^p, 0) \to (\mathbb{R}^p, 0)$ such that $H \circ \tilde{f} = F \circ h$. Let ψ be an element of $m_1^{(\delta-1)k\delta}$. Then, by Lemma 2.3, there must exist an element $\Psi \in m_{2p}^2$ such that $\psi = \Psi \circ (F \circ h, \tilde{f}) = \Psi \circ (H \circ \tilde{f}, \tilde{f})$. Thus, we have that $m_1^{(\delta-1)k\delta} \subset \tilde{f}^*m_p^2$. Therefore, by Lemma 2.1, we have that $m_1^{(\delta-1)k\delta} \subset f^*m_p^2$. $\square$

Lemma 2.5. *The inequality $\delta \leq r - 1$ holds.*

[*Proof of Lemma 2.5*]. Suppose that $\delta > r$. Then, since the map-germ $\widetilde{f}$ is r-$\mathcal{A}$-determined and $f_j(x) = o(x^\delta)$ for $j = 2, \ldots, p$, $\widetilde{f}$ must be $\mathcal{A}$-equivalent to the constant map-germ 0, which yields a contradiction. Next, suppose that $\delta = r$. Then, by the same reason as in the case that $\delta > r$, $\widetilde{f}$ must be $\mathcal{A}$-equivalent to $g(x) = (x^\delta, 0 \ldots, 0)$, which implies that g must be finitely $\mathcal{A}$-determined. However, the assumption $\delta > 2$ implies that g cannot be finitely $\mathcal{A}$-determined by Mather-Gaffney's geometric characterization of finite determinacy ([5], see also Theorem 2.1 of [7]). Thus, we have that $\delta \leq r - 1$. □

By now the proof of the assertion 2 of Theorem 1.1 is straightforward as follows. By Lemma 2.5 and the inequality $k\delta \leq r + \delta - 1$, we have that

$$m_1^{2r^2 - 6r + 4} = m_1^{2(r-2)(r-1)} \subset m_1^{(\delta-1)(r+\delta-1)} \subset m_1^{(\delta-1)k\delta}.$$

Hence, by lemma 4 and the same argument as in the proof of the assertion 1 of Theorem 1.1, we see that f is $(2r^2 - 6r + 3)$-$\mathcal{L}$-determined. □

References

1. V. I. Arnold, Simple singularities of curves, *Proc. Steklov Inst. Math.* **226** (1999) 20–28.
2. V. I. Arnold, *Arnold's problems*, Springer-Verlag Phasis, Berlin Heidelberg New York Moscow, 2005.
3. V. I. Arnold, S. M. Gusein-Zade and A. N. Varchenko, *Singularities of Differentiable Maps* I, *Monographs in Mathematics* **82** Birkhäuser, Boston, 1985.
4. J. W. Bruce, T. Gaffney and A. A. du Plessis, On left equivalence of map germs, *Bull. London Math. Soc.* **16** (1984) 303–306.
5. T. Gaffney, *Properties of finitely determined germs,* Ph.D. thesis Brandeis University, 1975.
6. J. J. Sylvester, Mathematical questions with their solutions, *Educational Times* **41** (1884) 21.
7. C. T. C. Wall, Finite determinacy of smooth map-germs, *Bull. London Math. Soc.* **13** (1981) 481–539.

T. Nishimura
Department of Mathematics
Faculty of Education and Human Sciences
Yokohama National University
Yokohama 240-8501
Japan
takashi@edhs.ynu.ac.jp

21

Regularity of the transverse intersection of two regular stratifications

PATRICE ORRO AND DAVID TROTMAN

Abstract

We give a general theorem stating that transversely intersecting regular stratified sets have regularly stratified intersection (and union) for a large class of regularity conditions. Such a result was previously known only for Whitney regular stratified sets and for weakly Whitney stratified sets.

1. Introduction

It is often useful to know that the transverse intersection of two regularly stratified sets is again regular. That the transverse intersection of two Whitney regular stratified sets is again Whitney regular was apparently first published in 1976 by Chris Gibson [10] in the Liverpool notes on the topological stability of smooth mappings. In their book "Stratified Morse Theory" [11], perhaps following Teissier's account for complex analytic varieties in his La Rabida notes [19] where Teissier attributes the detailed proof given there to Denis Chéniot, Mark Goresky and Robert MacPherson cite the 1972 Comptes Rendus note by Chéniot [6] for a proof of this result, which is a mistake, as Chéniot does *not* prove the Whitney regularity of intersections or even discuss it; he is concerned with the frontier condition in the case of a complex variety, and moreover only for intersections with a smooth complex submanifold. This mistake in attribution has unfortunately been copied by many authors.

That the transverse intersection of two weakly Whitney regular stratified sets is again weakly Whitney regular was proved by Karim Bekka in his thesis [1]

2000 *Mathematics Subject Classification* 58A35 (primary).

and published in [3]. We recall that weak Whitney regularity is the simultaneous validity of Whitney's condition (a) and a metric condition (δ) first introduced by Bekka and Trotman in 1987 [2]. See [2], [3], [4] and papers of Ferrarotti [9] and the books of Pflaum [16] and Schurmann [17] for further properties of weakly Whitney regular stratifications.

For Kuo-Verdier regularity (condition (w)) we know of no reference other than our joint work [15] in 2002: the property of invariance does not even seem to have previously been stated except in the special case when one of the sets is a smooth submanifold [21]. Here we state and prove a very general theorem, previously used in our work on normal cone structure [15], which applies to many regularity conditions at once.

Very recently (2009) Malgorzata Czapla [7] of the Jagellonian University in Cracow proved a more general invariance property for certain pairs of (b)- or (w)-regular stratified sets, assuming something less than transversality, as preparation for a (b)- and (w)-regular definable triangulation theorem [8] for definable sets, with a proof distinct from Shiota's 2005 proof [18] of the existence of Whitney regular triangulations for semialgebraic sets, using techniques which were developed by Guillaume Valette [20] to obtain a bilipschitz version of Hardt's local semialgebraic triviality theorem.

2. Regular stratifications

2.1. Notations

Let X and Y be two submanifolds of $\mathbf{R}^n$ such that $0 \in Y \subset \overline{X}$, and let π be the local projection onto Y. Let $<, >$ be the scalar product of $\mathbf{R}^n$. Following Hironaka [12] we write $\alpha_{X,Y}(x)$ for the cosine of the angle between $T_x X$ and $T_{\pi(x)} Y$:

$$\alpha_{X,Y}(x) = \max\{< u, v >: u \in N_x X - \{0\}, \|u\| = 1, v \in T_{\pi(x)} Y, \|v\| = 1\},$$

and write $\beta_{X,Y}(x)$ for the cosine of the angle between $x\pi(x)$ and $T_x X$:

$$\beta_{X,Y}(x) = \max\{< u, (x\pi(x))/\|x\pi(x)\| >: u \in N_x X - \{0\}, \|u\| = 1\}.$$

For $v \in \mathbf{R}^n$, we write $\eta(v, B)$ for the distance between v and a plane B:

$$\eta(v, B) = \sup\{< v, n >: n \in B^\perp, \|n\| = 1\}.$$

Set $d(A, B) = \sup\{\eta(v, B) : v \in A, \|v\| = 1\}$,

$$R_{X,Y}(x) = \frac{\|x\| \alpha_{X,Y}(x)}{\|x\pi(x)\|}$$

and

$$W_{X,Y}(x, z) = \frac{d(T_x X, T_z Y)}{\|x\pi(x)\|}.$$

2.2. Definitions

A pair of strata (X, Y) is said to satisfy:

Whitney's condition (a) at 0 if, for x in X,

$$\lim_{x \to 0} \alpha_{X,Y}(x) = 0,$$

Whitney's condition (b) at 0 if, for x in X,

$$\lim_{x \to 0} \alpha_{X,Y}(x) = \lim_{x \to 0} \beta_{X,Y}(x) = 0,$$

Kuo's condition (r) at 0 if, for x in X,

$$\lim_{x \to 0} R_{X,Y}(x) = 0,$$

the Kuo-Verdier condition (w) at 0 if, for x in X and y in Y,

$$W_{X,Y}(x, y) \text{ is bounded near 0,}$$

the Bekka-Trotman condition (δ) at 0 if, for x in X and y in Y, the angle between the line xy and $T_x X$ is bounded above near 0 by a positive constant $\delta < \pi/2$.

A stratification Σ of a (closed) set A in a manifold M is said to be E-regular if at each point y_0 of each stratum Y there is a chart for M in which for each stratum X of Σ distinct from Y, the pair (X, Y) is E-regular at y_0.

In [15] we introduced a condition (r^e), of Kuo-Verdier type.

Definition 2.1. Let $e \in [0, 1]$. We say that (X, Y) satisfies condition (r^e) at $0 \in Y$ if for $x \in X$ the quantity $R_e(x) = \frac{\|\pi(x)\|^e \alpha_{X,Y}(x)}{\|x\pi(x)\|}$ is bounded near 0.

This condition (r^e) is invariant by C^2 diffeomorphisms. It is exactly (w) when $e = 0$, thus (w) implies (r^e) for all $e \in [0, 1]$. But unlike (w), condition (r^e) when $e > 0$ does not imply condition (a) : one constructs easily a counter-example consisting of a semi-algebraic surface in $\mathbf{R}^3$ obtained by pinching the half-plane $\{z \geq 0, x = 0\}$, with boundary the axis $0y = Y$, in a cuspidal region $\Gamma = \{x^2 + y^2 \leq z^p\}$, where p is an odd integer such that $p > 2e$, so that in Γ there exist sequences tending to 0 for which condition (a) is not satisfied. The reader is invited to check that this example is (r^e)-regular.

2.3. The theorem

Theorem 2.2. *The conditions (a), (b), (r), (w), $(a + \delta)$ and $(a + r^e)$ for $0 \leq e < 1$ are invariant by transverse intersection of two stratifications of class C^2. This means that if (A, Σ) and (A', Σ') are closed E-regular C^2 stratified subsets of some manifold M whose strata have only transverse intersections, then their intersection $A \cap A'$ is E-regularly stratified by $\Sigma'' = \{X \cap X' : X \in \Sigma, X' \in \Sigma'\}$, where E is one of the conditions (a), (b), (r), (w), $(a + \delta)$, or $(a + r^e)$ for $0 \leq e < 1$.*

Note. For conditions (a), (b), and $(a + \delta)$, we can assume the stratifications are merely of class C^1.

Proof. Consider two transverse planes A and B.

For $v \in \mathbf{R}^n$, the distance of v to B is

$$\eta(v, B) = \sup\{< v, n >: n \in B^{\perp}, \|n\| = 1\}.$$

The distance of v to $A \cap B$ is then

$$\eta(v, A \cap B) = \sup\{< v, n >: n \in A^{\perp} + B^{\perp}, \|n\| = 1\}.$$

Decompose $A^{\perp} + B^{\perp}$ as $I + U + V$ where $I = A^{\perp} \cap B^{\perp}$, and U (resp. V) is the orthogonal complement of I in $A^{\perp}$ (resp. $B^{\perp}$). Then

$$\eta(v, A \cap B) = \sup\{< v, \Sigma_{i=1}^{3} n_i >: n_1 \in I, n_2 \in U, n_3 \in V, \|\Sigma_{i=1}^{3} n_i\| = 1\}.$$

Let now Σ and Σ' be transverse stratifications.

Suppose that Σ and Σ' each satisfies a regularity condition in a neighbourhood U of each point of the type $d(T_x S, T_y T) \leq K_{S,T} \phi_{S,T}(x, y)$ for some fixed function $\phi_{S,T}$ depending on the regularity condition, and a positive constant $K_{S,T}$, where S and T are adjacent strata of one of the stratifications, $x \in U \cap S$ and $y \in U \cap T$.

Let X, Y be strata of Σ and let X', Y' be strata of Σ' such that $Y < X$ and $Y' < X'$, and let $x \in X \cap X'$, and $y \in Y \cap Y'$.

Each case will be deduced from the proof that we now give for $(a + r^e)$.

For $(a + r^e)$ the associated function is

$$\phi_{X,Y}(x, y) = \frac{\|xy\|}{\|\pi_Y(x)\|^e},$$

where π_Y denotes the local projection onto Y.

We use here that each stratum Y of Σ (resp. Σ') is a differentiable submanifold of class C^2, so that the tubular projection π_Y onto Y is well-defined [13].

Let U_1 be a neighbourhood of 0 such that $||\pi_{Y \cap Y'}(x)|| \leq 2||\pi_Y(x)||$ if $x \in U_1$, and U_2 a neighbourhood of 0 such that $||\pi_{Y \cap Y'}(x)|| \leq 2||\pi_{Y'}(x)||$ if $x \in U_2$ (we may replace 2 by a constant $K > 1$, but we use 2 to simplify notation).

Then, for $x \in U_1$,

$$\frac{||xy||}{||\pi_Y(x)||^e} \leq \frac{2^e ||xy||}{||\pi_{Y \cap Y'}(x)||^e},$$

we obtain the inequality

$$\phi_{X,Y}(x, y) \leq 2\phi_{X \cap X', Y \cap Y'}(x, y)$$

when $x \in X \cap U \cap U_1$ and $y \in Y \cap U \cap U_1$, because $0 \leq e < 1$.

Similarly, when $x \in X' \cap U' \cap U_2$ and $y \in Y' \cap U' \cap U_2$,

$$\phi_{X',Y'}(x, y) \leq 2\phi_{X \cap X', Y \cap Y'}(x, y).$$

By what precedes for $A = T_x X$ and $B = T_x X'$, if $v \in T_y Y \cap T_y Y'$, and $||v|| = 1$, and setting

$$J = \{(n_1, n_2, n_3) \in I \times U \times V : ||\Sigma_{i=1}^3 n_i|| = 1\},$$

then

$$\eta(v, A \cap B) = \sup\{< v, \Sigma_{i=1}^3 n_i > : (n_1, n_2, n_3) \in J\}$$

$$(*) \qquad \leq \sup\{(\Sigma_{i=1}^3 | < v, n_i > |) : (n_1, n_2, n_3) \in J\}$$

$$\leq \sup\{(\Sigma_{i=1}^2 ||n_i||) K_{XY} \phi_{XY}(x, y) + ||n_3|| K_{X'Y'} \phi_{X'Y'}(x, y) :$$

$$(n_1, n_2, n_3) \in J\}$$

$$\leq \sup\{(\Sigma_{i=1}^3 ||n_i||) : (n_1, n_2, n_3) \in J\} K \phi(x, y),$$

where $K = 2 max(K_{XY}, K_{X'Y'})$ and $\phi(x, y) = \phi_{X \cap X', Y \cap Y'}(x, y)$.

As

$$||\Sigma_{i=1}^3 n_i||^2 = ||n_1||^2 + ||n_2||^2 + ||n_3||^2 + 2\cos(n_2, n_3)||n_2|| ||n_3|| = 1,$$

and $\Sigma_{i=1}^3 ||x_i||^2 + 2a||x_2|| ||x_3|| = 1$ is compact for $||a|| \neq 1$, we have that

$$d(T_y Y \cap T_y Y', A \cap B) = \sup\{\eta(v, A \cap B) : v \in T_y Y \cap T_y Y', ||v|| = 1\}$$

$$\leq C\phi(x, y),$$

where $C = \sup\{(\Sigma_{i=1}^3 ||x_i||) : \Sigma_{i=1}^3 ||x_i||^2 + 2a||x_2|| ||x_3|| = 1, ||a|| < 1 - \epsilon\}$, and ϵ is given by the minimal angle of $T_x X$ and $T_{x'} X'$ on a neighbourhood of y; ϵ is nonzero by the (a)-regularity of (X, Y) and (X', Y') at y and the transversality of Y and Y' at y. We need finally that (a) holds for $\Sigma \cap \Sigma'$, using the inequality $(*)$. This can be left as an exercise for the reader, or see [10] or [19]. The theorem follows because $\phi(x, y) = \phi_{X \cap X', Y \cap Y'}(x, y)$. $\square$

References

1. K. Bekka, *Sur les propriétés topologiques des espaces stratifiés*, Thesis, University of Paris-Sud, Orsay 1988.
2. K. Bekka and D. Trotman, Propriétés métriques de familles Φ-radiales de sous-variétés différentiables, *C. R. Acad. Sci. Paris Sr. I Math.* 305 (1987), 389–392.
3. K. Bekka and D. Trotman, Weakly Whitney stratified sets, *Real and Complex singularities (eds. J. W. Bruce, F. Tari)*, Chapman and Hall Research Notes in Math. 412 (2000), 1–15.
4. K. Bekka and D. Trotman, Metric properties of stratified sets, *Manuscripta Math.* 111 (2003), 71–95.
5. H. Broderson and D. Trotman, Whitney (b)-regularity is strictly weaker than Kuo's ratio test for real algebraic stratifications, *Math. Scand.*, 45 (1979), 27–34.
6. D. Chéniot, Sur les sections transversales d'un ensemble stratifié, *C. R. Acad. Sci. Paris Sér. A-B* 275 (1972), A915–A916.
7. M. Czapla, *Invariance of regularity conditions under definable, locally Lipschitz, weakly bi-Lipschitz mappings*, preprint, May 2009, 24 pages.
8. M. Czapla, *Definable triangulations with regularity conditions*, preprint, May 2009, 17 pages.
9. M. Ferrarotti, Trivializations of stratified spaces with bounded differential,*Singularities (Lille, 1991)*, London Math. Soc. Lecture Note Ser., 201,Cambridge Univ. Press (1994), 101–117.
10. C. G. Gibson, Construction of canonical stratifications, *Topological stability of smooth mappings*, Lecture Notes in Mathematics, Springer Verlag, New York, 552 (1976), 9–34.
11. M. Goresky and R. MacPherson, *Stratified Morse theory*, Springer-Verlag, New York, 1988.
12. H. Hironaka, Normal cones in analytic Whitney stratifications, *Pub. Math. I.H.E.S.* 36 (1969), 127–138.
13. M. Hirsch, *Differential Topology*, Graduate Texts in Mathematics, Vol. 33, Springer-Verlag, New York, 1976.
14. T.-C. Kuo, The ratio test for analytic Whitney stratifications, *Proceedings of Liverpool Singularities Symposium I (ed. C.T.C. Wall)*, Lecture Notes in Math., Springer-Verlag, New York, 192 (1971), 141–149.
15. P. Orro and D. Trotman, Cône normal et régularités de Kuo-Verdier, *Bulletin de la Société Mathématique de France* 130 (2002), 71–85.
16. M. Pflaum, *Analytic and Geometric Study of Stratified Spaces*, Lecture Notes in Mathematics, vol. 1768, Springer-Verlag, New York, 2001.
17. J. Schürmann, *Topology of singular spaces and constructible sheaves*, Mathematics Institute of the Polish Academy of Sciences. Mathematical Monographs (New Series) 63, Birkhaüser Verlag, Basel, 2003.

18. M. Shiota, Whitney triangulations of semialgebraic sets, *Ann. Polon. Math.* 87 (2005), 237–246.
19. B. Teissier, Variétés polaires II: Multiplicités polaires, sections planes, et conditions de Whitney, *Algebraic Geometry Proceedings, La Rabida 1981*, Lect. Notes in Math. Springer-Verlag, New York, 961 (1982), 314–491.
20. G. Valette, Lipschitz triangulations, *Illinois J. of Math.* 49 (2005), 953–979.
21. J.-L. Verdier, Stratifications de Whitney et théorème de Bertini-Sard, *Inventiones Math.*, 36 (1976), 295–312.

Patrice Orro
Laboratoire de Mathématiques
(UMR 5127),
Université de Savoie,
Campus Scientifique,
73376 Le Bourget-du-Lac
France
patrice.orro@univ-savoie.fr

David Trotman
Laboratoire d'Analyse, Topologie et Probabilités (UMR 6632),
Université de Provence,
Centre de Mathématiques et Informatique,
39 rue Joliot-Curie, 13453 Marseille
France
david.trotman@cmi.univ-mrs.fr

22

Pairs of foliations on surfaces

FARID TARI

Dedicated to the memory of Professor Carlos Gutierrez

Abstract

We survey in this paper results on a particular set of Implicit Differential Equations (IDEs) on smooth surfaces, called Binary/Quadratic Differential Equations (BDEs). These equations define at most two solution curves at each point on the surface, resulting in a pair of foliations in some region of the surface. BDEs appear naturally in differential geometry and in control theory. The examples we give here are all from differential geometry. They include natural families of BDEs on surfaces. We review the techniques used to obtain local models of BDEs (formal, analytic, smooth and topological). We also discuss some invariants of BDEs and present a framework for studying their bifurcations in generic families.

1. Introduction

An implicit differential equation (IDE) is an equation of the form

$$F(x, y, p) = 0, \quad p = \frac{dy}{dx} \tag{1.1}$$

where F is a smooth (i.e., C^∞) or real analytic function in some domain in $\mathbb{R}^3$. If $F(q_0) = 0$ and $F_p(q_0) \neq 0$ at $q_0 = (x_0, y_0, p_0) \in \mathbb{R}^3$ (when not indicated otherwise, subscripts denote partial differentiation), equation (1.1) can be written locally in a neighbourhood of q_0 in the form $p = g(x, y)$. It can then be studied using the methods from the theory of ordinary differential equations.

2000 *Mathematics Subject Classification* 53A05, 58K05, 34A09, 53B30, 37G10, 37C10

When $F(q_0) = F_p(q_0) = 0$, there may be more then one solution curve of equation (1.1) through points in a neighbourhood U of (x_0, y_0). We deal here mainly with the case when $F_{pp}(q_0) \neq 0$, so there are at most two solution curves through each point in U. In this case, it follows from the division theorem that equation (1.1) can be expressed in a quadratic form

$$a(x, y)dy^2 + 2b(x, y)dxdy + c(x, y)dx^2 = 0 \qquad (1.2)$$

where a, b, c are smooth or analytic functions in some neighbourhood U of (x_0, y_0) not vanishing simultaneously at any point in U. Equation (1.2) is called a *Binary Differential Equation* (BDE) or *Quadratic Differential Equation*. The functions a, b, c are called the *coefficients of the BDE*.

It is also of interest to study BDEs at points where their coefficients vanish at a given point. We shall label these *Type 2 BDEs* and reserve the label *Type 1 BDEs* for those with coefficients not vanishing simultaneously at any point. There are some crucial differences between the two types of BDEs. For instance, Type 1 BDEs may have finite codimension in the set of all IDEs and can be deformed in this set. However, Type 2 BDEs are of infinite codimension in the set of all IDEs and are deformed in the set of all BDEs. Other differences will be highlighted in the paper.

The *discriminant* of a BDE is the set $\Delta = \{(x, y) \in U : (b^2 - ac)(x, y) = 0\}$. A BDE determines a pair of transverse foliations or no foliations away from the discriminant. Thus, all the important features of the equation occur on the discriminant. The discriminant, together with the pair of foliations determined by the BDE is called *the configuration* of the BDE. In all the figures in this paper, we draw one foliation in a continuous line and the other in a dashed line. The discriminant is drawn in thick black.

BDEs have a long history (see for example [25] and [62] for historical notes). They appear in, and have application to, control theory, partial differential equations and differential geometry. The examples in this paper are from differential geometry. For applications to control theory see [25, 51]. The paper is organised as follows:

§2: lines of curvature, asymptotic and characteristic curves are classical pairs of foliations on surfaces and are given by BDEs. We consider their configurations on a surface endowed with a Riemannian or a Lorentzian metric. We also discuss the case of surfaces endowed with a metric of mixed type.

§3: the examples in §2 provide a good motivation for seeking models of the configurations of BDEs at points on the discriminant. We review the techniques involved for finding such models and clarify the meaning of the word model (up to formal, analytic, smooth or topological equivalence). We also give a complete list of local singularities of topological codimension ≤ 2.

§4: the discriminant of a BDE is a plane curve. However, the deformation of its singularities cannot always be modelled by the $\mathcal{K}$-deformations of a plane curve singularity. We review some invariants associated to BDEs and review Bruce's symmetric matrices framework ([6]) for studying the singularities of the discriminant.

§5: we review briefly a method for studying the bifurcations of a BDE in generic families of BDEs.

§6: we highlight references where work on more general IDEs and homogeneous differential equations of a given degree is carried out. We also give a list of existing local topological models of BDEs of codimension > 2.

An important aspect of BDE which is omitted here is the study of their foliations near a limit cycle. This study is initiated in the pioneering work of Sotomayor and Gutierrez [64] where they obtained a formula for the derivative of the Poincaré return map at a limit cycle of the lines of principal curvature on a smooth surface in $\mathbb{R}^3$. The behaviour of the asymptotic and characteristic curves at a limit cycle is also studied in [32, 33].

2. Examples from differential geometry

Let S be a smooth surface. We start with the case when S is immersed in the Euclidean space $\mathbb{R}^3$ and denote by "." the scalar product in $\mathbb{R}^3$. Let $x : U \subset \mathbb{R}^2 \to \mathbb{R}^3$ be a local parametrisation of S, and let S^2 denote the unit sphere in $\mathbb{R}^3$. The Gauss map

$$N : x(U) \subset S \to S^2,$$

assigns to each point $p = x(u, v)$ the normal vector $N(p) = (x_u \times x_v/\|x_u \times x_v\|)(u, v)$ to S at p.

The shape operator $A_p = -d_p N : T_p S \to T_p S$ (or the Weingarten map) has the following properties: it is a self-adjoint operator, i.e., it is a linear operator with $A_p(w_1).w_2 = w_1.A_p(w_2)$, for any pair of vectors in $T_p S$; it has always two real eigenvalues κ_1, κ_2 called *the principal curvatures*; it has two orthogonal eigenvectors (when $\kappa_1 \neq \kappa_2$) called the *principal directions*. The integral curves on S of the principal directions line fields are called the *lines of principal curvature*. The points where $\kappa_1 = \kappa_2$ are referred to as *umbilic* points. For generic immersions, the umbilic points are isolated points on S.

Let $E = x_u.x_u$, $F = x_u.x_v$, $G = x_v.x_v$ denote the coefficients of the first fundamental form and $l = N.x_{uu}$, $m = N.x_{uv} = N.x_{vu}$, $n = N.x_{vv}$ those of the second fundamental form on S. Then, the equation of the lines of principal curvature is given by the BDE

$$(Gm - Fn)dv^2 + (Gl - En)dvdu + (Fl - Em)dv^2 = 0. \qquad (2.1)$$

The discriminant of equation (2.1) is the set of umbilic points. Away from such points the lines of curvature form a net of orthogonal curves. The coefficients of equation (2.1) vanish at umbilic points, so we have a BDE of Type 2 there. There are generically three distinct topological configurations of the lines of curvature at umbilic points (Figure 22.6, top three figures). The configurations were first drawn by Darboux and a rigorous proof was given in [7, 64]. The global behaviour of the lines of principal curvature on closed orientable surfaces in $\mathbb{R}^3$ was first studied in [64]. (For historical notes on the study of the lines of curvature see [62].)

Two directions $w_1, w_2 \in T_p S$ are *conjugate* if $A_p(w_1).w_2 = 0$. An *asymptotic* direction is a self-conjugate direction, that is $A_p(w).w = 0$. There are two asymptotic directions at a hyperbolic point and none at an elliptic point on the surface. The integral curves of the pair of asymptotic line fields are called the *asymptotic curves*. The equation of the asymptotic curves is given by the BDE

$$ndv^2 + 2mdvdu + ldu^2 = 0. \qquad (2.2)$$

The discriminant of equation (2.2) is the parabolic set of the surface. The asymptotic curves form a family of cusps at a generic parabolic point. Their configurations at a cusp of Gauss are given in [3, 4, 50] (Figure 22.2, last three figures) and a more general approach for studying the singularities of their equation at such points is given in [23, 24, 51, 70]. Generic global properties of these foliations including the study of their limit cycles are given in [33].

At elliptic points there is a unique pair of conjugate directions for which the included angle is extremal ([28]). These directions are called the *characteristic directions* and their integral curves are called the characteristic curves. Characteristic directions on surfaces in $\mathbb{R}^3$ are studied in [28, 55, 60] and more recently in [8, 15, 32, 58]. In [32] they are labelled harmonic mean curvature lines and are defined as curves along which the normal curvature is K/H, where K is the Gaussian curvature and H is the mean curvature of S. The equation of the characteristic curve is given by the BDE

$$\begin{aligned}(2m(Gm - Fn) - n(Gl - En))dv^2 + 2(m(Gl + En) - 2Fln)dvdu \\ + (l(Gl - En) - 2m(Fl - Em))du^2 = 0.\end{aligned} \qquad (2.3)$$

It is shown in [15] that the BDEs of the asymptotic, characteristic and principal curves are related. A BDE (1.2) can be viewed as a quadratic form and represented at each point in U by the point $(a(x, y) : 2b(x, y) : c(x, y))$ in the projective plane. Let Γ denote the conic of degenerate quadratic forms. To a point in the projective plane is associated a unique polar line with respect to Γ, and vice-versa. A triple of points is called a self-polar triangle if the polar line of any point of the triple contains the remaining two points. It turns out that, at non parabolic or umbilic points on M, the triple asymptotic, characteristic and principal curves BDEs form a self-polar triangle ([15]). In particular, any two of them determine the third one.

In [29] is constructed a natural 1-parameter family of BDEs, called *conjugate curve congruence*, that links the asymptotic curves BDE and the principal curves BDE on a smooth surface in $\mathbb{R}^3$. In [15], it is constructed a natural 1-parameter family of BDEs, called *reflected conjugate congruence*, linking the characteristic curves BDE and that of the principal curves.

Consider the projective space PT_pM of all tangent directions through a point $p \in M$ which is neither an umbilic nor a parabolic point. Conjugation gives an involution on PT_pM, $v \mapsto \bar{v} = C(v)$. There is another involution on PT_pM which is the reflection in either of the principal directions, $v \mapsto R(v)$.

Definition 2.1 ([29]). 1. Let $\Theta : PTM \to [-\pi/2, \pi/2]$ be given by $\Theta(p, v) = \alpha$, where α denotes the oriented angle between a direction v and the corresponding conjugate direction $\bar{v} = C(v)$. The conjugate curve congruence, for a fixed α, is defined to be $\Theta^{-1}(\alpha)$ and is denoted by $\mathcal{C}_\alpha$.

2. ([15]) Let $\Phi : PTM \to [-\pi/2, \pi/2]$ be given by $\Phi(p, v) = \alpha$, where α is the signed angle between v and $R(\bar{v}) = R \circ C(v)$. Then, the reflected conjugate curve congruence, for a fixed α, is defined to be $\Phi^{-1}(\alpha)$ and is denoted by $\mathcal{R}_\alpha$.

Proposition 2.2 ([29]). 1. *The conjugate curve congruence $\mathcal{C}_\alpha$ of a parametrised surface is given by the BDE*

$$
\begin{aligned}
(\sin\alpha(mG - nF) &- n\cos\alpha\sqrt{EG - F^2})dv^2 \\
&+ (\sin\alpha(lG - nE) - 2m\cos\alpha\sqrt{EG - F^2})dvdu \\
&+ (\sin\alpha(lF - mE) - l\cos\alpha\sqrt{EG - F^2})du^2 = 0.
\end{aligned}
\tag{2.4}
$$

2. ([15]) *The reflected conjugate congruence $\mathcal{R}_\alpha$ is given by the BDE*

$$
\begin{aligned}
\left((2m(mG - nF) - n(Gl - En))\cos\alpha + (nF - mG)\tfrac{2mF - lG - nE}{\sqrt{(EG - F^2)}} \sin\alpha \right) dv^2 \\
+ \left(2(m(lG + nE) - 2lnF)\cos\alpha + (nE - lG)\tfrac{2mF - lG - nE}{\sqrt{(EG - F^2)}} \sin\alpha \right) dvdu \\
+ \left((l(lG - nE) - 2m(lF - mE))\cos\alpha \right. \\
+ \left. (mE - lF)\tfrac{2mF - lG - nE}{\sqrt{(EG - F^2)}} \sin\alpha \right) du^2 = 0.
\end{aligned}
$$

$$
\tag{2.5}
$$

Remark 1. The concepts of asymptotic, principal and characteristic curves and of conjugate and reflected curve congruences can be associated to any self-adjoint operator on a Riemannian surface ([65]).

We turn now to the case of surfaces embedded in a non-Euclidean space. We consider a smooth surface S endowed with a Lorentzian metric, that is, a metric which is locally equivalent to $\lambda(u, v)(dv^2 - du^2)$. We shall refer to such surfaces as *timelike* surfaces. In view of Remark 1, we consider a self-adjoint operator A on S, so $A : TS \to TS$ is a smooth map, where TS is the tangent bundle of S and its restriction $A_p : T_pS \to T_pS$ is a self-adjoint operator. An example of this situation is provided by an immersed timelike surface in the de Sitter space $S_1^3 \subset \mathbb{R}_1^4$, where $\mathbb{R}_1^4$ denotes the Minkowski 4-space. Then, there is a natural Gauss map $\mathbb{E} : S \to S_1^3$ and its derivative is a self-adjoint operator on S ([48]).

Because the metric on S is not positive definite, A_p does not always have real eigenvalues. When it does, we label them A-principal curvature and the associated eigenvectors the A-principal directions. The A-lines of principal curvature are given by the BDE (2.1), where E, F, G are the coefficients of the first fundamental form on S and l, m, n are referred to as the coefficients of the A-second fundamental form and are given by the same formulae as those for surfaces in the Euclidean 3-space. The discriminant of the equation is now a curve which is generically either empty or smooth except at isolated points where it has a Morse singularity of type node. We label these singular points *timelike umbilic points* as A_p is a multiple of the identity at such points. The configurations of the A-lines of curvature at timelike umbilic points are those in Figure 22.6, second and third rows.

The concepts of A-asymptotic and A-characteristic directions and curves can also be defined and their equations are given by the BDEs (2.2) and (2.3) respectively ([49]). The local behaviour of these pairs of foliations is distinct from that of their counterpart on a surface in the Euclidean 3-space.

Surfaces that have a mixed type metric give rise to interesting problems. Suppose given a metric $ds^2 = a(u, v)dv^2 + 2b(u, v)dvdu + c(u, v)du^2$ on a smooth surface S, where the set $ac - b^2 = 0$ is a smooth curve on S. Suppose the metric ds^2 is Riemannian in the region $ac - b^2 > 0$, so it is equivalent to $\lambda(u, v)(dv^2 + du^2)$. It is Lorentzian in the region $ac - b^2 < 0$, so it is equivalent to $\lambda(u, v)(dv^2 - du^2)$. Miernowski [52] considered the problem of finding an analytic model of the metric at points on the curve $ac - b^2 = 0$. The problem reduces to finding analytic models of BDEs of Type 1 at points on their discriminant (see §3.1).

Given an immersed surface S in the Minkowski space $\mathbb{R}_1^3$, the restriction of the pseudo scalar product in $\mathbb{R}_1^3$ to S gives a metric on S which can be of mixed type. The asymptotic, characteristic and principal curves associated to the Gauss map of S are well defined on the Riemannian and the Lorentzian parts of S. Their extensions to the degenerate locus of the metric on S are studied in [49].

3. Classification

We denote by $\omega(x, y, dx, dy) = a(x, y)dy^2 + 2b(x, y)dydx + c(x, y)dx^2$ the quadratic form associated to the BDE (1.2) and also use ω to refer to the equation $\omega = 0$. The interest here is local, so we take a, b, c to be germs of functions $\mathbb{R}^2, 0 \to \mathbb{R}$. We consider the origin to be a point on the discriminant. For BDEs of Type 1, we can rotate the coordinate axes in the plane, set $p = dy/dx$ and take p in a neighbourhood of zero. However, for BDEs of Type 2, we take $(dx : dy) \in \mathbb{R}P^1$.

Definition 3.1. Two germs, at the origin, of BDEs ω_1 and ω_2 are respectively smoothly, analytically or formally equivalent if there exist germs $H = (h_1, h_2) :$ $\mathbb{R}^2, 0 \to \mathbb{R}^2, 0$ of a smooth, analytic or formal diffeomorphism and $r : \mathbb{R}^2, 0 \to$ $\mathbb{R}$ of a smooth, analytic or formal function not vanishing at 0 such that

$$\omega_2 = r.H^*\omega_1,$$

that is, $\omega_2(x, y, dx, dy) = r(x, y)\omega_1(h_1(x, y), h_2(x, y), dh_1(x, y), dh_2(x, y))$.

Two germs of BDEs are topologically equivalent if there exists a germ of a homeomorphism that takes the configuration of one to the configuration of the other.

The aim is to produce representatives (models, preferably in simple forms) of equivalence classes of the equivalence relations in Definition 3.1. A more realistic task is to produce models of germs of low codimensions, which we define as follows. We associate to a germ of a BDE $\omega = (a, b, c)$ the jet-extension map

$$\Phi : \mathbb{R}^2, 0 \to \quad J^k(2, 3)$$
$$(x, y) \mapsto j^k(a, b, c)|_{(x,y)}$$

where $J^k(2, 3)$ denotes the vector space of polynomial maps of degree $\leq k$ from $\mathbb{R}^2$ to $\mathbb{R}^3$, and $j^k(a, b, c)|_{(x,y)}$ is the k-jet of (a, b, c) at (x, y). (This is simply the Taylor expansion of order k of (a, b, c) at (x, y).)

Definition 3.2. A singularity of ω is of *codimension m* if the conditions that define it yield a semi-algebraic set V of codimension $m + 2$ in $J^k(2, 3)$, for any $k \geq k_0$.

3.1. Formal and analytic classifications

We can reduce, inductively on the k-jet spaces, the coefficients of a BDE to simpler forms by making polynomial changes of coordinates in the plane and multiply by invertible polynomial functions. If this process converges, i.e., the composite of all the changes of coordinates (resp. multiplicative polynomials) converges to an analytic diffeomorphism (resp. non zero analytic function), then the obtained germ of a BDE is an analytic model. Otherwise we have a formal model. The convergence problem is a complicated one, see for example [1] for the case of vector fields. However, even if the process is not convergent (which is the case in general), the reduction of the k-jet of a BDE to a simple form is very valuable in practice. The local topological behaviour of the solutions and the relevant invariants of a BDE do, in general, depend only on some initial terms of the coefficients of the BDE. Taking these in simpler forms makes the geometric interpretation of the conditions involved more apparent and the calculations more manageable.

The formal classification of some BDEs of Type 1 is dealt with in [16]. We can reduce the constant part of the BDE to one of the following cases

$$dy^2 + dx^2, \quad dy^2 - dx^2, \quad dy^2, \quad 0.$$

(We show below how this is done.) It is shown in [16] that a BDE with constant part equivalent to $dy^2 \pm dx^2$ is analytically equivalent to $dy^2 \pm dx^2$. The initial form dy^2 leads to the following representatives of orbits in the space of 1-jets:

$$dy^2 + xdx^2, \quad dy^2 - ydx^2, \quad dy^2.$$

We reproduce from [16] the case $dy^2 + xdx^2$ as an example of how the formal reduction is carried out. Suppose a BDE has 1-jet $dy^2 + xdx^2$ and assume that the k-jets of the coefficients of the BDE are $1 + a_k, 2b_k, x + c_k$, with a_k, b_k, c_k belonging to the set of homogeneous polynomials of degree k which we denote by H^k. We make changes of coordinates of the form

$$x = X + p(X, Y)$$
$$y = Y + q(X, Y)$$

with $p \in H^k$ and $q \in H^{k+1}$ and multiply by $1 + r(X, Y), r \in H^k$. Then,

$$dx = (1 + p_X)dX + p_Y dY, \quad dy = q_X dX + (1 + q_Y)dY,$$

and the k-jet of the transformed BDE is

$$(1 + a_k + r + 2q_Y, b_k + q_X + Xp_Y, x + c_k + p + 2Xp_X).$$

We are seeking r, p, q so that $a_k + r + 2q_Y = b_k + q_X + Xp_Y = c_k + p + 2Xp_X = 0$, i.e.,

$$r + 2q_Y = -a_k,$$
$$q_X + Xp_Y = -b_k,$$
$$p + 2Xp_X = -c_k.$$

This process produces the linear map

$$L_k : H^k \oplus H^{k+1} \oplus H^k \to \qquad H^k \oplus H^k \oplus H^k$$
$$(p, q, r) \qquad \mapsto (r + 2q_Y, p + 2Xp_X, q_X + Xp_Y)$$

The map L_k is surjective and furthermore $\bar{L}_k = L_k|_{q_{k+1}=0}$ is an isomorphism, where q_{k+1} denotes the coefficient of y^{k+1} in q. Therefore, the above linear system has a solution, which means that we can reduce the k-jet of the coefficients of the BDE to $(1, 0, x)$.

Proposition 3.3 ([16]). 1. *Suppose that a germ of a BDE has linear part $dy^2 + xdx^2$. Then, for any $k \geq 1$ we can change coordinates and multiply by a non zero function so that the germ of the transformed BDE has k-jet $dy^2 + xdx^2$.*

2. *Suppose the germ of a BDE has linear part $dy^2 + xdx^2$. Then, there exist an analytic change of coordinates which transforms the BDE to $\mu(x, y)(dy^2 + xdx^2)$, where μ is an analytic function not vanishing at the origin.*

The BDEs with 1-jet $dy^2 - ydx^2$ are also considered in [16]. The 2-jet can be put in the form $dy^2 - (y + \lambda x^2)dx^2$. It is shown in [16] that

Proposition 3.4 ([16]). *For almost all value of λ, a BDE with 2-jet $dy^2 - (y + \lambda x^2)dx^2$ is formally equivalent to $dy^2 - (y + \lambda x^2)dx^2$.*

As pointed in the §2, Miernowski [52] considered the problem of finding analytic models of a metric $ds^2 = a(u, v)dv^2 + 2b(u, v)dvdu + c(u, v)du^2$ of mixed type at points where $ac - b^2 = 0$. Suppose that the point in consideration is the origin. Miernowski showed that if the 1-jet of ds^2 is equivalent to $dv^2 + udu^2$, then the metric is analytically equivalent to $\mu(u, v)(dv^2 + udu^2)$ (compare Proposition 3.3). However, if the 1-jet of ds^2 is equivalent to $dv^2 - vdu^2$, then Miernowski proved that there is a functional modulus in the classification. As a corollary of his result, a BDE with 2-jet $dy^2 - (y + \lambda x^2)dx^2$ cannot be reduced to the form $\mu(x, y)(dy^2 - (y + \lambda x^2)dx^2)$ by analytic changes of coordinates.

We turn now to BDEs of Type 2. We have the following orbits in the 1-jet space where $\epsilon = \pm 1$:

- $(y, b_1 x + b_2 y, \epsilon y), b_1 \neq 0, 2b_1 + \epsilon \neq 0, b_1 \neq \frac{1}{2}(b_2^2 - \epsilon)$, and $b_1 \neq \pm b_2 - 1$
  when $\epsilon = +1$ ([17, 38])
- $(x + a_2 y, 0, y), a_2 > \frac{1}{4}$ ([66])
- $(y, \pm x + b_0 y, 0)$, $(y, y, 0)$, $(x + y, 0, 0)$, $(x, b_0 y, 0)$, $(x + y, -y, 0)$,
  $(y, 0, 0), (0, 0, 0)$ ([13]).

It turns out that there are no discrete orbits in the formal classification. It is shown in [17] that a BDE with 1-jet $(y, b_1 x + b_2 y, \epsilon y), \epsilon = \pm 1$ can be reduced, for almost all values of (b_1, b_2), by a formal diffeomorphism and multiplication by non zero formal power series to $(y, b_1 x + b_2 y + b(x, y), \epsilon y)$, where $b(x, y)$ is a formal power series with no constant or linear terms. (See also [42] for a similar result for the case $\epsilon = -1$.)

3.2. Smooth and topological classifications

We deal with BDEs of Type 1 and 2 separately.

BDEs of Type 1

The study of BDEs of Type 1 follows form the general study of IDEs. The IDE (1.1) defines a surface

$$M = \{(x, y, p) \in \mathbb{R}^3 : F(x, y, p) = 0\}$$

in the 3-dimensional space of 1-jets of functions endowed with the contact structure $\alpha = dy - p dx$. Consider the projection $\pi : M \to \mathbb{R}^2, \pi(x, y, p) = (x, y)$. Generically, M is a smooth surface (that is 0 is a regular value of F) and the restriction of π to M is either a local diffeomorphism, a fold or cusp map. The set of critical points of the projection is called the *criminant* of the IDE and is given by the equations $F = F_p = 0$. The set of critical values of the projection is called the *discriminant* of the IDE, and is obtained by eliminating p from the equations $F = F_p = 0$.

The multi-valued direction field defined by F in the plane lifts to a single-valued direction field on the surface. This direction field is determined by the vector field

$$\xi = F_p \frac{\partial}{\partial x} + p F_p \frac{\partial}{\partial y} - (F_x + p F_y) \frac{\partial}{\partial p}$$

(which is along the intersection of M with the contact planes in $\mathbb{R}^3$). It is of course tangent to M at (x, y, p) and projects to a line through (x, y) with slope p.

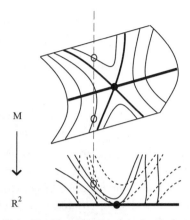

Figure 22.1. The lifted field ξ and the involution σ on M.

If π is a local diffeomorphism at (x, y, p), then the integral curves of ξ around (x, y, p) project to a family of smooth curves around (x, y).

Suppose that $\pi|_M$ has a fold singularity at (x, y, p), i.e., we can choose local coordinates in M and $\mathbb{R}^2$ for which π has the form (u, v^2). This means that $F = F_p = 0$ but $F_{pp} \neq 0$ at (x, y, p). Then, the IDE is locally a BDE at (x, y, p) and the discriminant is a smooth curve. (For BDEs of Type 1, the surface M is smooth if and only if the discriminant is smooth.) Every point in the plane near (x, y) which is not on the discriminant has two pre-images on M under π. This defines an involution σ on M near (x, y, p), which interchanges pairs of points with the same image under σ (Figure 22.1). The criminant is the set of fixed points of σ. Thus, locally at (x, y, p), we have a pair (ξ, σ) of a vector field and an involution on M. The classification (smooth or topological) of IDEs is the same as the classification (smooth or topological) of the pairs (ξ, σ). When ξ is regular, the IDE is smoothly equivalent to $dy^2 - x dx^2 = 0$ (i.e., $p^2 - x = 0$) ([1]). If ξ has an elementary singularity (saddle/node/focus), then the corresponding point in the plane is called a *folded singularity* of the BDE. At folded singularities, the equation is locally smoothly equivalent to

$$dy^2 + (-y + \lambda x^2)dx^2 = 0, \tag{3.1}$$

with $\lambda \neq 0, \frac{1}{16}$, provided that ξ is linearisable at the singular point; see [24, 25].

Normal forms at folded resonant saddles and nodes are given in [27]. At a degenerate elementary singular point of ξ of multiplicity $r \in \mathbb{N}$, $r > 1$, the equation is smoothly equivalent to

$$\left(\frac{dy}{dx} + \epsilon x^r + A x^{2r-1}\right)^2 = y$$

where $A \in \mathbb{R}$ and $\epsilon \in \{(\pm 1)^r\}$; see [26].

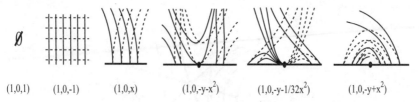

(1,0,1)   (1,0,-1)   (1,0,x)   (1,0,-y-x²)   (1,0,-y-1/32x²)   (1,0,-y+x²)

Figure 22.2. Topological models of stable singularities of BDEs of Type 1.

Suppose that $\pi|_M$ has a cusp singularity at (x, y, p), i.e., we can choose local coordinates in M and $\mathbb{R}^2$ for which π has the form $(u, v^3 + uv)$. This means that $F = F_p = F_{pp} = 0$ but $F_{ppp} \neq 0$ at (x, y, p). Then, the IDE is locally a cubic equation in p and the discriminant has a cusp singularity. Bruce [5] conjectured that the IDE has a functional modulus for smooth equivalence. Davydov [24] proved that the equation has a functional modulus even for the topological equivalence (see also §3.3).

We turn now to the topological equivalence. This can be treated in several ways. It can be done, as in [24, 25], by studying the pair (ξ, σ). Kuzmin [51] split the BDE into two ODEs and used the methods of ODE to analyse the behaviour of the solutions. In [9, 67, 69] we used the method of blowing up (see below). We give below the topological classification of the singularities of codimension ≤ 2.

There are three stable topological models (see [25] for references) at folded singularities: a folded saddle if $\lambda < 0$, a folded node if $0 < \lambda < \frac{1}{16}$ and a folded focus if $\frac{1}{16} < \lambda$ in equation (3.1); Figure 22.2, last three figures respectively. The labelling in the figures refers to the coefficients (a, b, c) of the model BDE. The first two figures in Figure 22.2 are models away from the discriminant, and the third at points on the discriminant corresponding to regular points of ξ.

Codimension 1 singularities are dealt with in [9, 25, 51, 67], see Figure 22.3. These occur when: the lifted field ξ has a saddle-node singularity (Figure 22.3, first figure, $\lambda = 0$ in equation (3.1)); the lifted field ξ has equal eigen values (Figure 22.3, second figure, $\lambda = 1/16$ in equation (3.1)); or when the discriminant has a Morse singularity, labelled Morse Type 1 singularities in [9] (Figure 22.3, last 4 figures). The Morse Type 1 singularities are distinguished by the type of the singularity of the discriminant, isolated point or node, and by the type of the folded singularities that appear in a generic deformation (two folded saddles or foci), see [9].

Codimension 2 singularities are classified in [69]. Degeneracy occurs in three ways: the discriminant is smooth and the lifted field has a degenerate elementary singularity of multiplicity 3 (Figure 22.4, first two figures); the

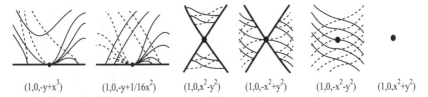

$(1,0,-y+x^3)$ $(1,0,-y+1/16x^2)$ $(1,0,x^2-y^2)$ $(1,0,-x^2+y^2)$ $(1,0,-x^2-y^2)$ $(1,0,x^2+y^2)$

Figure 22.3. Topological models of codimension 1 singularities of BDEs of Type 1.

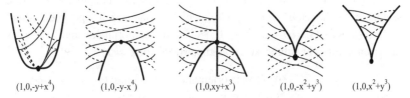

$(1,0,-y+x^4)$ $(1,0,-y-x^4)$ $(1,0,xy+x^3)$ $(1,0,-x^2+y^3)$ $(1,0,x^2+y^3)$

Figure 22.4. Topological models of codimension 2 singularities of BDEs of Type 1.

discriminant has a Morse singularity of type node and the unique direction determined by the IDE at the origin has an ordinary tangency with one of the branches of the discriminant (Figure 22.4, third figure); the discriminant has a cusp singularity with a limiting tangent transverse to the unique direction determined by the IDE (Figure 22.4, last two figures).

We have a general topological result about IDEs with $F = F_p = 0$ and $F_{pp} \neq 0$ at the origin. Such IDEs can be written in the form $\omega = dy^2 + f(x, y)dx^2 = 0$ ([13]). We say that ω is finitely topologically determined if there exists $k \in \mathbb{N}$ such that any BDE ω' with $j^k\omega' = j^k\omega$ is topologically equivalent to ω.

Theorem 3.5 ([69]). *A BDE $\omega = dy^2 + f(x, y)dx^2 = 0$ with $f(x, y)$ and $g(x) = f(x, 0)$ $\mathcal{K}$-finitely determined is finitely topologically determined.*

The hypotheses in Theorem 3.5 are equivalent to $m(\omega) < \infty$, where $m(\omega)$ is the multiplicity of the IDE (§4.1).

BDEs of Type 2

As pointed out in §3.1 there are no discrete local models under formal equivalence for BDEs of Type 2. To my knowledge, smooth equivalence has only been considered in one case in [42] (see Remark 2(2)). For topological equivalence there are several ways to proceed. We review here two techniques, another one can be found in [51].

One way to proceed when seeking topological models for BDEs of Type 2 is to consider a blowing-up of the singularities of the BDEs. This is first done in [64] where topological models of the lines of curvature at umbilic points on a smooth surface in $\mathbb{R}^3$ are sought. Guíñez ([39] and elsewhere) used this technique on BDEs whose discriminants are isolated points, labelled there positive quadratic equations. However, Guíñez's technique can be extended to deal with general BDEs ([57, 66, 68]). We highlight Guíñez's method below for the case when $j^1\omega = (y, b_1x + b_2y, -y)$.

Following the notation in [39], let $f_i(\omega)$, $i = 1, 2$ denote the foliation associated to the BDE $\omega = (a, b, c)$, which is tangent to the vector field $a\frac{\partial}{\partial u} + (-b + (-1)^i\sqrt{b^2 - ac})\frac{\partial}{\partial v}$. If ψ is a diffeomorphism and $\lambda(x, y)$ is a non-vanishing real valued function, then ([39]) for $k = 1, 2$,

1. $\psi(f_k(\omega)) = f_k(\psi^*(\omega))$, if ψ is orientation preserving;
2. $\psi(f_k(w)) = f_{3-k}(\psi^*(\omega))$, if ψ is orientation reversing;
3. $f_k(\lambda\omega) = f_k(\omega)$, if $\lambda(x, y)$ is positive;
4. $f_k(\lambda\omega) = f_{3-k}(\omega)$, if $\lambda(x, y)$ is negative.

We write $\omega = (\dot{y} + M_1(x, y), b_1x + b_2y + M_2(x, y), -y + M_3(x, y))$ and consider the directional blowing-up $x = u$, $y = uv$. (We also need to consider the blowing-up $x = uv$, $y = v$.) Then, the new BDE $\omega_0 = (u, v)^*\omega$ has coefficients

$$\bar{a} = u^2(uv + M_1(u, uv)),$$

$$\bar{b} = uv(uv + M_1(u, uv)) + u(b_1u + b_2uv + M_2(u, uv)),$$

$$\bar{c} = v^2(uv + M_1(u, uv)) + 2v(b_1u + b_2uv + M_2(u, uv)) - uv + M_3(u, uv).$$

We can write $(\bar{a}, \bar{b}, \bar{c}) = u(u^2A_1, uB_1, C_1)$ with

$$A_1 = v + uN_1(u, v),$$

$$B_1 = v^2 + b_2v + b_1 + u(N_2(u, v) + vN_1(u, v)),$$

$$C_1 = v(v^2 + 2b_2v + 2b_1 + \epsilon) + u(v^2N_1(u, v) + 2vN_2(u, v) + N_3(u, v)),$$

and $M_i(u, uv) = u^2N_i(u, v)$, $i = 1, 2, 3$.

The quadratic form $\omega_1 = (u^2A_1, uB_1, C_1)$ can be decomposed into two 1-forms, and to these 1-forms are associated the vector fields

$$X_i = (u^2A_1, -uB_1 + (-1)^i\sqrt{u^2(B_1^2 - A_1C_1)}), \quad i = 1, 2.$$

These vector fields are tangent to the foliations defined by ω_1. It is clear that we can factor out the term u in X_i, with an appropriate sign change when $u < 0$.

The vector fields

$$Y_i = (uA_1, -B_1 + (-1)^i \sqrt{B_1^2 - A_1 C_1}), \quad i = 1, 2$$

are then considered. Since the blowing up is orientation preserving if $u > 0$ and orientation reversing if $u < 0$, and we factored out u twice, it follows from the observation above (see [39]) that Y_1 corresponds to the foliation $\mathcal{F}_1$ of ω if $u > 0$ and to $\mathcal{F}_2$ if $u < 0$; while Y_2 corresponds to $\mathcal{F}_2$ if $u > 0$ and to $\mathcal{F}_1$ if $u < 0$.

One studies the vector fields Y_i in a neighbourhood of the exceptional fibre $u = 0$, and blows down to obtain the configuration of the integral curves of the original BDE. One can then proceed as in [9, 67, 69] to show that any two such configurations are homeomorphic.

Another way to proceed when seeking topological models for such BDEs of Type 2 is as follows (see for example [7] for the case $c = -a$ and in [11] for the general case). Consider the associated surface to the BDE

$$M = \{(x, y, [\alpha : \beta]) \in \mathbb{R}^2, 0 \times \mathbb{R}P^1 : a\beta^2 + 2b\alpha\beta + c\alpha^2 = 0\}.$$

As the coefficients of the BDE all vanish at the origin, the exceptional fibre $0 \times \mathbb{R}P^1$ is contained in M. The discriminant function $\delta = b^2 - ac$ plays a key role. When δ has a Morse singularity the surface M is smooth and the projection $\pi : M \to \mathbb{R}^2, 0$ is a double cover of the set $\{(x, y) : \delta(x, y) > 0\}$ ([11]; see also [6] for a general relation between the singularities of δ and those of M). We label these BDEs *Morse Type 2*. The bi-valued direction field defined by the BDE in the plane lifts to a single direction field ξ on M and extends smoothly to $\pi^{-1}(0)$. Note that the exceptional fibre $0 \times \mathbb{R}P^1 \subset \pi^{-1}(\Delta)$ is an integral curve of ξ. The closure of the set $\pi^{-1}(\Delta) - (0 \times \mathbb{R}P^1)$ is the *criminant* of the equation.

There is an involution σ on $M - (0 \times \mathbb{R}P^1)$ that interchanges points with the same image under the projection to $\mathbb{R}^2, 0$. It is shown in [11] that σ extends to M when the coefficients a, b, c are analytic. (In fact the result is true when the coefficients are smooth functions; see Remark 2 in [66].) Points on M are identified with their images by σ. A bi-valued field on the quotient space $M' = M/\sigma$ is then studied and models of the configurations of the integral curves of the BDE are obtained by blowing-down.

Consider the affine chart $p = \beta/\alpha$ (we also need to consider the chart $q = \alpha/\beta$), and set

$$F(x, y, p) = a(x, y)p^2 + 2b(x, y)p + c(x, y).$$

Then, the lifted direction field is parallel to the vector field

$$\xi = F_p \frac{\partial}{\partial x} + pF_p \frac{\partial}{\partial y} - (F_x + pF_y)\frac{\partial}{\partial p}.$$

The singularities of ξ on the exceptional fibre ($F = F_p = 0$) are given by the roots of the cubic

$$\phi(p) = (F_x + pF_y)(0, 0, p)$$
$$= a_2 p^3 + (2b_2 + a_1)p^2 + (2b_1 + c_2)p + c_1,$$

where $j^1 a = a_1 x + a_2 y$, $j^1 b = b_1 x + b_2 y$, $j^1 c = c_1 x + c_2 y$. The eigenvalues of the linear part of ξ at a singularity are $-\phi'(p)$ and $\alpha_1(p)$, where

$$\alpha_1(p) = 2(a_2 p^2 + (b_2 + a_1)p + b_1).$$

Therefore, the cubic ϕ and the quadratic α_1 determine the number and the type of the singularities of ξ (see [7, 11] for details).

The calculations simplify considerably when the 1-jet of the BDE is simplified. For instance, if α_1 and ϕ have no common roots or if ϕ has more than one root ([11, 38]): then one can take

$$j^1(a, b, c) = (y, b_1 x + b_2 y, \epsilon y), \epsilon = \pm 1.$$

(If α_1 and ϕ have a common root and ϕ has only one root, we can set $j^1(a, b, c) = (x + a_2 y, 0, y)$, $a_2 > \frac{1}{4}$, [66].)

The topological classification of codimension ≤ 2 singularities of BDEs of Type 2 is obtained using the above methods. We first observe that there are no topologically stable singularities of BDEs of Type 2. The discriminant of such BDEs are always singular and generic deformations within the set of BDEs remove the singularities.

The codimension 1 singularities are classified by the number and type of the singularities of ξ when (b_1, b_2) is away from some special curves in the (b_1, b_2)-plane ([7, 11, 39], see §3.1 and Figure 22.5). There are 3 topological models when the discriminant has an A_1^+ singularity and 5 when it has an A_1^--singularity (Figure 22.6). The bifurcations of these singularities in generic families are studied in [12], see also [51] for the case $\epsilon = -1$.

Codimension 2 singularities occur at generic points on the exceptional curves in Figure 22.5. These are classified in [66] using the blowing-up method; see Figure 22.7 for the models. The models in the first row in Figure 22.7 correspond to the case where ϕ and α_1 have one common root ($\epsilon = 1$, $b_1 = \pm b_2 - 1$), those in the second row to the case where ϕ has a double root ($2b_1 + \epsilon = 0$ or $b_1 = \frac{1}{2}(b_2^2 - \epsilon)$) and those in the third row to the case where the discriminant

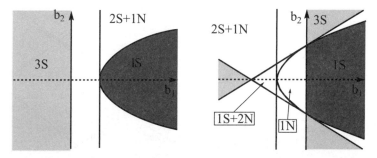

Figure 22.5. Partition of the (b_1, b_2)-plane, $\epsilon = -1$ left, $\epsilon = +1$ right. The labels refer to the number and type (S for saddle and N for node) of the singularities of ξ.

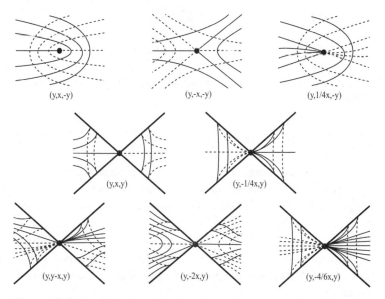

Figure 22.6. Topological models of codimension 1 singularities of Type 2 BDEs.

has a cusp singularity ($b_1 = 0$). (The first case in the second row in Figure 22.7 is also classified in [40].)

Remark 2. (1) For BDEs of Type 2, the second method is geometrical and works well when the surface M is smooth ([7, 11]). However, when M is singular, the involution σ presents some obstacles. One needs to show that σ extends to the exceptional fibre and this is not trivial. The first method is computational and the calculations are sometimes long and winding. (It is used

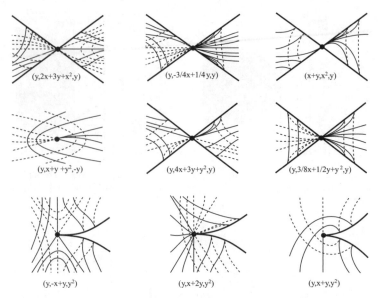

Figure 22.7. Topological models of codimension 2 singularities of Type 2 BDEs.

in [38, 39, 40, 41, 42, 57, 66, 68] to obtain topological models of BDEs with singular associated surface M.)

(2) There are classifications of some more degenerate singularities, motivated by geometric problems. We list the existing cases in the Appendix (§6). It is worth pointing out here that Gutierrez and Guíñez [42] showed that BDEs with 1-jet $(y, b_1x + b_2y, -y)$ are topologically 1-determined when the discriminant has a Morse singularity, i.e., when $b_1 \neq 0$. They also proved that any BDE with 1-jet $(y, b_1x + b_2y, -y)$ is smoothly equivalent to $(y, b_1x + b_2y + M_2(x, y), -y + M_3(x, y))$, where M_2, M_3 are smooth functions with zero 1-jets.

3.3. IDEs with first integrals

In [47], the authors studied germs of IDEs with independent first integral. An IDE is defined to be the surface $M = F^{-1}(0)$ in $PT^*\mathbb{R}^2$ endowed with its canonical contact structure given by the 1-form $\alpha = dy - pdx$. The surface M is supposed to be smooth, so is locally the image of a germ of an immersion $f : \mathbb{R}^2, 0 \rightarrow PT^*\mathbb{R}^2, z$. The IDE is then represented by the germ f.

The IDE has a first integral, that is, there exists a germ of a submersion $\mu : \mathbb{R}^2, 0 \rightarrow \mathbb{R}, 0$ such that $d\mu \wedge f^*\alpha = 0$ (this means that the integral curves of the lifted field ξ on M are images under f of the level sets of μ). As the solutions of the IDE in the plane are the images under $\pi \circ f$ of the level sets of μ, it is natural to consider the diagram $\mathbb{R}, 0 \xleftarrow{\mu} \mathbb{R}^2, 0 \xrightarrow{\pi \circ f} \mathbb{R}^2, 0$.

Consider in general a diagram (g, μ)

$$\mathbb{R}, 0 \xleftarrow{\mu} \mathbb{R}^2, 0 \xrightarrow{g} \mathbb{R}^2, 0$$

where g is a smooth map germ and μ is a germ of a submersion. The diagram (g, μ) is called an integrable diagram if there exists a germ of an immersion $f : \mathbb{R}^2, 0 \to PT^*\mathbb{R}^2, z$ such that $d\mu \wedge f^*\alpha = 0$ and $g = \pi \circ f$. Then (g, μ) is said to be induced by f.

Let $\pi : PT^*\mathbb{R}^2 \to \mathbb{R}^2$ be the natural projection. Two germs of immersions (IDEs) $f : \mathbb{R}^2, 0 \to PT^*\mathbb{R}^2, z$ and $f' : \mathbb{R}^2, 0 \to PT^*\mathbb{R}^2, z'$ are said to be equivalent if there exists germs of diffeomorphisms $\psi : \mathbb{R}^2, 0 \to \mathbb{R}^2, 0$ and $\phi : \mathbb{R}^2, \pi(z) \to \mathbb{R}^2, \pi(z')$ such that $\hat{\phi} \circ f = f' \circ \pi$, where $\hat{\phi} : PT^*\mathbb{R}^2, z \to PT^*\mathbb{R}^2, z'$ is the lift of ϕ.

The idea in [47] is to reduce the classification of IDEs with first integrals under the above equivalence to that of germs of integral diagrams. Two germs $(g, \mu), (g', \mu')$ of integral diagrams are equivalent if the diagram

$$\begin{array}{ccccc}
\mathbb{R}, 0 & \xleftarrow{\mu} & \mathbb{R}^2, 0 & \xrightarrow{g} & \mathbb{R}^2, 0 \\
\downarrow{\kappa} & & \downarrow{\psi} & & \downarrow{\phi} \\
\mathbb{R}, 0 & \xleftarrow{\mu'} & \mathbb{R}^2, 0 & \xrightarrow{g'} & \mathbb{R}^2, 0
\end{array}$$

commutes, with κ, ψ, ϕ germs of diffeomorphisms.

Suppose given two germs of IDEs f and f' with first integrals and with the set of critical points of $\pi \circ f$ and $\pi \circ f'$ nowhere dense. Then, ([47, Proposition 2.8]), f and f' are equivalent as IDEs if and only if the diagrams $(\pi \circ f, \mu)$ and $(\pi \circ f', \mu')$ are equivalent as integral diagrams.

A weaker equivalence relation of integral diagrams is introduced in [47]. Two germs $(g, \mu), (g', \mu')$ of integral diagrams are weakly equivalent if the diagram

$$\begin{array}{ccccc}
\mathbb{R}, 0 & \xleftarrow{(g, \mu)} & \mathbb{R}^2 \times \mathbb{R}, 0 & \xrightarrow{\pi_1} & \mathbb{R}^2, 0 \\
\downarrow{\phi} & & \downarrow{\Psi} & & \downarrow{\psi} \\
\mathbb{R}, 0 & \xleftarrow{(g', \mu')} & \mathbb{R}^2 \times \mathbb{R}, 0 & \xrightarrow{\pi_1} & \mathbb{R}^2, 0
\end{array}$$

commutes, with ϕ, Ψ, ϕ germs of diffeomorphisms and π_1 is the projection to the first component. Equivalent integral diagrams are weakly equivalent.

The authors in [47] used the theory of Legendrian singularities to classified generic integral diagrams under the weak equivalence. The word generic means the following. The set $Int(U, PT^*\mathbb{R}^2 \times \mathbb{R})$ of integral IDEs with first integral $(f, \mu) : U \subset \mathbb{R}^2 \to PT^*\mathbb{R}^2 \times \mathbb{R}$ is endowed with the Whitney C^∞-topology. A property is generic if the subset of (f, μ) that satisfy it is open and dense in $Int(U, PT^*\mathbb{R}^2 \times \mathbb{R})$.

Theorem 3.6 ([47], Theorem A). *For almost all differential equation germs with first integral (f, μ), the integral diagram is weakly equivalent to one of the germs in the following finite list:*

(1) $g = (u, v)$, $\mu = v$.

(2) $g = (u^2, v)$, $\mu = v - \frac{1}{3}u^3$.

(3) $g = (u, v^2)$, $\mu = v - \frac{1}{3}u$.

(4) $g = (u^3 + uv, v)$, $\mu = \frac{3}{4}u^4 + \frac{1}{2}u^2 v + v$.

(5) $g = (u, v^3 + uv)$, $\mu = v$.

(6) $g = (u, v^3 + uv^2)$, $\mu = \frac{1}{2}v^2 + v$.

An integral diagram (g, μ) is said to be of generic type if it is weakly equivalent to an integral diagram from the list in Theorem 3.6. Diagrams of generic type are then classified up to the "stronger" equivalence.

Theorem 3.7 ([47], Theorem B). *An integral diagram of generic type is equivalent to one of the following integral diagrams (g, μ)*

(1) $g = (u, v)$, $\mu = v$.

(2) $g = (u^2, v)$, $\mu = v - \frac{1}{3}u^3$.

(3) $g = (u, v^2)$, $\mu = v - \frac{1}{3}u$.

(4) $g = (u^3 + uv, v)$, $\mu = \frac{3}{4}u^4 + \frac{1}{2}u^2 v + \beta \circ g$,
   where $\beta(x, y)$ is a germ of a smooth function with $\beta(0) = 0$ and $\beta_y(0) = \pm 1$.

(5) $g = (u, v^3 + uv)$, $\mu = v + \beta \circ g$,
   where $\beta(x, y)$ is a germ of a smooth function with $\beta(0) = 0$.

(6) $g = (u, v^3 + uv^2)$, $\mu = \frac{1}{2}v^2 + \beta \circ g$,
   $\beta(x, y)$ is a germ of a smooth function with $\beta(0) = 0$ and $\beta_x(0) = 1$.

The cases (2) and (3) in Theorem 3.7 are first given in [22, 23]. In (4)–(6) Theorem 3.7, the discriminant is a cusp. We refer to [47] for further details and for the relation between Theorem 3.7, the previous classifications and Clairaut type equations.

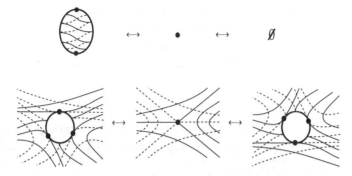

Figure 22.8. Bifurcations of a Morse Type 1 and a Morse Type 2 singularities.

4. Invariants

Consider the two BDEs

$$dy^2 + (x^2 + y^2)dx^2 = 0,$$
$$ydy^2 - 2xdxdy - ydx^2 = 0.$$

The first has a Morse Type 1 singularity and the second a Morse Type 2 singularity. Their discriminant, given by $x^2 + y^2 = 0$, has a Morse singularity of type A_1^+ (isolated point). Consider now the following 1-parameter deformations of these BDEs

$$dy^2 + (x^2 + y^2 + t)dx^2 = 0,$$
$$ydy^2 - 2xdxdy - (y + t)dx^2 = 0.$$

The discriminants in the first family $x^2 + y^2 + t = 0$ undergo the usual Morse transitions (Figure 22.8, first row). However, the discriminants in the second family $x^2 + y(y + t) = 0$ undergo transitions of type cone sections (Figure 22.8, second row). One can also show that for the first family, two folded singularities appear on the discriminant for $t < 0$ ([9], Figure 22.8). Three of these singularities appear on the discriminant of the second family for $t \neq 0$ ([18]). To explain these phenomena, an invariant of BDEs (multiplicity) is introduced in [13] and symmetric matrices are studied in [6].

4.1. The multiplicity of a BDE

We suppose here that the IDE (1.1) is given by an analytic function F and the coefficients of the BDE (1.2) are analytic functions (some of the results are also valid in the smooth category [21]). We can then complexify and denote by $\mathcal{O}(x, y, p)$ the ring of holomorphic function germs $\mathbb{C}^3, 0 \to \mathbb{C}$. We start with IDEs (1.1).

Definition 4.1 ([13]). A *singular point or zero* of the IDE given by $F(x, y, p) = 0$ is a zero of the canonical 1-form $dy - pdx$ on the criminant $F = F_p = 0$. The multiplicity of a singular point is the maximum number of zeros it can split up into under deformations of the equation $F = 0$ (including complex zeros).

Proposition 4.2 ([13]). (a) *The multiplicity of a singular point* $((x, y, p) = (0, 0, 0))$ *of the IDE* $F = 0$ *at a fold point of the projection corresponding to a zero of the vector field* ξ *is given by* $\dim_{\mathbb{C}} \mathcal{O}(x, y, p)/\mathcal{O}(x, y, p)\langle F, F_p, F_x + pF_y\rangle$.

(b) *The multiplicity of a non-fold singularity of the projection* $(x, y, p) \to (x, y)$ *is given by* $\dim_{\mathbb{C}} \mathcal{O}(x, y, p)/\mathcal{O}(x, y, p)\langle F, F_p, F_{pp}\rangle$ *provided that the vector field* ξ *is non-zero on the lift.*

(c) *If we have a non-fold singular point of the projection where the vector field* ξ *vanishes, then the multiplicity is the sum of the numbers occurring in* (a) *and* (b).

For BDEs of Type 1, the discriminant is smooth in a generic deformation, so the multiplicity is the number of folded singularities that occur in a generic deformation. If we assume that $a(0, 0) \neq 0$ and $p = 0$, then the multiplicity m of the BDE at $(0, 0, 0)$ is given by

$$m = m(\delta, a\delta_x - b\delta_y),$$

where $m(h, k)$ denotes $\dim_{\mathbb{C}} \mathcal{O}(x, y)/\mathcal{O}(x, y) \langle h, k\rangle$ ([13]).

In fact ([13]), any BDE of Type 1 can be transformed by changes of coordinates and multiplication by non-zero functions to one in the form $dy^2 + f(x, y)dx^2 = 0$. The multiplicity of the BDE is then given by

$$m = m(f, f_x) = \mu(f) + \mu(f(x, 0)) - 1,$$

where μ denotes the Milnor number of the function germ (which is the multiplicity of its Jacobian ideal).

If we consider the example at the beginning of this section, $f(x, y) = x^2 + y^2$, so $f_x(x, y) = 2x$ and $m = \dim_{\mathbb{C}} \mathcal{O}(x, y)/\mathcal{O}(x, y)\langle x^2 + y^2, 2x\rangle = 2$. This explains why we have two folded singularities appearing in the deformation (Figure 22.8, first row)

We turn now to BDEs of Type 2 where we complexify the coefficients.

Definition 4.3 ([13]). The multiplicity of a BDE of Type 2 is defined to be the (maximum) number of non-degenerate singular points of the perturbed equations within the set of BDEs, where this is finite.

We observe that if we deform a BDE of Type 2 in the set of all IDEs, then its multiplicity is infinite. Consider, for example, the BDE $yp^2 + 2xp - y = 0$ which has multiplicity 3 using Definition 4.3 (see below). If we view it as an IDE and consider the deformation $F = tp^n + yp^2 + 2xp - y = 0$, then the equations $F = F_p = F_{pp} = 0$ have a zero at the origin of multiplicity n. Therefore, using Definition 4.1, the multiplicity of the above BDE is infinite. (The key point here is that one cannot use the division theorem to reduce the BDE to an IDE of a fixed degree.)

Proposition 4.4 ([13]). *The multiplicity of a BDE is given by*

$$m = \tfrac{1}{2}m(\delta, a\delta_x^2 - 2b\delta_x\delta_y + c\delta_y^2)$$
$$= m(\delta, a\delta_x - b\delta_y) - m(a, b)$$
$$= m(\delta, b\delta_x - c\delta_y) - m(b, c).$$

The formula in Proposition 4.4 is also valid for BDEs of Type 1. For example, when $a(0, 0) \neq 0$, $m(a, b) = 0$ and we recover the formula for the multiplicity of a BDE of Type 1 given by $m(\delta, a\delta_x - b\delta_y)$.

If we consider the second example at the beginning of this section, we have $a = y$, $b = -x$, $c = -y$, so the multiplicity $m = m(x^2 + y^2, 4yx) - m(y, x) = 4 - 1 = 3$, which explains why we have three folded singularities appearing in the deformation of the BDE (Figure 22.8, second row).

4.2. The singularities of the discriminant

To a BDE with coefficients (a, b, c) is associated the family of symmetric matrices

$$S(x, y) = \begin{pmatrix} a(x, y) & b(x, y) \\ b(x, y) & c(x, y) \end{pmatrix}.$$

The discriminant of the BDE is precisely the determinant of S. Bruce classified in [6] families of symmetric matrices up to an equivalence relation that preserves the singularities of the determinant. Let $S(n, \mathbb{K})$ denote the space of $n \times n$-symmetric matrices with coefficients in the field $\mathbb{K}$ of real or complex numbers. A family of symmetric matrices is a smooth map germ $\mathbb{K}^r, 0 \to S(n, \mathbb{K})$. Denote by $\mathcal{G}$ the group of smooth changes of parameters in the source and parametrised conjugation in the target. Thus, two smooth map-germs A, B are $\mathcal{G}$ equivalent if $B = X^t(A \circ \phi^{-1})X$, where ϕ is a germ of a diffeomorphism $\mathbb{K}^r, 0 \to \mathbb{K}^r, 0$ and $X : \mathbb{K}^r, 0 \to GL(n, \mathbb{K})$. A list of all the $\mathcal{G}$-simple singularities of families of symmetric matrices is obtained in [6]. For more on symmetric matrices see [10, 36, 37].

The 2×2 matrix associated to a BDE of Type 1 is $\mathcal{G}$-equivalent to one in the form

$$\begin{pmatrix} 1 & 0 \\ 0 & f(x, y) \end{pmatrix}.$$

It turns out that, in this case, the $\mathcal{G}$-action reduces to the action of the contact group $\mathcal{K}$ on the ring of function germs $f : \mathbb{K}^2, 0 \to \mathbb{K}, 0$ ([6]). This explains, for instance, why we get the usual Morse transitions in the discriminants of the family of BDEs $dy^2 + (x^2 + y^2 + t)dx^2 = 0$ at the beginning of this section (Figure 22.8, first row).

For BDEs of Type 2, the $\mathcal{G}$-action does not reduce to the action of the contact group $\mathcal{K}$. For the example at the beginning of this section, the matrix $\begin{pmatrix} y & -x \\ -x & -y \end{pmatrix}$ of the BDE is 1-$\mathcal{G}$-determined and a versal $\mathcal{G}$-deformation is given by $\begin{pmatrix} y & -x \\ -x & -y + t \end{pmatrix}$. The zero sets of the determinants of these matrices undergo transitions of type cone sections (Figure 22.8, second row).

It is worth observing that the action $\mathcal{G}$ models the singularities of the discriminant of a BDE as well as its deformations in generic families of BDEs. The action does not preserve the pair of foliations determined by the BDE. Nevertheless, it provides important information when studying families of BDEs (see §5). All the key local information about the pair of foliations determined by the BDE occurs on the discriminant. It is also worth mentioning that all the $\mathcal{G}$-invariants associated the matrix of a BDE are invariants of the BDE.

4.3. The index of a BDE

The index of a BDE with discriminant an isolated point is defined as the index of one direction field determined by the BDE at the singular point. In [19, 20, 21], Challapa gave the following definition of the index of a BDE at a singular point (with discriminant not necessary an isolated point) when the coefficients are real analytic functions. Consider a family of BDEs $(a(x, y, t), b(x, y, t), c(x, y, t))$. The family is called a good perturbation if the discriminant δ_t is a regular curve for $t \neq 0$ and the BDEs for $t \neq 0$ fixed have only folded singularities. Challapa showed that such good perturbations exist. He defined the index of a folded saddle to be $K(S) = -1/2$ and the index of a folded node and focus to be $K(N) = K(F) = 1/2$ and gave the following definition of the index of an analytic BDE.

Definition 4.5. Let $\omega = (a, b, c)$ be a germ, at the origin, of an analytic BDE and ω_t be a good perturbation of ω. The index of ω at the origin is defined by

$$I(\omega) = \sum_i K_{\delta_t}(z_i) + \sum_{\delta^t(u_i)<0} \text{index}_{u_i} \nabla\delta_t$$

where $\nabla\delta_t$ denotes the gradient of δ_t, z_i are non-degenerate singular points of ω_t and u_i are the critical points in the negative part of δ_t (i.e., $\nabla\delta_t(u_i) = 0$ and $\delta_t(u_i) < 0$).

Challapa shows that the index I is independent of the choice of a good deformation, it is invariant under analytic changes of coordinates and it satisfies the Poincaré-Hopf formula.

4.4. Cr-invariant of asymptotic curves at folded singularities

Consider a surface S immersed in $\mathbb{R}^3$. Uribe-Vargas ([71]) produced an invariant of the folded singularities of the asymptotic curves BDEs on S. As asymptotic curves capture the contact of S with lines, one can also consider S immersed in an affine or projective 3-space. Recall that the discriminant of the BDE of the asymptotic curves is the parabolic set P of S. The flecnodal curve F is a curve on which an asymptotic direction has higher contact with the surface. In general the flecnodal curve is a smooth curve on S and is tangent to the parabolic set at the cusp of Gauss/godron point (i.e., at the folded singularities of asymptotic BDE). The flecnodal curve can also be captured using Legendre duality on the BDE of the asymptotic curves, see [14]. There is another curve D on S, called the *conodal curve*. It is the closure of the locus of points of contact of S with its bitangent planes. This curve is in general tangent to P and F at a cusp of Gauss.

Let g denotes the cusp of Gauss/godron point. Consider $\pi : PT^*S \to S$ endowed with the canonical contact structure and the Legendrian lifts L_D, L_F, L_P consisting of the contact elements of S tangent to D, F, P around g. Consider also L_g the fibre over g of π. The four Legendrian curves are tangent to the same contact plane Π and their tangent directions determine four lines l_D, l_F, l_P, l_g through the origin in Π.

Definition 4.6 ([71]). The cr-invariant $\rho(g)$ of a godron g is defined as the cross-ratio of the lines $l_D, l_F, l_d P$ and l_g of Π: $\rho(g) = (l_F, l_D, l_P, l_g)$.

Uribe-Vargas used ρ to obtain a classification of the configurations of the curves D, F, P at a cusp of Gauss; see [71] for more details.

5. Bifurcations

We consider in this section families of germs of BDEs. Two germs of families of BDEs $\tilde{\omega}$ and $\tilde{\tau}$, depending smoothly on the parameters t and s respectively, are said to be *locally fibre topologically equivalent* if, for any of their representatives, there exist neighbourhoods U and W of 0 in respectively the phase space (x, y) and the parameter space t, and a family of homeomorphisms h_t, for $t \in W$, all defined on U such that h_t is a topological equivalence between $\tilde{\omega}_t$ and $\tilde{\tau}_{\psi(t)}$, where ψ is a homeomorphism defined on W. (The map h_t is not required to be continuous in t.)

We associate to a germ of an r-parameter family of BDEs $\tilde{\omega} = (\tilde{a}, \tilde{b}, \tilde{c})$ the jet-extension map

$$\Phi: \quad \mathbb{R}^2 \times \mathbb{R}^r, (0, 0) \quad \rightarrow \quad J^k(2, 3)$$
$$((x, y), t) \quad \mapsto \quad j^k(\tilde{a}, \tilde{b}, \tilde{c})_t|_{(x,y)}$$

where $J^k(2, 3)$ denotes the vector space of polynomial maps of degree $\leq k$ from $\mathbb{R}^2$ to $\mathbb{R}^3$, and $j^k(\tilde{a}, \tilde{b}, \tilde{c})_t|_{(x,y)}$ is the k-jet of $(\tilde{a}, \tilde{b}, \tilde{c})$ at (x, y) with t fixed.

The singularity type of the BDE $\tilde{\omega}_0$ determines a semi-algebraic set V in $J^k(2, 3)$ of codimension, say, m. The family $\tilde{\omega}$ is said to be a *generic family* if the map Φ is transverse to V in $J^k(2, 3)$. A necessary condition for genericity is of course $r \geq m$. It follows from Thom's Transversality Theorem that the set of generic families is residual in the set of smooth map germs $\mathbb{R}^2 \times \mathbb{R}^r, 0 \rightarrow \mathbb{R}^3, 0$.

The *bifurcation set* of a generic family is the set of parameters t where the associated BDE has a singularity of codimension ≥ 1 at some point $p \in U$. This gives a stratification $\mathcal{S}$ of the parameter space consisting of following strata: the origin (if the singularity at $t = 0$ is isolated), and local and semi-local singularities of codimension s, $1 \leq s \leq m - 1$. The singularity of $\tilde{\omega}_0$ is local, but semi-local singularities can appear in $\tilde{\omega}_t$, for $t \neq 0$. The semi-local singularities are very hard to deal with, and there is so far no general approach to deal with them. Each case is dealt with separately. There is a result in [69] which is worth mentioning here.

Lemma 5.1 ([69]). *There are no Poincaré-Andronov (Hopf) bifurcations on the lifted field ξ of an IDE (1.1) at a regular point on the criminant.*

When studying bifurcations of a BDE $\tilde{\omega}_0$, the aim is to show that any two generic families of $\tilde{\omega}_0$ are (fibre) topologically equivalent. The strategy we adopted in [66, 69] is the following.

– Obtain a model for the BDE at $t = 0$ (using the methods in §3.2).
– Reduce the N-jet of the family to a normal form (using the formal reduction technique in §3.1).

- Obtain a condition for the family to be generic.
- Show that the bifurcation sets of generic families are homeomorphic.
- Obtain the configuration of the discriminant in each stratum of $\mathcal{S}$ (using the symmetric matrices framework §4.2).
- Show that the number of singularities, their type and their position on the discriminant are constant in each stratum of $\mathcal{S}$. (The results in [56] are of use here.)
- Show that the configurations of the integral curves have a constant topological type in each stratum of $\mathcal{S}$.

Models of generic families of BDEs with local codimension 2 singularities and their bifurcations in the families are given in [66, 69]. Singularities of codimension 1 are dealt with in [9, 18, 51, 67]. Some degenerate cases are studied in [54].

6. Appendix

In §2 the pairs of foliations asymptotic/characteristic/principal curves are defined on an immersed surface. These pairs (or some of them) are also considered on algebraic surfaces or surfaces with a cross-cap singularity. They are given by BDEs whose singularities are of higher codimension ([34, 57, 63, 68]).

When $F = F_p = F_{pp} = 0$ at the origin, the solution curves of the IDE form a web (see §3.3). A classification of hexagonal analytic 3-webs $p^3 + ap^2 + bp + c = 0$ is given in [2]. Other types of n-web occur in differential geometry. For example, the asymptotic curves on surfaces in $\mathbb{R}^5$ are given by a quintic differential equation $p^5 + a_1 p^4 + a_2 p^3 + a_3 p^2 + a_4 p + a_5 = 0$, where a_i, $i = 1 \ldots 5$, are smooth functions in (x, y) ([53, 59]).

Systems of IDEs $F_1(t, x, p) = \ldots = F_n(t, x, p) = 0$ with $x = (x_1, \ldots, x_n)$, $p = (p_1, \ldots, p_n)$, $p_i = dx_i/dt$ are considered in [61].

Homogeneous differential equations of degree greater than 2 are also considered. In [44] are defined lines of curvature on surface in $\mathbb{R}^4$ that are given by a quartic differential equation $a_0 dy^4 + a_1 dy^3 dx + a_2 dy^2 dx^2 + a_3 dy dx^3 + a_4 dx^4 = 0$, where a_i, $i = 1 \ldots 4$, are smooth functions in (x, y). The coefficients all vanish at some special points on the surface. The configuration of the solution curves at such points is given in [44].

In [31] Fukui and Nuño-Ballesteros studied equations of degree n which have n real solutions away from some isolated (singular) points where all the coefficients vanish. They defined the index of such BDEs at a singular point and proved a Poincaré-Hopf type theorem (see also [30]). They also gave a classification of the configuration of the n-web around generic singular points.

Below are some topological models of singularities of BDEs of Type 2 with codimension higher than 2 (see Figure 22.9). The models are of BDEs with discriminant having a given $\mathcal{K}$-singularity type. However, they do not form an exhaustive list of the topological types of BDEs with discriminant having that $\mathcal{K}$-singularity type. The models are topologically determined by the k-jet of the BDE, where k is the highest degree of the coefficients of the equation. These are as follows.

– The discriminant has an A_1^--singularity, α_1 and ϕ have two common roots, [68]:

$$(y, -x + y^2, y), \ (y, -x + xy, y).$$

– The discriminant has an A_2-singularity, $j^1 w \sim (x, by, 0)$, [57]:

$$(x, -y, x^2), \ (x, y, x^2).$$

– The discriminant has an A_3-singularity, $j^1 w \sim (0, x + y, 0)$, [41] and [43] respectively:

$$(y, x^2, -y), \ (y^2, x + y, -y^2).$$

– The discriminant has an A_3-singularity, $j^1 w \sim (0, b_0 x, y)$, [68]:

$$(\pm y^3, b_0 x + b_2 y^2, y).$$

The topological type is constant in open regions determined by some exceptional curves in the (b_0, b_2)-plane (see [68]).

– The discriminant has an A_3-singularity, $j^1 w \sim (\alpha x + y, \pm x, 0)$, [49]:

$$(y, x, \pm y^3), \ (y, -x, \pm y^3).$$

– The discriminant has an $X_{1,2}$-singularity [68]:

$$(x^2 + y^4, -xy, -x^2 + 2y^2 + y^3), \ (x^2 + y^4, -xy, -x^2 + 2y^2 + xy^2).$$

– The discriminant has an $Y_{5,6}$-singularity [57]:

$$(x^2, -xy, 2y^2 - x^3).$$

– Under some conditions, a BDE with discriminant an isolated point is topologically equivalent to its principal part defined by Newton polyhedra [45].

Remark 3. Some of the configurations in Figure 22.9 are topologically equivalent to those of less degenerate singularities. For instance, those in the second

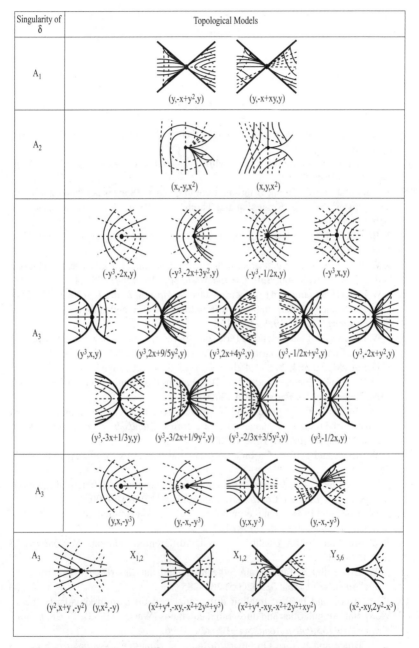

Figure 22.9. Some topological models of Type 2 BDEs with singularities of codimension > 2.

row are equivalent to a folded node and saddle respectively (Figure 22.2). However, their topological codimensions are distinct and so are their bifurcations.

References

1. V. I. Arnold, *Geometrical methods in the theory of ordinary differential equations*, Springer, Berlin, 1983.
2. S. I. Agafonov, On implicit ODEs with hexagonal web solutions. *J. Geom. Anal.* **19** (2009) 481–508.
3. T. Banchoff and R. Thom, Erratum et compléments: "Sur les points paraboliques des surfaces" by Y. L. Kergosien and Thom, C. R. Acad. Sci. Paris Sér. A-B **291** (1980) A503–A505.
4. T. Banchoff, T. Gaffney and C. McCrory, *Cusps of Gauss Mappings*. Pitman, New York, 1982.
5. J. W. Bruce, A note on first order differential equations of degree greater than one and wavefront evolution. *Bull. London Math. Soc.* **16** (1984) 139–144.
6. J. W. Bruce, On families of symmetric matrices. *Moscow Math. J.* **3** (2003) 335–360.
7. J. W. Bruce and D. Fidal, On binary differential equations and umbilics. *Proc. Royal Soc. Edinburgh* **111A** (1989) 147–168.
8. J. W. Bruce, G. J. Fletcher and F. Tari, Zero curves of families of curve congruences. Real and Complex Singularities (São Carlos, 2002), Ed. T. Gaffney and M. Ruas, Contemp. Math., **354**, Amer. Math. Soc., Providence, RI, June 2004.
9. J. W. Bruce, G. J. Fletcher and F. Tari, Bifurcations of implicit differential equations. *Proc. Royal Soc. Edinburgh* **130A** (2000) 485–506.
10. J. W. Bruce, V. V. Goryunov and V. M. Zakalyukin, Sectional singularities and geometry of families of planar quadratic forms. *Trends in singularities*, 83–97, Trends Math., Birkhauser, Basel, 2002.
11. J. W. Bruce and F. Tari, On binary differential equations. *Nonlinearity* **8** (1995) 255–271.
12. J. W. Bruce and F. Tari, Generic 1-parameter families of binary differential equations of Morse type. *Discrete Contin. Dynam. Systems* **3** (1997) 79–90.
13. J. W. Bruce and F. Tari, On the multiplicity of implicit differential equations. *J. of Differential Equations* **148** (1998) 122–147.
14. J. W. Bruce and F. Tari, Duality and implicit differential equations. *Nonlinearity* **13** (2000) 791–811.
15. J. W. Bruce and F. Tari, Dupin indicatrices and families of curve congruences. *Trans. Amer. Math. Soc.* **357** (2005) 267–285.
16. J. W. Bruce and F. Tari, Implicit differential equations from singularity theory viewpoint. Singularities and differential equations (Warsaw, 1993), 23–38, *Banach Center Publ.* **33** *Polish Acad. Sci., Warsaw,* 1996.
17. J. W. Bruce and F. Tari, On binary differential equations. *Nonlinearity* **8** (1995) 255–271.
18. J. W. Bruce and F. Tari, Generic 1-parameter families of binary differential equations. *Discrete Contin. Dyn. Syst.* **3** (1997) 79–90.

19. L. S. Challapa, Index of quadratic differential forms. *Contemporary Mathematics* **459** (2008) 177–191.
20. L. S. Challapa, Invariants of Binary Differential Equations. *J. Dyn. Control Syst.*, to appear.
21. L. S. Challapa, Indice de equações diferenciais binárias. PhD thesis, USP-São Carlos, Brazil, 2006.
22. M. Cibrario, Sulla reduzione a forma delle equationi lineari alle derviate parziale di secondo ordine di tipo misto. *Accademia di Scienze e Lettere, Instituto Lombardo Redicconti* **65** (1932) 889–906.
23. L. Dara, Singularités génériques des équations differentielles multiformes. *Bol. Soc. Brasil Math.* **6** (1975) 95–128.
24. A. A. Davydov, Normal forms of differential equations unresolved with respect to derivatives in a neighbourhood of its singular point. *Functional Anal. Appl.* **19** (1985) 1–10.
25. A. A. Davydov, *Qualitative control theory*. Translations of Mathematical Monographs **142** AMS, Providence, RI, 1994.
26. A. A. Davydov and L. Ortiz-Bobadilla, Smooth normal forms of folded elementary singular points. *J. Dynam. Control Systems* **1** (1995) 463–482.
27. A. A. Davydov and E. Rosales-González, Smooth normal forms of folded resonance saddles and nodes and complete classification of generic linear second order PDE's on the plane. International Conference on Differential Equations (Lisboa, 1995), 59–68, *World Sci. Publishing*, 1998.
28. L. P. Eisenhart, *A Treatise on the Differential Geometry of Curves and Surfaces.* Ginn and Company, 1909.
29. G. J. Fletcher, *Geometrical problems in computer vision.* Ph.D thesis, Liverpool University, 1996.
30. T. Fukui and J. J. Nuño-Ballesteros, Isolated roundings and flattenings of submanifolds in Euclidean spaces. *Tohoku Math. J.* **57** (2005) 469–503.
31. T. Fukui and J. J. Nuño-Ballesteros, Isolated singularities of binary differential equations of degree n. Preprint.
32. R. Garcia and J. Sotomayor, Harmonic mean curvature lines on surfaces immersed in $\mathbb{R}^3$. *Bull. Braz. Math. Soc.* **34(2)** (2003) 303–331.
33. R. Garcia and J. Sotomayor, Structural stability of parabolic points and periodic asymptotic lines, Workshop on Real and Complex Singularities (São Carlos, 1996), Mat. Contemp. **12** (1997) 83–102.
34. R. Garcia, C. Gutierrez and J. Sotomayor, Lines of principal curvature around umbilics and Whitney umbrellas. *Tohoku Math. J.* **52** (2000) 163–172.
35. R. Garcia and J. Sotomayor, Codimension two umbilic points on surfaces immersed in $\mathbb{R}^3$. *Discrete Contin. Dynam. Systems* **17** (2007) 293–308.
36. V. Goryunov and D. Mond, Tjurina and Milnor numbers of matrix singularities. *J. London Math. Soc.* **72** (2005) 205–224.
37. V. Goryunov abd V. Zakalyukin, Simple symmetric matrix singularities and the subgroups of Weyl groups A, D, E. Dedicated to Vladimir I. Arnold on the occasion of his 65th birthday. *Mosc. Math. J.* **3** (2003) 507–530, 743–744.
38. V. Guíñez, Positive quadratic differential forms and foliations with singularities on surfaces. *Trans. Amer. Math. Soc.* **309** (1988) 447–502.

39. V. Guíñez, Locally stable singularities for positive quadratic differential forms. *J. Differential Equations* **110** (1994) 1–37.

40. V. Guíñez, Rank two codimension 1 singularities of positive quadratic differential forms. *Nonlinearity* **10** (1997) 631–654.

41. V. Guíñez and C. Gutierrez, Rank-1 codimension one singularities of positive quadratic differential forms. *J. Differential Equations* **206** (2004) 127–155.

42. C. Gutierrez and V. Guíñez, Positive quadratic differential forms: linearization, finite determinacy and versal unfolding. *Ann. Fac. Sci. Toulouse Math.* **5** (1996) 661–690.

43. C. Gutiérrez, J. Sotomayor and R. Garcia, Bifurcations of umbilic points and related principal cycles. *J. Dynam. Differential Equations* **16** (2004) 321–346.

44. C. Gutierrez, I. Guadalupe, R. Tribuzy and V. Guíñez, Lines of curvatures on surface in $\mathbb{R}^4$. *Bol. Soc. Bras. Mat.* **28** (1997) 233–251.

45. C. Gutierrez, R. D. S. Oliveira and M. A. Teixeira, Positive quadratic differential forms: topological equivalence through Newton polyhedra. *J. Dyn. Control Syst.* **12** (2006) 489–516.

46. C. Gutierrez and M. A. S. Ruas Indices of Newton non-degenerate vector fields and a conjecture of Loewner for surfaces in R^4. (English summary) Real and complex singularities, 245–253, Lecture Notes in Pure and Appl. Math. **232** Dekker, New York, 2003.

47. A. Hayakawa, G. Ishikawa, S. Izumiya, K. Yamaguchi, Classification of generic integral diagrams and first order ordinary differential equations. *Internat. J. Math.* **5** (1994) 447–489.

48. S. Izumiya, Timelike hypersurfaces in de Sitter space and Legendrian singularities. *J. Math. Sciences.* **144** (2007) 3789–3803.

49. S. Izumiya and F. Tari, Self-adjoint operators on surfaces with singular metrics. To appear in *J. Dyn. Control Syst.*.

50. Y. Kergosien and R. Thom, Sur les points paraboliques des surfaces, C. R. Acad. Sci. Paris Sér. A-B **290** (1980) A705–A710.

51. A. G. Kuz'min, Nonclassical equations of mixed type and their applications in gas dynamics. *International Series of Numerical Mathematics*, 109. Birkhauser Verlag, Basel, 1992.

52. T. Miernowski, Forme normales d'une métrique mixte analytique réelle générique. *Ann. Fac. Sci. Toulouse Math. Sér.* 6 **4** (2007) 923–946.

53. D. K. H. Mochida, M. C. Romero-Fuster and M. A. S. Ruas, Inflection points and nonsingular embeddings of surfaces in $\mathbb{R}^5$. *Rocky Mountain Journal of Maths* **33** (2003) 995–1010.

54. A. C. Nabarro and F. Tari, Families of surfaces and conjugate curve congruences. *Adv. Geom.* **9** (2009) 279–309.

55. R. Occhipinti, Sur un systeme de lignes d'une surface. *L'emseignement Mathematiques*, **16** (1914) 38–44.

56. R. D. S. Oliveira and F. Tari, On pairs of regular foliations in the plane. *Hokkaido Math. J.* **31** (2002) 523–537.

57. J. Oliver, On pairs of geometric foliations of a parabolic cross-cap. To appear in *Qual. Theory Dyn. Syst.*

58. J. Oliver, On the characteristic curves on a smooth surface. Preprint, 2009.

59. M. C. Romero-Fuster, M. A. Ruas and F. Tari, Asymptotic curves on surfaces in $\mathbb{R}^5$. *Commun. Contemp. Math.* **10** (2008) 309–335.

60. L. Raffy, Sur le réseau diagonal conjugué. *Bull. Soc. Math. France* **30** (1902) 226–233.

61. A. O. Remizov, Generic singular points of implicit differential equations. *Vestnik Moskov. Univ. Ser. I Mat. Mekh.* **5** (2002) 10–16, **71**; translation in *Moscow Univ. Math. Bull.* **57(5)** (2002) 9–15 (2003).

62. J. Sotomayor and C. Gutierrez, Structurally stable configurations of lines of curvature and umbilic points on surfaces. Monographs of the Institute of Mathematics and Related Sciences, 3. IMCA, Lima, 1998.

63. J. Sotomayor and R. Garcia, Lines of curvature on algebraic surfaces. *Bull. Sci. Math.* **120** (1996) 367–395.

64. J. Sotomayor and C. Gutierrez, Structurally stable configurations of lines of principal curvature. Bifurcation, ergodic theory and applications (Dijon, 1981), 195–215, Astérisque, 98-99, Soc. Math. France, Paris, 1982.

65. F. Tari, Self-adjoint operators on surfaces in $\mathbb{R}^n$. *Differential Geom. Appl.* **27** (2009) 296-306.

66. F. Tari, Two-parameter families of binary differential equations. *Discrete Contin. Dyn. Syst.* **22** (2008) 759–789.

67. F. Tari, Geometric properties of the solutions of implicit differential equations. *Discrete Contin. Dyn. Syst.* **17** (2007) 349–364.

68. F. Tari, On pairs of geometric foliations on a cross-cap. *Tohoku Math. J.* **59** (2007) 233–258.

69. F. Tari, Two-parameter families of implicit differential equations. *Discrete Contin. Dyn. Syst.* **13** (2005) 139–162.

70. R. Thom, Sur les équations différentielles multiformes et leurs intégrales singulières. *Bol. Soc. Brasil. Mat.* **3** (1972) 1–11.

71. R. Uribe-Vargas, A projective invariant for swallowtails and godrons, and global theorems on the flecnodal curve. *Mosc. Math. J.* **6** (2006) 731–768, 772.

Farid Tari
Department of Mathematical Sciences
Durham University
Science Laboratories
South Road
Durham DH1 3LE
UK
email: farid.tari@durham.ac.uk

23

Bi-Lipschitz equisingularity

DAVID TROTMAN

Abstract

Much of the recent work of both Terry Gaffney and Maria Ruas
has centred on problems of equisingularity. I discuss recent results
on bilipschitz equisingularity including some important results
obtained by my former student Guillaume Valette, in particular
a bilipschitz version of the Hardt semiagebraic triviality theorem
and the resolution of a conjecture of Siebenmann and Sullivan
dating from 1977.

1. An oft-heard slogan

Stratifications are often used in singularity theory via the slogan that follows:

*"Given an analytic variety (or semialgebraic set or subanalytic set) take
some Whitney stratification. Then the stratified set is locally topologically trivial
along each stratum by the first isotopy theorem of Thom-Mather."*

Statements like this have been made hundreds of times, after the publication
in 1969 of René Thom's foundational paper "Ensembles et morphismes stratifiés" [35], and John Mather's 1970 Harvard notes on topological stability [19].
(A detailed published proof of the Thom-Mather isotopy theorem, somewhat
different to those of Thom and Mather, can be found in the write-up of the
1974-75 Liverpool Seminar published by Springer Lecture Notes in 1976, in
the second chapter written by Klaus Wirthmüller [7].)

It seems to be not yet so well-known that in the above statements one may
replace "Whitney stratification" by "Mostowski stratification" and "locally

2000 *Mathematics Subject Classification* 32S15 (primary), 14P10, 14P15, 32B15, 32B20, 32S60,
58A35 (secondary).

topologically trivial" by "locally bi-Lipschitz trivial", despite the fact that the corresponding theorems were published by Tadeusz Mostowski over twenty years ago in 1985 [22]. In this paper I shall describe the background to these and related developments as well as some recent improvements due to Guillaume Valette.

2. Existence of Whitney and Verdier stratifications

Recall that Hassler Whitney proved in 1965 that every complex analytic variety, and real analytic variety, admits a Whitney (b)-regular stratification [45, 46]. Also in 1965, Stanisław Łojasiewicz [17] published a proof of the existence of Whitney stratifications for the more general semianalytic sets, which include semialgebraic sets. (Łojasiewicz's classic account [17] of his theory of semianalytic sets is now available in a new TeX format due to Michel Coste as a downloadable pdf file on Coste's home page.)

Independently, a further generalisation was obtained by Thom [35], Heisuke Hironaka [12, 13], and then Robert Hardt [8, 9] who each gave existence theorems applying to what came to be called subanalytic sets. Thom's 1969 paper [35] refers to PSA sets, for "Projections of Semi-Analytic" sets. These were called subanalytic sets by Hironaka from 1972 on, and this terminology was adopted by the research community.

Jean-Louis Verdier [44] improved these existence theorems for Whitney stratifications for subanalytic sets in 1976 by showing, using Hironaka's resolution of singularities, that one can strengthen Whitney's condition (b) to find a subanalytic stratification satisfying what Verdier called condition (w), the 'w' standing for Whitney. Verdier remarked in his paper and in a seminar at Paris 7 in the spring of 1976 that he knew of no subanalytic example where (b) held and (w) did not. To answer this question in August 1976 I constructed the first semialgebraic example, while attending the Nordic Summer School on Real and Complex Singularities in Oslo [36]. During the following year I discovered a very simple real algebraic example, $\{y^4 = t^4 x + x^3\}$, which I put into my 1977 Warwick Ph. D. thesis [37], and published in a 1979 joint paper [3] with Hans Brodersen, then a Ph. D. student of Per Holm and Andrew du Plessis. Hans had independently discovered similar examples.

Two more distinct proofs of Verdier's existence theorem were obtained by the Cracow School in the mid 1980s using Łojasiewicz's theory of normal partitions, avoiding resolution of singularities. These were due firstly to Zofia Denkowska and Kristina Wachta [6], and secondly to Łojasiewicz, Jacek Stasica and Wachta [18]. As well as Łojasiewicz's classic notes [17] mentioned above, a very useful source for the methods of the Cracow School is the recently published book "Ensembles sous-analytiques à la Polonaise" [5], written by Denkowska and Stasica in 1984, and widely distributed as a manuscript since then.

3. O-minimal structures

This notion was developed in the late 1980s after it was noticed that many proofs of geometric properties of semialgebraic sets and maps could be carried over verbatim for subanalytic sets and maps.

Definition 3.1. An *o-minimal structure* on the real line $\mathbf{R}$ is a sequence $S = (S_n)_{n \in \mathbf{N}}$ such that for each n:

(1) S_n is a boolean algebra of subsets of $\mathbf{R}^n$, that is, S_n is a collection of subsets of $\mathbf{R}^n$, $\emptyset \in S_n$, and if A, $B \in S_n$, then $A \cup B \in S_n$, and $\mathbf{R}^n - A \in S_n$;

(2) $A \in S_n \Rightarrow A \times \mathbf{R} \in S_{n+1}$ and $\mathbf{R} \times A \in S_{n+1}$;

(3) $\{(x_1, \ldots, x_n) \in \mathbf{R}^n : x_i = x_j\} \in S_n$ for $1 \le i < j \le n$;

(4) $A \in S_{n+1} \Rightarrow \pi(A) \in S_n$, where $\pi : \mathbf{R}^{n+1} \to \mathbf{R}^n$ is the usual projection map;

(5) $\{r\} \in S_1$ for each $r \in \mathbf{R}$, and $\{(x, y) \in \mathbf{R}^2 : x < y\} \in S_2$;

(6) the only sets in S_1 are the finite unions of intervals and points. ("Interval" always means "open interval", with infinite endpoints allowed.)

A set $A \subset \mathbf{R}^n$ is said to be a *definable set* if $A \in S_n$. A map $f : A \to \mathbf{R}^m$ is said to be a *definable map* if its graph is definable.

If for every definable map $f : \mathbf{R} \to \mathbf{R}$ there exist $d \in \mathbf{N}$ and $K > 0$ such that $|f(x)| \le x^d$ for all $x > K$, the structure S is said to be *polynomially bounded*.

A theorem of Chris Miller [21] states that an o-minimal structure is not polynomially bounded if and only if the exponential function is definable.

Examples of (polynomially bounded) o-minimal structures are

- the semilinear sets,
- the semialgebraic sets (by the Tarski-Seidenberg theorem),
- the global subanalytic sets, i.e. the subanalytic sets of $\mathbf{R}^n$ whose (compact) closures in $\mathbf{P}^n(\mathbf{R})$ are subanalytic (using Gabrielov's complement theorem).

An important theorem of Wilkie [47] proves the o-minimality of the structure generated by graphs of polynomials and the graph of the exponential function. This was an old conjecture of Tarski.

The results in the previous section were extended ten or so years ago to the more general class of definable sets in arbitrary o-minimal structures (defined above) by Ta Lê Loi, who first proved the existence of Whitney stratifications in his Cracow thesis and then proved the existence of Verdier stratifications for such sets [16].

4. Local triviality along strata

Now we describe the developments concerning the local topological triviality of Whitney stratifications.

Definition 4.1. Let a closed subset Z of a smooth manifold M be stratified, with the strata indexed by a set Σ, so that $Z = \cup\{X : X \in \Sigma\}$. A stratified vector field $v = \{v_X : X \in \Sigma\}$ is said to be *rugose* if for each point $p \in Z$, where p belongs to the stratum Y say, there is a constant $C > 0$ and a neighbourhood U of p in M such that for each point $y \in Y \cap U$, and each point $x \in Z \cap U$, if X denotes the stratum of Σ containing x, then $\|v_X(x) - v_Y(y)\| < C|x - y|$ [44].

Note that this is close to being Lipschitz with constant C on Z, but is weaker: for v to be Lipschitz one would need to allow y to belong to X and to any other strata incident to Y, not just to Y.

A stratified homeomorphism $h : Z \to Z$ is similarly said to be *rugose* if for each point $p \in Z$, belonging to a stratum Y say, there is a constant $C > 0$ and a neighbourhood U of p in Z such that for each point $x \in U$ belonging to some stratum X, and each point $y \in Y \cap U$, $|h(x) - h(y)| < C|x - y|$.

If a homeomorphism h is controlled according to the definition of Mather, $|h(x) - h(y)| = |x - y|$, so that h is rugose.

The work of Thom [34, 35] and Mather [19], [20] established what is referred to as the Thom-Mather first isotopy theorem (or lemma!): every Whitney (b)-regular stratified set is locally topologically trivial along strata. This is proved by integrating controlled vector fields. The local topological trivialisation is given by a controlled stratified homeomorphism.

Verdier proved an analogue of this for (w)-regular stratifications [44]. His proof is somewhat simpler than Mather's and involves integrating stratified *rugose vector fields*, whose existence is actually *equivalent* to the stratification being (w)-regular [3]! It follows easily that the stratified homeomorphisms realising the resulting local topological trivialisation are also rugose. However note that the homeomorphisms produced by the Thom-Mather theorem are *already* rugose, as they are controlled [19], [20].

There is something strange about our slogan (*). The hypothesis is in terms of real or complex analytic sets, or semialgebraic, or subanalytic sets, while the conclusion is topological, or at best C^∞ stratified. Is there a way to construct a stratified homeomorphism which is analytic, or at least subanalytic? Or, if we start with a real algebraic variety or a semialgebraic set, is there a way to construct a semialgebraic local trivialisation? There are in fact positive answers to these questions. First there is a semialgebraic version of the Thom-Mather isotopy theorem due to Michel Coste and Masahirio Shiota [4].

Theorem 4.2 (Coste-Shiota, 1992). *Every semialgebraic set with a given semialgebraic Whitney (b)-regular stratification is locally semialgebraically trivial along the strata without further refinement.*

The proof uses model theory and the theory of real closed fields. An immediate consequence is the existence of locally semialgebraically trivial stratifications of semialgebraic sets. The methods allow one to obtain a similar result for subanalytic sets, where the local topological trivialisation is now subanalytic. Similar results are valid for definable sets in general o-minimal structures. More details are given in Shiota's 1997 monograph [31].

There is an earlier result due to Hardt [10] which implies the existence of locally semialgebraically trivial stratifications of semialgebraic sets using the existence of semialgebraic triangulations.

Theorem 4.3 (Hardt, 1980). *Every semialgebraic set admits a semialgebraic triangulation which is locally semialgebraically trivial along the simplexes.*

Whitney stratifying Hardt's semialgebraic triangulation of a semialgebraic set will give the existence of *some* Whitney stratification which is locally semialgebraically trivial along strata. But this method will produce many more strata than given by the Coste-Shiota theorem, which can even be applied to the minimal semialgebraic Whitney stratification for example.

5. Mostowski's Lipschitz stratifications

Thom suggested, in his November 1964 Bourbaki seminar [35] describing Whitney's work on stratifications of analytic varieties [45, 46], that the vector fields yielding local topological triviality of Whitney stratifications (conjectured by Whitney in his contribution to the Morse Jubilee volume [45], although Whitney's conjecture, still unproven, required a much stronger notion of local triviality) might even be made Lipschitz, so that local bi-Lipschitz triviality would follow. Thom's conjecture was verified by Tadeusz Mostowski in his

1985 habilitation [22]. An earlier positive result for the special case of families of plane curves had been obtained by Frédéric Pham and Bernard Teissier in 1969 [29, 30].

Mostowski [22] defined a notion of "Lipschitz" stratification (see the appendix below) via a set of equisingularity conditions, stronger than Verdier's (w)-regularity, and he proved that every complex analytic variety admits such a Lipschitz stratification. He showed further that these Lipschitz stratifications are locally bi-Lipschitz trivial along strata by integrating stratified Lipschitz vector fields. Adam Parusinski then proved such theorems for every real analytic variety [25], then for all semianalytic sets [26], and finally in 1994 for every subanalytic set [28]. It is NOT true however that every definable set in an o-minimal structure admits a Lipschitz stratification, as shown by the following simple example (due to Parusinski) where the local bi-Lipschitz type varies continuously along a line. Note that the set X in Example 1 is not definable in *any* polynomially bounded o-minimal structure, by Miller's dichotomy [21] mentioned in section 3.

Example 5.1. In $R^3 = \{(x, y, z)\}$ let $X(z) = \{y = 0\} \cup \{y = x^z\} \subset \{R^2 \times z\}$. Then the bi-Lipschitz types of $X(z)$ and $X(z')$ are distinct if $z \neq z'$, both > 1. Hence *no* locally bi-Lipschitz trivial stratification exists.

Example 5.2. Here is a semialgebraic example [14], due to Satoshi Koike, showing that (w)-regularity does not imply local bi-Lipschitz triviality. Let $V = \{y^2 = z^2 x^2 + x^3, x \geq 0\} \subset R^3$. Let Y denote the z-axis, and let $X = V - Y$. Then the pair (X, Y) is (w)-regular, but the bi-Lipschitz type of the germ of V at the point $(0, 0, 0) \in Y$ is distinct from the bi-Lipschitz type of the germ of V at points $(0, 0, z) \in Y$ if $z \neq 0$.

Note. In this article, the Lipschitz property is defined with respect to the *extrinsic* or *outer* metric, i.e. the metric induced by that of the ambient space. The recent papers of Birbrair, Fernandes and Neumann (cf. [1, 2] for example) refer to the *inner* metric, where one restricts to distance defined as the lower bound of arcs on the embedded stratified set. In their paper in these proceedings [2] they compare the two metrics and consider cases where the two metrics are equivalent, i.e. when one has a so-called *normal embedding*. The bi-Lipschitz type does not change along the z-axis in Example 2 if one uses the inner metric, and does not change continuously along the z-axis in Example 1.

Question 1. Do definable sets in polynomially bounded o-minimal structures admit Lipschitz stratifications in the sense of Mostowski ?

There is currently no analogue of the result of Coste and Shiota [4] for Lipschitz stratifications so as to produce local semialgebraic bi-Lipschitz triviality
along strata by isotopy.

However in 2005 Guillaume Valette [38], [39] proved a Lipschitz version of
Hardt's semi-algebraic triangulation theorem, so proving that semi-algebraic
sets admit semialgebraic stratifications (actually triangulations) which are
locally semi-algebraically bi-Lipschitz trivial. The proof combines techniques
of Hardt [10] and the model theory methods of Coste and Shiota [4],
together with a version of a preparation theorem due to Lion, Rolin and
Parusinski.

Restratifying as above, using a semialgebraic version of the existence of
L-stratifications, we would obtain the existence of a (Mostowski-type) Lipschitz stratification of a given semialgebraic set, which would be in addition
locally semialgebraically bi-Lipschitz trivial along the strata. I expect that
such a semialgebraic existence theorem for Lipschitz stratification of a given
closed semialgebraic set can be proved, by adapting Parusinski's semianalytic
proof [26].

Valette's proof applies to definable sets in *any* polynomial bounded o-
minimal structure (in particular to globally subanalytic sets), using a preparation theorem due to van den Dries and Speissegger.

6. Consequences of Valette's triangulation theorem

1. A germ at O of a semialgebraic set X in R^n has a link L_X whose metric
   type (its equivalence class defined by semialgebraic bi-Lipschitz homeomorphism) is well-defined, i.e. it is independent of the semialgebraic dis-
   tance function $\rho : (U, O) \to ([0, \epsilon), 0)$ such that $\rho^{-1}(0) = O$, $\rho(x) \equiv ||x||$,
   which defines the link [40].

   Two germs of semialgebraic sets X and Y in R^n have the same metric
   type if and only if their links L_X and L_Y have the same metric type. and
   two such germs X and Y of the same metric type can be related by a
   semialgebraic bi-Lipschitz homeomorphism preserving the distance to the
   origin [40].
2. The set of metric types of germs of real analytic sets is countable [41].
   This resolves a famous 1977 conjecture of Larry Siebenmann and Dennis
   Sullivan [32].
3. Two polynomially bounded o-minimal structures M_1 and M_2 over R with
   the same set of exponents define the same metric types, i.e. every definable
   set in R^n of M_1 is bi-Lipschitz homeomorphic to a definable set in R^n of

M_2 [35]. As an example, every global subanalytic set in R^n is bi-Lipschitz homeomorphic to a semialgebraic set in R^n.

4. Multiplicity mod 2 is a semialgebraic bi-Lipschitz invariant of real algebraic hypersurfaces in R^n with isolated singularities such that the tangent cones at these singularities have isolated singularities [42].

7. Appendix. Definition of Lipschitz stratifications

Tadeusz Mostowski in 1985 [22] introduced certain conditions (L_1), (L_2), (L_3) on a stratification, which strengthen the Kuo-Verdier condition (w), and which imply the possibility of extending Lipschitz vector fields and are indeed characterised by the existence of Lipschitz stratified extensions of Lipschitz vector fields on strata [28] with specified Lipschitz constants (see below).

Here are the definitions, which are necessarily somewhat complicated.

Definition 7.1 (cf. Mostowski [22]). Let $Z = Z_d \supset \cdots \supset Z_\ell \neq \emptyset$ be a closed stratified set in $\mathbf{R}^n$. Write $\overset{\circ}{Z}_j = Z_j - Z_{j-1}$.

Let $\gamma > 1$ be a fixed constant. A *chain* for a point $q \in \overset{\circ}{Z}_j$ is a strictly decreasing sequence of indices $j = j_1, j_2, \ldots, j_r = \ell$ such that each $j_s (s \geq 2)$ is the greatest integer less than j_{s-1} for which

$$dist(q, Z_{j_s-1}) \geq 2\gamma^2 dist(q, Z_{j_s}).$$

For each j_s, $1 \leq s \leq r$, choose $q_{j_s} \in \overset{\circ}{Z}_{j_s}$ such that $q_{j_1} = q$ and $|q - q_{j_s}| \leq \gamma dist(q, Z_{j_s})$.

If there is no confusion one calls $\{q_{j_s}\}_{s=1}^r$ a *chain* of q.

For $q \in \overset{\circ}{Z}_j$, let $P_q : \mathbf{R}^n \to T_q(\overset{\circ}{Z}_j)$ be the orthogonal projection to the tangent space and let $P_q^\perp = I - P_q$ be the orthogonal projection to he normal space $(T_q(\overset{\circ}{Z}_j)^\perp$.

A stratification $\Sigma = \{Z_j\}_{j=\ell}^d$ of Z is said to *be a Lipschitz stratification*, or to *satisfy the (L)-conditions*, if for some constant $C > 0$ and for every chain $\{q = q_{j_1}, \ldots, q_{j_r}\}$ with $q \in \overset{\circ}{Z}_{j_1}$ and each k, $2 \leq k \leq r$,

$$| P_q^\perp P_{q_{j_2}} \cdots P_{q_{j_k}} | \leq C | q - q_{j_2} | / d_{j_k-1}(q) \qquad (L1)$$

and for each $q' \in \overset{\circ}{Z}_{j_1}$ such that $| q - q' | \leq (1/2\gamma) d_{j_1-1}(q)$,

$$| (P_q - P_{q'}) P_{q_{j_2}} \cdots P_{q_{j_k}} | \leq C | q - q' | / d_{j_k-1}(q) \qquad (L2)$$

and

$$| P_q - P_{q'} | \le C \, | q - q' | \, / d_{j_1 - 1}(q) \qquad (L3).$$

Here $dist(-, Z_{\ell-1}) \equiv 1$, by convention.

For calculations of Lipschitz stratifications in the case of families of plane curves see my paper with Dwi Juniati [15]. For a proof that for subanalytic stratifications, (L) implies (L^*) (i.e. the condition is preserved by families of generic plane sections, in the notation introduced by Bernard Teissier) see my paper with Juniati and Valette [14].

According to Parusinski [28], an equivalent criterion to the L-conditions is that there exists a constant $C > 0$ such that for every set $W \subset Z$ such that $Z_{j-1} \subset W \subset Z_j$ for some $j \in \{\ell, \dots, d\}$, every Lipschitz stratified vector field v on W with Lipschitz constant L, such that v is bounded on $W \cap Z_\ell$ by K, extends to a Lipschitz stratified vector field on Z with Lipschitz constant $C(L + K)$.

It is not hard to show that for a given Lipschitz stratification $\exists \, C > 0$ such that $\forall x \in \overset{\circ}{Z}_j$, $\forall y \in \overset{\circ}{Z}_k$, where $k < j$,

$$|P_x^\perp P_y| \le \frac{C|x - y|}{dist(y, Z_{k-1})},$$

so that because $|P_x^\perp P_y| = d(T_y \overset{\circ}{Z}_k, \overset{\circ}{Z}_j)$, (w)-regularity follows, with a precise estimation for the local constant (which can tend to infinity as y approaches Z_{k-1}).

Theorem 7.2 (Mostowski 1985 [22]). *Every complex analytic set admits a complex analytic Lipschitz stratification. Moreover such Lipschitz stratifications are locally bi-Lipschitz trivial along strata.*

Theorem 7.3 (Parusinski 1994 [26]). *Every subanalytic set admits a subanalytic Lipschitz stratification. Moreover such Lipschitz stratifications are locally bi-Lipschitz trivial along strata.*

Note however that there is in general *no canonical* Lipschitz stratification attached to a given subanalytic set, even in the special cases of real algebraic varieties or complex algebraic varieties. See Mostowski's habilitation dissertation for details [22]. There *are* canonical Whitney (equivalently Verdier) stratifications of complex analytic varieties by the theory of Teissier ([33], see also [11]) which identifies canonical Whitney strata as precisely the loci of equimultiplicity of the different polar varieties. So it seems difficult to give algebraic criteria for Lipschitz regularity in general, in the spirit of Terry Gaffney's work using integral closures of modules. Mostowski did however give equivalent algebraic criteria for (L)-regularity in the case of complex surfaces in [23]. He also explained the tight relation of the polar varieties of a complex analytic variety X and the stratifications of X with locally trivialising Lipschitz vector fields in [24].

The definition of Lipschitz stratification is equivalent to being able to extend vector fields in a certain way, by specifying constants. Whitney, Verdier [3] and Bekka regularity can each be characterised in terms of extending vector fields without specifying constants. Is there some simpler variant of Mostowski's definition with nice geometric and algebraic characterisations? Of course the essential requirement of any such equisingularity criterion would be that the condition be generic and imply local bi-Lipschitz triviality.

References

1. L. Birbrair, A. Fernandes and W. D. Neumann, Bi-Lipschitz geometry of weighted homogeneous surface singularities, *Math. Ann.* **342** (2008), 139–144.
2. L. Birbrair, A. Fernandes and W. D. Neumann, On normal embedding of complex algebraic surfaces, *Proceedings of the 10th International Workshop on Real and Complex Singularities*, arXiv:0901.0030.
3. H. Brodersen and D. J. A. Trotman, Whitney (b)-regularity is weaker than Kuo's ratio test for real algebraic stratifications, *Math. Scand.* **45** (1979), 27–34.
4. M. Coste and M. Shiota, Thom's first isotopy lemma: a semialgebraic version, with uniform bound, *Real algebraic and analytic geometry*, De Gruyter (1995), 83–101.
5. Z. Denkowska and J. Stasica, *Ensembles sous-analytiques à la Polonaise*, Hermann, Paris, 2007.
6. Z. Denkowska and K. Wachta, Une construction de la stratification sous-analytique avec la condition (w), *Bull. Polish Acad. Sci. Math.* **35** (1987), 401–405.
7. C. G. Gibson, K. Wirthmüller, A. A. du Plessis and E. J. N. Looijenga, *Topological stability of smooth mappings*, Lecture Notes in Mathematics **552**, Springer (1976).
8. R. M. Hardt, Homology and images of semianalytic sets, *Bull. Amer. Math. Soc.* **80** (1974), 675–678.
9. R. M. Hardt, Stratification of real analytic mappings and images, *Invent. Math.* **28** (1975), 193–208.
10. R. M. Hardt, Semi-algebraic local-triviality in semi-algebraic mappings, *Amer. J. Math.* **102** (1980), 291–302.
11. J.-P. Henry and M. Merle, Limites de normales, conditions de Whitney et éclatement d'Hironaka, *Proceedings of Symposia in Pure Mathematics*, Volume **40**, Arcata 1981–Singularities, Part 2, American Mathematical Society, Providence, Rhode Island, 1983, 575-584.
12. H. Hironaka, Introduction aux ensembles sous-analytiques, *Singularités à Cargèse, 1972, Astérisque*, Nos. **7-8**, Soc. Math. France, Paris, 1973, 13–20.

13. H. Hironaka, Subanalytic sets, *Number theory, algebraic geometry and commutative algebra, in honor of Yasuo Akizuki*, Kinokuniya, Tokyo, 1973, pp. 453–493.

14. D. Juniati, D. Trotman and G. Valette, Lipschitz stratifications and generic wings, *J. London Math. Soc. (2)* **68** (2003), 133–147.

15. D. Juniati and D. Trotman, Determination of Lipschitz stratifications for the surfaces $y^a = z^b x^c + x^d$, *Singularités Franco-Japonaises (eds. J.-P. Brasselet, T. Suwa), Sémin. Congr.,* **10**, Soc. Math. France, Paris (2005), 127–138.

16. T. L. Loi, Verdier and strict Thom stratifications in o-minimal structures, *Illinois J. Math.* **42** (1998), 347–356.

17. S. Lojasiewicz, *Ensembles semianalytiques*, IHES notes, 1965.
    [Available from http://perso.univ-rennes1.fr/michel.coste/Lojasiewicz.pdf]

18. S. Lojasiewicz, J. Stasica and K. Wachta, Stratifications sous-analytiques. Condition de Verdier, *Bull. Polish Acad. Sci. Math.* **34** (1986), no. 9-10 (1987), 531–539.

19. J. N. Mather, *Notes on topological stability*, Harvard University, 1970.

20. J. N. Mather, Stratifications and mappings, *Dynamical systems (Proc. Sympos., Salvador, 1971)*, Academic Press, New York (1973), 195–232.

21. C. Miller, Exponentiation is hard to avoid, *Proc. Amer. Math. Soc.* **122** (1994), no. 1, 257–259.

22. T. Mostowski, Lipschitz stratifications, *Dissertationes Math.* vol. **243** (1985), 46 pp.

23. T. Mostowski, Tangent cones and Lipschitz stratifications, *Singularities (Warsaw, 1985)*, Banach Center **20**, PWN, Warsaw (1988), 303–322.

24. T. Mostowski, A criterion for Lipschitz equisingularity, *Bull. Polish Acad. Sci. Math.* **37** (1989), 109–116.

25. A. Parusi?ski, Lipschitz stratification of real analytic sets, *Singularities (Warsaw, 1985)*, Banach Center Publ., **20**, PWN, Warsaw (1988), 323–333.

26. A. Parusinski, Lipschitz properties of semi-analytic sets, *Ann. Inst. Fourier (Grenoble)* **38** (1988), 189–213.

27. A. Parusinski, Lipschitz stratification. *Global analysis in modern mathematics (Orono, ME, 1991; Waltham, MA, 1992)*, Publish or Perish, Houston, Texas (1993), 73–89.

28. A. Parusinski, Lipschitz stratification of subanalytic sets, *Ann. Sci. École Norm. Sup.(4)* **27** (1994), 661–696.

29. F. Pham and B. Teissier, *Fractions lipschitziennes d'un algèbre analytique complexe et saturation de Zariski*, Centre Math. École Polytech., Paris, 1969.

30. F. Pham, Fractions lipschitziennes et saturation de Zariski des algèbres analytiques complexes, *Actes du Congrès International des Mathématiciens (Nice, 1970), Tome 2*, Gauthier-Villars, Paris, 1971, 649–654.

31. M. Shiota, *Geometry of subanalytic and semialgebraic sets*, Progress in Mathematics **150**, Birkhäuser Boston, Inc., Boston, MA, 1997.

32. L. Siebenmann and D. Sullivan, On complexes that are Lipschitz manifolds, *Proceedings of the Georgia Topology Conference, Athens, Ga., 1977, Geometric topology*, Academic Press, New York-London (1979), 503–525.

33. B. Teissier, Variété polaires II, Multiplicités polaires, sections planes, et conditions de Whitney, *Alg. Geometry (La Ràbida, 1981)*, Lecture Notes in Math., **961**, Springer, Berlin (1982), 314–491.

34. R. Thom, Propriétés différentielles locales des ensembles analytiques, *Séminaire Bourbaki, 1964/65*, nÌ **281**, 12 p.
35. R. Thom, Ensembles et morphismes stratifiés, *Bull. Amer. Math. Soc.* **75** (1969), 240–284.
36. D. J. A. Trotman, Counterexamples in stratification theory: two discordant horns, *Real and complex singularities (Proc. Ninth Nordic Summer School/NAVF Sympos. Math., Oslo, 1976, edited by Per Holm)*, Sijthoff and Noordhoff, Alphen aan den Rijn, 1977, 679-686.
37. D. J. A. Trotman, *Whitney stratifications : faults and detectors*, Warwick thesis, 1978.
38. G. Valette, A bilipschitz version of Hardt's theorem, *C. R. Acad. Sci. Paris, Sér. I* , t. **340** (2005), 895–900.
39. G. Valette, Lipschitz triangulations, *Illinois J. Math.* **49** (2005), 953–979.
40. G. Valette, The link of the germ of a semi-algebraic metric space, *Proc. Amer. Math. Soc.* **135** (2007), 3083–3090.
41. G. Valette, On metric types that are definable in an o-minimal structure, *J. Symbolic Logic* **73** (2008), 439–447.
42. G. Valette, Multiplicity mod 2 as a semi-algebraic bi-Lipschitz invariant, to appear in *Discrete and Computational Geometry*.
43. L. van den Dries, *Tame Topology and O-minimal Structures*, London Mathematical Society Lecture Note Series **248**, Cambridge University Press, Cambridge, 1998.
44. J.-L. Verdier, Stratifications de Whitney et théorème de Bertini-Sard, *Inventiones Math.* **50** (1976), 273–312.
45. H. Whitney, Local properties of analytic varieties, *Differential and Combinatorial Topology (edited by Stewart S. Cairns), (A Symposium in Honor of Marston Morse)*, Princeton Univ. Press, Princeton, N. J. (1965) 205–244.
46. H. Whitney, Tangents to an analytic variety, *Ann. of Math. (2)* **81** (1965), 496–549.
47. A. J. Wilkie, Model completeness results for expansions of the ordered field of real numbers by restricted Pfaffian functions and the exponential function. *J. Amer. Math. Soc.* **9** (1996), 1051–1094.

D. Trotman

Laboratoire d'Analyse, Topologie et Probabilités (UMR 6632)

Centre de Mathématique et Informatique

Université de Provence

39 rue Joliot-Curie

13453 Marseille Cedex 13

France

trotman@cmi.univ-mrs.fr

24

Gaffney's work on equisingularity

C. T. C. WALL

Abstract

A survey of equisingularity theory focussed on Terry Gaffney's work.

The article begins with an account of the early history of equisingularity. Next I develop notation, particularly for polar varieties; recall the theory of integral closures of ideals, show how Gaffney generalised this to integral closures of modules, and list a variety of applications he has made.

The invariants available are classical and Buchsbaum-Rim multiplicities of modules, polar multiplicities and Segre numbers of ideals, and generalisations to modules. Some of the main theorems are of the form: the constancy of certain numerical invariants of a family imply equisingularity of the family (usually in the form of Whitney triviality). Many of the proofs use results showing that constancy of some invariants implies an integral dependence relation. One notable paper gives a sufficient condition for topological triviality of families of maps.

Introduction

The classification of singularities of plane curves was achieved in 1932 by Brauner [2], Burau [6], [7] and Zariski [60]: it yields an easily stated, necessary and sufficient condition for topological equivalence, which clearly does not imply analytic equivalence. Probably the simplest example is the case of 4

2000 *Mathematics Subject Classification* 32S15 (primary), 14B05, 13H15, 32S10, 32S50 (secondary).

concurrent lines $xy(x + y)(x + ty) = 0$ with t an invariant of analytic, but not of topological equivalence.

This situation presents the problem of creating a theory of equivalence of families of objects (e.g. algebraic varieties or morphisms) which will say when the members of the family are essentially the same. One needs a definition allowing some flexibility but with which calculations can be made. This is the problem of equisingularity, which lies at the heart of singularity theory.

Terry Gaffney has made major contributions to this, many of which appear in Proceedings of earlier São Carlos meetings. His philosophy is to seek invariants depending on members of the family whose constancy implies equisingularity.

In this article, I seek to describe Terry's work in this area. To put this in perspective, I also give an account of the earlier work which led up to it. This seems appropriate, as Terry has always sought to give full credit to others whose work or influence has contributed to his results. I am indebted to David Trotman, Andrew du Plessis and particularly to Terry himself for comments on earlier versions of this article.

I include a complete list of Gaffney's papers, which are cited with a G, e.g. as [G21], preceding the general bibliography.

1. Early results on equisingularity

The origins of the differential theory of equisingularity lie in attempts to classify singularities of differentiable mappings. This began with pioneering work of Whitney in the 1940s and 1950s [51], [52], [53], [54] classifying generic singularities in particular dimensions. A discussion in general was given by Réné Thom [44] in 1959, in particular conjecturing that topologically stable maps were C^∞−dense in all dimensions.

Thom announced new ideas at a lecture in Zürich in 1960 (see [45]) (where the writer had the good fortune to be present). This contains definitions of stratifications, mention of regularity and a statement that "Whitney has proved that real algebraic sets admit regular stratifications", the apparatus (tubes, local retractions, carpeting functions), and went on with corresponding ideas for C^∞−mappings. Thom had amazingly good geometric intuition; not only completely new ideas, but a good idea for what might be true and provable. It was often left to others to flesh out his ideas to obtain clear proofs. Gaffney is perhaps his true successor in having excellent geometric intuition, but he finds proofs with help from collaborators.

Details followed a few years later. In [58], Whitney studied the behaviour of the tangent plane $T_x X$ at a smooth point $x \in X$ as x tends to a smooth point y_0 on a subvariety Y of X and formulated conditions on (X, Y) at y_0:

(A) if there is a sequence $x_i \in X$ such that $x_i \to y_0$ and $T_{x_i} X$ tends to a limit L, then $T_y Y \subset L$,

(B) if there is also a sequence $y_i \in Y$ such that $y_i \to y_0$ and the unit vector in the direction $y_i x_i$ (in the ambient space) tends to a limit v, then $v \in L$.

In fact, it was soon realised that (B) implies (A); however, (A) remains an important condition. Whitney defined stratifications, called a stratification regular if these conditions hold at all points, and proved that any complex analytic variety has a regular stratification. For the real semi-analytic case, a proof of existence of regular stratifications was given, also in 1965, in notes [27] by Łojasiewicz, which established basic facts about semi-analytic sets, including his famous inequalities.

Thom's 'first isotopy lemma' states that a regularly stratified set is locally topologically trivial along strata. Proofs were given by Thom [47] and in widely circulated lecture notes by Mather [31] in 1970. These involve the construction of controlled vector fields, and their integration.

A useful variant of the regularity conditions was given by Verdier [49], anticipated in part by Hironaka [14] and the c-cosecance of Teissier [38]. For linear subspaces $A, B \subset \mathbb{R}^N$, define

$$\delta(A, B) := sup_{u \in A^\perp \setminus \{O\}, \ v \in B \setminus \{O\}} \frac{|(u, v)|}{\|u\| . \|v\|};$$

thus $\delta(A, B) = 0 \Leftrightarrow A \supseteq B$. We can re-state the Whitney (A) condition as $\delta(T_{x_i} X, T_{y_0} Y) \to 0$ as $x_i \to y_0$, i.e. as $\|x_i - y_0\| \to 0$. Now say that (X^o, Y) satisfies the Verdier condition W at y_0: if

(W) there exist a neighbourhood U of y_0 and $C > 0$ such that, for all $y \in U \cap Y$, $x \in U \cap X^o$, we have $\delta(T_x X, T_y Y) \leq C \|x - y\|$.

Verdier established that subanalytic sets admit stratifications satisfying this condition, that it is stronger than Whitney's condition (B); also that when it holds, one obtains controlled vector fields satisfying a condition he terms 'rugose' (which is stronger than continuity but weaker than Lipschitz). This can be generalised to 'the strict Whitney condition A with exponent r' by replacing the right hand side of the inequality by $C \|x - y\|^r$ (see e.g. [41]).

In [45] Thom also enunciated a 'second isotopy lemma' giving a sufficient condition for topological triviality of a family of C^∞−mappings. This was used in [46] to obtain deep results about singularities in general.

First he stated that any polynomial map is stratifiable, meaning that there are stratifications of source and target such that the map submerses each stratum of the source on a stratum of the target. He said that a map presents blowing-up if, writing for X a stratum of the source with image X', and $q(X)$ for $\dim X - \dim X'$, there exist strata $Y \subset \overline{X}$ with $q(Y) > q(X)$.

In [47], he defined a relative form of (A), now known as Thom regularity. If f is a stratified map, the Thom condition holds for (X, Y) relative to f at $y_0 \in Y$: if

(A_f) for any sequence $x_i \in X$ with $x_i \to y_0$ and $Ker(Tf|T_{x_i}X) \to L$ we have $Ker(Tf|T_yY) \subseteq L$.

Note that we can re-state this as: $\delta(Ker(Tf|T_xY), Ker(Tf|T_{x_i}X)) \to 0$ as $x_i \to y_0$. There is also a strict condition, first considered in [12],

(W_f) there exist a neighbourhood U of y_0 and $C > 0$ such that, for all $y \in U \cap Y$, $x \in U \cap X^o$, we have $\delta(Ker(Tf|T_yY), Ker(Tf|T_xX)) \leq C\|x - y\|$.

The second isotopy lemma now refers to a stratified map $f : X \to Y$ and a further $\pi : Y \to T$ such that, for each stratum S of Y, $\pi|S$ is a submersion. If also the maps are proper and A_f holds at all points, f is locally trivial, so the topological type of $f|(\pi \circ f)^{-1}(t)$ is independent of t. Proofs, similar to those of the first isotopy lemma, were sketched in [47] and in [31].

Clearly a necessary condition for f to possess a Thom regular stratification is that f does not exhibit blowing up. A useful construction of Thom regular maps is given in [9, §2]: here make the stronger hypothesis that the restriction $f| \sum(f)$ (where $\sum(f)$ denotes the critical set of f) is proper and finite-to-one. Then a stratification of $f(\sum(f))$ is a critical value stratification c.v.s. (called partial stratification in [9]) if, for all strata U, $f^{-1}(U) \cap \sum(f)$ is smooth and f induces a local isomorphism of it on U, and for all pairs U, V of strata, $f^{-1}(V) \cap \sum(f)$ and $f^{-1}(V) \setminus \sum(f)$ are Whitney regular over $f^{-1}(U) \cap \sum(f)$. Whitney's arguments yield the existence of a c.v.s. provided the spaces and maps are semialgebraic. Given a c.v.s., we can stratify the target of f by the strata of the c.v.s. and their complement and stratify the source by the strata just listed: then f is a stratified map which satisfies Thom regularity.

A general proof that a proper, complex analytic map which does not exhibit blowing up admits a Thom regular stratification, and even one satisfying W_f, was finally given by Henry, Merle and Sabbah in [12]: the proof uses the technique of polar varieties. The real analytic case was discussed in [15], and can also be treated by taking real parts of the stratifications of [12]. The writer has been unable to find a reference for existence of a Thom stratification in the real semi-analytic case.

We turn to the algebro-geometric approach to equisingularity. From 1964, when Hironaka [13] established the resolution of singularities in characteristic zero, interest began to shift from resolving singularities to classifying them. Zariski created a theory of equisingularity for families of curves, and hence for a variety along a smooth subvariety of codimension 1, and proposed a general definition by induction: roughly speaking, he required equisingularity of the discriminant of a generic projection to a subspace one dimension lower. The simplest non-trivial case is when we have a smooth point of Y, which has codimension 2, so a transverse slice meets X in a plane curve. The theory for this case was developed in considerable detail in the series [61] in 1965–68. In this case, Zariski's definition of equisingularity is equivalent to Whitney's, as holds more generally when the curve is not required to be planar.

In 1971 in [62], Zariski compared his definition with others and posed a number of searching and motivating questions, notably the famous problem of topological invariance of multiplicity. What is the relation between different conditions? When do equivalent singularities lie in a 1-parameter family? Do equivalent varieties have the same multiplicity? More generally, is equisingularity preserved under taking generic hyperplane sections? or under taking the discriminant of a generic projection? We will see that the techniques of projection and of generic hyperplane sections are built in to the theory as it has been developed by Gaffney.

The rather simple example $z^3 + tx^4z + x^6 + y^6 = 0$ (due to Briançon and Speder [3]) is Whitney equisingular but not Zariski equisingular; moreover, as Zariski shows in [64] in 1977, the blowup along the t-axis fails to be equisingular at $t = 0$, which led Zariski to reject Whitney equisingularity as a good notion.

A general discussion of the equisingularity notions as known in 1974 was also given by Teissier in [38]. He starts with Zariski's work, which was his own inspiration, and compares equisingularity in Zariski's sense, in Whitney's sense, and topological local triviality, and he too formulated a number of questions and conjectures.

Finally in 1979 Zariski proposed [65] a modified version of his definition which, unlike earlier ones, is clearly invariant under local analytic equivalence; here he also constructs a stratification.

Zariski's work in [61] also led him to the notion of saturation, which he developed in [63] in 1971–75. Here he gives a more algebraic form to equisingularity for the plane curve case. In the introduction to Vol IV of Zariski's works, Teissier and Lipman record that later work inspired by equisaturation led to the study of Lipschitz equisingularity. Talks were presented on Lipschitz equivalence at this conference by Birbrair & Neumann, and by Valette giving new insight on this concept: it does not at present seem likely to lead to a workable theory of equisingularity in general.

A rather different equisingularity notion, of 'blow analytic equivalence', was proposed by Kuo [25]. While this has had some success, it too does not seem likely to lead to a general theory, as it is not clear that if a family of varieties over T is blow analytically trivial over the open subsets T_1, T_2 of T it must also be trivial over $T_1 \cup T_2$.

Milnor's 1968 book [33] was enormously influential, and focussed attention on isolated singularities of hypersurfaces. Let $\{f_t : (\mathbb{C}^N, 0) \to (\mathbb{C}, 0) \mid t \in \mathbb{C}\}$ be a 1-parameter family of functions defining hypersurfaces X_t, each with an isolated singularity, with union $\mathbf{X} \subset \mathbb{C}^N \times \mathbb{C}$. There is an obvious numerical requirement for a family to be equisingular: constancy of the Milnor number $\mu(f_t)$.

For this case, major developments appeared at the Cargèse conference in 1972. Speder [36] proved that Zariski equisingularity implies the Whitney conditions. Lê [22] spoke on his result with Ramanujam that a μ−constant deformation is topologically trivial (except possibly if $N = 3$).

There is also a major paper by Bernard Teissier [37], which is the real starting point of modern equisingularity theory. He discussed integral closures, introduced and studied the sequence $\mu^*(X_t)$ of Milnor numbers of generic linear sections of X_t, and showed that a μ^*−constant family is Whitney equisingular. This result was completed shortly afterwards by a proof [4] of Briançon and Speder that Whitney equisingularity implies μ^*−constant and their example [3] of a family with μ, but not μ^*, constant: for $z^5 + tzy^6 + xy^7 + x^{15} = 0$, $\mu^{(3)} = 15$ is constant but $\mu^{(2)}$ is not, and the family is topologically trivial.

In [G30] Gaffney and Massey described (with hindsight) a somewhat simplified version of Teissier's argument, summarised in three steps:

Whitney regularity of $F = \{f_t\}$ is implied by W-regularity, the condition that the derivative $\partial F/\partial t$ belong to the integral closure $\overline{\mathfrak{m}_N \cdot J_z F}$, where $J_z F = \langle \{\partial F/\partial z_i\} \rangle$;

this condition holds at a dense open set and ("the PSID"), provided the multiplicity $m(\mathfrak{m}_N \cdot J_z F)$ is constant, holds on a closed set; and

$m(\mathfrak{m}_N \cdot J_z F)$ is a linear combination with positive coefficients of the upper semicontinuous invariants $\mu^{(i)}(X_t)$.

With the success of Teissier's theory, one would like to extend it as far as possible. To achieve this, the following are needed: an extension of the theory of integral closure of ideals; invariants of equisingularity, corresponding to μ^* above; a calculus for working with these invariants; and a generalisation of the PSID. Gaffney has obtained many results of all of these types, and I will try to summarise them. The next sections are devoted respectively to integral closure; invariants and formulae for them; and criteria for equisingularity.

2. Notations

I now fix notation for the rest of this article, for simplicity of exposition; though some of Gaffney's results were obtained in greater generality than I give below. The reader should be warned that though this notation is based on Gaffney's, it differs from his in many cases. We have a complex analytic variety-germ $(X, 0) \subset (\mathbb{C}^N, 0)$ (for brevity, I will restrict almost entirely to the complex analytic case, though Gaffney also has many results in the real case); we assume X equidimensional, of dimension d, and generically reduced. Write $\sum X$ for the singular set and $X^o := X \setminus \sum X$. We may suppose X given as $F^{-1}(0)$ for $F : \mathbb{C}^N \to \mathbb{C}^p$; take co-ordinates $\{z_i\}$ on $\mathbb{C}^N$.

When we wish to study families we take $T = \mathbb{C}^s$ as parameter space, with co-ordinates $\{t_j\}$, let $\mathbf{X} \subset \mathbb{C}^N \times T$ be given as $F^{-1}(0)$ for $F : \mathbb{C}^N \times T \to \mathbb{C}^p$, write $\pi : \mathbb{C}^N \times T \to T$ for the projection (and its restriction to $\mathbf{X} \to T$), $\mathbf{X}^o$ for the set of points where $\mathbf{X}$ is smooth and π submersive. Write also $X_t := \mathbf{X} \cap \pi^{-1}(t)$, and suppose each X_t as in the preceding paragraph. We write T for $\{0\} \times T \subset \mathbf{X}$ and study Whitney equisingularity of $\mathbf{X}$ over T along T.

For any k, $\mathcal{O}_k$ denotes the ring of germs of holomorphic functions at $0 \in \mathbb{C}^k$; $\mathfrak{m}_k$ denotes its maximal ideal. We write $\mathcal{O}_X$ for the sheaf of holomorphic functions on X, $\mathcal{O}_{X,x}$ for the sheaf of germs at $x \in X$, and $\mathfrak{m}_{X,x}$ for its maximal ideal. I will normally use roman letters to denote rings and modules and calligraphic ones for sheaves.

First we suppose each X_t has an isolated singular point at O; later we relax this. Also we first have the hypersurface case $p = 1$, then the complete intersection case where F is a submersion at a generic point, then the general case.

Sometimes regularity conditions A_f and W_f are considered relative to a further map $f : (X, O) \to (\mathbb{C}, 0)$ or $f : (\mathbf{X}, T) \to (\mathbb{C}, 0)$: these are non-trivial even if $p = 0$ so $\mathbf{X} = \mathbb{C}^N \times T$. We denote the zero locus of f by Z (or $\mathbf{Z}$).

Whitney's and other related conditions are defined in terms of the limiting behaviour of tangent spaces to X. Thus we are led to the study of the Nash blowup $N(X)$, the closure of the set of pairs $(x, T_x X)$ where $x \in X^o$, and the conormal space $C(X)$, the closure of the set of pairs (x, H) where $x \in X^o$ and $H \in P^{N-1}$ is a hyperplane containing the tangent space to X at x. If X has codimension 1 these coincide, but for a subset of higher codimension, while early work used the Nash blowup, it was shown by Henry and Merle [11] that the conormal space was more convenient and gave better results. For example, polar varieties and polar multiplicities are defined below by pulling back from linear subspaces of projective space; to use the Nash blowup, we would have to study instead subvarieties of Grassmannians. In the case of a family, we have the relative conormal, which can be defined as the closure of the set of pairs (x, H) with $x \in \mathbf{X}^o$ and H a hyperplane tangent to $\mathbf{X}$ at x and containing (a parallel of) T. We denote this by $C_T \mathbf{X}$ and can regard it as a subspace of $\mathbf{X} \times P^{N-1}$.

Given an ideal $\mathcal{I} = \langle g_1, \ldots, g_q \rangle \triangleleft \mathcal{O}_X$, the blowup $B_{\mathcal{I}}(X)$ is defined to be the closure in $X \times P^{q-1}$ of the graph of $X \setminus V(\mathcal{I}) \to P^{q-1}$ defined by $z \mapsto (g_1(z), \ldots, g_q(z))$, with projection $b_I : B_{\mathcal{I}}(X) \to X$; we write $D_{\mathcal{I}}$ for the exceptional divisor. Equivalently, we can form the Rees algebra (see [35]) $R(\mathcal{I}) := \bigoplus_{n \geq 0} \mathcal{I}^n$; then its graded ideals define $B_{\mathcal{I}}(X) = Proj(R(\mathcal{I}))$ (where *Proj* denotes the analytic homogeneous spectrum). If X has codimension 1, the Jacobian ideal is $J(F) = \langle \partial F/\partial z_1, \ldots, \partial F/\partial z_N \rangle$ and $C(X)$ coincides with its blowup $B_{J(F)}(X)$.

If X has codimension greater than 1, $C(X)$ is not a blowup of X: it has dimension $N - 1$. To extend the techniques to this case, Gaffney introduced the following. Write $\mathcal{E} := \mathcal{O}_X^p$ for the free module, $J_M F \subset \mathcal{E}$ for the (Jacobian) submodule generated by the columns of the Jacobian matrix $JF = (\partial F_i/\partial z_j)$, which has generic rank $N - d$; write $S\mathcal{E}$ for the symmetric algebra on $\mathcal{E}$ and $\mathcal{R} J_M F$ for the (Rees) subalgebra generated by $J_M F$. Then $P = Proj(S\mathcal{E})$ has dimension $d + p - 1$, and the image of $Proj(\mathcal{R} J_M F)$ in P can be identified with $C(X)$.

In general, for $L \subset \mathbb{C}^N \times T$ a linear subspace, write $J_M F_L$ for the submodule of $J_M F$ generated by the $\partial F/\partial v$ for $\partial/\partial v$ tangent to L (in the case $p = 1$, $J F_L$ for the subideal of $J F$). Thus in the case of a family $F : \mathbb{C}^N \times T \to \mathbb{C}^p$, $J_M F_T$ is generated by the columns $\partial F_i/\partial z_j$ for z_j the co-ordinates in $\mathbb{C}^N$. The relative conormal $C_T \mathbf{X}$ is now the image of $Proj(\mathcal{R} J_M F_T)$ in P.

Gaffney [G28] comments that the exceptional divisors of these blowups record behaviour of the limiting tangent hyperplanes, which are relevant for the Whitney conditions: more precisely, the fibre over O of the exceptional divisor of $J_M F_T$ records the limits as $(x, t) \to (0, 0)$ of limiting tangent hyperplanes to X_t; in the module case, the fibre of the conormal need not be a divisor, but does still record the limits.

3. Integral closures

Let I be an ideal in a ring R (we write $I \triangleleft R$). We say that x is integral over I if there exist elements $a_r \in I^r$ with $x^k + \sum_1^k a_r x^{k-r} = 0$. The set of such elements x is called the integral closure of I and denoted $\overline{I}$: it is an ideal in R. The proof uses the fact that x is integral over I if and only if there is a faithful finitely generated R−module M with $x M \subseteq I \cdot M$.

The integral closure has marvellous properties. An excellent reference is the beautiful set of lecture notes [24] of Monique Lejeune-Jalabert and Bernard Teissier. These were, it seems, motivated by a study of a section of Hironaka's big paper [13].

For R a complete local ring, and I a proper ideal, I defines a function $\bar{v}_I$ on R by setting $v_I(x) := \sup\{n \in \mathbb{N} \mid x \in I^n\}$ and $\bar{v}_I(x) := \lim_{k \to \infty} v_I(x^k)/k$. This is an order function, i.e. $\bar{v}_I(x + y) \geq \inf(\bar{v}_I(x), \bar{v}_I(y))$, $\bar{v}_I(xy) \geq \bar{v}_I(x) + \bar{v}_I(y)$, $\bar{v}_I(0) = \infty$ and $\bar{v}_I(1) = 0$. Then x is integral over I if and only if $\bar{v}_I(x) \geq 1$.

If $\bar{v}_I(x) > 1$, we say that f is *strictly dependent* on $\mathcal{I}$, and write $f \in \mathcal{I}^\dagger$: $\mathcal{I}^\dagger$ also is an integrally closed ideal. Although strict dependence had been used previously, the definition of $\mathcal{I}^\dagger$, and of a corresponding notion for modules, are due to Gaffney [G27].

We can regard $\bar{I}$ as the largest ideal equivalent to I: sometimes we would prefer a smallest. If $J \subset I$ and $\bar{J} = \bar{I}$, then J is said to be a *reduction* of I; a *minimal reduction* is one with the minimal number of generators. For I of finite codimension in $\mathcal{O}_k$, we can take the ideal generated by k general elements of I.

Now let X be a reduced complex analytic space, $\mathcal{I} \lhd \mathcal{O}_X$ the coherent sheaf of ideals defining a nowhere dense analytic subspace Y, $x \in Y$, $\mathcal{I}_x$ the germ of $\mathcal{I}$ at x; $f \in \mathcal{O}_X$. The following are equivalent ([24], see also [41]):

(i) (algebraic condition) $f \in \overline{\mathcal{I}_x}$;
(ii) (evaluation) for some f.g. faithful $\mathcal{O}_{X,x}$−module $\mathcal{M}_x$, $f.\mathcal{M}_x \subset \mathcal{I}_x.\mathcal{M}_x$;
(iii) (valuative criterion) for every arc, i.e. map-germ $\phi : (\mathbb{C}, 0) \to (X, x)$, we have $f \circ \phi \in \phi^*\mathcal{I}_x.\mathcal{O}_1$;
(iv) (growth condition) for V a neighbourhood of x in X, and $\{g_i\}$ generators of $\Gamma(V, \mathcal{I})$, there exist $C \in \mathbb{R}^+$ and a neighbourhood of x on which $|f(y)| \leq C \sup_i |g_i(y)|$.

Moreover, the $\overline{\mathcal{I}_x}$ are the stalks of a coherent sheaf $\overline{\mathcal{I}}$.

If $\mathcal{I} \subset \mathcal{J}$ there is a natural map $B_{\mathcal{J}}(X) \to B_{\mathcal{I}}(X)$. This map is finite if and only if $\overline{\mathcal{I}} = \overline{\mathcal{J}}$, i.e. I is a reduction of J.

We also need the normalised blowup, which we denote by $\tilde{b}_{\mathcal{I}} : \tilde{B}_{\mathcal{I}}(X) \to X$, with exceptional divisor $\tilde{D}_{\mathcal{I}}$; write $\mathcal{I}^*$ for the pullback of $\overline{\mathcal{I}}$. To construct it, take the normalisation $\tilde{X}$ of X, the pullback $\mathcal{I}'$ to it of $\mathcal{I}$, the integral closure $\overline{\mathcal{I}'}$ of $\mathcal{I}'$, and then $\tilde{B}_{\mathcal{I}}(X) = B_{\overline{\mathcal{I}'}}(\tilde{X})$.

Suppose X compact, or more generally that we consider germs along some compact subset $K \subset X$. Then $\tilde{D}_{\mathcal{I}}$ has only finitely many irreducible components D_α. The ideal $\mathcal{I}^*$ is supported on $\bigcup_\alpha D_\alpha$; in the neighbourhood of a smooth point of D_α, it can only be a power $\mathcal{I}_\alpha^{n_\alpha}$ of the ideal $\mathcal{I}_\alpha$ defining D_α.

Thus if $f \in \mathcal{I}$, the lift of f to $\tilde{B}_{\mathcal{I}}(X)$ must belong to each $\mathcal{I}_\alpha^{n_\alpha}$. But much more is true. Suppose X normal, $\mathcal{I} \lhd \mathcal{O}_X$ an invertible ideal defining a Cartier divisor $D = \bigcup_\alpha D_\alpha$, $f \in \Gamma(X, \mathcal{O}_X)$. Then

$$f_x \in \overline{\mathcal{I}_x} \text{ for all } x \in X \Leftrightarrow \text{ for each } \alpha \text{ we can find } x_\alpha \in D_\alpha \text{ with } f_{x_\alpha} \in \mathcal{I}_{\alpha,x}^{n_\alpha}.$$

It follows that in (iii) above, we only need one arc ϕ_α for each component D_α of D. This has numerous consequences: for example we can define fractional powers by letting $f \in \mathcal{I}^{\lceil p/q \rceil}$ if, for some N, $f^{qN} \in \overline{\mathcal{I}}^{pN}$: it follows that we can take all fractions to have denominator the least common multiple of the n_α; this is also a denominator for the Łojasiewicz exponent.

We can now give a global version of the above set of equivalent conditions to $f \in \overline{\mathcal{I}}$, referring to germs at K:

(i) $f \in \overline{\mathcal{I}_x}$ $(\forall x \in K)$;
(ii) for every proper $\pi : X' \to X$ with image containing K and $\mathcal{I}.\mathcal{O}_{X'}$ invertible, there is an open neighbourhood U' of $\pi^{-1}(K)$ with $f.\mathcal{O}_{U'} \in \mathcal{I}.\mathcal{O}_{U'}$;
(iii) for each α, $f \circ \phi_\alpha \in \phi_\alpha^* \mathcal{I}_x.\mathcal{O}_1$;
(iv) there exist an open neighbourhood U of K in X, generators $\{g_i\}$ of $\Gamma(U, \mathcal{O}_X)$ and $C \in \mathbb{R}^+$ such that $|f(x)| \leq C \cdot \sup\{|g_i(x)|\}$ for all $x \in U$.

Integral closures of ideals are the key to the study of equisingularity for families of hypersurfaces. To generalise to subvarieties of larger codimension, Gaffney begins [G21] by introducing the integral closure of a module or, more accurately, of a module given as a submodule of a free module. An earlier, purely algebraic treatment of an equivalent concept had been given by Rees [35], generalising results in [66]; but Gaffney takes the valuative criterion as definition. He notes that this also gives a useful notion in the real case, but we will restrict to the complex case.

For $h \in \mathcal{O}_{X,x}^p$ and $\mathcal{M}_x \subset \mathcal{E}_x = \mathcal{O}_{X,x}^p$ a submodule, the following are equivalent:

(i) $h \in \overline{\mathcal{M}_x}$ in the sense of Rees,
(ii) for some faithful $\mathcal{I}_x \lhd \mathcal{O}_{X,x}$ we have $\mathcal{I}_x \cdot h \subseteq \mathcal{I}_x \cdot \mathcal{M}_x$.
(iii) for all germs $\phi : (\mathbb{C}, 0) \to (X, x)$, we have $h \circ \phi \in \phi^*(\mathcal{M}_x) \cdot \mathcal{O}_1$ (as above, it suffices to check for rather few arcs ϕ),
(iv) For each choice $\{s_i\}$ of a set of generators for $\mathcal{M}_x$ there is a neighbourhood U of x such that for each $\phi \in \Gamma(Hom(\mathbb{C}^p, \mathbb{C}))$ there exists $C > 0$ such that, for all $z \in U$ we have $|\phi(z)h(z)| \leq C \sup_i |\phi(z)s_i(z)|$.

Moreover, if $\mathcal{M}$ is a coherent sheaf of submodules of $\mathcal{O}_X^p$, there is a unique coherent sheaf $\overline{\mathcal{M}}$ with $\overline{\mathcal{M}_x} = \overline{\mathcal{M}_x}$ for all $x \in X$.

Integral closures of modules can be reduced, to some extent, to those of ideals. Suppose X irreducible, and write $\mathcal{H} = \mathcal{M} + \mathcal{O}_X \cdot h$. Then [G21, 1.7] $h \in \overline{\mathcal{M}}$ if and only if $\wedge^k \mathcal{H} \subseteq \overline{\wedge^k \mathcal{M}}$, where k is the largest integer with $\wedge^k \mathcal{H} \neq 0$. Here $\wedge^k$ refers to exterior powers, or rather to their images in $\wedge^k \mathcal{E}$.

We can now define $\mathcal{M}$ to be a reduction of $\mathcal{N}$ if $\mathcal{M} \subset \mathcal{N}$ and $\overline{\mathcal{M}} = \overline{\mathcal{N}}$. According to [21, (2.6)], this is equivalent to the natural map $Proj(\mathcal{R}\mathcal{M}) \to Proj(\mathcal{R}\mathcal{N})$ being a finite map. Again a minimal reduction is one with the minimum number of generators; if $\mathcal{M}$ has finite codimension in $\mathcal{O}_{X,x}^p$, a general set of $d + p - 1$ elements generates a minimal reduction.

Strict dependence in the module case was introduced in [G27], see also [G29, §3]. We say that h is strictly dependent on $\mathcal{M}$, and write $h \in \mathcal{M}^\dagger$ if, for all germs $\phi : (\mathbb{C}, 0) \to (X, x)$, we have $h \circ \phi \in \phi^*(\mathcal{M}_x) \cdot \mathfrak{m}_1$. Then $\mathcal{M}^\dagger$ is a module, and if $\mathcal{M}$ is a reduction of $\mathcal{N}$, $\mathcal{M}^\dagger = \mathcal{N}^\dagger$.

Gaffney has shown the flexibility of the concept of integral closure of modules by applying it to obtain simplified and strengthened versions of a variety of interesting results. Most relevant to equisingularity theory is a recasting of Whitney regularity conditions using integral closures. Observe that regularity gives a condition on tangent hyperplanes of $\mathbf{X}$ at a series of points converging to T. In the hypersurface case, the tangent hyperplane to $\mathbf{X}$ is given by $t \partial F / \partial t + \sum_i z_i \partial F / \partial z_i = const$. The Whitney A condition requires $\partial F / \partial t$ to be 'smaller' than the $\partial F / \partial z_i$. Recall that JF_T denotes the ideal $\langle \partial F / \partial z_i \rangle \lhd \mathcal{O}_{N+s}$ generated by the $\partial F / \partial z_i$. For $\dim(T) = 1$, we have the following characterisations:

$\partial F / \partial t \in \mathfrak{m}_N \cdot JF_T$ is Mather's criterion for triviality under right equivalence.

$\partial F / \partial t \in (JF_T)^\dagger$ is Gaffney's [G27] criterion for A-regularity.

$\partial F / \partial t \in \overline{\mathfrak{m}_N \cdot JF_T}$ is equivalent [41] to the W condition of Verdier, but care is needed: for Mather's condition, these are ideals in $\mathcal{O}_{N+1}$, for the others, in $\mathcal{O}_X$.

For the general case $F : \mathbb{C}^N \times T \to \mathbb{C}^p$, we have:

A-regularity holds if and only if for all tangent vectors $\partial / \partial t$ to T, $\partial F / \partial t \in J_M F^\dagger$, i.e. $J_M F_T \subseteq J_M F^\dagger$ [G29];

W-regularity holds if and only if for all tangent vectors $\partial / \partial t$ to T, $\partial F / \partial t \in \overline{\mathfrak{m}_N \cdot J_M F}$, i.e. $J_M F_T \subseteq \overline{\mathfrak{m}_N \cdot J_M F}$ [G21], [G25];

There are corresponding assertions for A_f and for W_f with F replaced by (F, f) ([G33, 2.1], [G40, 2.8]). Simpler versions for the case when F is absent are given in [G27] for A_f and in [G28] for W_f.

We now mention briefly several other papers of Gaffney giving applications of integral closure of modules.

In [G27], Gaffney shows that a hyperplane H is a limiting tangent plane to X if and only if $J_M F_H$ is *not* a reduction of $J_M F$. In [G29] he shows that $\mathbf{X}$ is A-regular over T if and only if every limiting tangent hyperplane contains T (which gives the above condition for A-regularity); and that $J_M F_T$ is a reduction of $J_M F$ if and only if *no* hyperplane containing T is a limiting tangent hyperplane. Relativisations of these equivalences are given in [G29,§5].

In [G31], Gaffney obtains a necessary and sufficient condition for a polynomial map $F : \mathbb{C}^N \to \mathbb{C}^p$ to be non-characteristic over t_0 at infinity. He gives an inequality which generalises Malgrange's condition, and shows that this is equivalent to the non-characteristic condition, and to an inequality $\partial F'/\partial t_i \in \overline{J_M F'}$ at t_0, where F' is essentially F referred to affine co-ordinates with t_0 at the origin, and that this implies local topological triviality.

In [G34] he studies finite determinacy with respect to left equivalence. He recalls that if F is an injective immersion outside T, the condition for a rugose trivialisation is W regularity, i.e. $J_M F_T \subseteq \overline{\mathfrak{m}_N \cdot J_M F}$. For left equivalence, one must consider double points, so for any ideal $I \lhd \mathcal{O}_n$, define $I_D \lhd \mathcal{O}_{2n}$ to be the ideal generated by the $h(z) - h(w)$ for $h \in I$. Then he obtains the necessary and sufficient condition $(\mathfrak{m}_n^{k+1})_D \subset (f^*\mathfrak{m}_p)_D^\dagger$ for $f : \mathbb{C}^n \to \mathbb{C}^p$ to be k-determined with respect to rugose trivialisation.

In [G43] (with Trotman and Wilson, and generalising [G24]) he studies the t^r condition, due originally to Thom [46] in the real case, and refined by Kuo and Trotman [26] and Trotman and Wilson [48]. One says that $\mathbf{X}$ is (t^r) regular over T at O if every C^r-submanifold Q of dimension N, transverse to T at O, is transverse to $\mathbf{X}$ near O. We say $\mathbf{X}$ is (t^r) regular for P over T at O if this holds for all Q with the same r-jet as P (here we may regard Q as the graph of a map $\mathbb{R}^N \to T$). With this form of the condition it is the condition on jets that is crucial, not the degree of differentiability, so the condition is also non-vacuous in the complex case. There are also variants depending on the degree of differentiability of Q; P only enters via its r-jet. Then provided $r > 0$ (the results for $r = 0$ are slightly different),

$X \backslash \sum_F$ is t^r for P if and only if $\mathfrak{m}_N^r \cdot J M_t F \subseteq \overline{\mathfrak{m}_N \cdot J_M F_P + I(P) J M_t(F)}$,
and
X is t^r for P if and only if $X \backslash \sum_F$ is so and $\mathfrak{m}_N^r \cdot \mathcal{O}_\Sigma \subseteq \overline{I(P)\mathcal{O}_\Sigma}$.

A genericity theorem, which states that the multiplicity $m(J_M F_P; \mathbf{X} \cap P)$ takes its generic value among all transversals with a given $(r - 1)$-jet if and only if t^r holds for P, ties this to other results.

4. Invariants

The main theorems about equisingularity are stated in terms of numerical invariants, mostly multiplicities. The development of the theory of these invariants is an important part of the story. In fact, few of the definitions are solely due to Gaffney: his role has been that of someone with ideas and suggestions, provoking others (most notably, David Rees, Steve Kleiman and Anders Thorup), to make his ideas precise and to prove his conjectures.

Algebraic definitions of multiplicity start with an ideal I of finite colength in a sufficiently nice ring R of dimension d (e.g. a noetherian local ring $\mathcal{O}_X$ with dim $X = d$); then for k large, the length (in our case, dimension over $\mathbb{C}$) of R/I^k is a polynomial in k with leading term $m\frac{k^d}{d!}$; and m is defined to be the multiplicity $m(I)$. If $I \subseteq J$, then [34] $m(I) \leq m(J)$, and $m(I) = m(J) \Leftrightarrow \overline{I} = \overline{J}$. Gaffney points out that this shows that multiplicities 'control' integral closures; it follows that they control Whitney conditions. Moreover, I admits a reduction with d generators, and if I itself has d generators, $m(I) = \dim(R/I)$.

This is easily generalised to the mixed multiplicities $m_{i,j}(I, J)$ $(i + j = d)$ of two ideals. For k and ℓ large enough, it is shown in [37, Chap I, §2] that the length of $R/I^k J^\ell$ is a polynomial of degree d in k and ℓ, whose leading term we denote by $\sum_{i=0}^d m_{i,d-i}(I, J)\frac{k^i}{i!}\frac{\ell^{d-i}}{(d-i)!}$. It can be shown that $m_{i,j}(I, J)$ is equal to the multiplicity of the ideal generated by i general elements of I and j elements of J. The multiplicity of the product ideal is given [37] by $m(I \cdot J) = \sum_{i=0}^d \binom{d}{i} m_{i,d-i}(I, J)$.

The concept generalises to the Buchsbaum-Rim multiplicity [5] of a submodule M of finite colength of a free module $E := R^p$. Here we take the (graded) symmetric algebra $S(E)$ over R generated by E, the Rees subalgebra $R(M)$ generated by M, and consider its colength $\dim(S_n(E)/R_n(M))$ in grade n. For n large enough, this is polynomial of degree $d + p - 1$ in n; if the leading term is $m(M)\frac{n^{d+p-1}}{(d+p-1)!}$, then $m(M)$ is an integer defined to be the multiplicity.

The following were conjectured by Gaffney and proved by Rees and Kirby [19]: if $M \subseteq N$, then $m(M) \leq m(N)$, and $m(M) = m(N) \Leftrightarrow \overline{M} = \overline{N}$: in this latter case, M is said to be a reduction of N. Moreover, by [35] or [G25], M admits a reduction with $d + p - 1$ generators (it suffices to take $d + p - 1$ generic elements of M), and if M admits $d + p - 1$ generators, we have ([5], [G25]) the simple formula $m(M) = \dim(E/M) = \dim(R/ \wedge^p (M))$. In our usual setup, the function $m(J_M F_t)$ is upper semicontinuous [G29, Prop 1.1].

One can define mixed multiplicities for two modules M, N each of finite colength as in case of ideals: take i generic elements from M and j generic elements of N where $i + j = d + p - 1$, and compute the Buchsbaum-Rim multiplicity of the module they generate. There is a product formula [19] which, in the geometric case, is

$$m(I \cdot M) = \binom{m}{d}m(I) + \sum_{j=0}^{d-1} \binom{m}{j}m(M|S_j),$$

where S_j is the quotient of $\mathcal{O}_X$ by j generic elements of I.

We can also consider the relative multiplicity of a submodule $N \subset M \subset E$ of finite colength in M: not surprisingly, with increased technicalities.

Geometrical invariants are obtained from polar varieties and their generalisations. Polar curves of plane curves were used in the mid 19th century, for example in the proof of Plücker's formulae counting singularities of plane curves. The use of polar varieties in equisingularity theory begins with [38], where Teissier states that their use was advocated by Thom. For X as above, and $0 \le k < d$, take a linear subspace $L \subset \mathbb{C}^N$ of codimension $d - k + 1$, defining $p_L : \mathbb{C}^N \to \mathbb{C}^{d-k+1}$. Then the polar variety is defined to be the closure $P_k = P_k(X, L)$ of the set of $x \in X^o$ critical for $p_L|X$ (the notation is chosen so that P_k has codimension k). It can be shown that for a dense Z-open[†] set of subspaces L of codimension $d - k + 1$, the multiplicity at O of $P_k(X, L)$ is independent of L: this is defined to be the polar multiplicity $m_k(X)$. The extreme cases are $k = 0$, with $P_0 = X$, so $m_0(X)$ is the multiplicity of X at 0, and $k = d$, with P_d discrete, so $m_d(X) = 0$.

When we have a family $\mathbf{X}$, we still choose a generic $L \subset \mathbb{C}^N$, now defining $p_L : \mathbb{C}^N \times T \to \mathbb{C}^{d-k+1} \times T$, and define the relative polar variety to be the closure $P_{T,k}$ of the set of $x \in \mathbf{X}^o$ critical for p_L, and $m_{T,k}(\mathbf{X})_t$ to be the multiplicity of $P_{T,k} \cap (\mathbb{C}^N \times \{t\})$ at $0 \times \{t\}$. We may not assume that this is equal to $m_k(X_t)$ since the notion of genericity for the subspace L is different in the two cases. In the extreme case $k = d$, $P_{T,d}$ is finite over d, so its closure meets T at points with $m_{T,d}(\mathbf{X})_t > 0$. We see that the absence of such points is important for equisingularity.

More generally, if $\mathcal{I} \triangleleft \mathcal{O}_X$ is an ideal with q generators and $\Lambda \subset P^{q-1}$ is a linear subspace of codimension k, the polar variety $P_k(X, \Lambda)$ is the image in X of $B_{\mathcal{I}}X \cap (X \times \Lambda)$. Again, for a dense Z-open set of subspaces Λ of codimension k, the multiplicity at O of $P_k(X, \Lambda)$ is independent of Λ, and is defined to be the polar multiplicity $m_k(X, \mathcal{I})$. In the case when X is the zero set of $F : \mathbb{C}^N \to \mathbb{C}$ and $\mathcal{I}$ is the Jacobian ideal JF, we see (taking Λ as the annihilator of L) that $P_k(X, JF)$ is the same as $P_k(X)$. As above, if $\mathcal{I} \triangleleft \mathcal{O}_{\mathbf{X}}$ there is a relative version $m_{T,k}(\mathcal{I})_t$.

[†] Here and below we write Z-open to mean open in the Zariski topology.

If $\mathcal{I} \subset \mathcal{J}$ and $\overline{\mathcal{I}} = \overline{\mathcal{J}}$, the natural map $B_{\mathcal{J}}(X) \to B_{\mathcal{I}}(X)$ is finite, so we can take the same Λ for both $\mathcal{I}$ and $\mathcal{J}$, and use the projection formula to see that $m_k(X, \mathcal{I})$ depends only on the integral closure of $\mathcal{I}$.

If the polar variety is a curve, then so is its strict transform under the simple blowup of the origin in $\mathbb{C}^N$, and $m_1(X)$ is the intersection number of the preimage with the exceptional divisor. In general we can intersect with generic hyperplanes to cut down to a curve, then do the same. Equivalently, take the preimage in the blowup, and intersect with $d - k - 1$ hyperplanes and with the exceptional divisor. This leads to an interpretation of polar multiplicities as intersection numbers, which is due to Kleiman and Thorup [21].

The projection $B_{\mathrm{m}\cdot\mathcal{I}}(X) \to X$ factors through both $B_{\mathrm{m}}(X)$ and $B_{\mathcal{I}}(X)$. Over $B_{\mathcal{I}}(X)$ we have the line bundle induced from the universal bundle on P^{q-1}: write ℓ_I for its first Chern class and D_I for the corresponding exceptional divisor; and use the same notations for their pullbacks to $B_{\mathrm{m}\cdot\mathcal{I}}(X)$. Over $B_{\mathrm{m}}(X)$ we have the line bundle coming from the universal bundle on $P(\mathbb{C}^N)$, with first Chern class ℓ_{m} and divisor D_{m}. Then $P(X, L)$ corresponds to the intersection ℓ_I^k and $m_k(X, \mathcal{I})$ is given by $\ell_I^k \cdot \ell_{\mathrm{m}}^{d-k-1} \cdot D_{\mathrm{m}}$ or, in the notation of intersection theory,

$$m_k(X, \mathcal{I}) = \int \ell_I^k \cdot \ell_{\mathrm{m}}^{d-k-1} \cdot D_{\mathrm{m}}.$$

For the application to equisingularity for hypersurfaces, we take $\mathcal{I}$ to be the Jacobian ideal. If X has non-isolated singularity, this does not have finite codimension. To allow for this case, in [G28] Gaffney & Gassler define further invariants. For $L \subset P^q$ of codimension k, the polar variety of $\mathcal{I}$ is the projection to X of $P(X, \mathcal{I}) := B_{\mathcal{I}}(X) \cap (X \times L)$ and we now define the Segre cycle as the image of $Q(X, \mathcal{I}) := D_I \cap B_{\mathcal{I}}(X) \cap (X \times L)$. Its multiplicity is constant for L in a dense Z-open set, defining the Segre number $se_k(X, \mathcal{I})$. This too can be defined as an intersection number

$$se_k(X, \mathcal{I}) = \int \ell_I^{k-1} \ell_{\mathrm{m}}^{d-k-1} D_I \cdot D_{\mathrm{m}} (1 \leq k \leq d - 1),$$

$$se_d(X, \mathcal{I}) = \int \ell_I^{d-1} D_I.$$

Thus

$$se_k(X, \mathrm{m}_N \cdot \mathcal{I}) = \int h^{k-1} \ell_{\mathrm{m}}^{d-k-1} D \cdot D_{\mathrm{m}} (1 \leq k \leq d - 1),$$

$$se_d(X, \mathrm{m}_N \cdot \mathcal{I}) = \int h^{d-1} D$$

so substituting $h = \ell_l + \ell_m$ leads to the product rules

$$se_k(X, \mathfrak{m}_N \cdot \mathcal{I}) = \sum_{i=1}^{k} \binom{k-1}{i-1} se_i(X, \mathcal{I})(1 \leq k \leq d-1),$$

$$se_d(X, \mathfrak{m}_N \cdot \mathcal{I}) = \sum_{i=0}^{d-1} \binom{d}{i} m_i(X, \mathcal{I}) + \sum_{i=1}^{d} \binom{d-1}{i-1} se_i(X, \mathcal{I}).$$

These generalise the Lê numbers of Massey [29], which are defined in the case $p = 1$ as $\lambda_k(F_t) := se_k(X_t, J F_T)$. Massey showed that the reduced Euler characteristic $\tilde{\chi}^k(t)$ of the Milnor fibre of $F_t | L^k$ (for L^k a generic linear subspace of dimension k) is given by $\tilde{\chi}^k(t) = m_k(F_t) + \sum_{i=0}^{k}(-1)^i \lambda_{k-i}(F_t)$. It is also shown in [G28, Cor 4.5] that if $\mathcal{I} \lhd \mathcal{O}_X$ induces $\mathcal{I}_t$ on X_t, while the $se_i(X_t, \mathcal{I}_t)$ individually need not be upper semicontinuous in t, the sequence $(se_1, \ldots, se_d)$ is, provided sequences are ordered lexicographically.

Somewhat similar definitions for the case of modules are made in [G29], following [21]. Suppose $\mathcal{M} \subset \mathcal{E} = \mathcal{O}_X^p$ has finite colength (or, in the relative case, that $Supp(\mathcal{E}/\mathcal{M})$ is finite over T: in fact all the definitions are as easily made over T, this is what is needed for the application). Set $P := Proj(S\mathcal{E})$, and $P' := Proj(\mathcal{R}\mathcal{M})$, where $\mathcal{R}\mathcal{M}$ is the Rees algebra. Let B be the blowup of P by the sheaf of ideals on $S\mathcal{E}$ generated by $\mathcal{M}$, with exceptional divisor D. Let ℓ_E, ℓ_M denote the classes on B induced from the first Chern classes of the tautological sheaves on P and P'. Now define Segre classes by $se_i := \int \ell_M^{i-1} \ell_E^{r-i} [D]$, where $r = d + p - 1$. In fact, $se_i = 0$ for $i < d$.

Now set $e^j := \sum_{i=1}^{r-j} se_i$ (so that $e^{p-1} = se_d$, $e^0 = \sum_i se_i$). Then the e^j are all upper semicontinuous (the se_i need not be). With this definition, the Kleiman-Thorup multiplicity is e^0.

More generally, suppose given two submodules $\mathcal{M} \subset \mathcal{N} \subset \mathcal{E}$, with $\overline{\mathcal{N}}/\mathcal{M}$ of finite length, or equivalently, $\overline{\mathcal{M}} = \overline{\mathcal{N}}$ in a punctured neighbourhood of x. Write $\rho(\mathcal{M})$ for the ideal on $Proj(\mathcal{R}\mathcal{N})$ generated by $\mathcal{M}$, $B_\mathcal{M}$ for the normalised blowup of $Proj(\mathcal{R}\mathcal{N})$ along $\rho(\mathcal{M})$, and $D_{\mathcal{M},\mathcal{N}}$ for the exceptional divisor. Over $B_\mathcal{M}$ there are two canonical line bundles, one coming from $\mathcal{M}$ one from $\mathcal{N}$: denote their first Chern classes by ℓ_M, ℓ_N. Then, following Kleiman and Thorup [21], Gaffney defines the multiplicity of the pair $\mathcal{M} \subset \mathcal{N}$ as

$$m(\mathcal{M}, \mathcal{N}) := \sum_{j=0}^{d+r-2} \int \ell_M^{d+r-2-j} \ell_N^j D_{\mathcal{M},\mathcal{N}}.$$

This generalises the Buchsbaum-Rim multiplicity $m(\mathcal{M}) = m(\mathcal{M}, \mathcal{E})$, and satisfies additivity: if $\mathcal{L} \subset \mathcal{M} \subset \mathcal{N}$ then $m(\mathcal{L}, \mathcal{N}) = m(\mathcal{L}, \mathcal{M}) + m(\mathcal{M}, \mathcal{N})$.

However, the analogues of the polar and Segre multiplicities do not have such good properties as in the ideal case, as is shown by a counterexample in [G37, §4].

5. Criteria for equisingularity

We now discuss the main developments in chronological order, and begin in 1973 with Teissier's work [37] on isolated hypersurface singularities. The plan of his proof is as follows. Since the Whitney conditions hold generically, the inclusion $\partial F/\partial t \in \overline{\mathfrak{m}_N \cdot JF_T}$ which characterises them holds at an open set of points in T. The key step is now the "Principle of Specialisation of Integral Dependence" (PSID) which shows that for any g and any ideal $\mathcal{I} \lhd \mathcal{O}_{\mathbf{X}}$ such that the $\mathcal{I}_t \lhd \mathcal{O}_{X,t}$ have finite codimension and the multiplicity $m(\mathcal{I}_t)$ is constant along T, the set of $t \in T$ at which the germ $h \in \overline{\mathcal{I}_t}$ is closed.

The rough idea is as follows (for a fuller, but still very short and geometric account see [G28, pp 700–701]). Let $\mathcal{I} \lhd \mathcal{O}_{\mathbf{X}}$; blow up along $\mathcal{I}$; let D be the exceptional divisor of the blowup. If $m(\mathcal{I}_t)$ is constant, the projection $D \to T$ is equidimensional. Now whether or not $g \in \overline{\mathcal{I}}$ depends on the valuations $v_i(g)$ corresponding to the components D_i of D. Since the projection is equidimensional, the list of v_i is independent of t; since also the $v_i(g)$ are semicontinuous the result follows.

The invariants μ^* appear as follows. First, the relation of multiplicities to Milnor numbers for icis (due to Lê [23] and Greuel [10]) gives $m(JF_T) = \mu^{(N)}(X_t) + \mu^{(N-1)}(X_t)$. Next the formula for the multiplicity of a product of ideals gives $m(\mathfrak{m}_N \cdot J) = \sum_{i=0}^{N} \binom{N}{i} m_{i,N-i}(\mathfrak{m}_N, J)$. But the mixed multiplicity $m_{i,N-i}(\mathfrak{m}_N, J)$ is equal to the multiplicity of the restriction of J to a generic codimension i subspace, and hence to $\mu^{(i+1)}(X_t) + \mu^{(i)}(X_t)$. Thus $m(\mathfrak{m}_N \cdot JF_T) = \sum_{i=0}^{N} \binom{N}{i}(\mu^{(i+1)}(X_t) + \mu^{(i)}(X_t))$. It follows, since the $\mu^{(i)}$ are semicontinuous, that $\mu^*(X_t)$ is independent of t if and only if $m(\mathfrak{m}_N \cdot JF_T)$ is.

By 1980, Teissier had obtained a general criterion for regularity, which appeared in [41]. He developed the theory of (relative) polar multiplicities, in essentially our standard situation. Then his main theorem states that Whitney regularity is equivalent to constancy of polar multiplicities. More precisely, for $\mathbf{X}$ reduced complex analytic of pure dimension d and T a smooth subspace, the following are equivalent:

(i) $(m_0\mathbf{X}, m_1\mathbf{X}, \ldots, m_{d-1}\mathbf{X})_t$ is constant along T;

(ii) (if $\dim(T) = 1$) $\mathbb{C}^N \times \{0\}$ is transverse to all limits of tangent spaces of $\mathbf{X}^o$ and $(m(\mathbf{X})_t, m_{T,1}(\mathbf{X})_t, \ldots, m_{T,d}(\mathbf{X})_t)$ is constant along T;

(iii) $(\mathbf{X}^o, T)$ satisfies at t the Whitney A and B conditions;
(iv) $(\mathbf{X}^o, T)$ satisfies at t the strict Whitney A condition with exponent 1 and strict Whitney B with exponent > 0.

It follows that for any stratification of a complex analytic space, constancy along strata of polar multiplicities is necessary and sufficient for Whitney regularity, or for Verdier regularity. Teissier deduced that a complex analytic variety has a unique minimal regular stratification.

In Terry's great 1993 paper [G22] these results are refined and extended to obtain sufficient conditions for equisingularity of a family of maps. Jim Damon, using largely algebraic arguments, had previously proved in [8], unifying several earlier results:

If $F_0 : \mathbb{C}^N \to \mathbb{C}^p$ is $\mathcal{A}$–finite, i.e. finitely determined with respect to right-left-equivalence, any polynomial unfolding of non-negative weight is topologically trivial.

Gaffney aimed to improve this by replacing the weight condition by geometrical conditions on the unfolding. Now if F_0 is $\mathcal{A}$–finite, any (multi-)germ of F_0 outside the origin is stable. We require the same to hold for any (multi-)germ outside T of the unfolding $F : \mathbb{C}^N \times T \to \mathbb{C}^p \times T$ of F_0: a criterion on F for this to hold follows from Damon's work. Hence differentiable triviality is à priori guaranteed at points outside T. Adding a simplicity hypothesis, we may suppose that only finitely many $\mathcal{K}$–classes (strata) occur. It thus remains to prove, for each stratum in source or target, Whitney regularity over T: we then have a c.v.s. and the desired result will follow from the second isotopy lemma.

For this we have the above criterion of Teissier for regularity in terms of polar multiplicities. However, these are defined in terms of the whole variety $\mathbf{X} \subset \mathbb{C}^N \times T$ (or $\mathbf{Y} \subset \mathbb{C}^p \times T$). Gaffney's aim was to find a condition depending only on the individual fibres X_t. To replace relative polar multiplicities of $\mathbf{X}$ by polar multiplicities of X_t, two main issues arise. First, one needs to show that each $\mathbb{C}^N \times \{t\}$ is transverse to limiting tangent planes of strata. Gaffney succeeds here by using the fact that the stratification is regular in the complement of T and $m_{T,d}(\mathbf{X})$ vanishes along T. He then uses delicate arguments to show that $m_i(P_i(F, t))$ is constant if and only if $m_i(X_t)$ is so for $0 \le i < d$ and, again, $m_{T,d}(\mathbf{X}) = 0$ along T.

Secondly, since $d = \dim X_t$, $m_d(X_t)$ is not defined, so we need a replacement to control $m_{T,d}(\mathbf{X})$: this is the main difficulty. Let $\mathcal{Q}$ be a $\mathcal{K}$–equivalence class and $\mathcal{S}(\mathcal{Q})$ the corresponding stratum: write $d := \dim(\mathcal{S}(\mathcal{Q})) - \dim T$. To define the d–stable multiplicity $m_d(F_t, \mathcal{Q})$, take a versal unfolding $G : \mathbb{C}^n \times T \to \mathbb{C}^p \times T$, denote projection on T by π, and pick a general hyperplane $L \subset \mathbb{C}^p$

defining a polar variety $P := P_d(\overline{Q(G)}, \pi)$. Define $m_d(F, Q)$ as the multiplicity of the ideal $\mathfrak{m}_s \cdot \mathcal{O}_P \lhd \mathcal{O}_P$: this can be proved independent of all choices. Now to avoid 'coalescing', i.e. an arc in $\mathcal{S}(Q)$ converging to a point of T, it is enough to require $m_0(F, Q)$ constant.

This paper has led to a whole industry of obtaining more explicit, and much shorter lists of invariants in low dimensions whose constancy guarantees equisingularity. Many such results have been reported at São Carlos. In [G22], Gaffney gave a detailed study of the cases $\mathbb{C}^2 \to \mathbb{C}^2$ and $\mathbb{C}^2 \to \mathbb{C}^3$; his results were improved by Houston [16]. Jorge Peréz treated map-germs $\mathbb{C}^3 \to \mathbb{C}^3$ in his thesis [18]. In [G38], Gaffney and Vohra dealt with maps $\mathbb{C}^n \to \mathbb{C}^2$. Corank 1 maps $\mathbb{C}^n \to \mathbb{C}^{n+1}$ were treated by Houston [17]. In addition, in several of the equisingularity results (for spaces) below, constancy of the invariants controls some strata below the top dimension.

In successive papers generalising Teissier's first equisingularity theorem, certain themes reappear. In the original argument, the fact that Whitney conditions hold generically, which Whitney proved geometrically using his wing lemma, was used. Later Teissier replaced this by his 'idealistic Bertini theorem' [40]. In later papers, Gaffney and collaborators used a transversality theorem of Kleiman [20] for this purpose. There are also successive versions of the PSID. In each case, the general outline of the argument is the same: first we blow up, then examine the geometry. In each case we have **X** as usual, $h \in \overline{\mathcal{I}}$ (for ideals) or $h \in \overline{\mathcal{M}}$ (for modules) on X_t for a dense set of points $t \in T$, and can conclude the same at all points provided certain multiplicities are constant.

In 1996 Gaffney [G25] extends Teissier's results on isolated hypersurface singularities to the ICIS case. In fact in [G22] he had already shown that, in this case, Whitney regularity was equivalent to constancy of the polar invariants $m_k(X_t)$, so it follows from properties of the invariants discussed in §4 that, for ICIS germs, the following are equivalent:

(i) (X, T) is Whitney regular,
(ii) $m_k(X_t)$ is constant $(0 \le k \le d)$
(iii) for all tangent vectors $\partial/\partial t$ to T, $\partial F/\partial t \in \overline{\mathfrak{m}_N \cdot J F_T}$
(iv) $m(\mathfrak{m}_N \cdot J F_T)_t$ is constant.

He also proves $m(\mathfrak{m}_N \cdot J_M F) = \sum_{j=0}^{d} \binom{N-1}{j} m_{d-j}(X)$. The argument in this paper does not refer directly to a PSID, but such a result applicable to this case was later obtained in [G29]. Again the Lê-Greuel formula allows restatement in terms of $\mu^{(i)}$.

Three of Gaffney's major papers appeared in 1999. We referred above to the introduction and development by Gaffney & Gassler [G28] of the Segre numbers of an ideal. Next they obtain a PSID. Given a family in our usual notation and ideal $\mathcal{I} \lhd \mathcal{O}_X$, if h has germ at t in $\overline{\mathcal{I}_t}$ for a dense open set of t, and the Segre numbers $se_k(\mathcal{I}_t)$ are constant for $1 \leq k \leq d$, then $h \in \overline{\mathcal{I}}$. They apply this version of the PSID to the general hypersurface $\mathbf{Z}$ in $\mathbf{X} = \mathbb{C}^N \times T$, and show that the W_f condition holds, and hence the smooth part of $\mathbf{Z}$ is Whitney regular over T, provided the polar multiplicities $m_*(f_t)$ and Segre numbers $se_*(f_t)$ are constant, or equivalently, m_* and the reduced Euler characteristics $\overline{\chi}^*$ are constant. Also in this situation, the codimension 1 strata of $\sum(\mathbf{Z})$ are Whitney regular over T. Even more surprisingly, the converse holds: if $\mathbf{Z}$ admits a Whitney stratification with T a stratum, then W_f is satisfied.

Gaffney & Kleiman [G29] prove a version of PSID for modules $\mathcal{M}$ of finite codimension: more precisely, the support S of $\mathcal{E}/\mathcal{M}$ is finite over T. The required condition is that $m(\mathcal{M})$ is constant. It is also shown in the ICIS case in [G29] that A-regularity holds if the Segre numbers $se_*(J_M F)$ are constant: in fact they state the result in terms of the partial sums $e^j = \sum_1^{r-j} se_i$. This condition is not necessary for A-regularity, and they offer an interesting example to show that no condition depending only on the members of the family can be both necessary and sufficient: the families $z_1^2 - z_2^3 + z_2^2 t^b = 0$ are A-regular if $b \geq 2$ but not if $b = 1$, but have the same sets of members. They also obtain a number of results on A_f, which lead them to conjecture that A_f holds when the multiplicity or the Milnor numbers are defined and constant.

Gaffney and Massey [G30] prove the equivalence (in the complete intersection case) of:

(i) W_f holds,
(ii) both $\mathbf{X}$ and $\mathbf{Z}$ are Whitney regular over T, and
(iii) the X_t and Z_t all have only isolated singularities and the sequences $\mu^*(X_t)$ and $\mu^*(Z_t)$ are both constant in t.

If we just know that the X_t and Z_t have isolated singularities and $\mu(X_t)$ and $\mu(Z_t)$ are both constant in t, then A_f holds. In [G33] the relation between Milnor numbers and multiplicities is used to reduce the condition to constancy of $m(\mathrm{m} \cdot J_M(F, f))$.

It is also shown that the conditions W_f and A_f usually imply local analytic triviality if the target dimension of f exceeds 1, so are not of interest in this case.

6. Recent work

The pattern of nearly all the equisingularity theorems is to relate failure of regularity to jumps in dimension of exceptional sets arising in blowups and thence to non-constancy of invariants such as polar multiplicities. In [G37] Gaffney gives examples where the invariants previously studied (the Segre numbers) are constant and yet the dimension jumps. He argues that this is because strata of different dimensions are making contributions, and so he is led to a new approach.

The basic idea is as follows. Suppose X as usual and $\mathcal{M} \subset \mathcal{E} = \mathcal{O}_X^p$ a submodule. Decompose the support of $\mathcal{E}/\mathcal{M}$ in X into its irreducible components V_α and denote by $\mathcal{F}_k$ the union of the V_α of codimension $\geq k$. We think of these as defining a stratification, though the V_α do not need to be disjoint, or even to intersect nicely. A further complication is that rather than using the components of the support S of $\mathcal{E}/\mathcal{M}$ in X, we must form the preimage of S under π, take its components, and define the V_α as the projections of these back down to X.

Define the hull $\mathcal{H}_i(\mathcal{M})$ of $\mathcal{M}$ as the set of elements integrally dependent on $\mathcal{M}$ in codimension i: $h \in \mathcal{H}_i(\mathcal{M}) \iff \forall z \notin \mathcal{F}_{i+1}, h \in \overline{\mathcal{M}}_z$. This can be refined by considering germs at $x \in X$ of both $\mathcal{H}_i(\mathcal{M})$ and $\mathcal{F}_{i+1}$.

Then $\mathcal{F}_{i+1}$ is the projection to X of the cosupport of $\rho(\overline{\mathcal{M}})\mathcal{R}(\mathcal{H}_i(\mathcal{M}))$, thought of as a sheaf of modules on $Proj(\mathcal{R}(\mathcal{H}_i(\mathcal{M})))$.

For each α, we choose a smooth point z_α of V_α and a slice S_α at that point. Then if V_α has codimension $i + 1$, the multiplicity $m(\mathcal{M}|S_\alpha, \mathcal{H}_i(\mathcal{M})|S_\alpha, z_\alpha)$ is defined, and depends only on α. Denote it by $m_\alpha(\mathcal{M})$.

The above extends naturally to families $p : \mathcal{X} \to T$ and modules $\mathcal{M} \subset \mathcal{E}$ over $\mathcal{X}$, but we cannot necessarily identify the fibre of the cosupport of $\mathcal{M}$ over $t \in T$ with the cosupport of the restriction of $\mathcal{M}$ to X_t. This point makes for technical problems, and forces the proofs to go by induction on the codimension of the V_α.

The formal treatment appears in [G36]. Start with $\mathcal{M} \subset \mathcal{N} \subset \mathcal{E} = \mathcal{O}_X^p$, each of generic rank r: remark that though the cosupports of $\mathcal{M}$ and $\overline{\mathcal{M}}$ in $\mathcal{E}$ are the same, this is not the case for their cosupports in $\mathcal{N}$. Denote by π_X the projection to X of $Proj(\mathcal{R}(\mathcal{M}))$ or similars. Gaffney constructs a sequence

$$\mathcal{M} \subset \overline{\mathcal{M}} = \mathcal{H}_d(\mathcal{M}) \subset \mathcal{H}_{d-1}(\mathcal{M}) \ldots \mathcal{H}_0(\mathcal{M}) \subset \mathcal{E},$$

with each $\mathcal{H}_i(\mathcal{M})$ integrally closed, the components of the cosupport of $\rho(\overline{\mathcal{M}})\mathcal{R}(\mathcal{H}_i(\mathcal{M}))$ project to sets of codimension at least $i + 1$, and $\mathcal{H}_i(\mathcal{M})$ is as small as possible subject to this.

Proceed by induction on i (I omit numerous details) with induction basis:

$$e_0(\mathcal{M}) := p - r, \quad \mathcal{H}_0(\mathcal{M}) := \{h \in \mathcal{E} \mid e_0(\mathcal{M} + h \cdot \mathcal{O}_X) = e_0(\mathcal{M})\}.$$

Now assume $e_{k-1}(\mathcal{M})$ and a coherent sheaf $\mathcal{H}_{k-1}(\mathcal{M})$ defined; consider the cosupport CS_k of $\rho(\mathcal{M})$ on $Proj(\mathcal{R}(\mathcal{H}_{k-1}(\mathcal{M})))$ induced by the inclusion $\mathcal{R}(\mathcal{M}) \subset \mathcal{R}(\mathcal{H}_{k-1}(\mathcal{M}))$; let A_k index the components C_α of codimension k

of $\pi_X(CS_k)$. As part of the inductive construction, there are none of lower codimension; we can also expect these C_α to coincide with the V_α of codimension k. Now set

$$e_k(\mathcal{M}) : = \sum_{\alpha \in A_k} m(C_\alpha)m(\mathcal{M}|S_\alpha, \mathcal{H}_{k-1}(\mathcal{M})|S_\alpha),$$

$$\mathcal{H}_k(\mathcal{M}) : = \{h \in \mathcal{E} \mid e_k(\mathcal{M} + h \cdot \mathcal{O}_X) = e_k(\mathcal{M})\},$$

where S_α is a slice transverse to C_α at a smooth point.

As well as the hulls, this yields new invariants $e_i(\mathcal{M})$. An important first result: if $\mathcal{M} \subset \mathcal{N}$ and $e_i(\mathcal{M}) = e_i(\mathcal{N})$ for $0 \le i \le d$ then $\mathcal{N} \subset \overline{\mathcal{M}}$.

The central result of this work to date is called by Gaffney the 'multiplicity-polar' theorem. Suppose $\mathcal{M} \subset \mathcal{N} \subset \mathcal{E}$ such that the support C of $\overline{\mathcal{N}/\mathcal{M}}$ is finite over T. Then for each $t \in T$, $C \cap \pi^{-1}(t)$ is a finite set of points x_i, at each of which $m(\mathcal{M}_t, \mathcal{N}_t, x_i)$ is defined: define

$$m(\mathcal{M}, \mathcal{N})_t := \sum_{x \in C \cap \pi^{-1}(t)} m(\mathcal{M}, \mathcal{N}, x).$$

Recall that the polar variety $P_k(\mathcal{M})$ is defined as $\pi_X(Proj(\mathcal{R}\mathcal{M}) \cap (X \times L))$, where L is a generic plane of codimension $r + k - 1$. Then the multiplicity polar theorem states

$$m(\mathcal{M}, \mathcal{N})_t - m(\mathcal{M}, \mathcal{N})_{gen} = mult_t \, P_d(\mathcal{M}) - mult_t \, P_d(\mathcal{N}),$$

where *gen* is a generic point of T. A proof in the special case of ideals is given in [G37]; the general proof appears in [G46]. The proof is motivated by a review of the definition of Segre numbers of a module in terms of a sequence of polars of different codimensions.

In his São Carlos paper [G37], Gaffney considers the case of isolated singularities, and deduces the following version of the PSID: for $\mathcal{M} \subset \mathcal{N}$ as above, if $h \in \mathcal{N}$ and $h_t \in \overline{\mathcal{M}_t}$ for a dense open set of t, then provided $m(\mathcal{M}, \mathcal{N})_t$ is constant, we have $h \in \overline{\mathcal{M}}$. The key point of the proof is to study the dimensions of the fibres over T of the preimage in $Proj(\mathcal{R}\mathcal{M})$ of the locus of points where $\mathcal{M}$ is not free.

He deduces a criterion for Whitney regularity for the general case when the X_t have isolated singularities and $\sum \mathbf{X} = T$: W-regularity holds if and only if $m(\mathcal{M}, \mathcal{N})_t + mult_t \, P_d(\mathcal{N})$, with $\mathcal{M} = \mathfrak{m}_N \cdot J_M(F_t)$ and $\mathcal{N} = \mathcal{H}_0(J_{M_z}(F))$, is independent of t. In the relative case when also the Z_t have isolated singularities, the condition for A_f to hold is obtained from this by taking $\mathcal{M} = J_M(F, f)$ and $\mathcal{N} = \mathcal{H}_0(J_M(F, f))$.

Terry has also used the multiplicity polar theorem to obtain a version of the PSID in which the multiplicities required to be constant are the $m_\alpha(M_t) + mult_{z_\alpha} P_{i+1}(H_i(M))$.

As well as its theoretical value, Gaffney shows how to use the theorem to obtain numerous effective calculations of numerical invariants; in [G39] for a family of hypersurfaces whose singular locus has dimension 1, or other constant dimension; also for map-germs $\mathbb{C}^2 \to \mathbb{C}^3$; and in [G44] for isolated singularities; in particular a calculation of MacPherson's Euler obstruction.

Bibliography of Terry Gaffney's papers

G 1. *Properties of finitely determined germs*, thesis, Brandeis, 1976.

G 2. On the order of determination of a finitely determined germ, Invent. Math. **37** (1976) 83–92.

G 3. A note on the order of determination of a finitely determined germ, Invent. Math. **52** (1979) 127–130.

G 4. (with R. Lazarsfeld) On the ramification of branched coverings of P^n, Invent. Math. **59** (1980) 53–58.

G 5. (with A. A. du Plessis) More on the determinacy of smooth map-germs, Invent. Math. **66** (1982) 137–163.

G 6. (with J. W. Bruce) Simple singularities of mappings $(C, 0) \to (C^2, 0)$, J. London Math. Soc. **26** (1982) 465–474.

G 7. (with T. Banchoff and C. McCrory) *Cusps of Gauss mappings*, Research Notes in Math. **55**, Pitman, 1982.

G 8. (with J. N. Damon) Topological triviality of deformations of functions and Newton filtrations, Invent. Math. **72** (1983) 335–358.

G 9. The structure of $T\mathcal{A}(f)$, classification and an application to differential geometry, pp 409–427 in *Singularities* (ed. P. Orlik), Proc. Symp. Pure Math. **40(1)**, Amer. Math. Soc., 1983.

G 10. (with L. C. Wilson) Equivalence theorems in global singularity theory, pp 439–447 in *Singularities* (ed. P. Orlik), Proc. Symp. Pure Math. **40(1)**, Amer. Math. Soc., 1983.

G 11. Multiple points and associated ramification loci, pp 429–437 in *Singularities* (ed. P. Orlik), Proc. Symp. Pure Math. **40(1)**, Amer. Math. Soc., 1983.

G 12. The Thom polynomial of $\overline{\sum}^{1111}$, pp 399–408 in *Singularities* (ed. P. Orlik), Proc. Symp. Pure Math. **40(1)**, Amer. Math. Soc., 1983.

G 13. (with L. C. Wilson) Equivalence of generic mappings and C^∞ normalization, Compositio Math. **49** (1983) 291–308.

G 14. (with J. W. Bruce and A. A. du Plessis) On left equivalence of map germs, Bull. London Math. Soc. **16** (1984) 303–306.

G 15. (with H. Hauser) Characterizing singularities of varieties and of mappings, Invent. Math. **81** (1985) 427–447.

G 16. (with T. Banchoff and C. McCrory) Counting tritangent planes of space curves, Topology **24** (1985) 15–23.

G 17. New methods in the classification theory of bifurcation problems, pp 97–116 in *Multiparameter bifurcation theory (Arcata, 1985)*, Contemp. Math. **56**, Amer. Math. Soc., 1986.

G 18. Multiple points, chaining and Hilbert schemes, Amer. J. Math. **110** (1988), 595–628.

G 19. (with D. Mond) Cusps and double folds of germs of analytic maps $C^2 \to C^2$, J. London Math. Soc. **43** (1991) 185–192.

G 20. (with D. Mond) Weighted homogeneous maps from the plane to the plane, Math. Proc. Camb. Phil. Soc. **109** (1991) 451–470.

G 21. Integral closure of modules and Whitney equisingularity, Invent. Math. **107** (1992) 301–322.

G 22. Polar multiplicities and equisingularity of map germs, Topology **32** (1993) 185–223.

G 23. Punctual Hilbert schemes and resolutions of multiple point singularities, Math. Ann. **295** (1993) 269–289.

G 24. Equisingularity of plane sections, t_1 condition and the integral closure of modules, pp 95–111 in *Real and complex singularities (São Carlos III, 1994)* (ed. W. L. Marar), Pitman Res. Notes Math. **333**, 1995.

G 25. Multiplicities and equisingularity of ICIS germs, Invent. Math. **123** (1996) 209–220.

G 26. (with A. A. du Plessis and L. C. Wilson) Map-germs determined by their discriminants, pp 1–40 in *Stratifications, singularities and differential equations, I (Marseille, 1990; Honolulu, 1990)* (eds. D. Trotman and L. C. Wilson), Travaux en Cours **54**, Hermann, 1997.

G 27. Aureoles and integral closure of modules, pp 55–62 in *Stratifications, singularities and differential equations, II (Marseille, 1990; Honolulu, 1990)* (eds. D. Trotman and L. C. Wilson), Travaux en Cours **54**, Hermann, 1997.

G 28. (with R. Gassler) Segre numbers and hypersurface singularities, J. Alg. Geom. **8** (1999) 695–736.

G 29. (with S. Kleiman) Specialization of integral dependence for modules, Invent. Math. **137** (1999) 541–574.

G 30. (with D. Massey) Trends in equisingularity theory, pp 207–248 in *Singularity theory (Liverpool, 1996)* (eds. J. W. Bruce and D. Mond), London Math. Soc. Lecture Notes **263**, Cambridge Univ. Press, 1999.

G 31. Fibers of polynomial mappings at infinity and a generalized Malgrange condition, Compositio Math. **119** (1999) 157–167.

G 32. Plane sections, W_f and A_f, pp 16–32 in *Real and complex singularities (São Carlos V, 1998)* (eds. J. W. Bruce and F. Tari), Res. Notes Math. **412**, Chapman & Hall/CRC 2000.

G 33. (with S. Kleiman) W_f and integral dependence, pp 33–45 in *Real and complex singularities (São Carlos V, 1998)* (eds. J. W. Bruce and F. Tari), Res. Notes Math. **412**, Chapman & Hall/CRC 2000.

G 34. $\mathcal{L}^0$-equivalence of maps, Math. Proc. Camb. Phil. Soc. **128** (2000) 479–496.

G 35. The theory of integral closure of ideals and modules: applications and new developments. With an appendix by Steven Kleiman and Anders Thorup, pp 379–404 in *New developments in singularity theory (Cambridge, 2000)* (eds. D. Siersma, C. T. C. Wall and V. Zakalyukin), NATO Sci. Ser. II Math. Phys. Chem. **21**, Kluwer, 2001.

G 36. Generalized Buchsbaum-Rim multiplicities and a theorem of Rees, Comm. Algebra **31** (2003), no. 8 (Special issue in honor of Steven L. Kleiman), 3811–3827.

G 37. Polar methods, invariants of pairs of modules and equisingularity, pp 113–135 in *Real and complex singularities (São Carlos VII)* (eds. T. Gaffney and M. A. S. Ruas), Contemp. Math. **354**, Amer. Math. Soc., 2004.

G 38. (with R. Vohra) A numerical characterization of equisingularity for map germs from n-space, ($n \geq 3$), to the plane, J. Dyn. Syst. Geom. Theor. **2** (2004), 43–55.

G 39. The multiplicity of pairs of modules and hypersurface singularities, pp 143–168 in *Real and complex singularities (São Carlos VIII)* (eds. J.-P. Brasselet and M. A. S. Ruas), Trends Math., Birkhäuser, Basel, 2007.

G 40. Nilpotents, Integral Closure and Equisingularity conditions, pp 23–33 in *Real and Complex Singularities, (São Carlos IX)* (eds. M. J. Saia and J. Seade), Contemp. Math. **459**, Amer. Math. Soc., 2008.

G 41. Invariants of $D(q, p)$ singularities, pp 13–22 in *Real and Complex Singularities, (São Carlos IX)* (eds. M. J. Saia and J. Seade), Contemp. Math. **459**, Amer. Math. Soc., 2008.

G 42. (with J. Fernandez de Bobadilla) The Lê numbers of the square of a function and their applications, J. London Math. Soc. **77** (2008) 545–567.

G 43. (with D. Trotman and L. C. Wilson.) Equisingularity of sections, (t^r) condition, and the integral closure of modules, J. Alg. Geom. **18** (2009) 651–689.

G 44. The Multiplicity Polar Theorem and Isolated Singularities, J. Alg. Geom. **18** (2009) 547–574.

G 45. (with M. Vitulli) Weak subintegral closure of ideals. Math arXiv:0708.3105 25 pp. To appear in Advances in Math.

G 46. The Multiplicity-Polar Theorem. Math.CV/0703650 21 pp.

G 47. Complete Intersections with non-isolated singularities and the $\mathcal{A}_f$ condition, pp 85–93 in *Singularities I: Algebraic and Analytic Aspects, in honor of the 60th Birthday of Lê Dung Tráng.* (eds. J.-P. Brasselet, J. L. Cisneros-Molina, D. Massey, J. Seade, and B. Teissier), Contemp. Math. **474**, Amer. Math. Soc., 2008.

References

1. V. I. Arnold, Singularities of smooth mappings, Russian Math. Surveys **23i** 1-43 (1968), Reprinted as pp 3–45 in *Singularity Theory, London Math. Soc Lecture Notes* **53** Cambridge Univ. Press, 1981.

2. K. Brauner, Zur Geometrie der Funktionen zweier komplexen Veränderlichen II–IV, *Abh. Math. Sem. Hamburg* **6** (1928) 1–54.

3. J. Briançon and J.-P. Speder, La trivialité topologique n'implique pas les conditions de Whitney, *Comptes Rendus Acad. Sci. Paris* **280**A (1975) 365.

4. J. Briançon and J.-P. Speder, Les conditions de Whitney impliquent μ constant, *Ann. Inst. Fourier* **26** (1976) 153–163.

5. D. Buchsbaum and D. S. Rim, A generalized Koszul complex II: depth and multiplicity, *Trans. Amer. Math. Soc.* **111**(1963) 197–224.

6. W. Burau, Kennzeichnung der Schlauchknoten, *Abh. Math. Sem. Hamburg* **9** (1932) 125–133.

7. W. Burau, Kennzeichnung der Schlauchverkettungen, *Abh. Math. Sem. Hamburg* **10** (1934) 285–297.

8. J. N. Damon, Finite determinacy and topological triviality I, *Invent. Math.* **62** (1980) 299–324.

9. C. G. Gibson, K. Wirthmüller, A. A. du Plessis and E. J. N. Looijenga, *Topological stability of smooth mappings*, Lecture Notes in Math. **552**, Springer-Verlag, 1976.

10. G.-M. Greuel, Der Gauss-Manin Zusammenhang isolierter Singularitäten vollständiger Durchschnitten, *Math. Ann.* **214** (1975) 235–266.

11. J. P. G. Henry and M. Merle, Limites de normales, conditions de Whitney et éclatement d'Hironaka, pp 575–584 in *Singularities* (ed. P. Orlik), *Proc. Symp. in Pure Math.* **40(1)**, Amer. Math. Soc., 1983.

12. J. P. G. Henry, M. Merle and C. Sabbah, Sur la condition de Thom stricte pour un morphisme analytique complexe, *Ann. Sci. Ec. Norm. Sup.* **17** (1984) 227–268.

13. H. Hironaka, Resolution of singularities of an algebraic variety ove a field of characteristic zero, *Ann. of Math.* **79** (1964) 109–326.

14. H. Hironaka, Normal cones in analytic Whitney stratifications, *Publ. Math. IHES* **36** (1969) 27–138.

15. H. Hironaka, Stratification and flatness, pp 199–265 in *Real and complex singularities* (Proc. Ninth Nordic Summer School, Oslo, 1976) (ed. Per Holm) Sijthoff and Noordhoff, 1977.

16. K. Houston, Disentanglements and Whitney equisingularity, *Houston J. Math.* **33** (2007) 663–681.

17. K. Houston, On equisingularity of images of corank 1 maps, pp 201–208 in *Real and complex singularities (São Carlos VIII)* (eds. J.-P. Brasselet and M. A. S. Ruas), Trends Math., Birkhäuser, Basel, 2007.

18. V. H. Jorge Peréz, Polar multiplicities and equisingularity of map-germs from $\mathbb{C}^3$ to $\mathbb{C}^3$, *Houston J. Math.* **29** (2003) 901–923.

19. D. Kirby and D. Rees, Multiplicities in graded rings I: the general theory, in *Commutative algebra: syzygies, multiplicities and birational algebra* (eds. William J. Heinzer, Craig L. Hunecke and Judith D. Sally), *Contemp. Math.* **159**, Amer. Math. Soc., 1994.

20. S. L. Kleiman, On the transversality of a general translate, *Compositio Math.* **28** (1974) 287–297.

21. S. L. Kleiman and A. Thorup, A geometric theory of the Buchsbaum-Rim multiplicity, *J. Algebra* **167** (1994) 168–231.

22. D. T. Lê, Un critère d'équisingularité, pp 183–192 in *Singularités à Cargèse*, *Astérisque* **7-8**, Soc. Math. France, 1973.

23. D. T. Lê, Calcul du nombre de Milnor d'une singularité isolée d'intersection complète, *Funct. Anal. Appl.* **8** (1974) 45–52.

24. M. Lejeune-Jalabert and B. Teissier, *Clôture intégrale des idéaux et équisingularité*, Notes issued by Institut Fourier, Grenoble, 1974.

25. T.-C. Kuo, On classification of real singularities, *Invent. Math.* **82** (1985) 257–262.

26. T.-C. Kuo and D. Trotman, On (w) and (t^s)-regular stratifications, *Invent. Math.* **92** (1988) 633–643.

27. S. Łojasiewicz, *Ensembles semi-analytiques*, Notes issued by IHES, Bures-sur-Yvette, 1965.

28. B. Malgrange, *Ideals of differentiable functions*, Oxford Univ. Press, 1966.

29. D. Massey, *Lê cycles and hypersurface singularities*, Lecture Notes in Math. **1615**, Springer-Verlag, 1995.
30. J. N. Mather, Stability of C^∞−mappings I: The division theorem, *Ann. of Math.* **87** (1968) 89–104; II Infinitesimal stability implies stability, Ann. of Math. **89** (1969) 254–291; III Finitely determined map-germs, Publ. Math. IHES **35** (1969) 127-156; IV Classification of stable germs by $\mathbb{R}$−algebras, Publ. Math. IHES **37** (1970) 223–248; V Transversality, Advances in Math. **4** (1970) 301–335; VI The nice dimensions, Lecture Notes in math. **192**, Springer-Verlag 1971, 207–253.
31. J. N. Mather, *Notes on topological stability*, Notes issued by Harvard Univ., 1970.
32. J. N. Mather, Stratifications and mappings, pp 195-232 in *Proc. conference on dynamical systems* (ed. M. M. Peixoto) Academic Press, 1973.
33. J. W. Milnor, *Singular points of complex hypersurfaces*, Princeton Univ. Press, 1968.
34. D. Rees, A-transforms of ideals, and a theorem on multiplicities of ideals, *Proc. Camb. Phil. Soc.* **57** (1961) 8–17.
35. D. Rees, Reduction of modules, Math. *Proc. Camb. Phil. Soc.* **101** (1987) 431–449.
36. J.-P. Speder, L'équisingularité à la Zariski implique les conditions de Whitney, pp 41–46 in *Singularités à Cargèse*, Astérisque **7-8**, Soc. Math. France, 1973.
37. B. Teissier, Cycles évanescents, sections planes, et conditions de Whitney, pp 285–362 in *Singularités à Cargèse*, Astérisque **7-8**, Soc. Math. France, 1973.
38. B. Teissier, Introduction to equisingularity problems, pp 593–632 in *Algebraic geometry (Arcata 1974)* (ed. R. Hartshorne), *Proc. Symp. in Pure Math.* **29**, Amer. Math. Soc., 1975.
39. B. Teissier, Variétés polaires. I. Invariants polaires des singularités d'hypersurfaces, *Invent. Math.* **40(3)** (1977) 267–292.
40. B. Teissier, The hunting of invariants in the geometry of discriminants, pp 565–678 in *Real and complex singularities* (Proc. Ninth Nordic Summer School/NAVF Sympos. Math., Oslo, 1976), Sijthoff and Noordhoff, 1977.
41. B. Teissier, Variétés polaires. II. Multiplicités polaires, sections planes, et conditions de Whitney. *Algebraic geometry (La Ràbida, 1981)*, Lecture Notes in Math. **961**, Springer-Verlag 1982, 314–491.
42. R. Thom, Les singularités des applications différentiables, *Ann. Inst. Fourier* **6** (1955-56) 43-87. (see also Sém. Bourbaki 134).
43. R. Thom, 1956, Un lemme sur les applications différentiables, *Bol. Soc. Mat. Mex.* **1** 59-71.
44. R. Thom and H. I. Levine, *Singularities of differentiable mappings*, Bonn Math. Schrift **6**, 1959. Reprinted in Lecture Notes in Math. **192**, Springer-Verlag 1971, 1–89.
45. R. Thom, La stabilité topologique des applications polynomiales, *L'Ens. Math.* **8** (1962) 24–33.
46. R. Thom, 1964, Local topological properties of differentiable mappings, pp 191-202 in *Differential Analysis*, Oxford Univ. Press.
47. R. Thom, Ensembles et morphismes stratifiés, *Bull. Amer. Math. Soc.* **75** (1969) 240–284.
48. D. Trotman and L. C. Wilson, Stratifications and finite determinacy, *Proc. London Math. Soc.* **78** (1999) 334–368.

49. J.-L. Verdier, Stratifications de Whitney et théorème de Bertini-Sard, *Invent. Math.* **36** (1976) 295–312.
50. J. Wahl, Equisingular deformations of plane algebroid curves, *Trans. Amer. Math. Soc.* **193** (1974) 143–170.
51. H. Whitney, The general type of singularity of a set of $(2n - 1)$ smooth functions of n variables, *Duke Math. J.* **10** (1943) 161–172.
52. H. Whitney, The self-intersections of a smooth n-manifold in $2n$-space, *Ann. of Math.* **45** (1944) 220–246.
53. H. Whitney, The self-intersections of a smooth n-manifold in $(2n - 1)$-space, *Ann. of Math.* **45** (1944) 247–293.
54. H. Whitney, On singularities of mappings of euclidean spaces I: mappings of the plane into the plane, *Ann. of Math.* **62** (1955) 374–410.
55. H. Whitney, Elementary structure of real algebraic varieties, *Ann. of Math.* **66** (1957) 545–556.
56. H. Whitney, Singularities of mappings of euclidean spaces, pp 285–301 in *Symposium Internacional de Topología algebraica*, UNAM & UNESCO, 1958.
57. H. Whitney, Local properties of analytic varieties, pp 205–244 in *Differential and combinatorial topology* (ed. S. S. Cairns) Princeton, 1965.
58. H. Whitney, Tangents to an analytic variety, *Ann. of Math.* **81** (1965) 496–549.
59. H. Whitney and F. Bruhat, Quelques propriétés fondamentales des ensembles analytiques-réels, *Comm. Math. Helv.* **33** (1959) 132–160.
60. O. Zariski, On the topology of algebroid singularities, *Amer. J. Math.* **54** (1932) 453–465. (in Works, vol III)
61. O. Zariski, Studies in equisingularity I-III, *Amer. J. Math.* **87** (1965) 507–536, 972–1006, **90** (1968) 961–1023.
62. O. Zariski, Some open questions in the theory of singularities, *Bull. Amer. Math. Soc.* **77** (1971) 481–491.
63. O. Zariski, General theory of saturation and of saturated local rings I-III, *Amer. J. Math.* **93** (1971) 573–648, 872–964, **97** (1975) 415–502, see also *Astérisque* **7-8** (1973) 21–29.
64. O. Zariski, The elusive concept of equisingularity and related questions, pp 9–22 in *Algebraic geometry: the Johns Hopkins centennial lectures* (supplement to Amer. J. Math.), Johns Hopkins Press, 1977.
65. O. Zariski, Foundations of a general theory of equisingularity on r–dimensional algebraic and algebroid varieties, of embedding dimension $r + 1$, *Amer. J. Math.* **101** (1979) 453–514.
66. O. Zariski and P. Samuel, *Commutative algebra, vol. II*, van Nostrand, Princeton, 1960.

C. T. C. Wall,
Department of Mathematical Sciences,
University of Liverpool,
Liverpool L69 7ZL,
England.
ctcw@liv.ac.uk

25

Singularities in algebraic data acquisition

Y. YOMDIN

Abstract

We consider the problem of reconstruction of a *non-linear* finite-parametric model $M = M_p(x)$, with $p = (p_1, \ldots, p_r)$ a set of parameters, from a set of measurements $\mu_k(M)$. In this paper $\mu_k(M)$ are always the moments $m_k(M) = \int x^k M_p(x) dx$. This problem is a central one in Signal Processing, Statistics, and in many other applications. Typically, models of the above type contain "geometric parameters" which make the problem strongly non-linear.

In this paper we discuss singularities that appear in one of the basic examples in the reconstruction problem. The model in this example is a linear combination of δ-functions of the form $g(x) = \sum_{i=1}^{n} A_i \delta(x - x_i)$, with the unknown parameters A_i, x_i, $i = 1, \ldots, n$. As one can expect, (near-) singular situations occur as the points x_j approach one another and collide.

1. Introduction

In this paper we consider the following problem: let a finite-parametric family of functions $M = M_p(x)$, $x \in \mathbb{R}^m$ be given, with $p = (p_1, \ldots, p_r)$ a set of parameters. We call $M_p(x)$ a model, and usually we assume that it depends on some of its parameters in a non-linear way (this is always the case with the "geometric" parameters representing the shape and the position of the model).

2000 *Mathematics Subject Classification* 94A12 (primary), 62J02, 14P10, 42C99 (secondary).
This research was supported by the ISF, Grant No. 264/05 and by the Minerva foundation.

The problem is:

How do we reconstruct in a robust and efficient way the parameters p from a set of "measurements" $\mu_1(M), \ldots, \mu_l(M)$?

In this paper μ_j will be the moments $m_j(M) = \int x^j M_p(x) dx$. This assumption is not too restrictive - see, for example, [8, 7].

The above problem is certainly among the central ones in Signal Processing (non-linear matching), Statistics (non-linear regression), and in many other applications.

A typical approach to the reconstruction of the model's parameters from (or "fitting" the model to) the measurements is to minimize the fitting discrepancy with respect to the parameters (non-linear regression). The non-linear minimization is usually a difficult task.

There is also a direct (and somewhat "naive") approach to the above problem: substitute the model function $M_p(x)$ into the measurements μ_j and compute explicitly the resulting "symbolic" expressions of $\mu_j(M_p)$ in terms of the parameters p. Equating these "symbolic" expressions to the actual measurement results, we produce a system of nonlinear equations on the parameters p which we consequently try to solve. An advantage of this approach is that we produce an explicit and usually "algebraic" system of equations, whose solution promises an exact solution of the fitting problem. The problem with this approach is that a straightforward solving of large scale non-linear systems is itself a difficult task. It is comparable (and sometimes essentially equivalent) in complexity to the non-linear minimization as above.

What makes this approach feasible is the following fact: in many situations the algebraic structure of the arising non-linear systems is very specific. It allows for application of strong tools of classical Analysis, like Moment theory and Padé approximations, which provide a "closed form solution" (up to solving certain *linear* systems of equations).

Many specific results in this direction were known for a long time. Recently a bunch of new results and methods appeared (see [1, 2, 9, 14, 7, 8, 11, 16, 17, 18, 19, 28, 23, 24] and references therein), which are sometimes unified under the name "Algebraic Sampling". Let us stress that this approach assumes availability of *a priori information* on the nature of the data to be matched: namely, the form of an appropriate model.

Recently another non-linear reconstruction method has been suggested, utilizing a priori information on the data to be recovered: Compressive Sampling ([4, 6]). This approach assumes a priori only an existence of a sparse representation of the signal in a certain (wavelets) basis, and as this it presents a rather general and "universal" tool. Algebraic Sampling usually requires more specific a priori assumptions on the structure of the signals, but it promises a better reconstruction accuracy. In fact, we believe that ultimately the Algebraic Sampling approach has a potential to reconstruct "simple signals with singularities" as good as smooth ones (see [2], [3] and references therein). Notice that one can show in general that the *non-linear* approach is unavoidable in accurate reconstruction of signals with singularities ([9]).

As for the "measurements" considered in this paper, certainly, the polynomial moments do not present the best choice for practical applications. Indeed, the monomials x^j are far away from being orthogonal (see, for example, [27]). However, the main "algebraic" features of the arising non-linear systems are more transparent for the moments, while they remain essentially the same for a much wider class of measurements (including various forms of the Fourier transform).

As usual, the non-linear systems arising in Algebraic Sampling may degenerate as the parameters vary. The analysis of the corresponding singularities is crucial for a robust solution of these systems.

The aim of the present paper is to discuss one specific (but rather basic) example of singularities that typically appears in the model reconstruction problem. We concentrate on the model of the form $g(x) = \sum_{i=1}^{n} A_i \delta(x - x_i)$, with the unknown parameters A_i, x_i, $i = 1, \ldots, n$. The corresponding system of equations ("Prony's system") appears in surprisingly many apparently unrelated situations. As one can expect, (near -) singular situations occur as the points x_i approach one another and collide. We investigate the behavior of the moment inversion in these near-singular situations. More specifically, we address the following problem: assuming that the measured moments are bounded, it is easy to see that the coefficients A_i may tend to infinity as some of the points x_i collide. In particular, even a small noise in the measurements may be strongly amplified in such situations. We prove that *if we express the vector of the unknowns $(A_1, \ldots, A_n)$ in the basis of "divided finite differences" then the coefficients in this representation remain uniformly bounded.* This suggests a more robust reconstruction procedure which can treat also the cases of (near-) collision of the points x_i.

The paper is organized as follows: in Section 2.1 we present in detail the reconstruction from the moments of linear combinations of δ-functions. We describe the "Prony system" arising and its connection with the Padé approximation of the moment generating function. Further we give in Section 2 an overview of some additional results in Algebraic Sampling where Prony-like systems appear. In Section 3 we describe typical singularities of solutions of the Prony system and prove our main results. In Section 4 a possible more robust rearrangement of the Prony system is briefly discussed.

2. Examples of moment inversion

2.1. Linear combination of δ-functions

Let $g(x) = \sum_{i=1}^{n} A_i \delta(x - x_i)$, $x_i \in \mathbb{R}$, $i = 1, \ldots, n$. For this function we have

$$m_k(g) = \int_0^1 x^k \sum_{i=1}^{n} A_i \delta(x - x_i) dx = \sum_{i=1}^{n} A_i x_i^k. \qquad (2.1)$$

So assuming that we know the moments $m_k(g) = m_k$, $k = 0, 1, \ldots, 2n - 1$, we obtain the following system of equations for the parameters A_i and x_i, $i = 1, \ldots, n$, of the function g:

$$\sum_{i=1}^{n} A_i x_i^k = m_k, \quad k = 0, 1, \ldots, 2n - 1. \qquad (2.2)$$

Notice that system (2.2) is linear with respect to the parameters A_i and non-linear with respect to the parameters x_i.

System (2.2) appears in many mathematical and applied problems. First of all, if we want to approximate a given function $f(x)$ by an exponential sum

$$f(x) \approx C_1 e^{a_1 x} + C_2 e^{a_2 x} + \cdots + C_n e^{a_n x},$$

then the coefficients C_i and the values $\mu_i = e^{a_i}$ satisfy a system of the form (2.2) with the right-hand side (the "measurements") being the values of $f(x)$ at the integer points $x = 1, 2, \ldots$ (see [10], Section 4.9). The method of solution of (2.2) which we give below, is usually called Prony's method ([21]).

System (2.2) appears also in error correction codes, in array processing (estimating the direction of signal arrival) and in other applications in Signal Processing (see, for example, [18, 7] and references therein).

In [8, 11, 18] system (2.2) appears in reconstruction of plane polygons from their complex moments. In [12, 13] it arises in reconstruction of "quadrature domains", and in [1, 2, 14, 23, 24, 28] Prony system appears in reconstruction of D-finite functions and linear combinations of shifts of a given function. These results are shortly described in Sections 2.2 - 2.5 below.

System (2.2) appears also in some perturbation problems in nonlinear model estimation.

We now give a sketch of the proof of solvability of (2.2) and of the standard solution method. We follow the lines of [18, 20]. See also a literature on Padé approximation, in particular, [20] and references therein.

Theorem 2.1. *A linear combination* $g(x) = \sum_{i=1}^{n} A_i \delta(x - x_i)$ *of n δ-functions with non-zero coefficients* $A_1, \ldots, A_n$ *can be uniquely reconstructed from its 2n moments* $m_0(g), \ldots, m_{2n-1}(g)$ *via solving system (2.2).*

Proof.   Notice first of all that the requirement of A_i being non-zero is essential: the position of a δ-function with a zero coefficient cannot be identified.

Representation (2.1) of the moments immediately implies the following result for the moments generating function $I(z) = \sum_{k=0}^{\infty} m_k(g)z^k$:

Proposition 2.2. *For* $g(x) = \sum_{i=1}^{n} A_i\delta(x - x_i)$ *the moments generating function* $I(z)$ *is a rational function with the poles at* $\frac{1}{x_i}$ *and with the residues at these poles* A_i:

$$I(z) = \sum_{i=1}^{n} \frac{A_i}{1 - zx_i}. \tag{2.3}$$

We see that the function $I(z)$ encodes the solution of system (2.2).

So to solve this system it remains to find explicitly the rational function $I(z)$ from the first $2n$ its Taylor coefficients $m_0, \ldots, m_{2n-1}$.

We use the fact that the Taylor coefficients of a rational function of degree n satisfy a linear recurrence relation of the form

$$\sum_{j=0}^{n} C_j m_{r+j} = 0, \; r = 0, 1, \ldots. \tag{2.4}$$

Since we know the first $2n$ Taylor coefficients $m_0, \ldots, m_{2n-1}$ we can write a *homogeneous linear system* with n equations and $n + 1$ unknown coefficients C_l:

$$\sum_{j=0}^{n} C_j m_{j+r} = 0, \; r = 0, 1, \ldots, n - 1. \tag{2.5}$$

One can show that the rank of system (2.5) is $n - 1$. In particular, this can be done using the bilinear form $\langle S, T \rangle = \int S(x)T(x)g(x)dx$ whose matrix in the monomial basis turns out to be the same Hankel matrix as of system (2.5) with the last column omitted. On the other hand, the matrix of $\langle S, T \rangle$ with respect to the basis consisting of functions W_i defined by $W_i(x_j) = \delta_{(i,j)}$ is diagonal with the entries $A_1, \ldots, A_n$. So if all A_i are non-zero, the form $\langle S, T \rangle$ is non-degenerate.

Consequently, we can find a non-zero solution $C_0, \ldots, C_n$ of (2.5), the last entry C_n of this solution is non-zero, and it is unique, up to a multiplication by a non-zero scalar. Now if we rewrite the recurrence relation (2.4) in the form

$$m_{r+n} = \sum_{j=0}^{n-1} B_j m_{r+j}, \; r = 0, 1, \ldots, \tag{2.6}$$

with $B_j = -\frac{C_j}{C_n}$, $j = 1, \ldots, n - 1$, we see that the coefficients B_j of (2.6) are uniquely determined by the first $2n$ moments $m_0, \ldots, m_{2n-1}$ of g. Now via (2.6), starting with the known initial moments, we uniquely reconstruct the generating function $I(z)$ (see below). The poles and residues of $I(z)$ provide us the solution of system (2.2). This solution is unique (formally speaking, up to an arbitrary permutation of the indices i) since, as it was shown above, the generating function $I(z)$ is uniquely determined by the first $2n$ moments $m_0, \ldots, m_{2n-1}$ of g. The existence of solution was, in fact, our assumption - we have explicitly assumed that $m_0, \ldots, m_{2n-1}$ were the moments of a function g of the form $g(x) = \sum_{i=1}^{n} A_i \delta(x - x_i)$. This completes the proof of Theorem 2.1. $\qquad\square$

The solution of the Prony system described above is, essentially, identical to the solution of the Padé approximation problem for $I(z)$ (see, for example, [20]). We shall use below the Padé approximation form of the Prony system, so let us present it here. It produces also an explicit form for $I(z)$.

In general, in a diagonal Padé approximation of degree l of $I(z)$ we look for a rational function $R_l(z) = \frac{P_{l-1}(z)}{Q_l(z)}$, where P_{l-1} and Q_l are polynomials in z of the degrees $l-1$ and l respectively, such that

$$I(z) - R_l(z) = cz^{2l} + \cdots . \tag{2.7}$$

In other words, the Taylor coefficients of $I(z)$ and of $R_l(z)$ must coincide up to the degree $2l - 1$. Writing (2.7) in the form $Q_l(z)I(z) - P_{l-1}(z) = \tilde{c}z^{2l} + \cdots$ we easily obtain that for $Q_l(z) = C_0 + C_1 z + \cdots + C_l z^l$ the coefficients C_j satisfy a system

$$\sum_{j=0}^{l} C_j m_{j+r} = 0, \ r = 0, 1, \ldots, l - 1. \tag{2.8}$$

In our case $I(z)$ is a rational function of degree n, so for $l = n$ we get

$$I(z) = R_n(z) = \frac{P_{n-1}(z)}{Q_n(z)}. \tag{2.9}$$

The polynomial $P_{n-1}(z) = D_0 + D_1 z + \cdots + D_{n-1} z^{n-1}$ can now be reconstructed via

$$D_0 = C_0 m_0, \ D_1 = C_0 m_1 + C_1 m_0, \ldots \tag{2.10}$$

Finally, we represent $I(z) = R_n(z)$ as a sum of partial fractions:

$$I(z) = R_n(z) = \sum_{i=1}^{n} \frac{A_i}{1 - zx_i}. \tag{2.11}$$

Because of the uniqueness of the reconstruction provided by Theorem 2.1 we conclude that (2.11) coincides with (2.3). System (2.8) for $l = n$ coincides with (2.5), and we see that the recurrence coefficients C_j are the coefficients of the denominator $Q_n(z)$ while $\frac{1}{x_i}$ are the roots of Q_n. In particular, we conclude that all the roots of $Q_n(z)$ are simple. The identity (2.11) shows that the Prony system is just another interpretation of the n-th Padé approximation $R_n(z)$ of $I(z)$ - through its poles and residues.

For an arbitrary sequence $m = (m_1, \ldots, m_{2n-1})$, not necessarily appearing as a moment sequence of a linear combination of δ-functions, system (2.8) always has nonzero solutions - this gives the denominator Q_n of the n-th Padé approximation R_n of $\tilde{I}(z) = \sum_{k=0}^{2n-1} m_k z^k$, while the numerator can be found via (2.10). However, if we do not assume explicitly that the sequence m is a moment sequence of $g(x) = \sum_{i=1}^{n} A_i \delta(x - x_i)$, the resulting rational fraction $R_n(z)$ may be of a smaller degree than n. Its denominator Q_n may also have some complex and some multiple roots. So we have to replace (2.11) with

$$\tilde{I}(z) = R_n(z) + cz^{2n} + \cdots = \sum_{i=1}^{q} \sum_{j=0}^{\mu_i} \frac{A_{i,j}}{(1 - zx_i)^j} + cz^{2n} + \cdots , \qquad (2.12)$$

for some $q \leq n$ and the multiplicities μ_i with $\mu = \sum_{i=0}^{q} \mu_i \leq n$, and we have to allow $A_{i,j}$ and x_i to be complex.

We see that the original Prony system (2.2) is not always solvable. However, a generalized system corresponding to (2.12) is solvable for each input m. Let us restate this as follows:

Theorem 2.3. *For each $m = (m_1, \ldots, m_{2n-1})$ there are $q \leq n$, the multiplicities μ_i with $\mu = \sum_{i=0}^{q} \mu_i \leq n$, the points $x_i \in \mathbb{C}$, $i = 1, \ldots, q$, and the coefficients $A_{i,j} \in \mathbb{C}$, $i = 1, \ldots, q$, $j = 0, \ldots, \mu_i$ such that for a linear combination of δ-functions and their derivatives*

$$g(x) = \sum_{i=1}^{q} \sum_{j=0}^{\mu_i} A_{i,j} \delta^j (x - x_i) \qquad (2.13)$$

we have $m_k = m_k(g) = \int x^k g(x) dx$, $k = 0, \ldots, 2n - 1$.

The main problem considered in this paper is the following: *what may happen with the coefficients A_i in the original Prony system (2.2) as the poles x_i approach one another and the function as in (2.1) degenerates into a function of the form (2.12)?*

In the rest of this section we discuss more examples of the moment inversion where Prony-like systems naturally appear.

2.2. Piecewise-polynomials and piecewise-solutions of linear ODE's

In [28] a method for reconstruction of piecewise-polynomials from samplings is suggested (which starts with a reconstruction of linear combinations of δ-functions and of their derivatives). It leads to "Prony-like" systems of equations on the unknown parameters. One can consider, as a natural generalization of piecewise-polynomials, the class A_D of piecewise-analytic functions g, each piece being annihilated by a linear differential operator D with polynomial coefficients. Such functions are also called "L-splines" (see [25, 26] and references therein). For piecewise-polynomials of degree d we have $D = \frac{d^{d+1}}{dx^{d+1}}$. In the case of a known operator D a reconstruction procedure based on an explicit determination of the operator $\tilde{D}$ with polynomial coefficients annihilating the moment generating function $I(z)$ has been described in [14]. In the case of the unknown operator D a reconstruction procedure has been proposed in [1, 2], based on an explicit construction of a difference-differential operator $\hat{D}$ annihilating g in a sense of distributions. Both approaches ultimately lead to certain Padé (Padé-Hermite) approximation problems.

2.3. Linear combinations of shifts of a known function

Reconstruction of this class of signals from sampling has been described in [28]. A rather similar problem of reconstruction from the moments has been studied in [2, 24]. Our method proposed in these papers is based on the following approach: we construct convolution kernels dual to the monomials. Applying these kernels, we get a Prony-type system of equations on the shifts and amplitudes.

More accurately, let $F(x) = \sum_{i=1}^{n} a_i f(x + x_i)$, with the unknown parameters a_i, x_i. We are given a finite number of moments m_k of the signal F: $m_k = m_k(F) = \int x^k F(x) dx$. We look for the dual functions $\psi_k(x)$ satisfying the convolution equation

$$\int f(t + x)\psi_n(t)dt = x^n \tag{2.14}$$

for each index n. To solve this equation we apply Fourier transform to both sides of (2.14). Assuming that $\hat{f}(\omega) \in C^\infty$, $\hat{f}(0) \neq 0$ we find (see [24]) that there is a unique solution to (2.14) provided by

$$\phi_n(x) = \sum_{k \leq n} C_{n,k} x^k, \tag{2.15}$$

where

$$C_{n,k} = \frac{1}{(\sqrt{2\pi})^d} \binom{n}{k} (-i)^{n+k} \left[\left. \frac{\partial^{n-k}}{\partial \omega^{n-k}} \right|_{\omega=0} \frac{1}{\hat{f}(\omega)} \right].$$

Now it is shown in [2, 24] that if we set the generalized polynomial moments M_s as $M_s = \sum_{k \leq s} C_{s,k} m_k$ then we obtain the following system of equations for the unknowns a_i, x_i:

$$\sum_{i=1}^{n} a_i x_i^s = M_s, \; s \geq 0. \tag{2.16}$$

This is once more a Prony system (2.2). Notice that the generalized moments M_s are expressed in a known way through the original measurements $m_k = m_k(F)$.

2.4. Reconstruction of polygons from complex moments

In [18, 11, 8] the problem of reconstruction of a planar polygon from its complex moments is considered. The complex moments of a function $f(x, y)$ are defined as

$$\mu_k(f) = \iint z^k f(x, y) dx dy, \; k = 0, 1, \ldots, \; z = x + iy. \tag{2.17}$$

Complex moments can be expressed as certain specific linear combinations of the real double moments $m_{kl}(f)$.

For a plane subset A its complex moments $\mu_k(A)$ are defined by $\mu_k(A) = \mu_k(\chi_A)$, where χ_A is the characteristic function of A.

Let P be a closed n-sided planar polygon with the vertices z_i, $i = 1, \ldots, n$. The reconstruction method of [18] is based on the following result of [5]:

Theorem 2.4. *There exist a set of n coefficients a_i, $i = 1, \ldots, n$, depending only on the vertices z_i, such that for any analytic function $\phi(z)$ on P we have*

$$\iint_P \phi''(z) dx dy = \sum_{i=1}^{n} a_i \phi(z_i).$$

The coefficients a_j, $j = 1, \ldots, n$ are given as $a_j = \frac{1}{2} \left(\frac{\bar{z}_{j-1} - \bar{z}_j}{z_{j-1} - z_j} - \frac{\bar{z}_j - \bar{z}_{j+1}}{z_j - z_{j+1}} \right)$.

Applying this formula to $\phi(z) = z^k$ we get

$$k(k-1)\mu_{k-2}(\chi_P) = \sum_{i=1}^{n} a_i z_i^k, \; k = 0, 1, \ldots, \tag{2.18}$$

where we put $\mu_{-2} = \mu_{-1} = 0$. So on the left-hand side we have shifted moments of P.

If we ignore the fact that a_j can be expressed through z_i and consider both a_j and z_i as unknowns, we get from (2.18) a system of equations

$$\sum_{i=1}^{n} a_i z_i^k = v_k, \ k = 0, 1, \ldots, \tag{2.19}$$

where v_k denotes the "measurement" $k(k-1)\mu_{k-2}(P)$. System (2.19) is identical to system (2.2) which appears in reconstruction of linear combination of δ-functions. One of the solution methods suggested in [18] is the Prony method described in Section 2.4 above. Another approach is based on matrix pencils. In [11, 8] an important question is investigated of polygon reconstruction from noisy data.

2.5. Quadrature domains

We introduce, following [12], a slightly different sequence of double moments: for a function $g(z) = g(x + iy)$ the moments $\tilde{m}_{kl}(g)$ are defined by

$$\tilde{m}_{kl}(g) = \iint z^k \bar{z}^l g(z) dx dy, \ k, l \in \mathbb{N}. \tag{2.20}$$

One defines the moment generating function $I_g(v, w) = \sum_{k,l=0}^{\infty} \tilde{m}_{kl}(g) v^k w^l$ and the "exponential transform"

$$\tilde{I}_g(v, w) = 1 - \exp\left(-\frac{1}{\pi} I_g(v, w)\right)$$

$$= \exp\left(-\frac{1}{\pi} \iint_{\Omega} \frac{g(z) dx dy}{(z - v)(\bar{z} - w)}\right) := \sum_{k,l=0}^{\infty} b_{kl}(g) v^k w^l.$$

Now (classical) quadrature domains in $\mathbb{C}$ are defined as follows:

Definition. A quadrature domain $\Omega \subset \mathbb{C}$ is a bounded domain with the property that there exist points $z_1, \ldots, z_m \in \Omega$ and coefficients c_{ij}, $i = 1, \ldots, m$, $j = 0, \ldots, s_i - 1$, so that for all analytic integrable functions $f(z)$ in Ω we have

$$\iint_{\Omega} f(x + iy) dx dy = \sum_{i=1}^{m} \sum_{j=0}^{s_i-1} c_{ij} f^{(j)}(z_i). \tag{2.21}$$

$N = s_1 + \cdots + s_m$ is called the order of the quadrature domain Ω.

The simplest example is provided by the disk $D_R(0)$ of radius R centered at $0 \in \mathbb{C}$: $\iint_{D_R(0)} f(x + iy) dx dy = \pi R^2 f(0)$. The results of Davis ([5]; Theorem 3.1 above) give another example in this spirit.

The following result ([13], [12], Theorem 3.1) provides a necessary and sufficient condition for $\Omega \subset \mathbb{C}$ to be a quadrature domain: let $\tilde{I}_\Omega(v, w) = \tilde{I}_{\chi_\Omega}(v, w)$ be the exponential transform of Ω.

Theorem 2.5. Ω *is a quadrature domain if and only if there exists a polynomial $p(z)$ with the property that the function $\tilde{q}(z, \bar{w}) = p(z)\bar{p}(w)\tilde{I}_\Omega(z, \bar{w})$ is a polynomial at infinity (denoted by $q(z, \bar{w})$). In that case, by choosing $p(z)$ of minimal degree, the domain Ω is given by $\Omega = \{z \in \mathbb{C}, \ q(z, \bar{z}) < 0\}$. Moreover, the polynomial $p(z)$ in this case is given by $p(z) = \prod_{i=1}^m (z - z_i)^{s_i}$, where z_i are the quadrature nodes of Ω.*

Now, the algorithm in [12] for reconstruction of a quadrature domain from its moments consists of the following steps:

1. Given the moments $\tilde{m}_{kl}(\Omega) = \tilde{m}_{kl}(\chi_\Omega)$ up to a certain order, construct the (truncated) exponential transform $\tilde{I}(v, w) = \sum_{k,l=0}^\infty b_{kl} v^k w^l$.
2. Identify the minimal integer N such that $\det(b_{k,l})_{k,l=0}^N = 0$. Then there are coefficients α_j, $j = 0, \ldots, N-1$, such that for $B = (b_{k,l})_{k,l=0}^N$ and $\alpha = (\alpha_1, \ldots, \alpha_{N-1}, 1)^T$ we have

$$B\alpha = 0. \tag{2.22}$$

   We solve this system with respect to α. Then the polynomial $p(z)$ defined above is given by $p(z) = z^N + \alpha_{N-1} z^{N-1} + \cdots + \alpha_0$.
3. Construct the function

$$R_\Omega(z, \bar{w}) = p(z)\bar{p}(w) \exp\left(\frac{-1}{\pi} \sum_{k,l=0}^{N-1} \tilde{m}_{kl}(\Omega) \frac{1}{z^{k+1}} \frac{1}{\bar{w}^{l+1}}\right)$$

   and identify $q(z, \bar{w})$ as the part of $R_\Omega(z, \bar{w})$ which does not contain negative powers of z and $\bar{w}$. Then the domain Ω is given by $\Omega = \{z \in \mathbb{C}, \ q(z, \bar{z}) < 0\}$.

Remark. Let us substitute into the definition of the quadrature domain (formula (2.21) above) $f(z) = z^k$. Assuming that all the quadrature nodes z_i are simple, we get for the complex moments $\tilde{m}_{k,0}(\Omega) = m_k(\Omega)$ the expression

$$m_k(\Omega) = \sum_{i=1}^m c_i z_i^k,$$

which is identical to (2.19) in reconstruction of planar polygons. So we can reconstruct the quadrature nodes z_i and the coefficients c_i from the complex moments only, and we get once more a complex system which is identical to (2.2). Allowing quadrature nodes z_i of an arbitrary order, we get a system corresponding to a linear combination of δ-functions and their derivatives (compare [28, 17]).

Notice that system (2.22) that appears in step 2 of the reconstruction algorithm above is very similar to system (2.5) in the solution process of (2.2).

3. Singularities in reconstruction of linear combinations of δ-functions

Our setting and notations in this section are the same as in Section 2.1. So we have $g(x) = \sum_{i=1}^{n} A_i \delta(x - x_i)$, and we assume that the points $x_i \in \mathbb{R}$, $i = 1, \ldots, n$, are pairwise distinct. However, we shall encounter functions of the form $h(x) = \sum_{i=1}^{q} \sum_{j=0}^{\mu_i} A_{i,j} \delta^{(j)}(x - x_i)$ as the limit cases.

There are several types of singularities that may occur in the reconstruction of $g(x)$ from its first $2n - 1$ moments. Some of the points x_i may "escape to infinity". This corresponds to the case where the rank of system (2.5) drops. A couple of x_i may collide and disappear (i.e. to become complex conjugate). If some of A_i are small, finding x_i becomes ill-posed. We shall consider here only the case where the points x_i remain bounded, real and distinct, but may approach one another.

More accurately, we assume that the moments $m_k(g)$, $k = 0, 1, \ldots, n - 1$ are bounded in absolute value by K, while the points x_i are contained in $[0, 1]$ and are pairwise distinct. The problem is to describe, under the above assumption, the possible structure of the coefficients A_j, stressing the case where the poles collide.

Easy examples (see Proposition 3.2 below) show that the coefficients A_i may tend to infinity under the above assumptions. This may present a serious problem to a robust solution of a non-linear system (2.2) (for example, if we use the continuation method). Informally, our result suggests that *it may be better to consider instead of the coefficients A_i themselves their representation in the basis of divided finite differences: as Theorem 3.1 below claims, the coefficients of such a representation remain uniformly bounded independently of the geometry of the points $x_1, \ldots, x_n$, assuming that the moments $m_0, \ldots, m_{n-1}$ are bounded.*

Let us recall the definition of the divided finite differences (see, for example, [10]). Let $X = \{x_1, \ldots, x_n\}$ be a set of points in $\mathbb{R}$, and let $Y = Y(x)$ be a function on X, $Y(x_i) = y_i$, $i = 1, \ldots, n$.

Definition. For $j = 0, 1, \ldots, n - 1$ the j-th divided finite difference $\Delta_j[X, Y]$ is defined as the sum

$$\Delta_j[X, Y] = \sum_{i=1}^{j+1} \frac{y_i}{(x_i - x_1) \ldots (x_i - x_{j+1})} = \sum_{i=1}^{j+1} \alpha_i^j y_i.$$

In particular, $\Delta_0[X, Y] = y_1$, $\Delta_1[X, Y] = \frac{y_2 - y_1}{x_2 - x_1}$.

We shall need the following property of the divided differences:

Proposition 3.1. *Assume that $f(x)$ is a C^n-function. Then the divided finite differences of f satisfy $\Delta_j[X, f] = \frac{1}{j!} f^{(j)}(\eta)$ for some $\eta \in [x_1, x_j]$.*

There is another convenient expression for the divided differences: denoting by δ_j the function (measure) $\delta_j = \sum_{i=1}^{j+1} \alpha_i^j \delta(x - x_i)$ we get for each function f that

$$\Delta_j[X, f] = \int f(x) \delta_j(x) dx.$$

The functions $\delta_j(x)$ are linear combinations of δ-functions with the coefficients tending to infinity as some of the points $x_1, \ldots, x_j$ approach one another. Still, their moments remain uniformly bounded:

Proposition 3.2. *For each $x_1, \ldots, x_n$ in $[0, 1]$ and for each k we have*

$$0 < m_k(\delta_j) \leq \binom{k}{j}.$$

Proof. Indeed,

$$m_k(\delta_j) = \int x^k \delta_j(x) dx = \Delta_j[X, x^k] = \frac{1}{j!} (x^k)^{(j)}(\eta_j) = \binom{k}{j} \eta_j^{k-j}$$

with $\eta_j \in [x_1, x_{j+1}] \subset [0, 1]$. $\square$

Theorem 3.1 below shows that the divided differences and their bounded linear combinations are, essentially, *the only* linear combinations of δ-functions with uniformly bounded moments.

Let V be the space of all functions on $X = \{x_1, \ldots, x_n\}$. For $j = 0, 1, \ldots, n - 1$ denote by $v_j = \{\alpha_1^j, \ldots, \alpha_{j-1}^j, 0, \ldots, 0\} \in V$ the corresponding discrete functions on X (where α_i^j are the divided differences coefficients in Definition 3.1). Clearly, the functions $v_0, v_1, \ldots, v_{n-1}$ form a basis of the space of all functions on X.

Let us now return to the function $g(x) = \sum_{i=1}^n A_i \delta(x - x_i)$. The vector of the coefficients $A(g) = \{A_1, \ldots, A_n\}$, considered as a function on $X = \{x_1, \ldots, x_n\}$, can be represented as a linear combination of the basis vectors $v_0, v_1, \ldots, v_{n-1}$: $A(g) = \sum_{j=0}^{n-1} c_j v_j$.

The following theorem provides a necessary and sufficient condition for the moments $m_k(g)$, $k = 0, 1, \ldots, n - 1$, of g to be uniformly bounded, independently of the geometry of the nodes x_i:

Theorem 3.3. *If the moments* $m_k(g)$, $k = 0, 1, \ldots, n-1$ *of the function* $g(x) = \sum_{i=1}^{n} A_i \delta(x - x_i)$ *are bounded in absolute value by K then the coefficients c_j in the representation $A(g) = \sum_{j=0}^{n-1} c_j v_j$ satisfy*

$$\sum_{j=0}^{n-1} |c_j| \leq C(n)K.$$

Conversely, if $|c_j| \leq K_1$, $j = 0, 1, \ldots, n-1$, then $|m_k(g)| \leq 2^k K_1$ for all k. Here $C(n)$ is a constant depending only on n.

Proof. In one direction the result is immediate: if $A(g) = \sum_{j=0}^{n-1} c_j v_j$ then, in the notations introduced above, we have $g(x) = \sum_{j=0}^{n-1} c_j \delta_j(x)$. Hence

$$m_k(g) = \sum_{j=0}^{n-1} c_j \int x^k \delta_j(x) dx = \sum_{j=0}^{n-1} c_j m_k(\delta_j).$$

Assuming $|c_j| \leq K_1$, $j = 0, 1, \ldots, n-1$, and bounding $m_k(\delta_j)$ via Proposition 3.2 we get $|m_k(g)| \leq K_1 \sum_{j=0}^{n-1} \binom{k}{j} \leq 2^k K_1$.

In the opposite direction, let us assume that $|m_k(g)| \leq K$ for $k = 0, 1, \ldots, n-1$. This implies that for each polynomial $S(x) = a_0 + a_1 x + \cdots + a_{n-1} x^{n-1}$ of degree $n-1$ we have $|\int S(x)g(x)dx| \leq K \|S\|$, where $\|S\| = \sum_{r=0}^{n-1} |a_r|$. From the representation $g = \sum_{j=0}^{n} c_j \delta_j$ as above we get $\int S(x)g(x)dx = \sum_{j=0}^{n-1} c_j \int S(x)\delta_j(x)$ which is the sum of the divided differences $\sum_{j=0}^{n-1} c_j \Delta_j[X, S(x)]$. Finally we get, via Proposition 3.1,

$$\left| \int S(x)g(x)dx \right| = \left| \sum_{j=0}^{n-1} \frac{c_j}{j!} S^{(j)}(\eta_j) \right| \leq K \|S\|. \tag{3.1}$$

Now we construct the auxiliary polynomial $S(x)$ of degree $n-1$ with the following properties:

1. For each $j = 0, \ldots, n-1$ the j-th derivative of $S(x)$ satisfies $|S^{(j)}(x)| \geq j!$ on $[0, 1]$.
2. The sign of $S^{(j)}(x)$ on $[0, 1]$ is the same as the sign of c_j in (3.1).

The polynomial S can be constructed as follows: first we take $S_0(x) = \pm x^{n-1}$, with the sign plus or minus chosen according to the sign of c_n. Next we put $S_1(x) = S_0(x) + cx^{n-2}$ with the coefficient c chosen in such a way that the properties 1 and 2 are valid for $S_1^{(n-2)}(x)$ on $[0, 1]$. Notice that the last derivative does not change. By induction we define $S_{r+1}(x) = S_r(x) + cx^{n-r-2}$ with the coefficient c chosen in such a way that the properties 1 and 2 are valid for $S^{(n-r-2)}(x)$ on $[0, 1]$. Notice that all the derivatives of the polynomial

$S_{r+1}(x)$ of orders higher than $n - r - 2$ are the same as for S_r. Finally we put $S(x) = S_{n-1}(x)$. By construction S has the properties 1 and 2 above.

The polynomial S constructed depends only on the sign pattern of $c_0, \ldots, c_{n-1}$. We denote the maximum of the norm $\|S(x)\|$ over all the sign patterns of c_j by $C(n)$.

Finally we substitute the constructed polynomial S into (3.1). We get

$$\sum_{j=0}^{n-1} |c_j| \le \left| \sum_{j=0}^{n-1} \frac{c_j}{j!} S^{(j)}(\eta_j) \right| \le C(n)K.$$

This completes the proof of Theorem 3.1. □

Corollary 3.4. *If the moments $m_k(g)$, $k = 0, 1, \ldots, n - 1$ of $g(x) = \sum_{i=1}^{n} A_i \delta(x - x_i)$ are bounded in absolute value by K then for each $k \ge n$ we have*

$$|m_k(g)| \le 2^k C(n)K.$$

Proof. By the first part of Theorem 3.1 the coefficients c_j in the representation $A(g) = \sum_{j=0}^{n-1} c_j v_j$ satisfy $\sum_{j=0}^{n-1} |c_j| \le C(n)K$. In particular, for each j we have $|c_j| \le C(n)K$. Now the required bound follows directly from the second part of Theorem 3.1. □

Remark 1. Although our model $g(x) = \sum_{i=1}^{n} A_i \delta(x - x_i)$ has $2n$ "degrees of freedom", by Corollary 3.1 it is enough to bound just its n first moments to guarantee that the rest are uniformly bounded, independently of the geometry of the nodes x_i. This fact can be considered as a quantitative version of the following property of the Tailor coefficients of a rational function $R_n(z) = \frac{P_{n-1}(z)}{Q_n(z)}$ of degree n (which also has $2n$ degrees of freedom): vanishing of its first n Taylor cocfficients implies its identical vanishing. This is an immediate consequence of the *existence* of a recurrence relation (2.1), although to find explicitly the coefficients B_j of this recurrence we need $2n$ Taylor coefficients.

Remark 2. The setting of our main problem is somewhat different from a typical approach of classical Moment Theory. The Hamburger and Stieltjes moment problems assume positivity of the measure. The Hausdorff moment problem assumes bounded variation of the measure (see, for example, [20] for the classical case and [15] for a multi-dimensional setting). In contrast, our problem is meaningful exactly for the measures $g(x) = \sum_{i=1}^{n} A_i \delta(x - x_i)$ with unbounded variation, while the uniform bound on the moments is implied by a cancellation of the positive and negative parts. Positivity assumption would immediately imply a bound on A_i: the first of the equation in the Prony system is

$$A_1 + \cdots + A_n = m_0,$$

so in case of positive A_i they all are bounded by m_0.

Remark 3. It is not necessary to assume in Theorem 3.1 that the points $x_1, \ldots, x_n$ are pairwise distinct. Extending the finite difference to the case of multiple points (taking at such points the derivatives of an appropriate order – see [10]) we can include into consideration the functions of the form

$$h(x) = \sum_{i=1}^{q} \sum_{j=0}^{\mu_i} A_{i,j} \delta^j (x - x_i).$$

Theorem 3.1 remains valid with no changes, and its proof in the extended situation can be obtained by passing to the limit as some of the points x_i collide.

4. Noisy measurements

In this paper we do not consider the problems of a practical implementation of the reconstruction algorithms. So we just discuss shortly some important issues of such an implementation related to the "collision singularities". As it was shown above, as some of the nodes x_i in $g(x) = \sum_{i=1}^{n} A_i \delta(x - x_i)$ approach one another the amplitudes A_i may tend to infinity while all the moments (measurements) remain uniformly bounded. Let us assume now that the moments are measured with a certain error (noise):

$$m_k = \hat{m}_k + n_k,$$

where m_k are the measured values of the moments, $\hat{m}_k$ are their "true" values and n_k are the noise components. We can assume that the noise components n_k are uniformly bounded. Still, for almost colliding configurations of the nodes *this noise may be strongly amplified*. The Prony systems (2.2) with the nodes fixed becomes a linear Vandermonde system

$$\sum_{i=1}^{n} A_i x_i^k = m_k, \ k = 0, 1, \ldots, n - 1$$

with respect to the amplitudes A_i, and its determinant tend to zero as the nodes approach one another.

A natural suggestion may be to represent the amplitudes vector $A(g) = (A_1, \ldots, A_n)$ in the base of divided differences, as in Section 3 above:

$$A(g) = \sum_{j=0}^{n-1} c_j v_j,$$

and to take c_j, $j = 0, \ldots, n-1$ as the new unknowns. As it was shown in the proof of Theorem 3.1, the moments of g are expressed as

$$m_k(g) = \sum_{j=0}^{n-1} c_j \int x^k \delta_j(x) dx = \sum_{j=0}^{n-1} c_j m_k(\delta_j).$$

Here $m_k(\delta_j)$ are the successive divided differences of x^k, and a new linear system for c_j takes a form

$$\sum_{j=0}^{n-1} c_j m_k(\delta_j) = m_k(g), \ k = 0, \ldots, n-1. \tag{4.1}$$

Theorem 3.1 guarantees that the solutions of (4.1) remain bounded for the right hand side bounded, independently of the configuration of the nodes. In particular, solving (4.1) does not amplify the noise. We plan to present implementation details separately.

References

1. D. Batenkov, *Moment inversion of piecewise D-finite functions*, arXiv: 0901.4665, Inverse Problems **25**(10): 105001, October 2009.
2. D. Batenkov, N. Sarig, Y. Yomdin, *An "algebraic" reconstruction of piecewise-smooth functions from integral measurements*, arXiv: 0901.4659, to appear in the Proceeding of the SAMPTA conference, Luminy, May 2009.
3. D. Batenkov, Y. Yomdin, Algebraic Fourier reconstruction of piecewise smooth functions, arXiv Math. 1005.1884, May 2010.
4. E. J. Candeś. *Compressive sampling*. Proceedings of the International Congress of Mathematicians, Madrid, Spain, 2006. Vol. III, 1433–1452, Eur. Math. Soc., Zurich, 2006.
5. P. J. Davis, *Plane regions determined by complex moments*, J. Approximation Theory **19** (1977) 148–153.
6. D. Donoho, *Compressed sensing*. IEEE Trans. Inform. Theory **52**(4 (2006) 1289–1306.

7. P. L. Dragotti, M. Vetterli and T. Blu, *Sampling Moments and Reconstructing Signals of Finite Rate of Innovation: Shannon Meets Strang-Fix*, IEEE Transactions on Signal Processing, **55(5)** Part 1 (2007) 1741–1757.

8. M. Elad, P. Milanfar, G. H. Golub, *Shape from moments—an estimation theory perspective*, IEEE Trans. Signal Process. **52(7)** (2004) 1814–1829.

9. B. Ettinger, N. Sarig. Y. Yomdin, *Linear versus non-linear acqusition of step-functions*, J. of Geom. Analysis **18(2)** (2008) 369–399.

10. F. B. Hildebrand, Introduction to Numerical Analysis, Second Edition, Dover PUblications, New York, 1987.

11. G. H. Golub, P. Milanfar, J. Varah, *A stable numerical method for inverting shape from moments*, SIAM J. Sci. Comput. **21(4)** (1999/00) 1222–1243 (electronic).

12. B. Gustafsson, Ch. He, P. Milanfar, M. Putinar, *Reconstructing planar domains from their moments.* Inverse Problems **16(4)** (2000) 1053–1070.

13. B. Gustafsson, M. Putinar, *Linear analysis of quadrature domains. II*, Israel J. Math. **119** (2000) 187–216.

14. V. Kisunko, Cauchy type integrals and a D-moment problem. C.R. Math. Acad. Sci. Soc. R. Can. **29(4)** (2007) 115–122.

15. S. Kuhlmann, M. Marshall, *Positivity, sums of squares and the multidimensional moment problem,* Trans. Amer. Math. Soc. **354(11)** (2002) 4285–4301 (electronic).

16. I. Maravić, M. Vetterli, *Sampling and reconstruction of signals with finite rate of innovation in the presence of noise*, IEEE Trans. Signal Process. **53(8)** part 1 (2005) 2788–2805.

17. I. Maravić, M. Vetterli, *Exact sampling results for some classes of parametric nonbandlimited 2-D signals*, IEEE Trans. Signal Process. **52(1)** (2004) 175–189.

18. P. Milanfar, G. C. Verghese, W.C. Karl, A. S. Willsky, *Reconstructing Polygons from Moments with Connections to Array Processing*, IEEE Transactions on Signal Processing, **43(2)** (1995) 432–443.

19. P. Milanfar, W. C. Karl, A.S. Willsky, *A Moment-based Variational Approach to Tomographic Reconstruction*, IEEE Transactions on Image Processing, **5(3)** (1996) 459–470.

20. E. M. Nikishin, V. N. Sorokin, *Rational Approximations and Orthogonality*, Translations of Mathematical Monographs, **92**, AMS, 1991.

21. R. de Prony, Essai experimentale et analytique, *J. Ecol. Polytech. (Paris)*, **1(2)** (1795) 24–76.

22. M. Putinar, *Linear analysis of quadrature domains.*, Ark. Mat. **33(2)** (1995) 357–376.

23. N. Sarig, Y. Yomdin, Signal Acquisition from Measurements via Non-Linear Models, C. R. Math. Rep. Acad. Sci. Canada Vol. **29(4)** (2007), 97–114.

24. N. Sarig and Y. Yomdin, *Reconstruction of "Simple" Signals from Integral Measurements*, in preparation.

25. K. Scherer, L. L. Schumaker, *A dual basis for L-splines and applications*, J. Approx. Theory **29(2)** (1980) 151–169.

26. L. L. Schumaker, *Spline functions: basic theory.* Third edition, *Cambridge Mathematical Library*, Cambridge University Press, Cambridge, 2007. xvi+582 pp.

27. G. Talenti, *Recovering a function from a finite number of moments*, Inverse Problems **3** (1987) 501–517.
28. M. Vetterli, P. Marziliano, T. Blu, *Sampling signals with finite rate of innovation*, IEEE Trans. Signal Process. **50(6)** (2002) 1417–1428.

Y. Yomdin
Department of Mathematics
Weizmann Institute of Science
Rehovot 76100
Israel
yosef.yomdin@weizmann.ac.il